ACS SYMPOSIUM SERIES **638**

Heterogeneous Hydrocarbon Oxidation

Barbara K. Warren, EDITOR
Union Carbide Corporation

S. Ted Oyama, EDITOR
Virginia Polytechnic Institute and State University

Developed from a symposium sponsored by
the Division of Colloid and Surface Chemistry,
and the Division of Petroleum Chemistry, Inc.

American Chemical Society, Washington, DC

Library of Congress Cataloging-in-Publication Data

Heterogeneous hydrocarbon oxidation / Barbara K. Warren, editor, S. Ted Oyama, editor

p. cm.—(ACS symposium series, ISSN 0097–6156; 638)

"Developed from a symposium sponsored by the Division of Colloid and Surface Chemistry, and the Division of Petroleum Chemistry, Inc." Symposium held in conjunction with the 211th National Meeting of the American Chemical Society, New Orleans, La., Mar. 24–29, 1996.

Includes bibliographical references and indexes.

ISBN 0–8412–3422–1

1. Hydrocarbons—Oxidation—Congresses. 2. Catalysis—Congresses. 3. Combustion—Congresses.

I. Warren, Barbara K., 1946– . II. Oyama, S. Ted, 1955– . III. American Chemical Society. Division of Colloid and Surface Chemistry. IV. American Chemical Society. Division of Petroleum Chemistry. V. American Chemical Society. Meeting (211th: 1996: New Orleans, La.) VI. Series.

QD305.H5H47 1996
547′.0104593—dc20 96–24782
CIP

This book is printed on acid-free, recycled paper.

PRINTED IN THE UNITED STATES OF AMERICA

Foreword

THE ACS SYMPOSIUM SERIES was first published in 1974 to provide a mechanism for publishing symposia quickly in book form. The purpose of this series is to publish comprehensive books developed from symposia, which are usually "snapshots in time" of the current research being done on a topic, plus some review material on the topic. For this reason, it is necessary that the papers be published as quickly as possible.

Before a symposium-based book is put under contract, the proposed table of contents is reviewed for appropriateness to the topic and for comprehensiveness of the collection. Some papers are excluded at this point, and others are added to round out the scope of the volume. In addition, a draft of each paper is peer-reviewed prior to final acceptance or rejection. This anonymous review process is supervised by the organizer(s) of the symposium, who become the editor(s) of the book. The authors then revise their papers according to the recommendations of both the reviewers and the editors, prepare camera-ready copy, and submit the final papers to the editors, who check that all necessary revisions have been made.

As a rule, only original research papers and original review papers are included in the volumes. Verbatim reproductions of previously published papers are not accepted.

ACS BOOKS DEPARTMENT

Contents

Novel Oxidants

Indexes

Preface

CATALYTIC HYDROCARBON OXIDATION PROCESSES include reactions for functionalization of organic molecules to make intermediates as well as reactions involved in total combustion of hydrocarbons to carbon oxides. The processes of selective oxidation and combustion share a number of characteristics, but they also represent opposing extremes of product selection. For both types of processes, a great deal of information about the underlying chemical and chemical engineering principles exists, and it can be used to improve current technologies and invent new ones.

This book provides a comparative study of both oxidation regimes and summarizes state-of-the-art advances in the areas of catalytic combustion of hydrocarbons, hydrocarbon oxidation with alternative oxidants, synthesis and characterization of materials, and related surface science studies. Topics covered include catalyst synthesis, catalyst characterization, kinetics and mechanisms, catalytically active sites, and advanced process applications.

Acknowledgments

We are grateful to the authors who cooperated in the preparation of this book and its rapid completion. We very much appreciate the contributions of all of the participants in the symposium entitled "Catalytic Heterogeneous Hydrocarbon Oxidation," cosponsored by the Divisions of Colloid and Surface Chemistry and Petroleum Chemistry, Inc., at the 211th National Meeting of the American Chemical Society in New Orleans, Louisiana, March 24–29, 1996. These participants made the symposium a success and this book possible.

We are also very grateful to have had financial support from Union Carbide Corporation, BP Chemicals, ARCO Chemical Company, Exxon Chemical Company, and the ACS Divisions of Petroleum Chemistry, Inc., and Colloid and Surface Chemistry to cover partial costs of this symposium. Finally, we acknowledge and thank the donors of the Petroleum

Research Fund, administered by the American Chemical Society, for partial support of this scientific meeting.

BARBARA KNIGHT WARREN
Union Carbide Corporation
3200 Kanawha Turnpike
P.O. Box 8361
South Charleston, WV 25303

S. TED OYAMA
Department of Chemical Engineering
Virginia Polytechnic Institute and State University
133 Randolph Hall
Blacksburg, VA 24061

May 7, 1996

Overview

Chapter 1

Factors Affecting Selectivity in Catalytic Partial Oxidation and Combustion Reactions

S. Ted Oyama

Departments of Chemical Engineering and Chemistry, Virginia Polytechnic Institute and State University, 133 Randolph Hall, Blacksburg, VA 24061–0211

Control of selectivity is a dominant issue for both partial and total oxidation of hydrocarbons. This chapter provides a critical overview of various factors that affect selectivity. It is suggested that both kinetic and thermodynamic aspects are important in determining selectivity in oxidation, and that selectivity is intimately tied to reactivity. Discussion is made of the role of different types of oxygen, the mode of adsorbate bonding, the occurrence of branching steps, the reducibility of the catalyst, the oxygen-metal bond strength, and the effect of structure. Since oxidation is a complex process many contrary views exist, and, where possible, these are presented and contrasted.

Control of selectivity is one of the central problems in catalytic hydrocarbon oxidation (*1*). The problem arises naturally because reactants undergoing reaction can be oxidized to various extents. This paper discusses various factors that affect selectivity in both partial oxidation and complete combustion. The discussion of combustion is restricted to purely surface phenomena, and excludes surface-initiated gas-phase processes. Where possible, comparisons will be made between partial oxidation and combustion.

The reactions of methane are diverse and provide a good illustration of the role of selectivity in oxidation.

Table 1
Products of Methane Oxidation

	Reactant	Products						
Species	CH_4	C_2H_6	C_2H_4	CH_3OH	HCHO	HCOOH	CO	CO_2
Oxidation state	-4	-3	-2	-2	0	+2	+2	+4
$-\Delta G_{oxid}^{700K}$ kJ/mol CH_4		69	74	91	295	503	579	801

0097–6156/96/0638–0002$15.00/0

In the course of the various transformations the formal oxidation state of carbon increases to values that depend on the chemical species formed. The degree of oxidation required depends on the process. For oxidative coupling the desired products are C_2H_6 and C_2H_4, for chemicals production the end products are CH_3OH, HCHO and HCOOH, in syngas formation the target is CO, while in catalytic combustion the objective is full oxidation to CO_2.

There has been limited success in the direct production of partial oxidation products from methane. Methane coupling to C_2 products has been studied over a number of catalysts and maximum yields are less than 40% (*2,3,4*). Oxygenates have been obtained over oxides of vanadium (*5,6*), molybdenum (*7,8*), and iron (*9*) as well as biological (P-450, and methane monooxygenase) (*10,11*) catalysts and yields have been even lower, less than 10%. Syngas ($CO + H_2$) has been produced in high selectivity (> 90%) at low contact times with noble metals supported on monoliths (*12,13,14*) and at high contact times on perovskite (*15*) and pyrochlore (*16*) oxides.

In the case of catalytic combustion (*17,18,19,20*) the problem of selectivity is different. In part the objective is to achieve full oxidation of methane to CO_2 and H_2O without the formation of partial oxidation products, but it is also essential to limit the formation of NO_x. One major application of catalytic combustion is in gas turbines (*21*), where catalysts are used to lower the temperature of reaction in the "precombustion region" (*22*), where most of the NO_x is generated. Another substantial application is in volatile organic compounds (VOC) abatement (*23*). Here the challenge is to eliminate small quantities of pollutants in an air stream. As in partial oxidation, both metals and oxides are used as catalysts in combustion.

Kinetics or Thermodynamics?

The network of reactions involving methane is complex, with many parallel and consecutive reactions. The solid lines depict the main expected routes for consecutive reactions, while the dashed lines show other possible pathways.

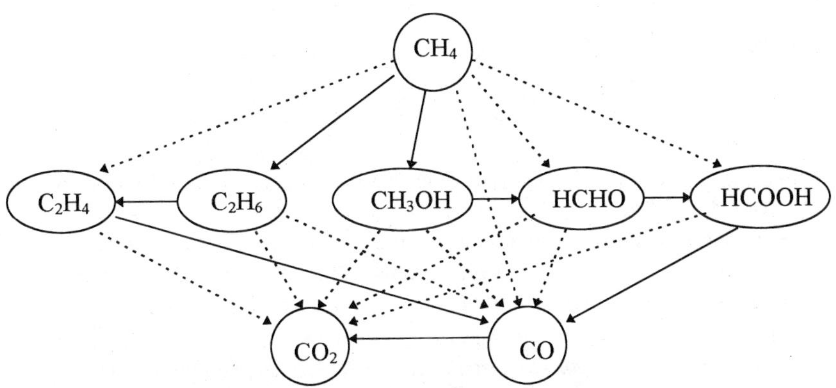

Figure 1
Reaction Network of Methane Oxidation

Any part of the network dealing with an intermediate can be simplified and described by a simple series of reactions:

$$A + O \xrightarrow{k_1} B$$

$$B + O \xrightarrow{k_2} C$$

In this sequence A is methane, O is oxygen, B is a partly oxidized species, C is the end member of the oxidation chain, CO_2, and k_1 and k_2 are effective rate constants. Depending on whether the objective is partial or total oxidation, the desired product is either B or C. As seen from the Table 1 the standard free energy of formation is highly negative for CO_2. Thermodynamically CO_2 formation is highly favored, and it is often stated that stopping at B requires kinetic control (7).

In the simplest case the rate of formation of B on a surface undergoing reduction and oxidation steps may be given by

$$r_1 = \frac{k_1 K_A(A) K_O(O)}{K_A(A) + K_B(B) + K_C(C) + K_O(O)} \quad (1)$$

The rate of formation of C is similarly given by

$$r_2 = \frac{k_2 K_B(B) K_O(O)}{K_A(A) + K_B(B) + K_C(C) + K_O(O)} \quad (2)$$

In these equations the small case *k*'s represent rate constants while the upper case *K*'s refer to equilibrium adsorption constants. The selectivity to B of the process is given by the relation giving the maximum value of (B).

$$r_1 - r_2 = \frac{d(B)}{dt} = k_1 K_A(A) K_O(O) - k_2 K_B(B) K_O(O) = 0 \quad (3)$$

This results in

$$\frac{(B)_{\max}}{(A)} = \frac{k_1 K_A}{k_2 K_B} \quad (4)$$

Equation (3) shows that selectivity is intimately connected with the rates of the reactions involved in the sequence. Equation (4) demonstrates that both *kinetic and thermodynamic* factors are involved in determining the selectivity to B. Since in general the intermediate product is more functionalized than the reactant, it is more reactive and is also more strongly adsorbed, so $k_2 > k_1$ *and* $K_B > K_A$. This is particularly true for alkane oxidations, and thus, obtaining a large selectivity to B is a challenge. For non-alkanes there are cases in which $K_A \gg K_B$, and high selectivity can be obtained, for example, in the oxidation of methanol to formaldehyde (*24*).

One way to improve selectivity to the intermediate product is by reducing the rate of the second step, but this is usually accompanied by a decrease in the rate of the first reaction. Although eventually a favorable selectivity may be achieved, this is usually at the cost of the overall rate. Table 2 shows an example.

Table 2
Comparison of Activity and Selectivity on an Oxide and a Metal

Catalyst	Turnover rate / s^{-1}	Selectivity / %			Reference
		CH_3CHO	$CH_2{=}CH_2$	CO_x	
7.7% V_2O_5/SiO_2	1.1×10^{-3}	12	68	20	(*25*)
Pt	6.0×10^{5}	–	–	100	(*26*)

Table 2 compares the performance of vanadium oxide and platinum in the oxidation of ethane. Turnover rates are based on sites titrated by selective chemisorption, using oxygen in the case of the vanadium oxide (*27*). It is seen that the metal is more active than the oxide by *over 8 orders of magnitude*, but produces mostly CO_2. Evidently, the price to pay for an increase in selectivity is a reduction in rate. This empirical observation has also been reported for a collection of methane oxidation catalysts where selectivity was found to decrease with conversion (*28*).

Electrophilic versus Nucleophilic Oxygen

The role of adsorbed oxygen in controlling selectivity has been championed by Bielanski and Haber (*29,30*). Adsorbed oxygen is classified into two types: electrophilic and nucleophilic, which differ in their mode of reaction. *Electrophilic* oxygen includes species like, O^- (oxide), O_2^- (superoxide), and O_2^{2-} (peroxide), which are believed to be responsible for deep oxidation. These species are electron deficient and are expected to react with the electron-rich regions of a hydrocarbon molecule, such as double bonds. *Nucleophilic* oxygen refers to the oxide ion, O^{2-}, often identified as lattice oxygen, which is believed to carry out partial oxidation. Because this oxygen has its full complement of electrons, it is expected to react with the portions of the molecule that are electron-poor.

Evidence for this difference in reactivity between the two types of oxygen is provided by the simultaneous measurement of the mass and charge of surface oxygen species during catalytic reactions (*31,32*). On catalysts such as Co_3O_4, which favor total oxidation, the type of oxygen detected was electrophilic (O^{2-}, O^-), while on catalysts like $Bi_2Mo_3O_{12}$, which carry out partial oxidation, the type of oxygen found was nucleophilic (O^{2-}). Unfortunately, direct measurements like these have not been applied to a variety of catalysts and conditions. Such measurements are very desirable. Nevertheless, the concept of reactivity control by the type of oxygen appears to be useful for qualitatively explaining selectivity in a variety of oxide systems (*33,34*).

Considerable care must be employed in the use of the above categorization. The concept does not hold in metallic systems (*e.g.* combustion catalysts), where ionic oxygen species are not present. Furthermore, in many oxide systems differing views have been proposed.

For the case of *selective oxidation*, Tagawa, *et al*., have concluded that O^- forms on the surface and abstracts a β-hydrogen from an adsorbed complex in the dehydrogenation of ethylbenzene (*35*). Szakás, *et al*., suggest that a mobile O^- radical is responsible for both partial and total oxidation of 1-butene and *n*-butene to maleic anhydride on V-P-O_x (*36*). Akimoto, *et al.*, have given evidence that O_2^- ions are involved in the oxidation of butadiene to maleic anhydride over supported molybdena catalysts (*37*). Yoshida, *et al.*, have reported the reaction of O_2^- on V_2O_5 with propylene and benzene to form aldehydes while the lattice oxygen shows little reactivity below 423 K (*38*). Gleaves, *et al.*, (*39*) suggest that adsorbed oxygen species are responsible for selective oxidation in vanadyl pyrophosphate, while Schiøtt, *et al.* (*40*) suggest that these are specifically peroxo oxygen species. In the methane coupling area many active forms of oxygen in the catalyst have been proposed, including O^-, O_2^- and O_2^{2-}. These are reviewed by Lee and Oyama (*41*).

In the case of *combustion* Arai and co-workers proved the participation of both adsorbed and lattice oxygen in the complete oxidation of methane (*42*).

Selectivity Control by the Mode of Adsorbate Bonding

So far not much has been said about the manner in which the hydrocarbon reactant is adsorbed on the surface. Yet, this is probably a crucial factor, since the stability of the surface intermediate will determine its lifetime on the surface, and its susceptibility to retrograde reactions leading to total oxidation.

On a transition metal oxide surface an activated hydrocarbon molecule has the option of being bound *to the metal or to oxygen*. (See scheme below.) When the hydrocarbon is attached to the surface through an oxygen atom by an ether-type linkage it can undergo reaction by two well-known pathways. An α-H elimination produces an aldehyde, while a β-H elimination produces an olefin. On the other hand, if the hydrocarbon is bonded directly to the transition metal (M) by an M-C bond, there are no ready elimination reactions. The intermediate is relatively stable and remains on the surface for a long time, allowing the possibility of deep oxidation.

There are several examples where differences in bonding result in different oxidation pathways. For methane oxidation on a variety of transition metal oxides Dowden, *et al.* (*43*) suggested that methoxy intermediates can be hydrated to methanol, whereas methylene species are oxidized to CO_x. In the case of ethane oxidation on supported vanadium oxide (*44*) it was found that the production rates of CH_3CHO and C_2H_4 do not vary substantially with catalyst structure whereas the formation rate of CO_x varies by a factor of 30. This suggests that the products are not formed from a common intermediate, but rather that two independent paths for the formation of the two product types exist. These are suggested to be

$$MO + C_2H_6 \rightarrow M{-}O{-}CH_2CH_3 \longrightarrow C_2H_4 + CH_3CHO$$

$$MO + C_2H_6 \rightarrow M{-}CH_2CH_3 \longrightarrow CO_x$$

The *oxygen-bonded* intermediate produces selective oxidation products, while the *metal-bonded* intermediate produces CO_x. Another example of this situation is found in the oxidation of propylene on supported molybdenum oxide (*45*). To test the effect of bonding, two model reactants are employed, allyl alcohol and allyl iodide, which are expected to form different intermediates.

$$MO + X\text{-}CH_2\text{-}CH{=}CH_2 \rightarrow \begin{cases} \underset{M}{/}\ O\text{-}CH_2\text{-}CH{=}CH_2 \longrightarrow OHC\text{-}CH{=}CH_2 \\ \underset{M}{/}\ CH_2\text{-}CH{=}CH_2 \longrightarrow CO_x \end{cases}$$

The intermediates were decomposed in a temperature programmed experiment, and the products were detected by mass spectrometry. Again the oxygen-bonded intermediate produced a selective oxidation product, in this case acrolein, whereas the metal-bonded intermediate produced CO_x.

Still another example of the importance of bonding is found in the oxidation of ethanol on silica-supported molybdenum oxide. Here the major products are the selective oxidation products, acetaldehyde and ethylene. Raman spectroscopy at *in situ* reaction conditions indicates that there are two different adsorbed ethoxide species associated with the formation of the two products (*46*).

$$CH_3CH_2OH + Mo{=}O \longrightarrow \underset{Mo}{/}\ O\text{–}CH_2CH_3 \longrightarrow CH_3CHO$$

$$CH_3CH_2OH + \underset{Mo\ \ Mo}{\overset{O}{/\ \backslash}} \longrightarrow \underset{Mo\ \ Mo}{\overset{O\text{–}CH_2CH_3}{/\ \backslash}} \longrightarrow CH_2{=}CH_2$$

The ethoxide species bound to the terminal oxygen group produces acetaldehyde, while the ethoxide species bound to the bridging oxygen group produces ethylene.

Selectivity Determining Step

Kung (*47*) has suggested the existence of a selectivity-determining step which determines the fate of a surface intermediate. For the following sequence

$$A \longrightarrow B^* \begin{cases} \xrightarrow{k_1} C \\ \xrightarrow{k_2} D \end{cases}$$

The selectivity-determining step is that involving the reaction of the intermediate, B*, and it includes the two irreversible reactions denoted by rate constants k_1 and k_2. The intermediate is an adsorbed species, as otherwise the network would reduce to the trivial case of two independent reactions. For the two reactions the rates of formation of C and D are

$$r_C = \frac{k_1(B^*)}{F} \qquad r_D = \frac{k_2(B^*)}{F}$$

where F is a common function of concentrations. The fractional selectivity to C can be defined as $S = r_C/(r_C + r_D) = k_1/(k_1 + k_2)$. Thus, provided that there are no other reactions, the selectivity can be related to the ratio of rate constants for the selectivity-determining step.

An example of where this concept arises is in the dehydrogenation of alkanes on vanadium oxide-containing catalysts (*48*). The intermediate B* in that case is an adsorbed alkyl or alkene.

$$\underset{\substack{|\\M}}{CH_2}CH_2R \text{ or } CH_2{=}\underset{\substack{|\\M}}{CH}R \; \begin{cases} \rightarrow \text{dehydrogenation products} \\ \rightarrow \text{oxygenates or COx} \end{cases}$$

The idea of a selectivity-determining step implies that there exists a *critical species* which can react in either of two ways, only one of which leads to a desired product.

Reducibility of Catalysts

Sachtler and De Boer (*49*) attempted to correlate the differences in selectivity among different catalysts for the oxidation of propylene to the reducibility of the catalysts. They hypothesized that in general the higher the reducibility, the higher would be the conversion and lower the selectivity. However, there were marked differences in the catalytic activity for samples of very similar reducibility. The discrepancies were attributed to different active site densities in these catalysts. Reducibility should be related to the heat of formation, but a number of workers were unsuccessful in correlating catalytic activity of oxides in hydrocarbon oxidation and heat of formation of the oxides (*50,51*). Subsequently, Sachtler, *et al.*, correlated oxidation activity and selectivity to the *differential* ($\delta\Delta H/\delta x$) increase of the heat of oxygen release, ΔH, with increasing reduction level, x (*52*). Kung, *et al.*, found that the selectivity for dehydrogenation of butane on V_2O_5/Al_2O_3 samples increased with increasing degree of reduction (47).

Electron Transfer

Related, but distinct from the concept of reducibility, is the idea of electron transfer. There are two facets to electron transfer: i) electron transfer from the reactant to the catalyst, and ii) electron transfer through the catalyst and eventually to oxygen. The first aspect was discussed by Sleight, who suggested that the conduction band of oxides acts as an electron sink for the electrons produced in the activation of hydrocarbons (*53,54*). The second aspect is an essential part of the Mars-van Krevelen redox mechanism. For example high electronic conductance in multicomponent oxides has been associated with good catalytic activity (*55*).

In cases where the active centers are isolated metal ions or clusters, the conduction band is replaced by the density of empty states. This was suggested by Weber (*56*) in his theoretical work on methanol oxidation on molybdenum oxide, which indicated that the hydrogen atom removed in the rate-determining step is a hydride which goes to the metal, not a proton which goes to oxygen. Morokuma's group (*57*) reached similar conclusions with their *ab initio* calculations in elimination reactions. These ideas were verified experimentally by Oyama and coworkers who studied ethanol oxidation on different supported molybdenum catalysts (46). They found the same activation energies but different preexponential factors, suggesting that the magnitude of the electronic partition function determined the rate. This partition function is associated with the density of empty states.

Metal-Oxygen Bond Strength

The importance of the reactivity of oxygen at the surface of oxides and metals has been explored by a number of investigators *(52,58-61)*. On *oxides* Balandin (*62,63*) suggested that the maximum hydrocarbon oxidation rate would occur at a heat of reaction, Q_o, for reoxidation of the catalyst that was one-half the heat, Q_r, of combustion of the hydrocarbon, $Q_o = \frac{1}{2}Q_r$. Moro-oka and Ozaki (*64*) found a correlation between the rate of oxidation of propylene to form mostly carbon dioxide and the metal-oxygen bond strength (Fig. 2). The latter was measured by the heat of formation of the oxide normalized by the number of oxygen atoms in the oxide (ΔH_o). They also found that the larger the ΔH_o, the higher the order in propylene, and the higher the oxygen coverage.

The group of Kung has correlated the differential heat of reoxidation measured calorimetrically and selectivity for oxidative dehydrogenation of butane on V_2O_5/Al_2O_3 samples (*65,66*). It was found that the selectivity is low when the heat is low and increases rapidly when the heat increases rapidly. The heat of reoxidation is a measure of the metal-oxygen bond strength, and suggests that the ease of removal of lattice oxygen is important in determining selectivity for dehydrogenation.

In combustion on *metals* it has been found that oxidation rate decreases with increasing strength of the metal-oxygen bond (*67*). This may be due to the fact that the metal catalysts operate at close to full coverage in oxygen, and the combustion reaction involves breaking metal-oxygen bonds.

Oxidation State of the Surface

The role of oxidation state is related to the concepts presented above of reducibility of the surface and metal oxygen bond strength. Many selective oxidation catalysts are found in intermediate states of oxidation. For example, in V-P-O catalysts for *n*-butane oxidation, the predominant oxidation state of vanadium is +4. Phosphorus is found to moderate the reoxidizability and the reducibility of the catalyst (*68*).

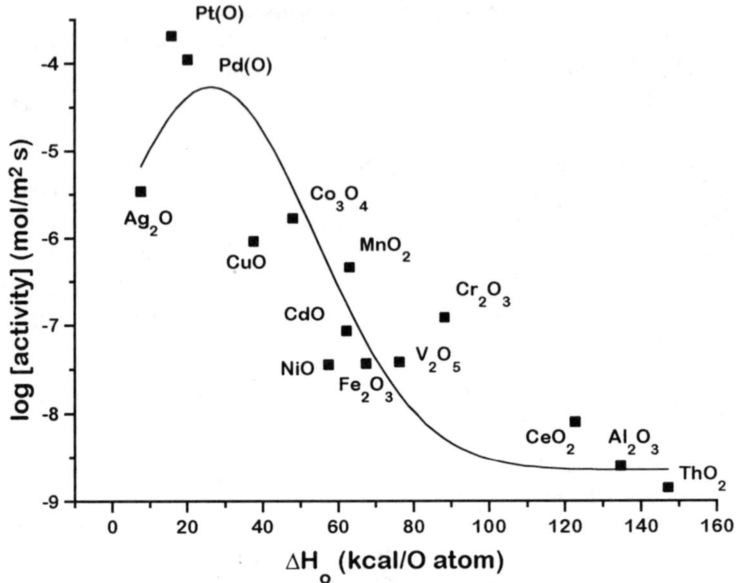

Figure 2.
Correlation between the Catalytic Activity of Propene Oxidation and the Heat of Formation (Adapted from Ref. 64)

Ultra-high vacuum studies of oxidation reactions have revealed that on most transition metals complete combustion is favored at high oxygen coverages (*69*). However, there are notable exceptions. In ethylene oxidation on silver it has been found that selectivity increases with oxygen coverage (Figure 3) (*70*). In C_2 and C_3 alkyl oxidations on rhodium higher oxygen concentrations lead to olefins or aldehyde/ketones while low oxygen concentrations produce CO_x (*71,72*). In this work, Bol and Friend have suggested that oxidation selectivity can be controlled by manipulating the oxygen coverage (*73*).

Surface Structure

Exposed Crystal Face

The effect of surface structure in selectivity in partial oxidation has been discussed extensively in the literature (*74,75*). Much work has been done with macroscopic crystallites of mm dimensions.

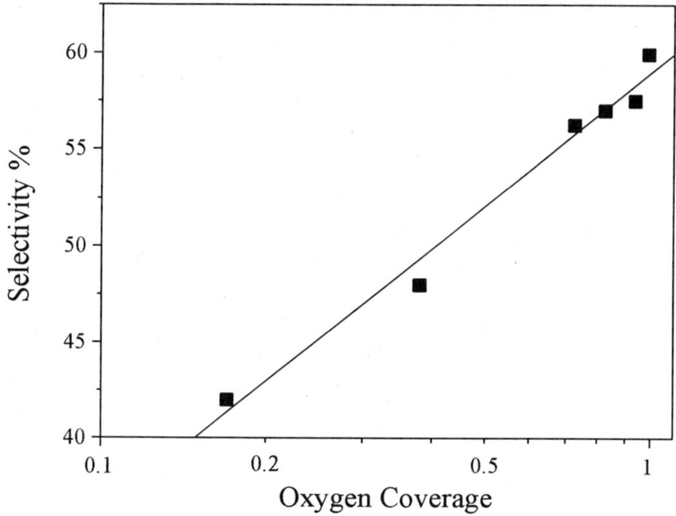

Figure 3
Dependence of Selectivity on Oxygen Coverage for Ethylene Oxidation on Silver (Adapted from Ref. 70)

Propylene oxidation on molybdenum oxide has been looked at in detail. There are two opposing views. One view, expounded by Volta, Tatibouët, Moraweck and Germain, suggests that the effect of structure is substantial and can lead to large differences in selectivity (0 to 100%, depending on exposed crystal face). Another position, taken by Oyama, acknowledges that there is an effect of structure, but that selectivity differences are not so extreme. The first group of researchers studied samples with different relative amounts of different crystal planes and related these to selectivity to different reaction products (*76,77*). This allowed the assignment of the formation of various products to specific crystal faces. It was concluded that the (100) side face of MoO_3 was responsible for acrolein formation and that the (010) basal face resulted in CO_x formation (*78,79*). This method was criticized for overanalyzing scant data, although the results appeared to be reproducible (*80*). In a refinement of the earlier studies it was concluded that stepped (1k0) faces of the {100} family were involved in acrolein formation (*81*).

These conclusions were contested in another study involving allyl halides and allyl oxalate (*82*) in which the (010) plane was suggested to be responsible for acrolein production. But these studies were carried out with uncontrolled water vapor partial pressure in a pulse mode, and the conclusions were based on a model involving sequential activation and reaction of propylene on different planes. Such desorption-adsorption behavior of propylene has not been observed.

Another study, comparing silica-supported molybdenum oxide with macroscopic crystallites, found considerable differences between samples indicating that the reaction was structure sensitive (*83*). The sites responsible for selective oxidation were suggested to be low-coordination molybdenum centers present in the supported samples and on the edges of crystallites. The results were consistent with the suggestion that (1k0) faces are responsible for selective oxidation.

A similar approach has been taken with crystals of $(VO)_2P_2O_7$ in *n*-butane oxidation as reviewed by Cavani and Trifirò (*84*). Work by Matsuura and Yamazaki (*85*), Volta, *et al.* (*86*), and Okuhara, *et al.* (*87*) indicates that the basal (100) plane is selective for producing maleic anhydride, while the side faces produce CO_x (Fig. 4).

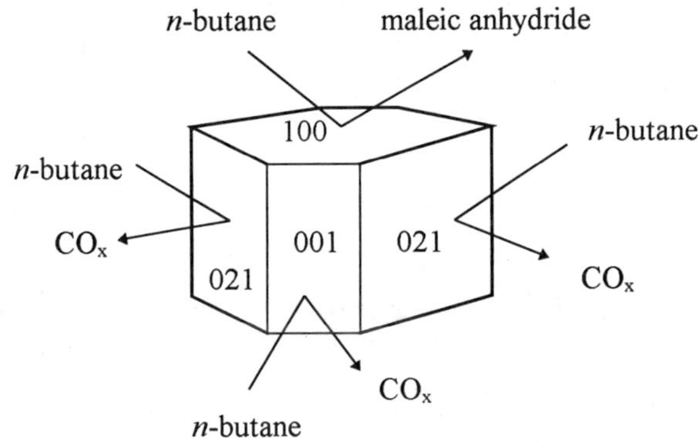

Figure 4
Selective and Non-selective Oxidation Sites on the Crystal Faces of $(VO)_2P_2O_7$. (Adapted from Ref. 85)

The effect of structure in *combustion* reactions has also been studied extensively. Briot and Primet found an increase of the turnover rate of methane combustion on supported Pt and Pd catalysts with aging and attributed it to a higher oxygen activity on *larger* crystallites (*88,89*). Hicks, *et al.* reached a similar conclusion in comparing supported Pt and Pd catalysts of different dispersions (*90,91*). Pd was reported to have larger structure-sensitivity. Work on Pt by Otto again confirmed these results (*92*).

Site Isolation

Callahan and Grasselli employed a model based on a grid of oxygen atoms (Fig. 5) to explain selectivity in propylene oxidation on copper oxide (*93*). They concluded that oxygen atoms must be distributed over the surface of a catalyst so as to limit the number of oxygen atoms at an active site (Fig. 5). The optimum number of oxygen atoms was just the number necessary to stoichiometrically obtain the desired product. A Monte Carlo calculation predicted that maximum selectivity could be achieved at 60-70% reduction. The number of oxygen atoms could be controlled by isolating the site in various manners, such as operating at an intermediate oxidation state, selecting metal oxides with an intrinsic limited grouping of oxygen atoms, or modifying the surface of an oxide with a reagent. An example of site isolation by structural means was provided in the U-Sb-O system (*94*).

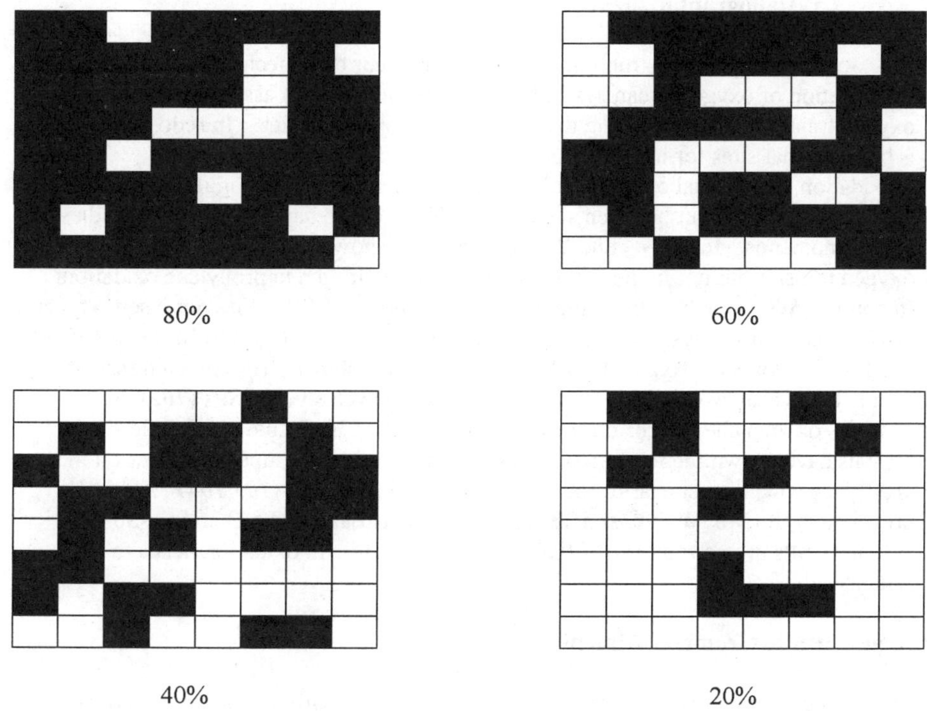

Figure 5
Oxidation of Reduced Surface Grid With Oxygen (Adapted from Ref. 93) The dark areas represent the portion of the surface covered with oxygen.

It has been found by $^{18}O_2$ labeling of $(VO)_2P_2O_7$ that the oxidation of *n*-butane involves only catalytic sites near the surface that do not exchange oxygen with the bulk (*95*). Vanadium sites on this surface are separated by phosphate groups requiring that all oxygens necessary for oxidation of the n-butane be in the close neighborhood of the adsorption site (84). This may limit overoxidation of the *n*-butane and explain the excellent selectivity obtained on V-P-O.

Defects

It has been proposed, based on x-ray diffraction line-broadening, that the most active $(VO)_2P_2O_7$ catalyst possess a defective structure. Defects have been suggested to increase the Lewis acidity of calcined V-P-O catalysts (*96*), or to facilitate formation of the active phase (84).

In general the relationship of defects to selectivity is not clear, and probably depends on the system. It has also been pointed out that defects are in equilibrium with gas-phase reactants and products, and so will change in concentration with reaction conditions (*97*).

Crystallographic shear

Long range reconstruction on oxides can occur by a mechanism involving organization of oxygen vacancies. The phenomenon has been associated with assisting oxygen transport and facilitating desorption of adsorbed species. In redox kinetics it is believed that sites for hydrocarbon oxidation are different from those for reoxidation, and crystallographic shear planes (CSPs) have been suggested to assist bulk oxygen movement between sites (*98,99,100*). There have been limited studies of this phenomenon. In $WO_{2.95}$ and $WO_{2.9}$ it has been shown that CSPs are involved in oxygen transfer but not in the initial abstraction of hydrogen in propylene oxidation. In contrast $WO_{2.72}$, which has a tunnel structure without CSPs, does not insert oxygen (*101*). In the Mo-O system during the oxidation of $Mo_{18}O_{52}$ (001) to MoO_3 at 670 K and low pressures of oxygen, RHEED and SEM have shown that oxygen transport occurs by a vacancy exchange mechanism involving exclusively CSPs (*102*). On $Mo_{18}O_{52}$ oxygen chemisorbs on the CSP boundaries. LEED studies of rutile single crystals have shown the formation of (1x3), (1x5), and (1x7) superstructures on the (100) face which are similar to the CSP structures in the bulk (*103,104*). Although prevalent in the Mo, W, Nb, and Ti systems, in industrial Mo-Te-O and Sn-Sb-O systems CSPs are not observed (*105*). Clearly, their role in oxidation needs to be established.

Reactant-Site Geometry Matching

For reactions where different portions of a hydrocarbon need to be activated the distance between the reaction points must correspond to the separation between active functions on the surface

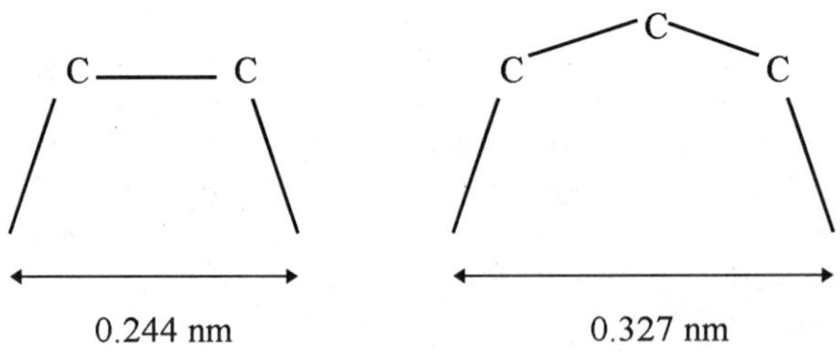

Figure 6
Schematic Representation of the Geometry in Ethyl and Propyl Species (Adapted from Ref. 48)

For example Kung and coworkers (48) have suggested that in the dehydrogenation of propane on vanadium-containing catalysts, the bound C_3 species (distance between end C atoms 0.327 nm) can readily bond with the two V ions in the active site of $(VO)_2P_2O_7$ (separation 0.319 nm) but can only do so with difficulty with those in $Mg_2V_2O_7$ (separation 0.339 nm) (Fig. 6). Thus, oxidation occurs on $(VO)_2P_2O_7$ while dehydrogenation occurs on $Mg_2V_2O_7$. The reaction of C_2 would require a different explanation (48).

Ziolkowski, *et al.*, have developed a crystallographic model of active sites which describes a detailed correspondence between the geometry and energetics of the different faces in $(VO)_2P_2O_7$ and the n-butane reactant (*106,107*). They suggest activation of the C_1 and C_4 positions by hydrogen abstraction to produce C-O bonded intermediates.

General Properties of Oxide Catalysts

Sokolovskii (1) has outlined a set of consistent requirements for oxide catalysts for combustion and selective oxidation. The results of his survey are summarized below.

Catalysts for *combustion* must provide a high rate of oxygen activation. To do so they must have a large number of sites for coordinating oxygen molecules, and have the ability to easily donate and accept electrons. Oxides of *3d* elements with an unfilled d shell exemplify materials suitable for combustion. Sokolovskii believes catalysts should also permit only a slow transformation of adsorbed oxygen to lattice oxygen.

At low temperatures the rate-determining process in combustion is the oxidation of surface species, and the rate will be higher the faster the oxygen is activated. At high temperatures the limiting process is the interaction of the reactant with the catalyst, and an inverse relation is observed between reaction rate and the initial heat of chemisorption of oxygen.

The properties of catalysts for *selective oxidation* depend on the polarity of the C-H bond to be activated. When the polarity is *low*, as in methane, catalysts have generally i) required high temperatures, ii) provided low oxygen mobility, iii) possessed a high degree of reduction, iv) provided nucleophilic sites, and v) accepted one electron (have low electrophilicity). When the polarity is *high*, as in acrolein, the catalyst have i) operated at low temperatures, ii) provided high oxygen mobility, iii) maintained a low level of reduction, iv) provided electrophilic sites, and v) accepted two electrons.

Summary

This chapter has given an overview of factors that control selectivity in catalytic partial oxidation and combustion reactions. Both oxide and metal systems were discussed.

First, it was shown that not only thermodynamic quantities, but also kinetic parameters, were important in determining selectivity. It was also conjectured that the price for a gain in selectivity was a decrease in rate. The possible roles of electrophilic versus nucleophilic oxygen were discussed, and it was concluded that electrophilic oxygen does not always lead to complete combustion. Instead of focusing on the type of oxygen species to explain reaction pathways, an alternative explanation was presented based on the mode of adsorbate bonding. It was suggested that when organic intermediates were bonded to the surface through an oxygen atom, partial oxidation products would be favored, but when direct metal-carbon bonds were formed, total oxidation products would be preferred. If instead of having different intermediates with different bonding properties, one common intermediate was formed with a choice of reaction pathways, then the reaction was deemed to involve a selectivity-determining step.

Lastly, other factors that control selectivity were discussed, with a concentration on catalyst properties. These included the reducibility of catalysts, the ease of electron transfer, the strength of metal-oxygen bonds, the oxidation state of the surface, the identity of the exposed crystal face, the isolation of active sites, the presence of defects, the occurrence of crystallographic shear, the need of reactant-site geometry matching, and the general properties of oxides.

Acknowledgment

This work was supported by the National Science Foundation, Division of Chemical and Thermal Systems (CTS-9311876). Discussions with B. K. Warren are gratefully acknowledged.

References Cited

1. Sokolovskii, V. D. *Catal. Rev.-Sci. Eng.* **1990**, 32, 1.
2. E. E. Wolf, Ed. *Methane Conversion by Oxidative Processes: Fundamental and Engineering Aspects*; Van Nostrand Reinhold: New York, 1992.
3. Amenomiya, Y.; Birss, V. I.; Goledzinowski, M.; Galuszka, J.; Sanger, A. R. *Catal. Rev.-Sci. Eng.* **1990**, 32, 163.
4. Otsuka, K.; Jinno, K.; Morikawa, A. *J. Catal.* **1986**, 100, 353.
5. Zhen, K. J.; Khan, M. M.; Mak, C. H.; Lewis, K. B.; Somorjai, G. A. *J. Catal.* **1985**, 94, 501.
6. Iwamoto, M. *Japan Kokai Tokkyo Koho* 58-92,630, **1983**.
7. Spencer, N. D.; Pereira, C. J. *J. Catal.* **1989**, 116, 399.
8. Pitchai, R.; Klier, K. *Catal. Rev.-Sci. Eng.* **1986**, 28, 13.
9. Kobayashi, T.; Nakagawa, K.; Tabata, K.; Haruta, M. *J. Chem. Soc. Chem. Commun.* **1994**, 1609.
10. P. R. Ortiz de Montellano, Ed. *Cytochrome P-450: Structure, Mechanism and Biochemistry*, Plenum: New York, 1985.
11. A. C. Rosenzweig; S. J. Lippard *Acc. Chem. Res.* **1994**, 27, 229.
12. Hickman, D. A.; Haupfear, E. A.; Schmidt, L. D. *Catal. Lett.* **1993**, 17, 223.
13. Torniainen, P. M.; Chu, X.; Schmidt, L. D. *J. Catal.* **1994**, 146, 1.
14. Huff. M.; Schmidt, L. D. *J. Catal.* **1994**, 149, 127.
15. Slagtern, Å.; Olsbye, U. *Appl. Catal. A* **1994**, 99, 108.
16. Ashcroft, A. T.; Cheetham, A. K.; Foord, J. S.; Green, M. L. H.; Grey, C. P.; Murrell, A. J.; Vernon, P. D. F. *Nature* **1990**, 344, 319.
17. Spivey, J. J. *Ind. Eng. Chem. Res.* **1987**, 26, 2165.

18. Zwinkels, M. F. M.; Järås, S. G.; Menon, P. G.; Griffin, T. A. *Catal. Rev.-Sci. Eng.* **1993**, 35, 319.
19. Pfefferle, L. D.; Pfefferle, W. C. *Catal. Rev.-Sci. Eng.* **1987**, 29, 219.
20. Prasad, R.; Kennedy, L. A.; Ruckenstein, E. *Catal. Rev.-Sci. Eng.* **1984**, 26, 1.
21. Ismagilov, Z. R.; Kerzhentsev, M. A. *Catal. Rev.-Sci. Eng.* **1990**, 32, 51.
22. Arai, H.; Machida, M. *Shokubai* **1991**, 33, 328.
23. Chu, W.; Windawi, H. *Chem. Eng. Prog.* March **1996**, 37
24. Cheng, W.-H. *J. Catal.* **1996**, 158, 477.
25. Oyama, S. T.; Middlebrook, A. M.; Somorjai, G. A. *J. Phys. Chem.* **1990**, 94, 5028.
26. Hiam, L.; Wise, H.; Chaikin, S. *J. Catal.* **1968**, 10, 272.
27. Oyama, S. T.; Went, G. T.; Lewis, K. B.; Bell, A. T.; Somorjai, G. A. *J. Phys. Chem.* **1989**, 93, 6786.
28. Dautzenberg, F. M.; Garten, R. L.; Klingman, G. *Fuels from Remote Natural Gas: Defining the Research and Development Challenge*, ACS Division of Industrial and Engineering Chemistry, 21st State-of-the-Art Symposium, June 15-18, 1986, Marco Island, Florida.
29. Bielanski, A.; Haber, J. *Catal. Rev. Sci. Eng.* **1979**, 19, 1.
30. Bielanski, A.; Haber, J. *Oxygen in Catalysis*; Marcel Dekker, Inc., New York, 1991.
31. Libre, J. M.; Barbaux, Y.; Grzybowska, B.; Bonnelle, J.-P. *React. Kinet. Catal. Lett.* **1982**, 20, 249.
32. Barbaux, Y.; Elamrani, A.; Bonnelle, J.-P. *Catal. Today* **1987**, 1, 147.
33. Keulks, G. W. *J. Catal.* **1970**, 19, 232.
34. Wragg, R. D.; Ashmore, P. G.; Hockey, J. A. *J. Catal.* **1971**, 22, 49.
35. Tagawa, T.; Hattori, T.; Murakami, Y. *J. Catal.* **1982**, 75, 66.
36. Szakács, S.; Wolf, H. ; Mink, G. ; Bertóti, I.; Wostneck, N.; Locke, B. ; Seeboth, H. *Catal. Today* **1987**, 1, 27.
37. Akimoto, M.; Echigoya, E. *J. Catal.* **1974**, 35, 278.
38. Yoshida, S.; Matsuzaki, T.; Ishida, S.; Tarama, K. *Proc. Int. Cong. Catal. 5th, 1972*, Miami Beach, Florida, **1973**, 21, 1049.
39. Gleaves, J. T.; Ebner, J. R.; Kuechler, T. C. *Catal. Rev.-Sci. Eng.* **1988**, 30, 49.
40. Schiøtt, B.; Jørgensen, K. A. *Catal. Today* **1993**, 16, 79.
41. Lee, J. S.; Oyama, S. T. *Catal. Rev. Sci. Eng.* **1988**, 30, 249.
42. Arai, H.; et al. *Appl. Catal.* **1986**, 26, 265.
43. Dowden, D. A.; Schnell, C. R.; Walker, G. T. *Proc. Int. Cong. Catal. 4th, 1968*; Moscow, **1970**, p. 201.
44. Oyama, S. T. *J. Catal.* **1991**, 128, 210.
45. Oyama, S. T.; Desikan, A. N.; Zhang, W. In *Catalytic Selective Oxidation*, Oyama S. T.; Hightower, J. W., Eds.; ACS Symposium Series 523; American Chemical Society, Washington DC 1993, p. 16.
46. Zhang, W.; Desikan, A.; Oyama, S. T. *J. Phys. Chem.* **1995**, 99, 14468.
47. Kung, H. H. *Adv. Catal.* **1994**, 40, 1.
48. Michalakos, P. M.; Kung, M. C.; Jahan, I.; Kung, H. H. *J. Catal.* **1993**, 140, 226.
49. Sachtler, W. M. H.; De Boer, N. D. *Proc. Int. Cong. Catal. 3rd. 1964*; Amsterdam, **1965**, 1, 252.

50. Roiter, V. A.; Golodets, G. I. *Proc. Int. Cong. Catal. 4th, 1968*; Moscow, **1970**, 1, 365.
51. Germain, J. E.; Laugier, R. *Bull. Soc. Chim. Fr.* **1972**, 541.
52. Sachtler, W. M. H.; Dorgelo, G. H.; Fahrenfort, J.; Voorhoeve, R. J. H. *Rec. Trav. Chim. Pays-Bas.* **1970**, 89, 460.
53. Lim, W. J.; Sleight A. W. *Ann. N.Y. Acad. Sci.* **1976**, 272, 22.
54. Sleight, A. W. In *Advanced Materials in Catalysis*; Burton, J. J.; Garten, R. L., Eds.; Academic Press: New York, 1977, p. 181.
55. Bonceblanc, H.; Millet, J. M. M.; Coudurier, G.; Védrine, J. C. In *Catalytic Selective Oxidation*, Oyama S. T.; Hightower, J. W., Eds.; ACS Symposium Series 523; American Chemical Society, Washington DC 1993, p. 262.
56. Weber, R. S.; *J. Phys. Chem.* **1994**, 98, 2999.
57. Koga, N.; Obara, S.; Kitaura, K.; Morokuma, K. *J. Am. Chem. Soc.* **1985**, 107, 7109.
58. Boreskov, G. K.; Popovskii, V. V.; Sazonov, V. A. *Proc. 4th Int. Cong. Catal., Moscow*, 1970, Nauka: 1970, p. 343.
59. Ilchenko, N. I.; Golodets, G. I. *J. Catal.* **1975**, 39, 73.
60. Moro-oka, Y.; Morikawa, Y.; Ozaki, A. *J. Catal.* **1967**, 7, 23.
61. Callahan, J. L.; Grasselli, R. K. *AIChE J.* **1962**, 9, 755.
62. Balandin, A. A. *Adv. Catal.* **1958**, 10, 96.
63. Balandin, A. A. *Adv. Catal.* **1969**, 19, 1.
64. Moro-oka, Y.; Ozaki, A. *J. Catal.* **1966**, 5, 116.
65. Andersen, P. J.; Kung. H. H. *J. Phys. Chem.* **1992**, 96, 3114.
66. Michalakos, P.; Kung, M. C.; Jahan, I.; Kung, H. H. *Prep. Amer. Chem. Soc. Div. Petrol. Chem.* **1992**, 37, 1201.
67. Golodets, G. I. *Heterogeneous Catalytic Reactions Involving Molecular Oxygen*; Elsevier: New York, 1983, p. 439.
68. Cavani, F.; Centi, G.; Trifirò, F. *La Chimica & La Industria, Milan*, **1992**, 74, 182.
69. Madix, R. J.; Roberts, J. T. In *Surface Reactions*, Madix, R. J., Ed.; Springer-Verlag: Berlin, 1994, p. 5.
70. Akella, L. M.; Lee, H. H. *J. Catal.* **1984**, 86, 456,
71. Xu, X.; Friend, C. M. *J. Am. Chem. Soc.* **1991**, 113, 8572.
72. Xu, X.; Friend, C. M. *J. Phys. Chem.* **1991**, 95, 10753.
73. Bol, C. W. J.; Friend, C. M. *J. Phys. Chem.* **1995**, 99, 11930.
74. Murakami, Y.; Inomata, M.; Miyamoto, A.; Mori, K. *Proc. 7th Int. Cong. Catal. Tokyo, 1980*, Seiyama, Y.; Tanabe, K., Eds.; Elsevier: Amsterdam, 1981, Part B, p. 1344.
75. Volta, J. C.; Portefaix, J. L. *Appl. Catal.* **1985**, 18, 1.
76. Volta, J. C.; Desquesnes, W.; Moraweck, B.; Coudurier, G. React. *Kin. Catal. Lett.* **1979**, 12, 241.
77. Volta, J. C.; Desquesnes, W.; Moraweck, B.; Tatibouët, J.-M. *Proc. 7th Int. Cong. Catal. Tokyo, 1980*, Seiyama, Y.; Tanabe, K., Eds.; Elsevier: Amsterdam, 1981, Part B, p. 1398.
78. Volta, J. C.; Tatibouët, J.-M.; Phichitkul, C.; Germain, J. E. *Proc. 8th Int. Cong. Catal. Berlin, 1984*, Vol. 4; Verlag Chemie: Berlin, 1984, p. 451.

79. Volta, J. C.; Moraweck, *J. Chem. Soc. Chem. Commun.* **1980**, 330.
80. Oyama, S. T. *Bull. Chem. Soc. Japan* **1988**, 61, 2585.
81. Abon, M.; Mingot, B.; Massardier, J.; Volta, J. C. *Preprints*, Division of Petroleum Chemistry, American Chemical Society Boston Meeting, April 22-27, 1990, Vol. 35, p. 44.
82. Brückman, K.; Grabowski, R.; Haber, J.; Mazurkiewicz, A.; Sloczynski, J.; Wiltowski, *J. Catal.* **1987**, 104, 71.
83. Desikan, A. N.; Oyama, S. T. In *Surface Science of Catalysis*, Dwyer, D. J.; Hoffmann, F. M., Eds.; ACS Symposium Series 482; American Chemical Society, Washington DC 1992, p. 260.
84. Cavani, F.; Trifirò, F. In *Catalysis, Specialist Periodical Report*; Chp. 7, Royal Society of Chemistry: London, 1995, p. 246.
85. Matsuura, I.; Yamazaki, M. *In New Developments in Selective Oxidation*, Centi, G.; Trifirò, F., Eds.; Elsevier: Amsterdam, 1990, p. 21.
86. Volta, J. C.; Bere, K.; Zhang, Y. J.; Olier, R. In *Catalytic Selective Oxidation*, Oyama, S. T.; Hightower, J. W., Eds., American Chemical Society: Washington D.C., 1993, p. 217.
87. Okuhara, T.; Inumaru, K.; Misono, M. . In *Catalytic Selective Oxidation*, Oyama, S. T.; Hightower, J. W., Eds., American Chemical Society: Washington D.C., 1993, p. 156.
88. Briot, P. et al. *Appl. Catal.* **1990**, 59, 141.
89. Briot, P.; Primet, M. *Appl. Catal.* **1991**, 68, 301.
90. Hicks, R. F., et al. *J. Catal.* **1990**, 122, 280.
91. Hicks, R. F., et al. *J. Catal.* **1990**, 122, 295.
92. Otto, K. *Langmuir* **1989**, 5, 1364.
93. Callahan, J. L.; Grasselli, R. K. *AIChE J.* **1963**, 9, 755.
94. Grasselli, R. K.; Suresh, D. D. *J. Catal.* **1972**, 25, 273.
95. Okuhara, T.; Misono, M. *Catal. Today* **1993**, 16, 61.
96. Centi, G.; Trifirò, F. *Appl. Catal. A* **1992**, 88, 115.
97. Haber, J., *Catalytic Hydrocarbon Oxidation*, Warren, B. K.; Oyama, S. T., Eds.; ACS Books, Washington, D.C., 1996, Chp. 2.
98. O'Keefe, M. *Fast Ion Transport in Solids*; North Holland: Amsterdam, 1973; p. 233.
99. Anderson, J. S. In *Problems in Nonstoichiometry*; A. Rabenau, Ed.; North Holland: Amsterdam, 1970; Chap. 1, p. 1.
100. Grasselli, R. K.; Bradzil, J. F.; Burrington, J. D. *Appl. Catal.* **1986**, 25, 335.
101. de Rossi, S.; Jacono, M. W.; Pepe, F.; Schiavello, M.; Tilley, R. J. D. *Z. Phys. Chem.* **1982**, 130, 109.
102. Floquet, H.; Betrand, O. *Surf. Sci.* **1988**, 198, 449.
103. Chung, Y. W.; Lo, W. J.; Somorjai, G. A. *Surf. Sci.* **1977**, 45, 531.
104. Firment, L. E. *Surf. Sci.* **1981**, 116, 205.
105. Gai, P. L.; Boyes, E. D.; Bart, J. C. J. *Philos. Mag. A* **1987**, 45, 531.
106. Ziolkowski, J.; Bordes, E.; Courtine, P. *J. Catal.* **1990**, 122, 126.
107. Ziolkowski, J.; Bordes, E.; Courtine, P. In *New Developments in Selective Oxidation*, Centi, G.; Trifirò, F., Eds.; Elsevier: Amsterdam, 1990, p. 625.

Chapter 2

Selectivity in Heterogeneous Catalytic Oxidation of Hydrocarbons

J. Haber

Institute of Catalysis and Surface Chemistry, Polish Academy of Sciences, ul.Niezapominajek, 30239 Cracow, Poland

Interaction of a hydrocarbon molecule with oxygen at the surface of an oxide catalyst results in a network of parallel and consecutive elementary steps, in which hydrogen is abstracted, nucleophilic oxygen is inserted, electrophilic oxygen reacts with π-electrons and C-C bonds are cleaved. All products of partial oxidation are derived by kinetic control of the reaction. The catalyst to be selective must accelerate those consecutive steps which lead to the desired product and depress all others. Active sites must be thus present at its surface, which: 1) - abstract hydrogen from the appropriate position in the molecule, 2) - inject or remove electrons from the surface intermediates, 3) - properly orient the reacting molecule(s) with respect to the surface, 4) - perform a nucleophilic addition of a surface oxide ion, 5) - activate and orient properly the oxygen molecule with respect to the hydrocarbon molecule in electrophilic oxidation steps, 6) - enable rapid diffusion of hydrogen and undergo a facile dehydroxylation, 7) - easily reoxidize by interaction with gas phase oxygen and show rapid diffusion of oxygen through the lattice to the reaction site. Defect structure of transition metal oxides is an important factor determining the selectivity in oxidation of hydrocarbons. The latter may be influenced by using oxygen transfer reactions in multicomponent oxide systems with different cationic redox pairs, selecting appropriate type of defect structure and making use of monolayer oxide catalysts, contact potentials at grain boundaries and gas-solid interactions resulting in surface reconstruction. The dynamic behaviour of oxides in oxidation reactions is analysed. The mechanism of selective oxidation of hydrocarbons on metal catalysts is shortly presented.

Selective catalytic oxidation is the simplest method to functionalize hydrocarbon molecules (1,2). The processes, in which hydrocarbons are oxygenated to form alcohols, aldehydes or acids or their anhydrides, are the basis of the modern petrochemical industry and constitute the largest category of catalytic organic reactions. Practically all monomers used in manufacturing artificial fibers and plastics are obtained by catalytic oxidation processes. Moreover, these reactions are responsible for two basic functions of the living cells: they provide energy for

0097–6156/96/0638–0020$15.00/0

endergonic cellular processes and transform dietary materials into cellular constituents. In chemical industry more than 60% of products obtained by catalytic routes are products of oxidation.

The achievement of high selectivity in oxidation of hydrocarbons is particularly challenging for both physical chemists and chemical engineers because the final result depends on many opposing factors (3), among which the following are the most important:

- in hydrocarbon oxidation processes thermodynamics favours the ultimate formation of carbon dioxide and water. Therefore, all products of partial oxidation are derived by kinetic control of the reaction;
- the hydrocarbon-oxygen mixture can usually react along many different pathways in the network of competing parallel and consecutive reactions and therefore the catalyst must strictly control the relative rates, accelerating the series of consecutive elementary steps leading to the desired product and hindering those, in which unwanted byproducts are formed;
- the C-H bonds in the initial reactant are usually stronger than those in the intermediate products, which makes these intermediates prone to rapid further oxidation;
- all oxidation processes are strongly exothermic and efficient heat removal must be ensured to control the temperature and prevent over-oxidation, appearance of hot-spots as well as catalyst damage.

As oxidation reactions are multistep processes and it is usually an intermediate appearing as a result of kinetic control, which is the wanted product, the integral selectivity as determined in the laboratory flow-reactor is a complex result of the operation of a catalyst in the given reactor in the given reaction conditions and therefore should be considered as reactor selectivity. It is the outcome of the intrinsic selectivity resulting from the reaction at the catalyst surface and subsequent transformations of the desorbed products in the course of their diffusion through the pores of the catalyst particle (particle selectivity) and later in the gas phase along the reactor. Thus, the reactor selectivity depends on the degree of conversion and on the type of the reactor used even if heat and mass transfer effects are eliminated. A typical example of selectivity-conversion dependence is shown in Figure 1, in which selectivity to maleic anhydride is plotted as a function of the conversion of butane (4). The dependence of the maleic anhydride yield on conversion shows a well defined maximum for a conversion of about 80%. A further increase in conversion drastically decreases the selectivity to around 20% for total conversion. The decrease of selectivity to maleic anhydride follows an increase of the ratio of residual concentration of oxygen to n-butane suggesting that at the down-stream end of the catalytic bed combustion of maleic anhydride takes place due to the overoxidation of the catalyst surface.

Therefore, when referring to selectivity data, the conversions at which they were determined should be always given, because otherwise the comparison of different catalysts is meaningless. If possible, in the attempts to correlate the performance of the catalyst with its surface structure and physico-chemical properties the intrinsic selectivity values should be preferably used.

Selectivity is a critical factor in scaling up an oxidation process to a commercial scale. It is obvious that capital and labour costs can be lowered by increasing the plant size. However, in larger plants the contribution of the cost of feedstock to the total costs is higher than in smaller plants so that the cost advantage resulting from improvement in selectivity constitutes an important incentive, specially in view of the rising prices of hydrocarbons. On the other hand, the increase in selectivity permits a reduction of the plant size for a specified capacity, and hence further reduction of the investment costs. Also, because the combustion to carbon oxides provides the main contribution to the heat release from the conversion of feedstocks and the major part of capital costs in oxidation processes is

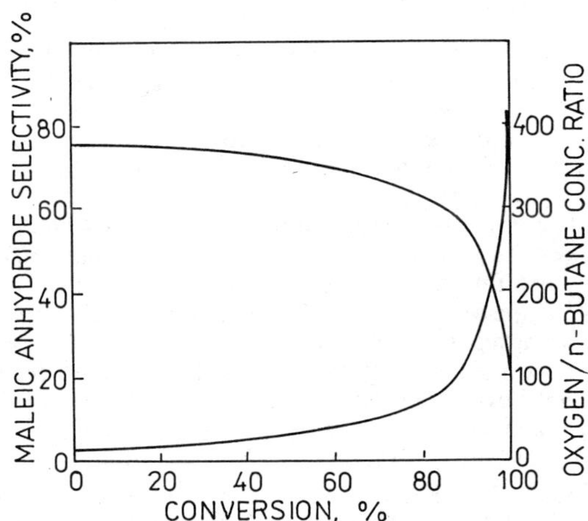

Figure 1. Selectivity to maleic anhydride and oxygen/butane residual ratio as a function of the conversion of butane at 300°C (3).

related to the efficient removal of heat from the reaction zone, an increase in selectivity results in additional reduction of capital costs.

Electrophilic and Nucleophilic Oxidation

As the hydrocarbon molecule is a singlet and the oxygen molecule is a triplet, either the hydrocarbon or the oxygen must be activated to get an efficient reaction. Molecular oxygen may be activated in different ways (5,6): by excitation to the singlet state or by transfer of electrons from the catalyst to the oxygen molecule to form molecular or atomic ion radicals O_2^- or O^-. All these forms are strongly electrophilic reactants. They may abstract hydrogen from the hydrocarbon molecule with the formation of alkyl radicals, which may start a chain reaction leading to total oxidation. In reactions with olefins or aromatics they attack the regions of highest electron density - the π-bond system (7). Peroxo- and superoxo-complexes are formed, which decompose by C-C bond cleavage to give oxygenated fragments or undergo combustion. These reactions may be classified as electrophilic oxidations (8). The second route of heterogeneous oxidation starts with the activation of the hydrocarbon molecule by abstraction of hydrogen from a given carbon atom, which becomes prone to nucleophilic addition of the oxide ion O^{2-}. It should be emphasized that the latter has no oxidizing properties, but is a nucleophilic reactant. The consecutive steps of hydrogen abstraction and oxygen addition may be then repeated to obtain selectively more oxygenated molecules. These reactions are classified as nucleophilic oxidation (8). The role of oxidizing agents in these steps of the reaction sequence is played by cations of the catalyst lattice. After the nucleophilic addition of the lattice oxygen ion the oxygenated product is desorbed, leaving a vacancy at the catalyst surface.

As an example, Figure 2 illustrates the reaction network of an alkane molecule at the surface of an oxide catalyst. An alkane molecule begins to interact with an oxide surface by forming weak hydrogen bonds with the surface O^{2-} ions and OH^- groups. When the surface oxide ions have enough strong basic properties, an attack on the C-H bond takes place with the abstraction of a proton to form a surface OH group and the generation of adsorbed alkyl groups, which may lose a second hydrogen atom and transform into olefins. When the surface contains also OH groups which show Brønsted acid properties, their protons may form hydrogen bonds with the π-bonds of the olefins and when the acid properties are strong enough the transfer of a proton from the surface to the olefin may take place resulting in the formation of a carbocation. This may start a whole network of reactions proceeding by a carbocation mechanism, e.g. isomerization, transalkylation, cracking etc.

The π-bond of the olefin, instead of interacting with a surface proton, may react with a transition metal cation, displaying the properties of a Lewis acid site, and may form a surface π-complex. When the basicity of surface oxide ions is strong enough, they may perform a nucleophilic attack on the C-H bond in the α-position, which results in the abstraction of hydrogen and formation of an allylic species (9). The latter may be bonded to the transition metal ion either side-on as the so called π-allyl, or end-on as the σ–allyl. An equilibrium between these two forms exists at the oxide surface, the latter being an intermediate in oxidative coupling to form dienes, the former undergoing a nucleophilic attack by another surface oxide ion resulting in the formation of an aldehyde in the case of an attack on a primary carbon atom of the hydrocarbon molecule, or ketone in the case of the secondary one (nucleophilic oxidation). Quantum-chemical modeling has shown that the reaction pathway followed by the reacting system depends on the type of hydrocarbon molecule and its orientation on approach to the surface (10,11,12,13).

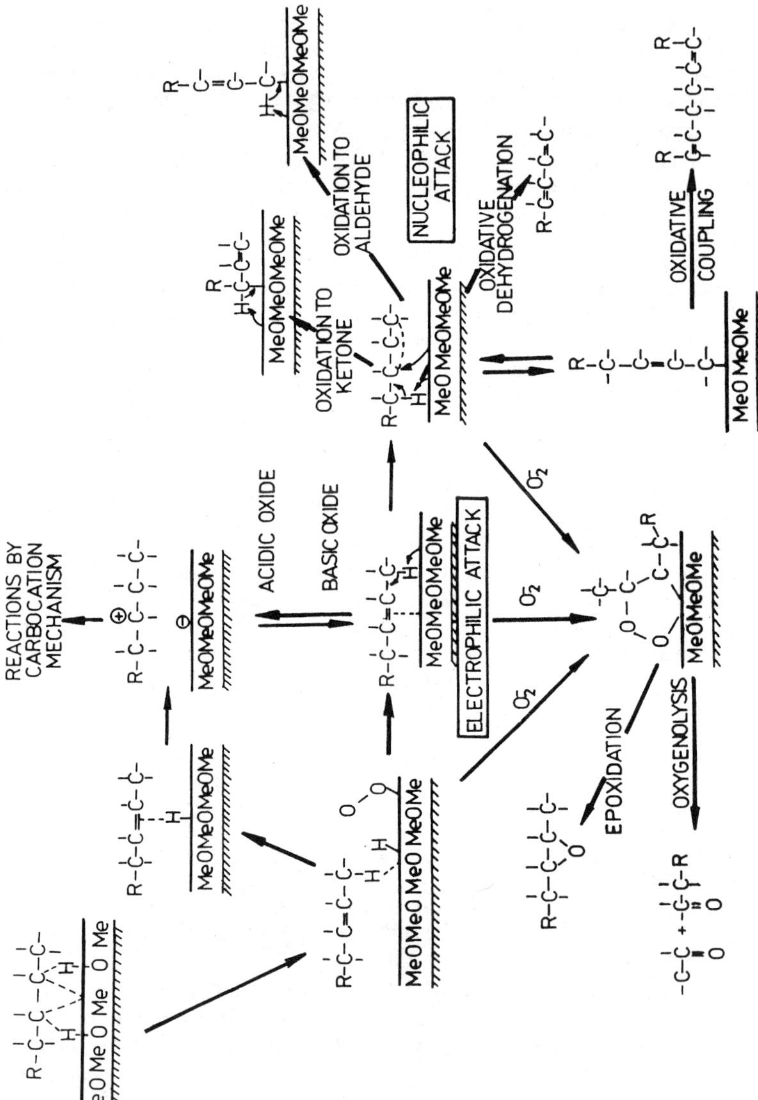

Figure 2. Reaction network of the oxidation of a hydrocarbon molecule at the surface of an oxide catalyst.

When e.g. alkylaromatic molecules are oriented side-on, with the ring-plane parallel to the surface, strong interactions develop between all carbon atoms and surface oxygen atoms, resulting in the destruction of the molecule and formation of CO. When the molecule approaches end-on pointing to the surface with its alkyl group, abstraction of hydrogen takes place followed by nucleophilic addition of surface oxygen atom and desorption of the oxygenated molecule (13). Similarly, in the case of electrophilic oxidation of benzene with adsorbed oxygen the product formed depends on the mutual orientation of the adsorbed reactants (11,12).

Many oxides contain surface vacancies that form F^--centres which may play the role of sites for activation of oxygen molecules to various reactive moieties of electrophilic character. Moreover, non-stoichiometric transition metal oxides are in dynamic equilibrium with the gas phase and the surface may be populated with transient electrophilic oxygen moieties as discussed below. These moieties may perform an electrophilic attack on any of the intermediates of the hydrocarbon reaction network resulting in oxygenolysis (electrophilic oxidation).

The presence of O_2^- or O^- species at the surface of an oxide may be detected by different techniques (14). One of the methods which permits the quantitative determination of the number of electrons transferred between the solid and the adsorbed layer is the measurement of the changes of work function in the course of adsorption (15). When changes of work function due to exposure to oxygen are followed upon temperature variation and the amount of oxygen adsorbed is simultaneously measured, the number of electrons localized per oxygen atom adsorbed may be determined and hence the type of oxygen species residing at the surface may be found. Results of such experiments (15) carried out with different oxides in different temperature ranges are summarized in Table I and compared with the catalytic properties of these oxides. They clearly show that whenever electrophilic oxygen species O_2^- and O^- are present at the surface, total oxidation is observed in the course of the catalytic oxidation of hydrocarbons.

Table I. Oxygen Species at Surfaces of Various Oxides (15).

Catalyst	*Temp. range (K)*	*Oxygen species*	*Catalytic behaviour*
Co_3O_4	293-423	O_2^-	Total oxidation
	573-673	O^-	Total oxidation
V_2O_5 and	293-393	O_2^-	Total oxidation of aromatics
V_2O_5-TiO_2	533-653	O^-	Total oxidation of aromatics
	653	O^{2-}	Selective oxidation of aromatics
$Bi_2Mo_3O_{12}$	538-673	O^{2-}	Selective oxidation of olefins

Thus, in order to prepare a selective catalyst, active sites must be generated at its surface, which accelerate the series of consecutive elementary steps that transform the reactant molecules into the desired final products, while all other active sites, which lead to the formation of undesired byproducts must be eliminated or deactivated.

Reoxidation of the Catalyst

After the desorption of the oxygenated product molecule an oxygen vacancy is formed and the active center at the surface is left in the reduced state. Before the next catalytic cycle can be repeated, the active center must be reoxidized. The reoxidation may be realised in three ways (16): i/ by incorporation of oxygen from the gas phase, ii/ by reoxidation at some other surface site and diffusion of oxygen ions through the bulk, iii/ by diffusion of the reduced lattice constituents to the nuclei of the new phase of lower valent oxide or metal and exposure of underlying fresh surface. Depending on the rate of such regeneration of the active center, its "dead time" may be shorter or longer. When the mobility of lattice oxygen is high the regeneration by mechanism (ii) operates very effectively, the dead time of active centers is very short, and the turn-over frequency on these centers very high (17). In order to accelerate the regeneration by this mechanism a mixture of oxides or a composite oxide are used, one redox pair participating in the oxidation of the hydrocarbon molecule and the second redox pair mediating the reoxidation of the catalyst by gas phase oxygen, the transport of oxide ions occurring by lattice or surface diffusion. The dramatic influence of the introduction of the redox pair Fe^{2+}/Fe^{3+} into the $Bi_2(MoO_4)_3$ catalyst in the oxidation of propene to acrolein is illustrated by the following data obtained at similar experimental conditions (18):

Catalyst	Selectivity to acrolein, (%)	Conversion of propene, (%)
$Bi_2(Mo_4)_3$	90.3	7.4
$M^{2+}{}_aM^{3+}{}_bBi_xMo_yO_z$	95.7	69.7

Functions of the Selective Catalyst

For catalysts to be active and selective in the oxidation of hydrocarbons they must be able to perform complex multifunctional operations and therefore must conform to a number of conditions (19). Namely, active sites must be present at the catalyst surface, which: 1) - activate the hydrocarbon molecule by abstraction of hydrogen from the appropriate position in the molecule, 2) - inject or remove electrons from the surface intermediates by providing energy levels of cationic redox pairs of appropriate redox potential, 3) - properly orient the reacting molecule in respect to the surface, 4) - perform a nucleophilic addition of a surface oxide ion which could be easily extracted from the surface in the desorption of the oxygenated intermediate, 5) - activate the oxygen molecule and orient it properly at the surface in respect to the hydrocarbon molecule in electrophilic oxidation steps, 6) - enable rapid diffusion of hydrogen and undergo a facile dehydroxylation, 7) - easily reoxidize by interaction with gas phase oxygen and show rapid diffusion of oxygen through the lattice to the reaction site, 8) - all other active sites which could accelerate the possible side reactions must be eliminated. Very often this last requirement refers to Lewis acid sites, whose presence might be detrimental e.g. in the case of the oxidation of hydrocarbon molecules with π-electron systems. Such molecules form stable surface π-complexes with Lewis acid sites and become strongly adsorbed at the surface of the catalyst. This prolongs their residence time and increases the probability of total oxidation. Such effect was observed in the oxidation of o-xylene to phthalic anhydride on V_2O_5/TiO_2 catalysts, for which the complete coverage of TiO_2 surface by a monolayer of VO_x to block its acid sites considerably improved the selectivity (20).

Defect Structure in Oxides and Oxygen Reactivity

Transition metal oxides used as catalysts in selective oxidation of hydrocarbons are non-stoichiometric compounds, their composition depending on the equilibrium between the lattice and its constituents in the gas phase. Changes in oxygen pressure cause changes in stoichiometry of the oxide, which may be accommodated by the crystal lattice in two ways (Figure 3): either by generation of point defects or by alteration of the mode of linkage between the oxide coordination polyhedra, resulting in the formation of extended defects (crystallographic shear). When non-stoichiometry is introduced by the presence of point defects, a series of equilibria is established at the surface on the pathway of lattice oxygen being transferred into the gas phase or in the reversed process of reoxidation of the lattice (Figure 4). When the temperature increases, the equilibrium shifts in the direction of higher dissociation pressure of the oxide and the surface becomes more and more populated with electrophilic oxygen species. When used as catalysts in oxidation of hydrocarbons, such oxides may show high selectivity to partial oxidation products at low temperatures, when conversion is very low,and on raising the temperature the conversion increases but selectivity rapidly decreases, the total oxidation becoming the predominant reaction pathway (21). It should be borne in mind that in the case of those oxides, in which the transition metal cation is not in its highest oxidation state, chemisorption of oxygen takes place at low temperatures, electrons being transfered from the oxide to adsorbed oxygen molecules with formation of higher valent cations and electrophilic oxygen species. This may be followed by surface reconstruction and formation of a monolayer of higher valent oxide at the surface of the lower valent oxide (21).

Different behaviour is shown by oxides, in which the change of stoichiometry is accommodated be the formation of shear structures. Because there is no vacancy formation on addition of nucleophilic oxygen ion from the oxide surface to the hydrocarbon molecule and desorption of oxygenated product, but instead a shear plane is formed or an existing one grows (7,16,22), on raising the temperature the activity increases with rising mobility of oxygen, but the selectivity to partial oxidation products remains very high because of the absence of electrophilic oxygen. This difference in behaviour is illustrated in Figure 5, in which the activity and selectivity in the oxidation of propene to acrolein as a function of the temperature of reaction is compared for the two classes of compounds. $Ni_3(PO_4)_2$ is an example of an oxysalt with point defects, and $NiMoO_4$ an example of an oxysalt forming shear structures.

As the phenomenon of crystallographic shear appears in transition metal oxides with anisotropic lattices, pronounced structure-sensitivity of catalytic properties is observed and the habit of crystallites of the catalyst may have strong influence on the selectivity of the reaction (23,24,25). This is illustrated in Figure 6, in which the selectivity in the oxidation of o-xylene to phthalic anhydride and CO_2 on V_2O_5 are plotted as the function of the habit of crystallites, described by a structural factor which represents the ratio of intensities of (110) to (001) reflections in the X-ray diagram (26).

Oxide Monolayer Catalysts

One of the methods of securing the exposure of a selected crystal plane of a transition metal oxide is to disperse the oxide in the form of a monolayer at the

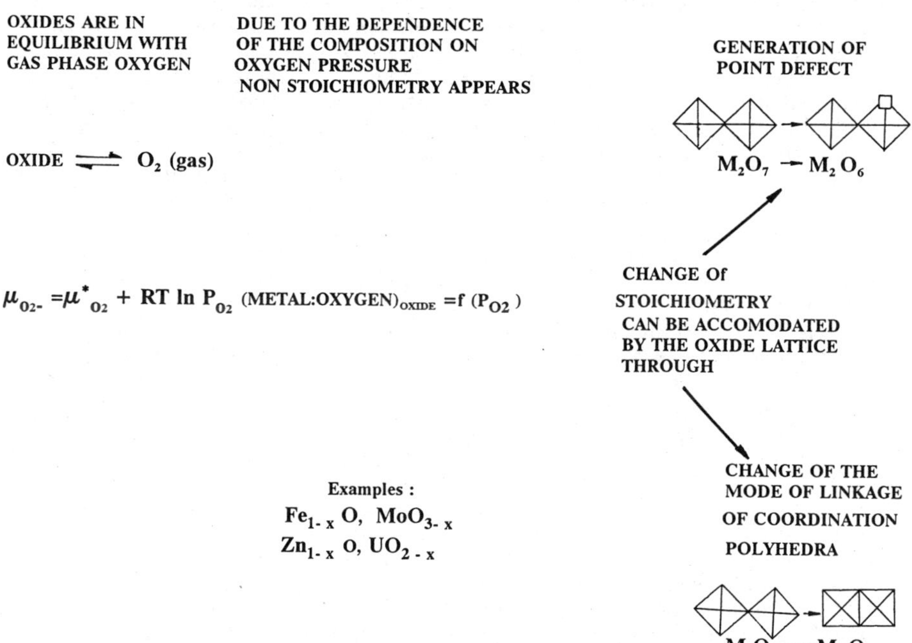

Figure 3. Defect structures in transition metal oxide.

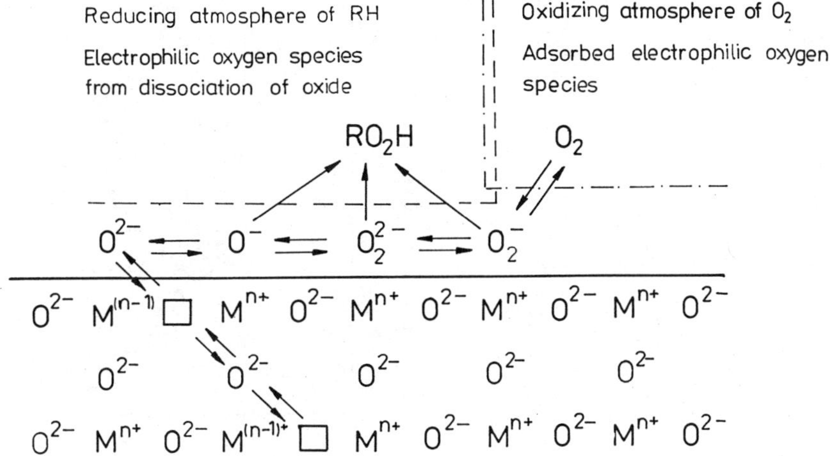

Figure 4. Formation of electrophilic oxygen species at the oxide surface.

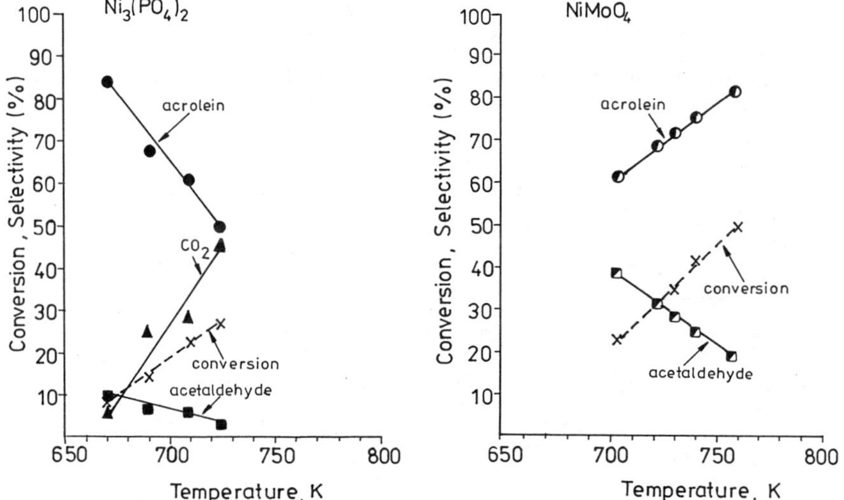

Figure 5. Conversion and selectivity of propene oxidation to acrolein, acetaldehyde and CO_2 on $NiMoO_4$ and $Ni_3(PO_4)_3$ as function of the reaction temperature.

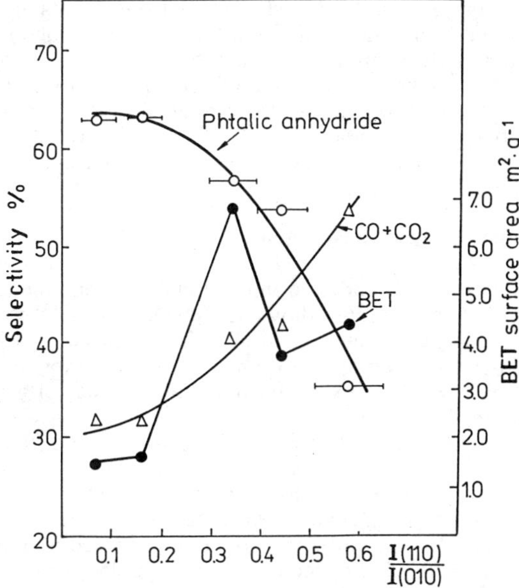

Figure 6. Selectivities in oxidation of o-xylene on V_2O_5 catalysts as function of the structural factor, describing the crystal habit. Its low value means the presence of thin platelets exposing mainly the basal (010) plane, high value - spherical particles exposing predominatly other planes [after (26)].

surface of another oxide, which plays the role of a support. Modification of the properties of an oxide monolayer may be due to the following factors (22):

- he oxide monolayer exposes only one crystal face, which may be different at surfaces of various supports due to the effect of epitaxy. Catalysts may be thus tailored to have surface structure optimal for the given reaction and eliminate parasitic reactions at other crystal planes;
- in the monolayer the molecular properties of dispersed isolated species may dominate over the collective properties of clusters permitting the obtention of the properties of the monolayer from those of the transition metal coordination complexes;
- due to the transfer of electrons between the monolayer and the support the occupancy of molecular orbitals of the active centers may be changed and their HOMO-LUMO character modified. When surface clusters are large enough to present collective properties, this transfer of electrons results in the appearance of contact potential modifying the position of the Fermi level and hence the chemical potential of electrons and the catalytic properties. As the position of this level is sensitive to the presence of additives in the lattice and particularly at the surface, the catalytic properties of the monolayer may be modified by doping. The correlation between the catalytic performance of different monolayers and their type, structure and chemical properties has recently been discussed in a number of papers (27).

Dynamics of the Oxide Catalyst Surface

In the last decade a vast amount of experimental evidence and the results of quantum-chemical calculations have shown that the surface of the catalyst, both oxide and metal, is in dynamic interaction with the gas phase (13,21,25,29,30). Adaptability to changes in external conditions is one of the most important properties of solid surfaces, responsible for many phenomena of great theoretical and practical importance.

Oxide systems may respond to changes in composition of the reacting catalytic mixture in three ways:

- changes in composition of the gas phase, which entail the modification of its redox potential, may induce a change of the steady-state degree of reduction of the catalyst. Surface defect equilibria at the oxide surface or in the whole bulk may be shifted and changes of concentration of a given type of site involved in the catalytic transformation may cause changes in catalytic properties;
- when the concentration of defects at the oxide surface surpasses a certain critical value, ordering of defects or formation of a new bidimensional surface phase may occur resulting often in a dramatic change in catalytic properties (28);
- when a redox mechanism operates in the catalytic reaction, the ratio of the rate of catalyst reduction and its reoxidation may be different for various phases and hysteresis in the dependence of catalytic properties on the composition of the gas phase may appear, these properties being then strongly influenced by the type of pretreatment. One of examples is the formation of the active phase of the vanadyl pyrophosphate catalyst for selective oxidation of butane to maleic anhydride by pretreatment of the precursor in an atmosphere of reactants (30). Recent experiments indicate that evolution of the active phase takes place in the course of the reduction of vanadyl phosphate precursor to vanadyl pyrophosphate by butane in the presence of maleic anhydride as the reaction product (31). This might be the reason why VPO catalysts after standard preparation attain their highest activity and selectivity only after many hours of time-on-stream.

Oxidation of Hydrocarbons on Metal Catalysts

The only elements which in the conditions of catalytic oxidation resist oxidation to metal oxides are noble metals. Therefore only in the case of these elements can one speak of catalytic oxidation on metal catalysts. With the exception of silver they catalyse total oxidation of hydrocarbons. It is generally agreed (32) that oxygen may be adsorbed on metal surfaces as: 1) - weakly adsorbed molecular oxygen, 2) - chemisorbed atomic oxygen, 3) - oxygen atoms penetrating into the outermost layers of metal crystals, thus forming "subsurface oxygen". On the platinum surface, in the temperature range of catalytic reactions (500-700 K) atomic oxygen predominates. It is a highly reactive radical which generates hydrocarbon radicals and initiates a chain reaction developing into the gas phase and resulting in rapid total oxidation. Platinum is therefore used for catalytic combustion of hydrocarbons. It is noteworthy that the platinum surface undergoes reconstruction in vacuum and returns to the initial structure under the influence of adsorbates such as CO or hydrocarbons (33). In the case of the e.g. Pt(110) surface, the sticking coefficient of oxygen is much higher on the unreconstructed surface than on the reconstructed one. Therefore the rate of the oxidation of adsorbate molecules is greater and they are swept away, the surface undergoing again reconstruction. Self-sustained oscillations of the kinetics of the catalytic reaction thus appear at certain conditions. Such oscillatory behaviour has been observed in the total oxidation of ethylene and propene on Pt and Rh catalysts prepared in different forms (34). Recently, spatially resolved measurements carried out with the help of the photoemission electron microscopy (PEEM) technique revealed the formation of spatiotemporal patterns such as propagating and standing waves as well as chemical turbulence, requiring some kind of communication between different regions of the surface, i.e. spatial selforganisation (35).

The only metal, which is known as a catalyst for selective oxidation, is silver. It is used in the catalytic epoxidation of ethylene to ethylene oxide and the oxidation of methanol to formaldehyde. Earlier it was assumed that oxygen at the surface of silver is adsorbed dissociatively until only single sites remain vacant. Molecular oxygen is adsorbed at these sites and reacts with ethylene from the gas phase to form ethylene oxide (36). Single oxygen atoms left at the surface are removed by reacting with ethylene to total oxidation products. More recently it was demonstrated (37) that the adsorbed oxygen atoms, which exchange with oxygen atoms in the subsurface layer, are the intermediates for the epoxidation reaction rather than the molecular oxygen species. The selectivity of ethylene epoxidation increases with increasing oxygen surface coverage and the epoxide is produced when the oxygen-to-silver atom ratio in the reactive surface layer surpasses the value of 0.5. This is possible only if some part of oxygen is present in the subsurface layer.

An interesting observation of the behaviour of rhodium was recently reported (38). When alkyl radicals, formed in situ by decomposition of alkyl iodides, were introduced on the oxygen-covered Rh(111) surface, they reacted along two parallel reaction pathways: either the surface oxygen atoms abstracted hydrogen from the alkyl radicals to form alkenes, or they bound these radicals forming transient alkoxides, which subsequently underwent dehydrogenation to form aldehydes or ketones. On clean Rh(111) surface the alkyl radicals adsorbed and decomposed with the formation of carbon deposit. This is another example of a catalyst, whose selectivity in hydrocarbon oxidation is determined by the oxygen coverage of the catalyst surface.

References

(1) Bielanski, A.; Haber J., *Oxygen in Catalysis*; Marcel Dekker Inc: New York, 1991.

(2) Pasquon, I. *Catal.Today* **1987**, *1*, 297.
(3) Haber, J. Stud.Surf.Sci.Catal. 72; Elsevier: Amsterdam, 1992; p.279.
(4) Centi, G.; Fornasari, G.; Trifiro, F. *Ind.Eng.Chem. Prod.Res.Dev.* **1985**, *24*, 32.
(5) Che, M.; Tench, A. J. *Adv.Catal.* **1982**, *31*, 78; **1983**, *32*, 1.
(6) Sojka, Z. *Catal.Rev.Sci.Eng.* **1995**, *37*, 461.
(7) Haber, J. In *Proc.8th Intern.Congr.Catal.,Berlin 1984*; Verlag Chemie-Dechema: 1984, vol.1; Plenary Lectures, p.85.
(8) Haber, J. In *The Role of Solid State Chemistry in Catalysis*; Grasselli, R. K.; J.F.Brazdil, J. F., Eds.; ACS Symp. Series No 279; Washington, 1985; p.3.
(9) Adams, C. R. In *Proc.3rd Intern.Congr.Catal., Amsterdam 1964*; North Holland Publ.Co.: Amsterdam, 1965; p.240;
(10) Haber, J.; Witko, M. *Catal.Lett.* **1991**, *9*, 297.
(11) Broclawik, E.; Haber, J.; Witko, M. *J. Molecular Catal.* **1984**, *26*, 249.
(12) Witko, M.; Broclawik, E.; Haber, J. *J. Molecular Catal.* **1986**, *35*, 179.
(13) Haber, J.; Witko, M. *Catal.Today* **1995**, *23*, 311.
(14) Che, M. In Stud.Surf.Sci.Catal. 75A; Elsevier: Amsterdam, 1993; p.31
(15) Libre, J. M.; Barbaux, Y.; Grzybowska, B.; Bonnelle, J. P. *React.Kinet.Catal.Lett.* **1982**, *20*, 249.
(16) Haber, J. *Pure&Appl.Chem.* **1978**, *50*, 923.
(17) Coulson, D. R.; Mills, P. L.; Kourtakis, K.; Lerou, J. J.; Manzer, L. E. In Stud.Surf. Sci.Catal. 72; Elsevier: Amsterdam, 1992; p.305.
(18) Grasselli, R. K. *Appl.Catal.* **1985**, *15*,127.
(19) Haber, J. In *Proc.8th Annual Science&Technology Seminar "Advances in Catalytic Technologies", Santa Clara 1991*; Catalytica Stud.Div.: Mountain View, 1992.
(20) Grzybowska-Swierkosz, B. *Materials Chem.Phys.* **1987**, *17*.
(21) Haber, J. *Materials Sci.Forum* **1988**, *25/26*, 17.
(22) Haber, J. In *Perspectives in Catalysis*; Thomas, J.M.; Zamaraiev,K. I., Eds.; Blackwell Scientific Publ.: Oxford, 1992; p.371.
(23) Germain, J. E. In Stud.Surf.Sci.Catal. 21; Elsevier: Amsterdam, 1985; p.355.
(24) Vedrine, J. C.; Coudurier, G.; Forissier, M.; Volta, J. C. *Catal.Today* **1987**, *1*, 261.
(25) Haber, J. In Stud.Surf.Sci.Catal. 48; Elsevier: Amsterdam, 1989; p.447.
(26) Gasior, M.; Machej, T. *J.Catal*.**1983**, *83*, 472.
(27) Deo, G.; Wachs, I. *J.Phys.Chem.* **1991**, *95*, 5889.
(28) Haber, J. In *Proc.3rd Intern.Conf.Chemistry and Uses of Molybdenum*; Barry, H. F.; Mitchell, P. C. H., Eds.; Climax Molybden. Co: An Arbor, 1980; p.114.
(29) Samorjai, G. *Surf.Sci* **1994**, *299/300*, 849.
(30) Cavani, F.; Trifiro, F. In Specialists Periodical Reports, Catalysis 11; The Royal Society of Chemistry: London, 1994; p.246.
(31) Hutchings, G. J.; Desmartin-Chomel, A.; Olier, R.; Volta, J. C. *Nature* **1994**, *368*, 41.
(32) Ponec, V.; Bond, G. C. *Catalysis by Metals and Alloys*; Stud.Surf.Sci.Catal. 95; Elsevier: Amsterdam, 1995.
(33) Ertl, G. *Adv.Catal.* **1990**, *37*, 213.
(34) Slinko, M. M.; Jaeger, N. I. *Oscillating Heterogeneous Catalytic Systems*; Stud.Surf.Sci.Catal. 86; Elsevier: Amsterdam, 1994.
(35) Ertl, G. *Surf.Sci.* **1994**, *299/300*, 742.
(36) Kilty, P. A.; Sachtler, W. M. H. *Catal.Rev.Sci.Eng.* **1974**, *10*, 1.
(37) van Santen, R. A.; Kuipers, H. P. E. *Adv.Catal.* **1987**, *35*, 265.
(38) Bol, C. W. J.; Friend, C. M. *J.Phys.Chem.* **1995**, *99*, 11930.

COMBUSTION

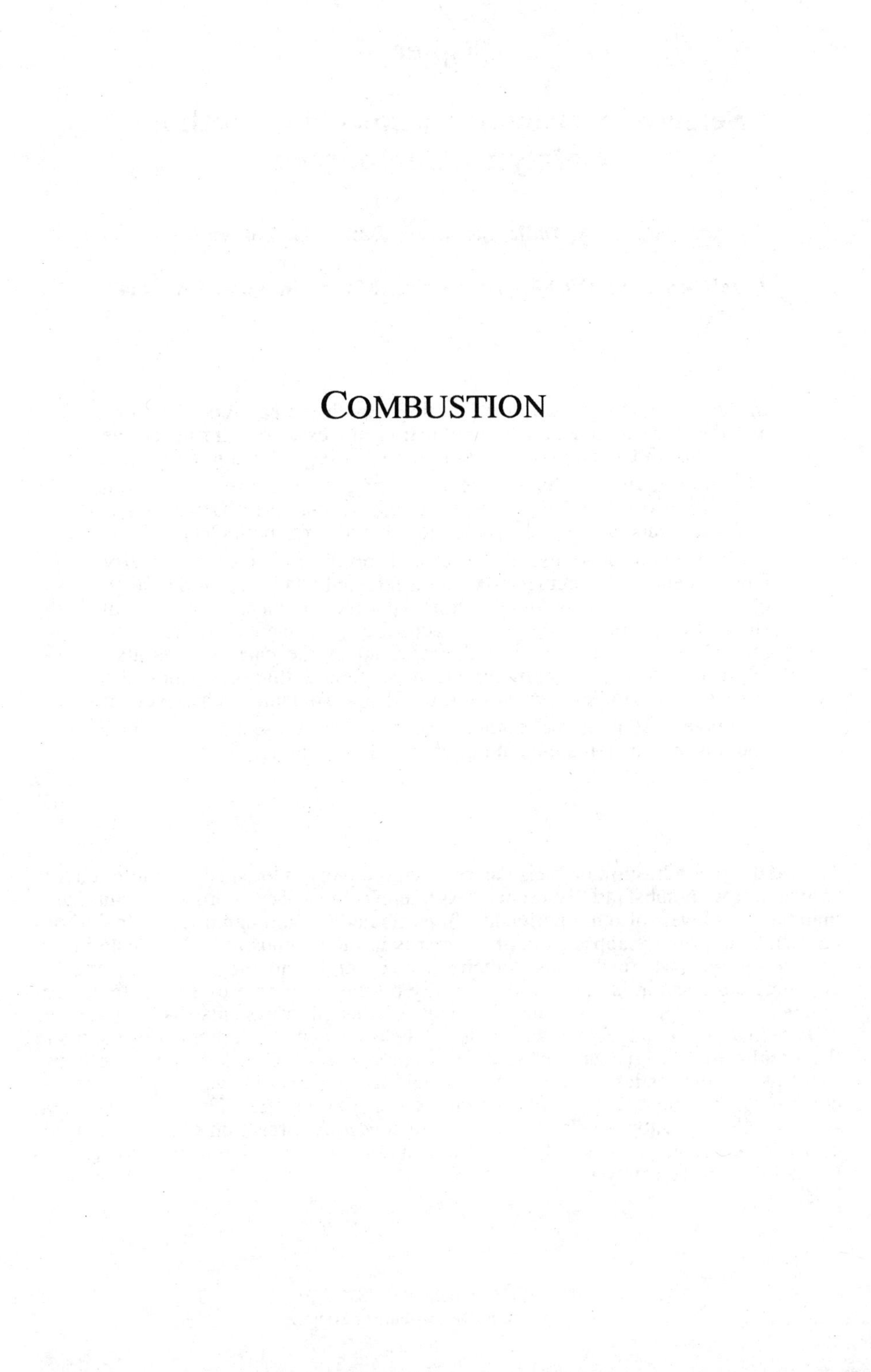

Chapter 3

Selectivity Considerations in Methane Catalytic Combustion

Ralph A. Dalla Betta and Daniel G. Löffler

Catalytica Inc., 430 Ferguson Drive, Mountain View, CA 94043

In the catalytic combustion of methane, both the catalytic reactions and the downstream radical combustion processes can influence the selectivity to CO and CO_2. Assuming CH_4 is oxidized to CO_2 via a CO intermediate, the CO level exiting a supported palladium monolithic catalyst was modeled assuming that the CO was a gas phase intermediate and using literature kinetics for both CH_4 and CO surface reaction rates. The model predicts a CO selectivity independent of the methane oxidation rate and CO levels in the range of 1 to 6 ppm. This agrees with experimental measurements that showed levels of 5 to 8 ppm at practical gas turbine conditions. The radical reaction process just downstream of the catalyst was also explored using a kinetic reaction sequence that describes the homogeneous radical combustion of CH_4. To achieve CO levels in the range of 10 ppm in a reaction time adequate for use in a gas turbine combustor, reaction temperatures above 1100°C are required.

The catalytic combustion of fuels has been vigorously explored as a route to energy conversion with substantially reduced NOx emissions. These processes must also maintain low levels of other pollutants, in particular CO and unburned hydrocarbons (UHC). The primary application of interest is in the combustion of methane in gas turbine engines where the gas velocity is very high and the residence time for complete combustion is very short, typically on the order of 5 to 30 ms. In current practical gas turbine systems, development is directed at systems that can achieve NOx below 5 ppm and CO and UHC levels below 10 ppm. A number of factors in the combustor design can influence the emissions of CO and UHC including inadequate combustion of the fuel, quenching in the cold wall regions of the combustor and quenching by injected cooling or dilution air. This paper will deal with the CO emission levels that arise from inadequate reaction of the fuel in the catalytic oxidation process and the subsequent homogeneous combustion process just downstream of the catalyst.

0097–6156/96/0638–0036$15.00/0

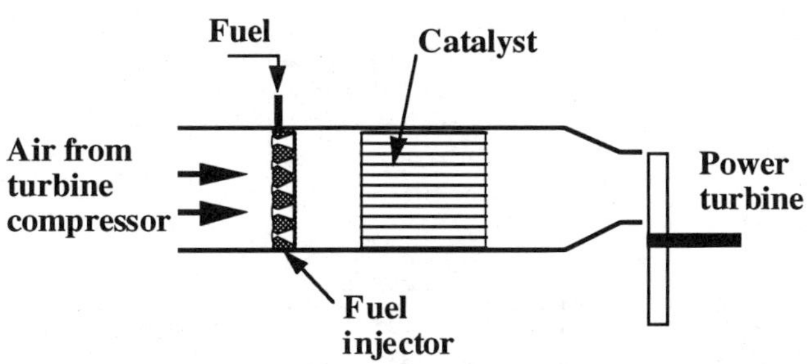

Figure 1 Schematic diagram of a gas turbine catalytic combustion system showing major components of a typical system.

A schematic diagram of a catalytic combustor system is shown in Figure 1. Hot, high pressure air produced by the gas turbine compressor flows into the combustor. The methane is injected into this air through a fuel injector to make a uniform fuel-air mixture that enters the catalyst. The fuel undergoes catalytic oxidation on a monolithic catalyst and any fuel or CO that is not reacted on the catalyst could be combusted in a homogeneous radical reaction in the region downstream of the catalyst. At the end of the combustor, the hot combustion gases enter the power turbine where work is extracted and the temperature of the combustion gas drops rapidly. This rapid temperature decrease quenches any further reaction and "freezes" the composition of the combustion gas stream, essentially locking the CO concentration at the level existing at this point.

The details of the mechanism of the catalytic reaction of methane on noble metal catalysts are not well understood. Even the basic question of whether the oxidation goes from CH_4 all the way to CO_2 without formation of gas phase CO is not unambiguously established. If CH_4 oxidation is a stepwise process from CH_4 to CO then to CO_2, the catalytic oxidation could produce significant amounts of CO which must be subsequently oxidized to CO_2 in the catalyst or in the downstream sections of the combustor or result in a high level of CO emissions.

Catalytic combustion processes can be divided into two basic types of systems. The first is a system in which most of the reaction and energy release occurs within the catalyst and the catalyst temperature is close to the combustor exit temperature. The second approach is one in which only a portion, approximately 30 to 70%, of the fuel is reacted within the catalyst and the remainder is reacted in the free space just downstream of the catalyst. These two processes are schematically compared in Figure 2. In the system described in Figure 2a, the fuel oxidation is essentially catalytic and any partial oxidation products produced by the catalyst will exit the combustor. These products could be further oxidized downstream of the catalyst if the temperature is high enough for radical homogenous reactions to be significant. However, in general these system are not operated at high temperatures since the temperatures above 1100°C result in rapid catalyst deterioration due to sintering and vaporization of catalyst components (*1*).

The system described in Figure 2b uses the catalyst to partially oxidize the fuel and subsequently completes the fuel oxidation in a radical homogenous oxidation process just downstream of the catalyst. In this case, it is the radical homogenous oxidation process that controls the level of partial oxidation products that will exit the combustor.

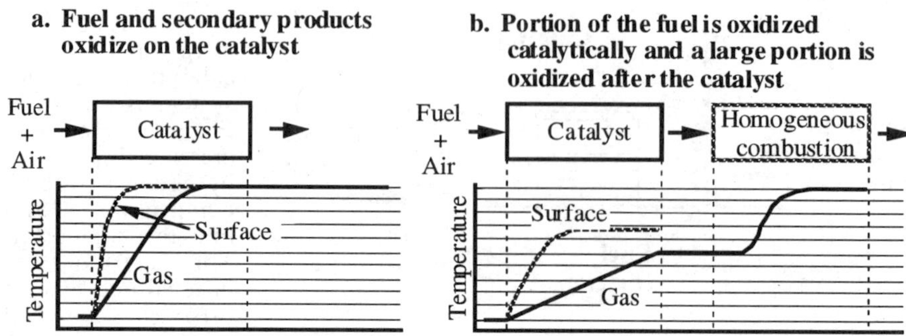

Figure 2. Reaction and heat release profiles for two catalytic combustion system designs.

Experimental

Catalyst Preparation. Catalysts were prepared using cordierite honeycomb monoliths of 50 mm OD by 50 mm length and with a cell density of 200 cells per square inch. These monoliths were coated with a zirconia sol prepared by ball milling a zirconia powder with a surface area of 50 m^2/g at a slurry concentration of 25%. The monoliths were dipped in the zirconia sol and blown out with an air stream, then dried and calcined in air to a final temperature of 950°C. The loading of zirconia on the cordierite monolith was typically 0.27 to 0.29 g ZrO_2/g cordierite. The zirconia coated cordierite monolith was impregnated with an acidic $Pd(NH_3)_2(NO_2)_2$ solution by dipping the monolith into the Pd solution, blowing out the excess and drying and calcining for 10 hours at a temperature of 950°C. Since the expected maximum operating temperature of the catalyst is 950°C, this 10 hour calcination should provide sufficient stability for a short term reactor test without significant sintering of the Pd. To achieve the desired concentration, the initial Pd solution was adjusted so that the solution uptake would give the required Pd loading. The Pd loading was 0.02 gPd/g cordierite.

High Pressure Combustion Test. Experimental measurements of CO production from Pd catalysts were obtained in the flow reactor shown in Figure 3. This reactor can operate up to pressures of 20 atm and air flow rates up to 10,000 standard liters/min (SLPM). For these experiments, the inlet air was heated with the electric heater shown in Figure 3. The flow path is 50 mm in diameter and the catalyst is installed in the flow path with thermocouples located as shown in Figure 3 to measure catalyst and gas temperatures. The catalyst substrate temperature was measured by inserting a thermocouple into one of the channels of the cordierite monolith and sealing the channel with a high temperature stable ceramic cement. The thermocouple leads exited the monolith and were connected to a feedthrough on the high pressure reactor wall. Both air and fuel were controlled by mass flow controllers and the system pressure was controlled by a PID closed loop controller at the reactor exit. The air, provided by a high pressure compressor, was dried prior to entering the reactor.

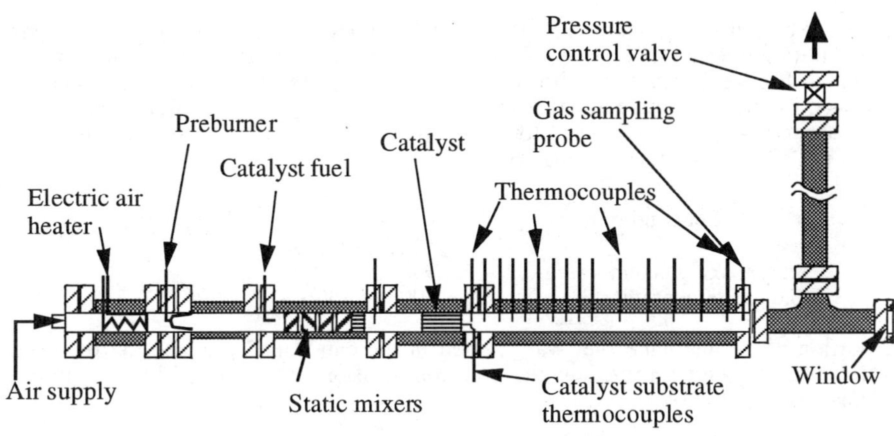

Figure 3 High Pressure catalytic combustion reactor system

The fuel was natural gas from the local distribution system and had a typical composition as follows:

methane	95.9
ethane	2.0
propane	0.1
butane	0.03
C_5 and higher	minor
carbon dioxide	1.42
nitrogen	0.48

A water cooled gas sampling probe can be located at any downstream position. For the experiments discussed here, the sampling probe was located as close as possible to the catalyst, 7 cm from the exit of the monolithic catalyst. The CO and CO_2 concentration was measured using non-dispersive infrared analyzers with calibrated ranges of 0-50 ppm for the CO and 0-5% for CO_2. CH_4 concentration was measured at the inlet to the reactor and at the exit of the catalyst using continuous flame ionization (FID) analyzers. Water was removed from the gas stream prior to these measurements. The reported concentrations were corrected to the composition present in the reactor stream by estimating the water concentration from the measured CO and CO_2 concentration in the product gas stream.

Kinetic Modeling of Homogeneous Methane Combustion

The homogenous radical combustion of methane has been extensively studied and a large data base of reactions and kinetic parameters has been developed to describe this process. One complete reaction model has been reported by Miller and Bowman in 1989 *(2)*. The complete mechanism involves 234 reactions and includes all radical species and many hydrocarbon fragments up to C_4. To permit rapid routine calculations, this kinetic model has been reduced to 143 reactions and 44 species by eliminating reactions that are not important in the temperature and excess oxygen regime explored in this work *(3)*. This kinetic model was used to calculate species concentration over a range of conditions typical of the catalytic combustion scheme shown in Figure 2b, that is, with partial oxidation of the fuel within the catalyst followed by homogenous radical reaction after the catalyst. The calculations were done assuming a plug flow reactor using the CHEMKIN simultaneous equation solver *(4)*. The conditions for this series of model calculations were:

Catalyst inlet conditions	450°C, 2.4 to 4.4% CH_4 in air
Pressure	12 atm absolute
Catalyst exit gas temperature	950°C
Final gas temperature	1000 to 1400°C

A portion of the methane fuel was reacted in the catalyst to produce the required outlet catalyst temperature and the gas composition input to the kinetic model calculation.

Figure 4 shows the concentration profile of several of the important species during the homogeneous radical reaction at a final adiabatic reaction temperature of 1300°C. At the catalyst exit temperature of 950°C, the radical homogeneous reaction process occurs quite rapidly, within ~2 ms. The methane, oxygen and other species react to build up a radical pool that accelerates the reaction of methane.

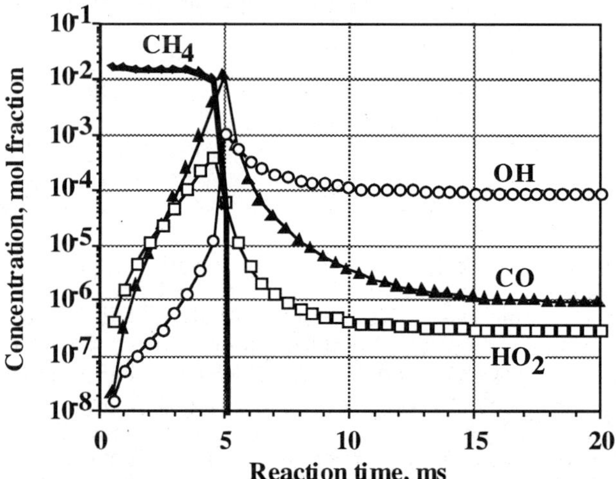

Figure 4. Kinetic model simulation of homogeneous radical reaction of CH_4 downstream of catalytic combustion catalyst.

The CO builds to a relatively high concentration, as much as 1%, and then oxidizes to CO_2. In the case of homogeneous radical reaction, CO is clearly an intermediate product and builds up to a level that is in the range of 25% of the original CH4 concentration. This calculation was done for a final gas temperature of 1300°C and at this temperature, the reaction of CO is very rapid and proceeds to a relatively low final CO concentration, approximately 1×10^{-6} mol fraction. This kinetic simulation was run at several final temperatures by reducing the amount of CH_4 at the catalyst inlet. The CO concentration profiles are shown in Figure 5. The CO reaction profile is strongly dependent on the final gas temperature. The reaction time required to achieve CO levels of ~10 ppm (10^{-5} mol fraction) is 2.5 ms at 1400°C and increases to 12 ms at 1100°C.

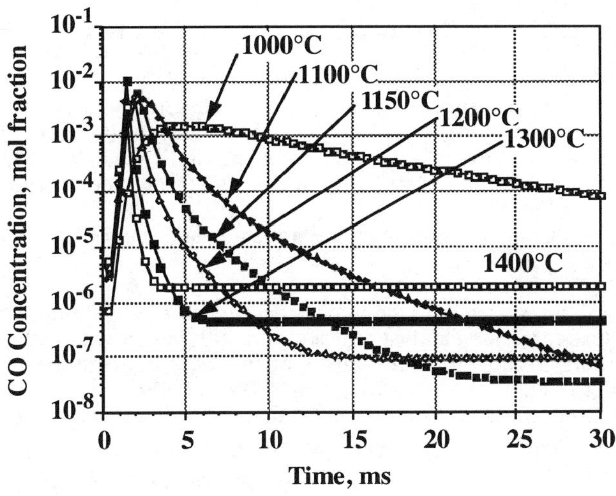

Figure 5. Calculated CO concentration profiles for post catalyst homogeneous reaction with 2.2 to 4.4% methane.

The slow oxidation of CO at low temperatures is shown in Figure 6. In this figure, the time required for CO reaction is taken as the reaction time to reach a level of 10 ppm (10^{-5} mol fraction) after the methane has been full reacted. The reaction time required to reach 10 ppm CO is shown to increase rather abruptly at temperatures below 1100°C, requiring 50 ms to obtain 10 ppm CO. Since the desired emission level is 10 ppm or lower, and the residence times in a typical gas turbine combustor are in the range of 10 to 30 ms, final reaction temperatures should be 1100°C or higher.

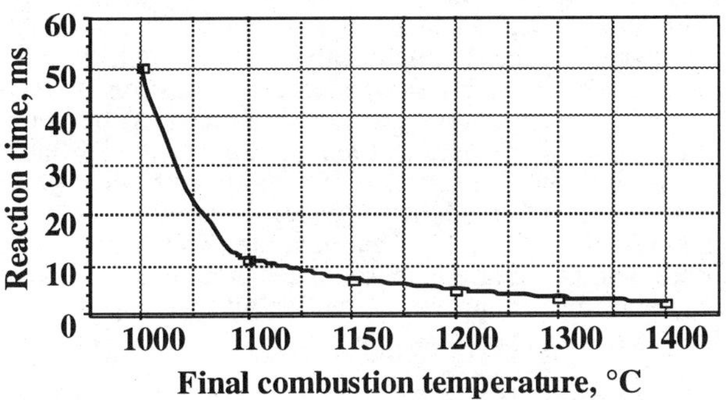

Figure 6. Calculated reaction time to achieve a CO concentration of $1x10^{-5}$ mol fraction for final combustor temperatures from 1000 to 1400°C.

These data suggest another interesting factor, that is, that the equilibrium CO level is dependent on the final temperature. This can be seen from Figure 5 where at 1400°C the CO concentration levels off at about $2x10^{-6}$ mol fraction while at lower temperatures the equilibrium CO level is substantially less. The effect of equilibrium on the achievable CO level is shown in more detail in Figure 6 where the equilibrium CO level at final combustor temperatures from 1000 to 1500°C is shown for pressures from 1 to 20 atm. These data were calculated using a standard energy minimization program combined with the JANAF thermochemical data base. At 1500°C and low pressures, levels of 10 ppm CO may be difficult to achieve. However, at temperatures below 1400°C, levels below 10 ppm should be quite accessible. If the target CO level is below 5 ppm and the system operates at low pressure, the combustor temperature should be 1300°C or lower.

These data establish several criteria for catalytic combustion systems using the design shown in Figure 2b:

- If significant oxidation of fuel occurs after the catalyst, then high post catalyst temperatures are required to rapidly oxidize the intermediate CO.
- Below a combustor temperature of 1100°C, CO oxidation by radical homogeneous chemistry is so slow that it would not be possible to achieve CO emissions targets of 10 to 20 ppm CO without unreasonably large combustor systems.
- The equilibrium CO level can be significant at combustor outlet temperatures above 1300°C and at low pressures.

The above discussion has concentrated on the effect of combustor temperature on CO reaction. In all model test cases the catalyst exit temperature was maintained at 950°C and as shown in Figure 5 and under the range of methane concentrations studies with combustor outlet temperatures ranging from 1000 to 1400°C, the delay time before rapid CH_4 reaction is relatively constant. Changes in the catalyst exit temperature by changing the fraction of CH_4 reacted within the catalyst will change the time required to achieve complete CH_4 reaction. This can be a substantial effect but was not dealt with here because it has a negligible effect on the CO reaction profile.

The homogeneous modeling calculations also provide insight to the requirements for a catalytic system such as that described schematically in Figure 2a. In such a system, the major portion of methane is reacted in the catalyst. If complete reaction is not obtained in the catalyst and post catalyst homogeneous combustion is to be used to further lower the CO level, then the outlet gas temperature must be above 1100°C. This would require that the catalyst operate at temperatures above 1100°C. At temperatures this high, sintering of high surface area supports and active components would be greatly accelerated. In addition, the vapor pressure of many catalytic materials becomes significant at these high temperatures and vaporization loss will become a significant problem *(1)*. For this reason, such combustion systems would probably only be workable at lower temperatures and would have to react all of the fuel and CO within the catalyst via surface reaction.

CO Selectivity on a Supported Pd Catalyst

The selectivity of methane oxidation to CO and CO_2 was measured at several gas flow rates and total pressures for 1.5% CH_4 in air at 600°C inlet temperature to the monolithic catalyst, corresponding to an adiabatic combustion temperature of 950°C. The catalyst temperature was measured with three thermocouples placed 5 mm inside the monolithic structure at the catalyst outlet.. A low level of CH_4 was selected to limit the maximum temperature of the catalyst. These data are shown in Table I.

Since the measurement of the CO concentration is of prime importance to the interpretation of these results, it is important to consider whether the sampling technique used here can lead to errors. One process that could lead to a low CO measurement is further reaction of CO by either homogeneous gas phase reaction while at high temperature before the gas temperature is quenched in the probe. A second mechanism of CO loss is by catalytic oxidation to CO_2 on the sampling probe walls. The gas temperature at the catalyst exit where the sample probe is located is in the range of 840 to 945°C. As can be seen in the results presented in Figure 5 above,

Table I. CO concentration from a Pd monolithic catalyst during CH_4 oxidation

Run	Air flow	Press	Gas temperature Inlet	Gas temperature Outlet	Catalyst temp (outlet)	Gas velocity	Residence time	CO mol fraction
	SLPM	atm	°C	°C	°C	m/s	s x 10^3	x10^6
1	6000	9.5	599	738	945	19.3	1.9	7
2	8000	9.5	600	707	940	25.3	1.4	8
3	6000	2.4	596	658	860	56.5	6.4	8
4	8000	3.1	596	668	840	75.7	4.8	9

at temperatures below 1000°C, homogeneous CO oxidation is very slow. At 950°C, reaction times of 10 to 20 ms would be required before significant CO would be oxidized. The gas is cooled to less then 300°C in 1 to 2 ms by a combination of expansion cooling and heat transfer from the probe walls. The possibility of significant CO oxidation on the probe walls is low since the probe is water cooled and the metal wall temperature is less then 100°C.

It is interesting to note that the CO concentration is approximately 7-9 x $1O^{-6}$ mol fraction (7-9 ppm) at all the conditions evaluated. The linear velocity varied from 19.3 m/s to 75.7 m/s, corresponding to residence times of 1.4 and 4.8 milliseconds, respectively. The space velocity for these runs varies from 2 to about $6x10^6$ hr^{-1}. From the methane conversion measured in the experimental runs, the selectivity of the catalytic reaction can be calculated. This is shown in Table II where CO_2 selectivity, the desired product, is defined as $[CO_2]/\{[CO_2]+[CO]\}$. The selectivity to CO_2 ranges from 99.2 to 99.8% for the four runs.

Table II. Measured methane conversion and CO and CO_2 selectivity from CH_4 reaction on Pd monolithic catalyst

Run	CH_4 conversion	CO mol fraction	Selectivity to CO_2
	%	x10^6	
1	19	7	0.998
2	15	8	0.996
3	9.1	8	0.994
4	7.9	9	0.992

Kinetic Modeling of Heterogeneous Methane Combustion

The concentration of CO within the channels of a monolithic catalyst can be estimated assuming a surface reaction in a system with significant mass transfer resistances. In excess oxygen, the methane oxidation reaction on a catalytic surface can be treated as a series reaction

$$CH_4 \xrightarrow{k1} CO \xrightarrow{k2} CO_2 \tag{1}$$

where k1 and k2 are the rate constants for the reaction of methane with oxygen to give CO, and for the oxidation of CO, respectively.

Consider a system consisting of an active catalytic surface with reactants diffusing from a homogeneous gas phase through a stagnant boundary layer to react on the catalytic surface, as shown in Figure 7.

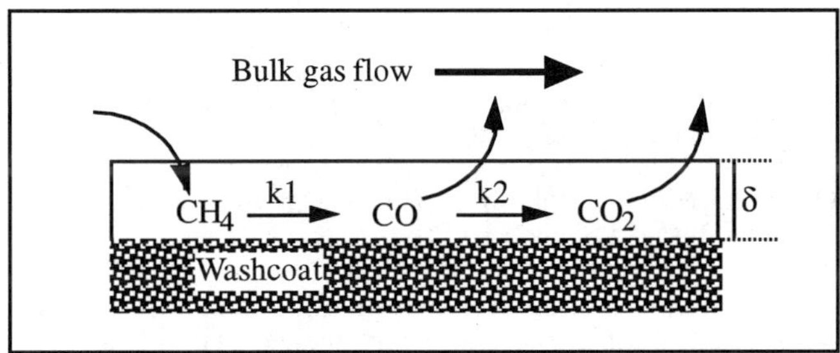

Figure 7. Sketch of the model and parameters used to simulate the surface reaction process.

Assuming the oxidation of methane is first order in methane, the rate of consumption of methane can be written

$$r = \eta \ k1 \ [CH_4]_0 \tag{2}$$

where η is an effectiveness factor

$$\eta = \frac{1}{1 + \frac{k1}{km}} \tag{3}$$

Here km is the mass transfer coefficient defined as the ratio of the molecular diffusion coefficient for methane to the boundary layer thickness:

$$km = \frac{D}{\delta} \tag{4}$$

Equation (2) shows that the rate of consumption of methane is a function of the reaction rate constant and the boundary layer thickness.

The products of the oxidation of methane are CO and CO_2. The selectivity to CO depends on the ability of this species to escape from the catalytic surface. Consequently, diffusion limitations will lower the selectivity of production of CO below that obtained under kinetically controlled conditions.

The selectivity for CO production is given by

$$S = \frac{CH_4 \text{ consumed} - CO_2 \text{ produced}}{CH_4 \text{ consumed}} \tag{5}$$

where

$$CH_4 \text{ consumed} = km\, \{[CH_4]_{\text{bulk gas}} - [CH_4]_{\text{surface}}\}$$

$$CO_2 \text{ produced} = k2[CO]_{\text{surface}}$$

The carbon atom mass balance can be written

$$CH_4 \text{ consumed} = CO \text{ produced} + CO_2 \text{ produced}$$

in symbols

$$km\, \{[CH_4]_{\text{bulk gas}} - [CH_4]_{\text{surface}}\} = km\{[CO]_{\text{surface}} - [CO]_{\text{bulk gas}}\} + k2[CO]_{\text{surface}} \tag{6}$$

After introducing the assumptions

$$[CH_4]_{\text{bulk gas}} >> [CH_4]_{\text{surface}}$$

$$[CO]_{\text{surface}} >> [CO]_{\text{bulk gas}}$$

equation (6) is reduced to

$$km\, [CH_4]_{\text{bulk gas}} = km[CO]_{\text{surface}} + k2[CO]_{\text{surface}} \tag{7}$$

and the amount of methane consumed is

$$CH_4 \text{ consumed} \approx km\, [CH_4]_{\text{bulk gas}} \tag{8}$$

Substituting (7) and (8) in (5) we obtain

$$S = \frac{km[CO]_{surface} + k2[CO]_{surface} - k2[CO]_{surface}}{km[CO]_{surface} + k2[CO]_{surface}} \quad (9)$$

$$S = \frac{1}{1 + \frac{k2}{km}} \quad (10)$$

With the assumptions above, the selectivity for production of CO depends only on the ratio k2/km. It does not depend on the rate constant of the oxidation of methane.

Using the model described above and the kinetic rates reported by Landry et al for CO oxidation on Pd (5) and the rate of CH_4 oxidation on Pd reported by Ribeiro et al (6), the expected CO concentration in the product stream from the experimental runs reported in Tables I and II above can be calculated. The results are reported in Table III.

Table III. Calculated CO Concentrations for operating conditions similar to those in experimental runs

Air flow	Pressure	CO mol fraction	Selectivity to CO_2
SLPM	atm	x 10^6	
6000	9.5	1.6	0.9993
8000	9.5	2.0	0.9993
6000	3	5.2	0.9956
8000	3	6.4	0.9953

In spite of the assumption that a gas phase CO intermediate was assumed, the model predicts a very low CO concentration exiting the catalyst at these very high space velocities. The low CO concentration is due to the very fast catalytic oxidation rate for CO at the temperatures used in this work. In addition, the model suggest a strong inverse pressure dependence. This results from the assumed first order dependence of CO oxidation which results in an increase in rate as the total pressure increases. The mass flux of CO through the boundary layer into the bulk flow, however, is independent of total pressure.

The calculated CO concentration presented in Table III can be compared with the measured values reported in Table II. The measured values are somewhat higher, with 7 to 8 ppm CO measured at 9.5 atm while the model calculated values are 1.6 to 2 ppm. At 3 atm pressure, the agreement is somewhat better, with a measured value of 8 to 9 ppm and a calculated value of 5 to 6 ppm. This level of agreement is quite good considering the difficulty in measuring such low levels of CO. The trends are predicted correctly, with CO concentration increasing as the gas velocity is increased (residence time decreased). The model predicted strong effect of pressure is not seen in the measured data. The reason for this discrepancy is not understood.

Conclusions

The selectivity of methane catalytic combustion processes to CO_2 must be very high since the required CO emissions level is very low, < 10 ppm. On supported Pd catalysts, the measured CO concentration at the outlet of the catalyst is very low, in the range of 7 to 9 ppm, surprising for a system operating at such high space velocities, 2-6x10^6 hr^{-1}. The kinetic model of the surface reaction shows that the probable cause of the low CO level is that CO oxidation on Pd is so fast that the CO reacts before it can diffuse into the bulk flow. The calculated selectivity to CO_2 can be very high, in excess of 99.9% at practical pressures with monolithic catalysts operating with residence times in the range of 1.5 ms. This is supported by the experimentally measured CO_2 selectivity of 99.8%.

For processes that utilize homogeneous oxidation of a portion of the methane, the selectivity to CO_2 can also be very high if the final combustion temperature is maintained above 1100°C. Below this temperature, the rate of CO oxidation is sufficiently slow that the time required to achieve a low CO level would require a large combustor volume and would not be viable in a gas turbine application. These results also show that a catalytic combustion process that is designed to react most of the methane within the catalyst cannot expect post catalyst homogeneous reactions to aid in the combustion of CO unless the gas temperature is above 1100°C. This would require the catalyst to operate at temperatures above 1100°C, placing severe thermal stresses on the catalyst system.

Acknowledgments

The authors would like to thank Thomas Rostrup-Nielsen for the homogeneous kinetic model calculations and George Voss for the high pressure reactor measurements.

References

1. Dalla Betta, R. A., *Catalysis Today* , Accepted for publication.
2. Miller, J. A. and Bowman, C.T., *Prog. Energy Combust. Sci.* **1989**, *15*, 287.
3. Bowman, C. T., Private communication.
4. Kee, R. J., Miller, J. A. and Jefferson, T. H., *"CHEMKIN:A General-Purpose, Problem-Independent, Transportable, Fortran Chemical Kinetics Code Package,"* Sandia National Laboratories Report SAND87-8251 **1987**.
5. Landry, S. M., Dalla Betta, R. A., Lu, J.P., and Boudart, M., *J. Phys. Chem.* **1990**, *94*, 1203.
6. Ribeiro, F. H., Chow, M. and Dalla Betta, R. A., *J. Catal.* **1994**, *146*, 537.

Chapter 4

Ignition and Extinction of Hydrogen–Air and Methane–Air Mixtures over Platinum and Palladium

F. Behrendt, O. Deutschmann, R. Schmidt, and J. Warnatz

Interdisciplinary Center for Scientific Computing, Universität Heidelberg, Im Neuenheimer Feld 368, D–69120 Heidelberg, Germany

Ignition and extinction of hydrogen-oxygen as well as methane-oxygen mixtures flowing towards resistively heated platinum or palladium foils are experimentally investigated. The ignition temperatures and the hysteresis observed between ignition and extinction are compared with time-dependent simulations. The simulation is performed for the coupled system of surface and gas phase utilizing elementary reaction mechanisms for both phases. Comparison of experimental and numerical results for ignition and extinction at platinum shows a very good agreement with respect to temperature as well as its temporal development during ignition. Due to deficiencies regarding kinetic data for the surface reactions on Pd, the results here do not compare as well as for Pt.

Extensive experimental and theoretical attention has been given to catalytic combustion in the past decade. The potential of heterogeneous processes for reducing emission of pollutants, improved ignition, and enhanced stability of flames has been recognized. Here, catalytic ignition as a sudden transition from a kinetically controlled system to one controlled by mass transport is of special interest *(1-4)*. Its description needs a detailed knowledge of both the elementary reaction steps at the gas-surface interface and the gas phase as well of the transport processes between surface and gas phase.

In the present work, mixtures of hydrogen or methane with oxygen (for some of the measurements diluted by nitrogen) at atmospheric pressure are utilized. These mixtures flow slowly through a tube towards a platinum or palladium foil, which is heated resistively. When the foil is heated to a sufficiently high temperature, heterogeneous reactions start and the reactants are consumed at the metal surface.

To simulate the catalytic ignition under the given conditions the above mentioned transition from kinetic to transport control requires a closely coupled solution of the governing equation for the gas phase and the surface processes *(9,15)*. The set of differential equations describing this coupled system is discretized and solved using a time exact solver *(10,11)*. This numerical code, originally developed for the simulation of gas phase combustion processes *(8)*, applies a detailed description of the elementary chemical processes occurring at the gas-surface interface, and couples them to the reactive flow in the surrounding gas phase.

0097–6156/96/0638–0048$15.00/0

The ignition temperature is calculated as a function of the fuel concentration, and is compared with the experimental results. The surface chemistry is described by a detailed reaction mechanism including adsorption, desorption, and surface reactions.

Experiment

The experimental setup is shown in Figure 1. The catalyst is mounted in a rectangular tube with a cross section of 28 x 38 mm^2 and a length of 16 cm. This catalyst is a polycrystalline foil of either platinum or palladium with a purity of 99.95 % for Pt and 99.9 % for Pd. The thickness of the Pt foil is 0.027 mm and that of the Pd foil 0.025 mm, the outer dimensions of both are 25 by 5 mm. Before each experiment the foil is cleaned following a procedure described by Keck et al. *(16)*. After several ignition/extinction cycles a irreversible deactivation of the foil is observed, and it has to be replaced.

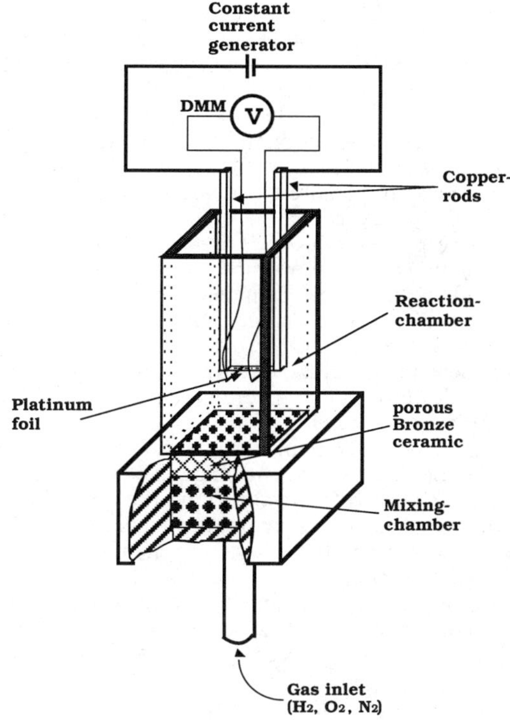

Figure 1. Experimental Setup

The foils are mounted on two copper rods so that the flow is directed towards the foils (stagnation point flow configuration). Through these rods the electrical current used to heat the foil is supplied. Additionally, two thin platinum wires are attached to the foil to measure the temperature dependent electrical resistivity of the foil, and thus, the temperature. Compared to the application of a spot-welded thermocouple this method is simpler to use, it is very responsive even to fast changes of the foil temperature, and does not introduce additional catalytically active material to the system.

After passing through a mixing chamber filled with glass spheres the gas mixture flows towards the foil. The fuel content of the mixture is defined as $\alpha = p_{fuel}/(p_{fuel}+p_{oxygen})$ and is varied between $0.25 < \alpha < 0.75$ for hydrogen and $0.28 < \alpha < 0.96$ for methane. The combined partial pressure of fuel and oxygen for mixtures diluted by nitrogen is introduced as $\delta = (p_{fuel}+p_{oxygen})/p_{total}$. The flow velocity is 8 cm/s.

Model and Simulation

The experiment is simulated using a one-dimensional stagnation point flow model with time and distance from the foil as independent variables. By confining the attention to the centre of the surface, edge effects can be neglected. Details of this approach are discussed by Warnatz et al. *(4)*.

The dependent variables (density, momentum, temperature, and mass fraction of each gas phase species) depend only on the distance from the foil. The system is closed by the ideal gas law. The boundary-value problem that has to be solved has been stated by Evans and Greif *(5)* and by Kee et al. *(6)*. This set of governing equations is analogous to that used for simulation of laminar counterflow diffusion-flames *(7,8)*.

The solution of the gas-phase problem is coupled to the surface properties and the surface reaction mechanism. Therefore, the surface boundary-conditions become more complex compared with the pure gas-phase problem. In a time-dependent formulation the mass fraction of a gas-phase species at the surface is determined by the diffusive and convective processes as well as the creation or depletion rate of that species by surface reactions. The temperature boundary-condition is derived from energy contributions at the interface. Included are the conductive, convective, and diffusive energy transport in the gas phase, the chemical heat release at the surface, radiation, and the resistive heating of the foil. Furthermore, the heat capacity of the platinum foil and the dependence of the specific resistance of the foil on temperature are taken into account. Details on the governing equations and boundary conditions can be found in Behrendt et al. *(9,15)*.

To solve this stagnation flow problem numerically, a computer code originally developed to simulate counterflow diffusion-flames is adapted. This code computes species mass fraction, temperature, and velocity profiles, and fluxes at the gas-surface interface as a function of time. The program accounts for finite-rate gas-phase and surface chemical kinetics. A simplified multicomponent molecular transport model is used.

The Navier-Stokes equations describing the gas phase together with the boundary conditions represent a differential-algebraic equation system. Discretization bases on a finite-difference scheme using a statically adapted non-equidistant grid. The resulting systems of algebraic and ordinary differential equations is solved by the extrapolation solver LIMEX *(10,11)*.

The chemistry is modelled by a set of elementary reactions in the gas phase as well as on the surface. The gas-phase reaction data are taken directly from modelling work on flame chemistry *(12,13)*. Its validity has been established through extensive studies of various combustion systems, and is applied here without modification. The surface rate data can be found in *(9)*. This set of detailed reaction steps including O and H atoms, OH, H_2O, CO, and CH_i (i = 0, 3) molecules is based on several publications *(17-18)*. The rate coefficients for the reverse reactions as well as the reaction enthalpies of the surface species are calculated from the thermochemical data for the H_2/O_2-system as stated in *(4)* extended for CO and CH_i (i = 0, 3).

The only adjustable parameter to the simulation is the heat loss to the supporting copper rods. This loss is determined experimentally using inert gas and accounted for in the simulation.

In a previous paper *(15)* experimental data on the ignition of methane-air mixtures published by Schmidt et al. *(17)* have been used to evalute the recation mechanism and the numerical code used here.

Results

Ignition of H_2-O_2 Mixtures. In Fig. 2 the surface temperature is plotted versus the current supplied to the foil.

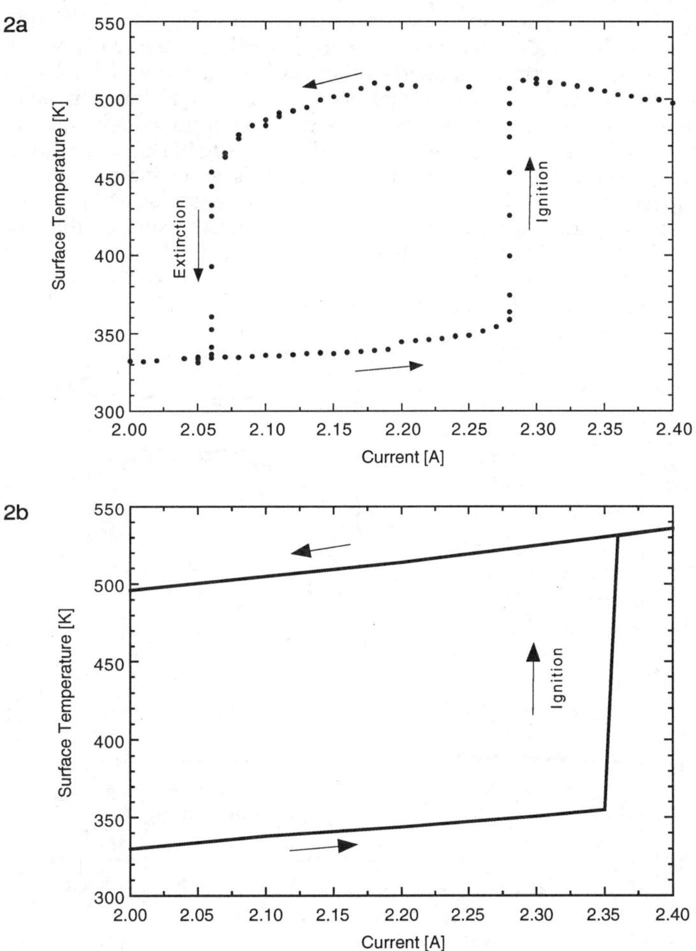

Figure 2. Experimental (top, 2a) and numerical (bottom, 2b) ignition and extinction curve for hydrogen and oxygen on platinum ($\alpha = 0.5$)

Before ignition, the increase of foil temperature is due to this resistive heating, while during and after ignition the heat released by surface reactions contributes significantly to the foil temperature.The point of ignition is clearly marked by the sudden rise of the foil temperature. Figure 2a shows a typical experimental ignition/extinction curve for a hydrogen-oxygen mixture on platinum.

Before ignition, the foil is covered by hydrogen keeping oxygen from adsorption, and no surface reactions take place (see Fig.3). Elevating the foil temperature by electrical heating above a certain value leads to a shift of the adsorption/ desorption equilibrium of hydrogen towards desorption, thus liberating surface sites for dissociative oxygen adsorption. The oxygen atoms react with hydrogen atoms forming OH radicals and finally water.

The heat released by the water formation leads to a further increase of the surface temperature, which in turn enhances the desorption of hydrogen. This self-accelerated process leads to ignition. During ignition the system undergoes a transition from kinetically controlled surface reactions to diffusion control, where the transport of reactants from the surrounding gas phase to the surface limits the reaction rate.

After ignition, the current supplied to the foil is reduced until finally extinction occurs for a current much lower than that one needed for ignition. The heat released at the foil maintains enough free surface sites for the continued adsorption of either reactant until finally the surface is fully covered by hydrogen again. Figure 2b shows the results of calculations using the same parameters as in the measurement. The ignition temperature in both cases is 355 K. In the calculation no extinction is observed for these parameters, while in the experiment the surface reactions extinguish again. The polycrystalline foil undergoes recrystallization during this heating and cooling cycle, an effect which cannot be accounted for in the simulation. Extinction is observed for calculations using smaller α.

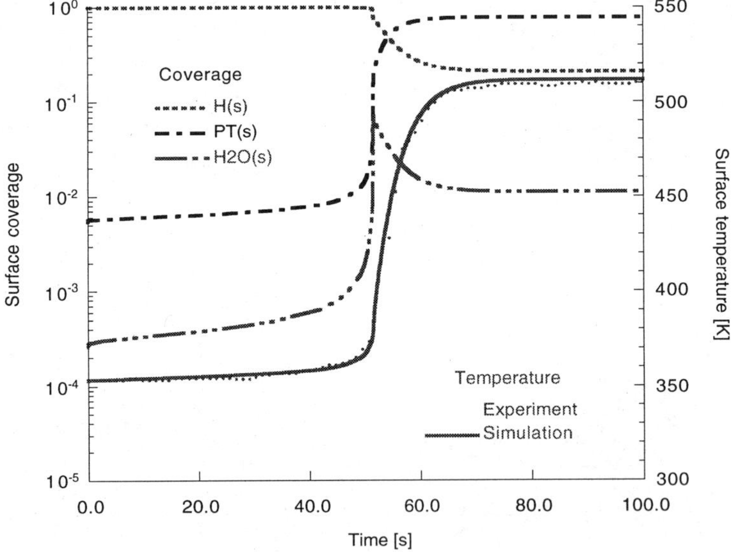

Figure 3. Time-resolved change of surface temperature and coverage for hydrogen oxidation on platinum ($\alpha = 0.5$)

The detailed time-resolved development of the surface temperature (experimental and calculated) and surface coverage (calculated only) is presented in Figure 3. Time zero here is the moment of last increase of the current through the foil. After ignition, most of the platinum sites are empty with about 15 % H atoms and 1 % water remaining.

In Figures 4 (platinum) and 5 (palladium) experimental results for the ignition of nitrogen-diluted mixtures of hydrogen and oxygen are compared with simulations. For all cases δ is kept at 0.059. In general, for a given value of α, the ignition temperature on platinum is higher than on palladium.

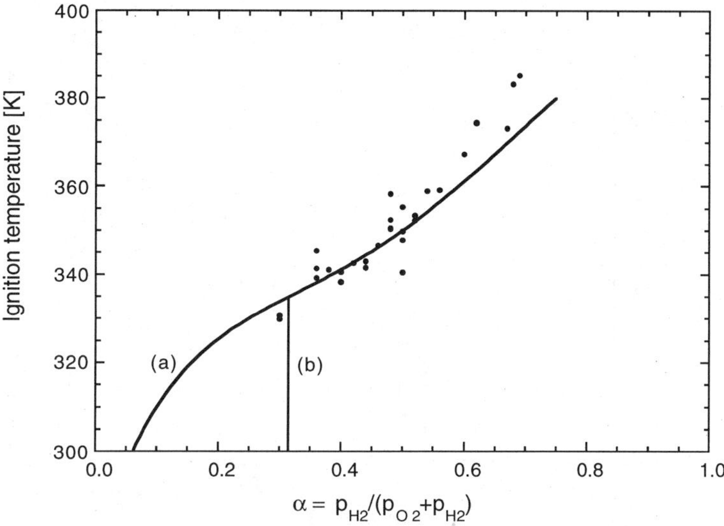

Figure 4. Measured and calculated ignition temperatures for hydrogen and oxygen diluted by nitrogen on Pt ($\delta = 0.059$) (for a and b see text)

The difference between both catalysts is more pronounced at the lower ignition limit. Below $\alpha = 0.3$ for platinum and $\alpha = 0.26$ for palladium significant differences are observed in experiment and simulation depending on the initial coverage of the metal surface, i.e., bistability is observed. Starting the calculation with a surface covered by hydrogen atoms (case a;using an initially pure hydrogen flow), ignition starts without external supply of heat for α smaller than 0.12 for platinum and smaller than 0.23 for palladium. Even at room temperature the desorption of hydrogen leads to a sufficient number of free sites allowing oxygen to adsorb, causing the onset of reactions at the surface. At low α, the lower sticking coefficient of oxygen is compensated for by the much higher concentration of oxygen in the gas phase.

Above $\alpha = 0.12$ and 0.23, respectively, hydrogen sticking becomes so efficient that external heating is required to create enough free sites, so that adsorption of oxygen remains competitive. The sticking coefficients for both oxygen and hydrogen are more than an order of magnitude larger for palladium than for platinum, thus extending the range of ignition without heating to higher values of α.

When the surface is initially covered by oxygen atoms (case b; gas flow of pure oxygen), ignition without external heating is observed below $\alpha = 0.32$ for both catalysts. Hydrogen has a higher sticking coefficient compared to oxygen and, at the same time, is more mobile on the surface. Consequently, on an oxygen-covered surface, even at room temperature, hydrogen will be able to adsorb and react to form water. This releases some heat, leading to more free sites and finally to ignition.

The discussion above conclusively shows the platinum and palladium surface to be covered initially by hydrogen for sufficient high α, i.e., α larger than 0.32 for the present conditions.

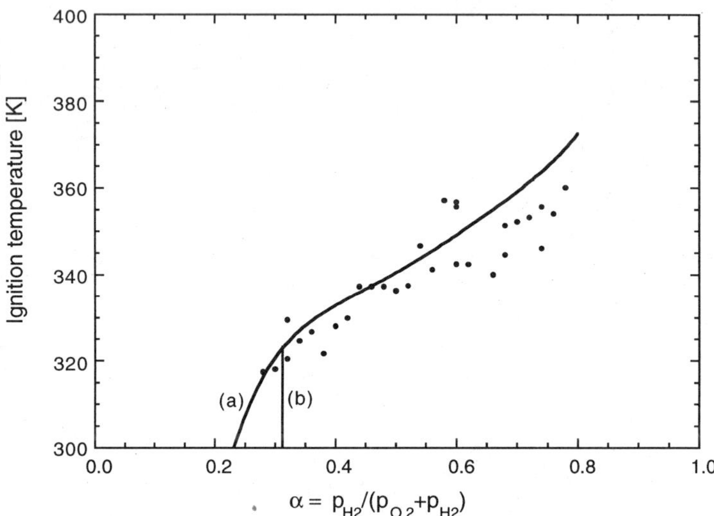

Figure 5. Measured and calculated ignition temperatures for hydrogen and oxygen diluted by nitrogen on Pd ($\delta = 0.059$) (for a and b see text)

Ignition of CH_4-O_2 Mixtures. Experimental ignition temperatures for mixtures of methane and oxygen diluted by nitrogen on platinum as a function of α are shown in Figure 6. The relative partial pressure of methane $\alpha = p_{methane}/ (p_{methane}+p_{oxygen})$ is varied between 0.28 and 0.96, while the dilution by nitrogen for all α is given by $\delta = (p_{methane}+p_{ooxygen})/p_{total} = 0.059$.

The ignition temperatures for methane-oxygen mixtures are much higher compared with hydrogen-oxygen mixtures. Remarkable is the reversed dependence of ignition temperature on α. While for hydrogen-oxygen mixtures the ignition temperature increases with increasing α, here a decrease of the ignition temperature is observed. In contrast to the hydrogen-oxygen system, the platinum surface is initially covered by oxygen instead by the fuel. So, the reaction is initiated by the desorption of oxygen, offering free sites for the adsorption of methane.

For small α, i.e., a high concentration of oxygen in the gas phase, adsorption of oxygen on these free sites is more likely than of methane. For higher temperatures more free sites are generated, resulting in a higher probability of adsorption of methane leading to ignition. With increasing α, the larger mole fraction of methane in the gas phase increases the chance of methane to adsorb, thus lowering the temperature needed for ignition.

Figure 7 shows a complete calculated ignition-extinction cycle for a methane-oxygen mixture at $\alpha = 0.5$. Compared to the same kind of cycle for a hydrogen-oxygen mixture as shown in Figure 2, the difference between ignition and extinction temperature is much larger. Additionally, for the oxidation of methane a reduction of electrical power supplied to the foil after ignition causes the surface temperature to decrease about 300 K before the system extinguishes.

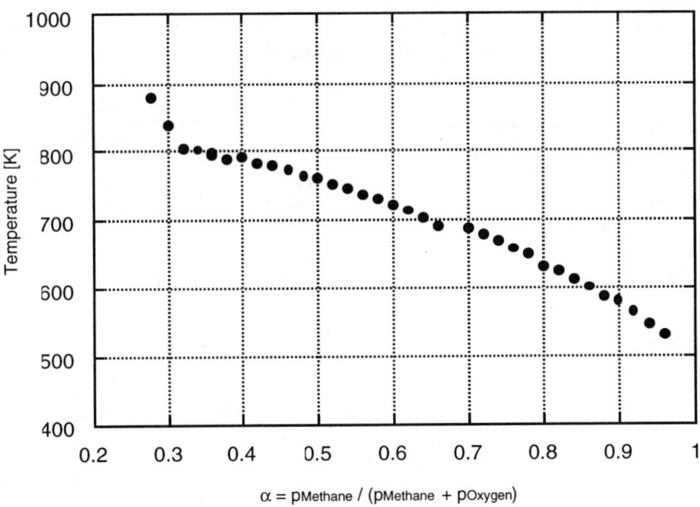

Figure 6. Experimental ignition temperatures for methane-oxygen mixtures diluted by nitrogen on platinum

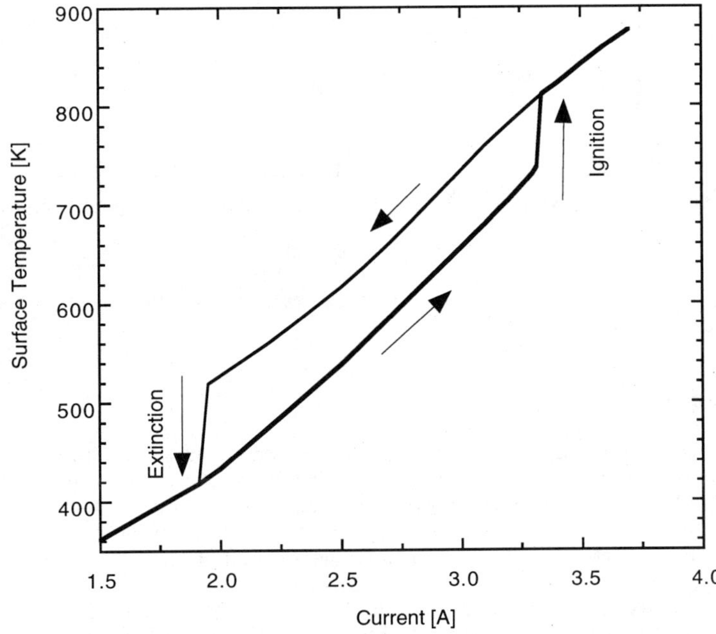

Figure 7. Numerical ignition and extinction curve for methane and oxygen on platinum ($\alpha = 0.5$)

In Figure 8 the calculated time-resolved surface coverage for platinum, oxygen, and CO as well as the temperature are plotted. At stated above, the surface is dominantly covered by O atoms before ignition. At ignition, enough free sites become available for methane to adsorb. The mechanism used here assumes a fast lost of H atoms from methane and subsequent oxidation of the C atom to CO following the work of Hickmann and Schmidt (*14*). After ignition, the surface is dominantly covered by CO (about 80 %). Reduction of the electrical power supplied to the foil does not extinguish the surface reactions until the final step of the reaction sequence, the oxidation of the CO, is suppressed.

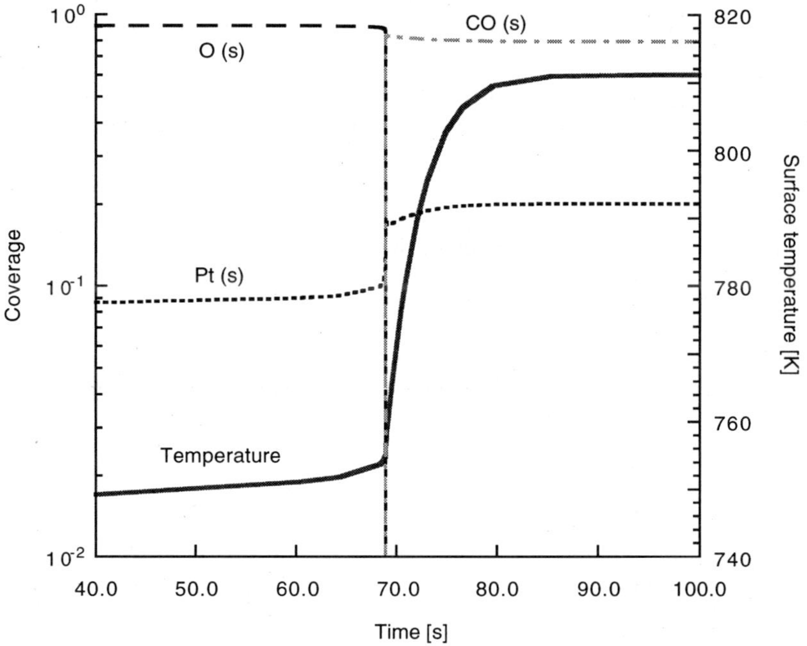

Figure 8. Time-resolved change of surface temperature and coverage for methane oxidation on platinum ($\alpha = 0.5$)

Conclusions

The present study shows that the ignition experiments for hydrogen-oxygen and methane-oxygen mixtures on two important catalysts, platinum and palladium, can be interpreted in terms of elementary reaction steps on the metal surface. After ignition, the catalytic system is controlled by a transport limitation for the reactants. Therefore, to describe catalytic ignition, the instationary simulation of the coupled surface/gas-phase system is required.

The analysis so far is based on the mean field approximation neglecting structural effects (e.g., island formation of adsorbates) on the surface. For higher temperatures this assumption holds due to the high mobility of the adsorbates on the surface. The influence of spatial inhomogenities on the surface reaction rates, and thus on ignition temperatures, will be investigated in future work.

Acknowledgment

The financial support of this work by the Deutsche Forschungsgemeinschaft (DFG) within the Sonderforschungsbereich 359 "Reaktive Strömung, Diffusion und Transport" is gratefully acknowledged.

Literature Cited

1. Cho, P. and Law, C. K., *Combust. Flame* **1986**, *66*, 159.
2. Williams, W. R., Stenzel, M. T., Song, X., and Schmidt, L. D., *Combust. Flame* **1991**, *84*, 277.
3. Rinnemo, M., Fassihi, M., and Kasemo, B., *Chem. Phys. Lett.* **1993**, *211*, 60.
4. Warnatz, J., Allendorf, M. D., Kee, R. J., and M. E. Coltrin, *Combust. Flame* **1994**, 96, 393.
5. Evans, G. H. and Greif, R., *ASME J. Heat Transfer* **1987**, *109*, 928.
6. Kee, R. J., Miller, J. A., Evans, G. H., and Dixon-Lewis, G., *Twenty-Second Symposium (International) on Combustion*, The Combustion Institute, Pittsburgh, PA, 1988, p 1479.
7. Dixon-Lewis, G., Giovangigli, V., Kee, R. J., Miller, J. A., Rogg, B., Smooke, M. D., Stahl, G., and Warnatz, J., *Progress in Astronautics and Aeronautics* (A. L. Kuhl, J.-C. Leyer, A.A. Borisov, W. A. Sirignano, eds.), AIAA, Washington, D.C., 1991, Vol. 131, pp 125-144.
8. Behrendt, F. and Warnatz, J., *Progress in Astronautics and Aeronautics* (A. L. Kuhl, J.-C. Leyer, A.A. Borisov, W. A. Sirignano, eds.), AIAA, Washington, D.C., 1991, Vol. 131, pp 145-160.
9. Deutschmann, O., Behrendt, F., and Warnatz, J., *Catal. Today* **1994**, *21*, 461.
10 Deuflhard, P., Hairer, E., and Zugck, J., *Num. Math.* **1987**, *51*, 501.
11 Deuflhard, P. and Nowak, U., *Progress in Scientific Computing* (P. Deuflhard, B. Enquist, B., eds.), Birkhaeuser, 1987, Vol. 7, p 37.
12 Warnatz, J., *Combustion Chemistry* (Gardiner, W. C., ed.), Springer, NY, 1984, Chapter 5.
13 Baulch, D. L., Cobos, C. J., Cox, R. A., Esser, C., Frank, P., Just, Th., Kerr, J. A., Pilling, M. J., Troe, J., Walker, R. J., and Warnatz, J., *J. Phys. Chem. Ref. Data* **1992**, *21*, 411.
14 Hickman, D. A. and Schmidt, L. D., *AIChE J.* **1993**, *39(7)*, 1164.
15 Behrendt, F., Deutschmann, O., Maas, U., Warnatz, J., *JVST A* **1995**, *13*, 1373.
16 Keck, K.-E., Kasemo, B., Högberg, T., *Surf. Sci.* **1986**, *167*, 313.
17 Williams, W. R., Stenzel, M. T., Song, X., Schmidt, L. D., *Combust. Flame* **1991**, *84*, 277.
18 Ljungström, S., Kasemo, B., Rosén, A., *Langmuir* **1994**, *10*, 699.

Chapter 5

A New Class of Uranium Oxide Based Catalysts for the Oxidative Destruction of Benzene and Butane Volatile Organic Compounds

Graham J. Hutchings[1], Catherine S. Heneghan[1], Ian D. Hudson[2], and Stuart H. Taylor[1]

[1]Leverhulme Centre for Innovative Catalysis, Department of Chemistry, University of Liverpool, P.O. Box 147, Liverpool L69 3BX, United Kingdom
[2]Company Research Laboratory, BNFL, Springfields Works, Salwick, Preston PR4 OXJ, United Kingdom

The oxidative destruction of two typical volatile organic compounds (VOCs), benzene and butane, have been investigated by uranium oxide based catalysts. A new range silica supported uranium oxide catalysts have also been synthesised and tested under industrially relevant conditions..
Catalytic activity studies have shown that U_3O_8 is an active catalyst for the destruction of both compounds at high gas hourly space velocity. Comparison with Co_3O_4, a known active combustion catalyst, showed that U_3O_8 was more active under the majority of our experimental conditions. The silica supported catalysts also showed high activity for benzene and butane destruction.. Doping the supported catalyst with chromium considerably enhanced the activity for butane destruction.
These studies have demonstrated that under commercially realistic operating conditions, catalysts based on uranium oxide show high activity for the oxidative destruction of benzene and butane, which are considered to be representative of two types of very different common volatile organic compounds.

In recent years it has become essential to reduce the emissions of volatile organic compounds (VOCs) from chemical installations. VOCs are a wide ranging class of compounds commonly occurring in commercial waste streams and constitute a major source of air pollution. Several VOC abatement technologies have been proposed, these include thermal oxidation and catalytic oxidation [*1*]. Thermal oxidation is carried out at high temperature, typically >1000°C, and is relatively insensitive to space

0097-6156/96/0638-0058$15.00/0

velocity and to the composition and concentration of the waste gases. Whereas, catalytic oxidation involves combustion at lower temperatures, typically 400-600°C, but is greatly affected by space velocity, waste gas composition and concentration. In general, it is expected that catalytic VOC destruction efficiency will increase with increased temperature and decreased space velocity [*2*]. The lower temperatures required for catalytic combustion results in a lower fuel demand and can therefore be more cost effective than a thermal oxidation process. The catalytic process also exerts more control over the reaction products and is less likely to produce toxic by-products, such as dioxins, which may be produced by thermal combustion. A combination of the two methods has been proposed and it is claimed that this process gives increased VOC destruction efficiency [*3*].

Catalytic oxidation using an air oxidant therefore provides a convenient route for VOC destruction. The relative ease of destruction of VOCs by catalytic oxidation varies according to the class of compound and follows the general order, alcohols > aldehydes > aromatics > ketones > acetates > alkanes > chlorinated alkanes [*3*]. Various catalysts have been proposed and these fall into two broad categories, noble metals and metal oxides [*4,5*]. Noble metal catalysts, which are often used in the supported form, show high intrinsic combustion activity, however, they are relatively expensive, susceptible to poisoning even at low levels and in some cases show poor stability [*5*]. The oxides of cobalt, copper, chromium, manganese and nickel have all been used for VOC destruction [*6-8*] and it would be expected that oxide catalysts could tolerate higher levels of poisons, however the activity shown by these oxides is generally lower than noble metal catalysts. It is therefore evident that if high activity oxide catalysts can be developed for VOC destruction these will be preferred.

Uranium oxides have previously been used as catalysts both in pure form and as catalysts components. The oxide U_3O_8 was shown to have appreciable activity for the oxidation of CO by molecular oxygen [*9*]. Whilst uranium oxides have also been used as major catalyst components in dual oxide systems, such as uranium-antimony for propylene ammoxidation [*10,11*] and uranium-bismuth catalysts for oxidative demethylation [*12*].

This study has been undertaken to investigate the efficiency of uranium oxide and uranium oxide based catalysts for the destruction of model VOCs in the vapour phase. The classes of compounds investigated in this study include aromatics and alkanes. These were chosen because aromatics have been ranked intermediate and alkanes low based on their ease of destruction [*3*]. The specific compounds benzene and butane were selected as typical compounds within these groups, and are also common VOC pollutants. A characterisation study has also been undertaken to investigate the nature of the supported uranium oxide species and the stability of uranium oxide under typical reaction conditions.

Experimental Details

Thermal Analysis Thermal gravimetric analysis (TGA) and differential scanning calorimetry (DSC) were used to investigate the nature of the chemical and physical transformations during the catalyst preparative calcination procedures. Analysis was performed using a Perkin Elmer series 7 thermal analysis system. Thermal gravimetric analysis was carried out in a stream of flowing air. Typically a 10-20 mg sample was used and this was heated from 40°C to 800°C at a linear heating rate of 40°C min^{-1}.

Differential scanning calorimetry experiments were performed in flowing nitrogen, air and static air atmospheres. Sample weights were in the range 10-20 mg and were encapsulated in aluminium sample pans with a perforated lid, to permit sample exposure to the atmosphere. Transformations were investigated in the temperature range from 40°C to 550°C with a linear temperature ramp rate of either 20°C min^{-1} or 40°C min^{-1}. The ramp rate selected was dependent on the requirement for increased resolution or sensitivity. However, the temperature for transformation remained independent of the chosen ramp rate.

Powder X-ray Diffraction. Powder X-ray diffraction patterns were recorded using a Phillips X-PERT MPD double goniometer system. The system consisted of a conventional goniometer and a second goniometer equipped with a high temperature Anton Parr XRK *in-situ* reaction chamber and a 15° scanning position sensitive detector (PSD). Both goniometers were housed on a common X-ray tube generating Cu Kα X-rays.

Analysis of catalysts ex-situ was performed over the angular range 5° < 2Θ >-75° with the source operated at 40 KeV and 50 mA. Patterns were processed using Phillips software and phases were identified by matching to standard samples and entries in the JCPDS powder diffraction file.

Studies *in-situ* used an X-ray source of lower intensity (30 KeV, 40 mA) in order not to saturate the PSD which was used in a scanning mode from 18° to 60° 2Θ. The *in-situ* reaction cell was designed so that gases flowed through the catalyst sample which was heated from ambient to 600°C. Experiments were carried out with a flow of dry air and an air stream containing ca 4% water. Typical analysis times at each temperature were in the region of 1.5 min.

Catalyst Preparation and Testing. The single oxide catalysts, U_3O_8 and Co_3O_4 (ex Johnson Matthey 99.999%) were used as supplied. A silica supported catalyst, denoted U/SiO_2, was prepared by an incipient wetness impregnation technique followed by drying at 100°C for 24 hours, the active loading was 10 mol% ($U:SiO_2$). A range of silica supported catalysts were also prepared by doping a 9 mol% uranyl nitrate precursor with 1 mol% loadings of cobalt, chromium, iron and copper via the nitrate solutions. These catalysts were calcined at 300°C for 1 hour and then for a further 3 hours either at 600°C or 800°C in static air.

Catalytic activity was determined in a stainless steel microreactor. Benzene was fed into the reactor system via a syringe pump whilst butane was introduced by a mass flow controller. Air was used as the oxidant and flow rates controlled using a mass flow controller. Typical reaction conditions used in these studies employed a 1% VOC concentration the balance being air. The catalysts were pelleted to a particle size range of 425-600μm and were secured in the reactor tube by plugs of silica wool. The catalyst bed was packed to a constant volume for all experiments, which were performed using a gas hourly space velocity of 70,000 h^{-1}.

Product and reactant analysis was carried out on-line using a Hiden quadrupole mass spectrometer and a Varian 3400 Gas Chromatograph with a thermal conductivity detector.

Results and Discussion

Thermal Analysis. The results obtained from DSC experiments in flowing air are shown for pure uranyl nitrate ($UO_2(NO_3)_2.6H_2O$) and the pre-calcination silica supported catalyst precursor in figure 1.

Pure uranyl nitrate showed four endotherms below 265°C, these being at 82°C, 113°C and 262°C, with a very broad peak at 145°C, these endotherms corresponded to the loss of water from the hydrated uranyl nitrate starting material. Peaks identified at 294°C and 307°C (shown in figure 1.), and 404°C and 518°C (not shown), were due to the endothermic decomposition of the nitrate ion. The product from this decomposition is UO_3 which reduces endothermically to $UO_{2.9}$ at 590°C and ultimately to U_3O_8 via an endothermic transition at 630°C [*13*]. In our experiments the furnace temperature was limited to 550°C by the use of aluminium sample pans and it was therefore not possible to observe these higher temperature transitions.

Studies to investigate the silica supported catalyst pre-cursor showed somewhat different results. The major feature was the endotherm at 244°, corresponding to the decomposition of the nitrate ion. The process of supporting uranyl nitrate on silica destabilised the nitrate, as it decomposed at a temperature approximately 55°C lower. The same effect has previously been observed and has been shown to extend to other supports such as Al_2O_3, TiO_2 and MgO [*13*]. Two other minor peaks were also evident, these were related to the loss of water and were located at 115°C and 162°C, which were at similar temperatures to those of the pure uranyl nitrate.

Thermal gravimetric results showing the % weight loss and the 1st derivative of this loss are shown in figures 2 and 3. TGA of the pure uranyl nitrate showed an overall weight loss of 41%, which was complete around 600°C. The mass loss at 81°C, and in the region of 150°C were due to water loss, whilst the mass loss at 279°C, 305°C, 381°C and 533°C were due to nitrate decomposition. The peaks at 381°C and 533°C were not shown in the DSC figures but are clearly visible in the TGA results. The peak temperatures for transformations obtained from DSC and TGA result were in good agreement.

The TGA of the supported catalyst pre-cursor showed an overall weight loss of 14%, considerably lower than the pure uranyl nitrate, as a consequence of the silica support. The overall weight loss was also completed around 600°C. The TGA peak at 249°C confirmed the destabilising effect of the support on the nitrate. A major TGA peak was observed at 201°C, this was not evident in the DSC studies, but it is this temperature region is likely to be due to the loss of water.

DSC and TGA studies were also carried out under static air and nitrogen atmospheres, no significant differences between these studies and the studies performed under flowing air were apparent. From the results presented in this section it was evident that nitrate decomposition was complete below 540°C and therefore the silica supported catalyst precursor was ultimately calcined at 600°C and 800°C.

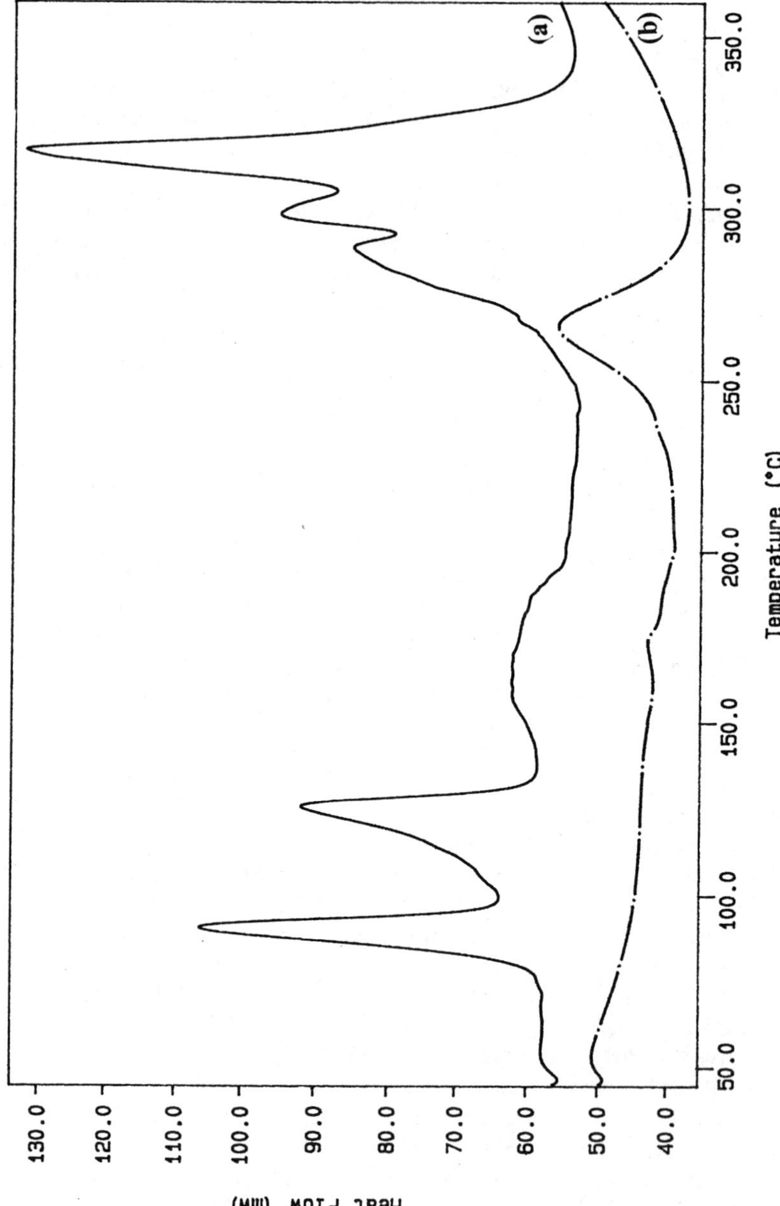

Figure 1. Differential scanning calorimetry in flowing air; (a) pure uranyl nitrate, (b) silica supported catalyst pre-cursor.

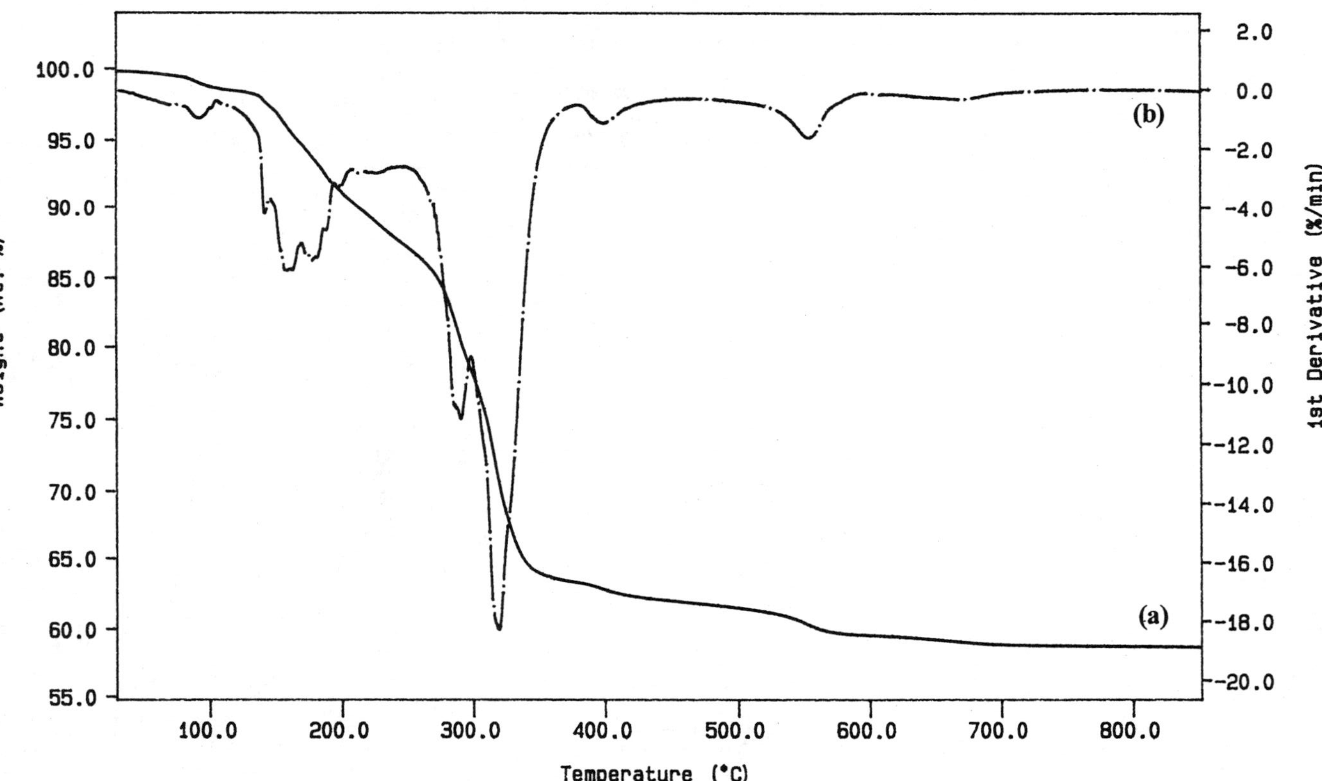

Figure 2. Thermal gravimetric analysis of $UO_2(NO_3)_2.6H_2O$ in flowing air; (a) % weight loss, (b) 1st derivative % weight loss.

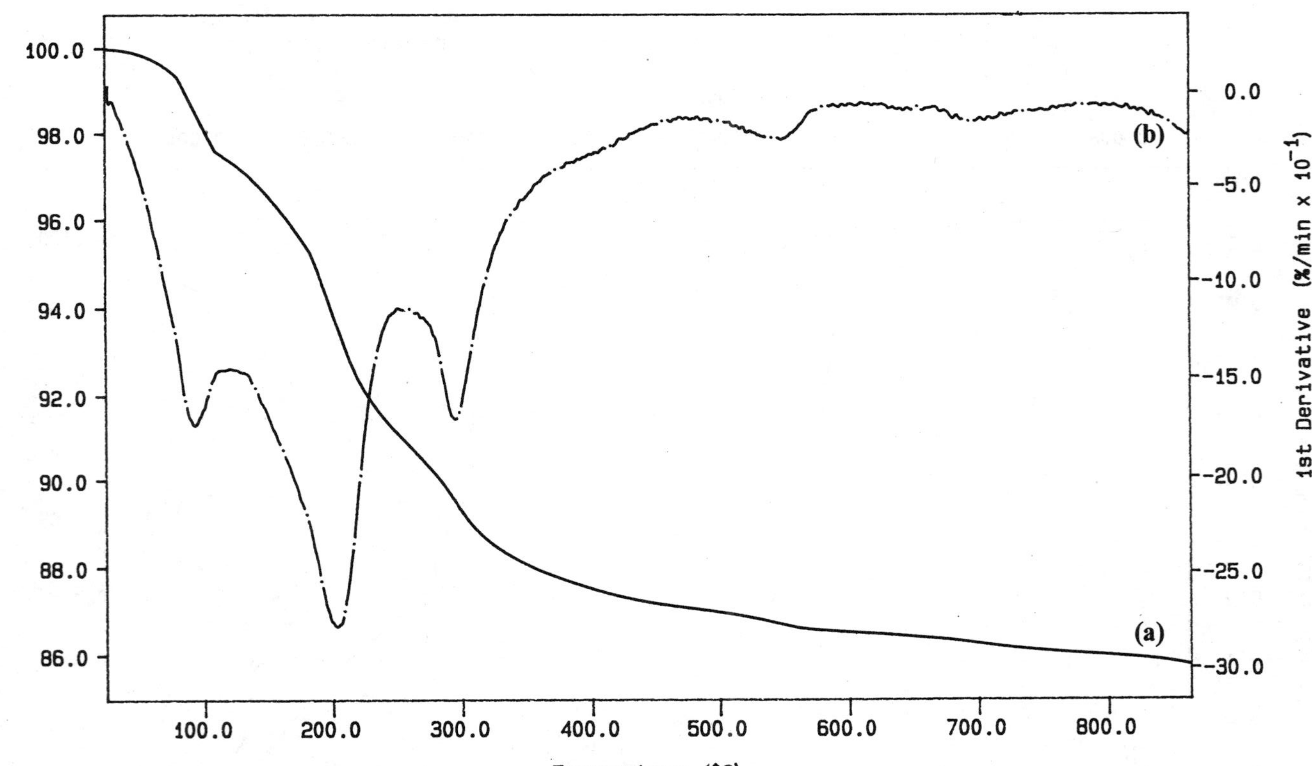

Figure 3. Thermal gravimetric analysis of silica supported catalyst pre-cursor in flowing air; (a) % weight loss, (b) 1st derivative % weight loss.

Powder X-ray Diffraction. Powder X-ray diffraction studies of the silica supported catalyst pre-cursor dried at 100°C showed that the dihydrate and hexahydrate uranyl nitrate phases were crystalline and intact on the support. The identification of these phases on the catalyst pre-cursor validates the use of the pure uranyl nitrate as a model compound for thermal analysis studies. A colour change from pale yellow to a darker mustard colour was observed on calcination of the catalyst pre-cursor at 600°C and XRD results showed no diffraction peaks. Such an observation does not necessarily indicate the absence of a crystalline uranium oxide crystalline phase as the considerable X-ray line broadening associated with a highly dispersed phase would produce similar results. The XRD pattern for the catalyst pre-cursor calcined at 800°C is shown along with that of a standard U_3O_8 reference in figure 4. The comparison clearly indicates that the supported phase produced on calcination at 800°C was U_3O_8. The supported U_3O_8 crystallite size was in the region of 120Å determined by line broadening analysis. No phase changes after use were detected for the U/SiO_2 catalyst, however line broadening analysis showed that the particle size increased to 150Å. The increase in a particle size was consistent with a sintering process at the higher reaction temperatures.

The silica supported catalysts doped during preparation with cobalt, chromium, iron and copper were also investigated by XRD. The $Cr/U/SiO_2$ catalyst showed a very similar diffraction pattern to the U/SiO_2 catalyst. The catalysts doped with cobalt, iron and copper all exhibited asymmetrically shaped diffraction peaks corresponding very closely to the d-spacing observed for U/SiO_2. These features can be best interpreted as a second minor peak, at a slightly higher diffraction angle, not fully resolved from the major diffraction peaks of U_3O_8. The origin of these secondary peaks cannot be unequivocally identified, but a possible explanation is that the dopant component ions may to some extent be incorporated into the U_3O_8 lattice, distorting the unit cell to produce a new set of diffraction peaks. The distortion caused by low levels of incorporation would probably be small and so the shift in peak position would also be small. The ionic radii of the dopant ions are all smaller than U^{5+} and U^{6+} in U_3O_8 and substitution into the lattice would result in a unit cell contraction which is concordant with our observations.

After use U_3O_8 was the major phase identified on the silica supported doped catalysts, peak widths were reduced after use mirroring the increase in particle size of the used U/SiO_2 catalyst. The decrease of peak width increased peak resolution confirming the asymmetric peak in the diffraction patterns of the unused catalysts doped with cobalt, iron and copper were in fact due to a new set of diffraction peaks.

In-situ XRD studies were carried out using U_3O_8 to investigate the phase stability under realistic operating conditions. Increasing the sample temperature from ambient to 600°C in a flowing air atmosphere showed that U_3O_8 was stable. A slight loss of resolution was observed as the temperature was increased however this was expected as peak broadening occurs due to increased thermal motion within the sample. The catalyst was maintained at 600°C and then cooled in 50°C steps down to 200°C with no detectable phase changes. A certain degree of confusion exists over the phase diagram of uranium oxides, a transition from the U_3O_8 phase to a mix of γ-UO_3 and U_3O_8 around 500°C has been reported [*14*] but no evidence for this bulk transformation was observed.

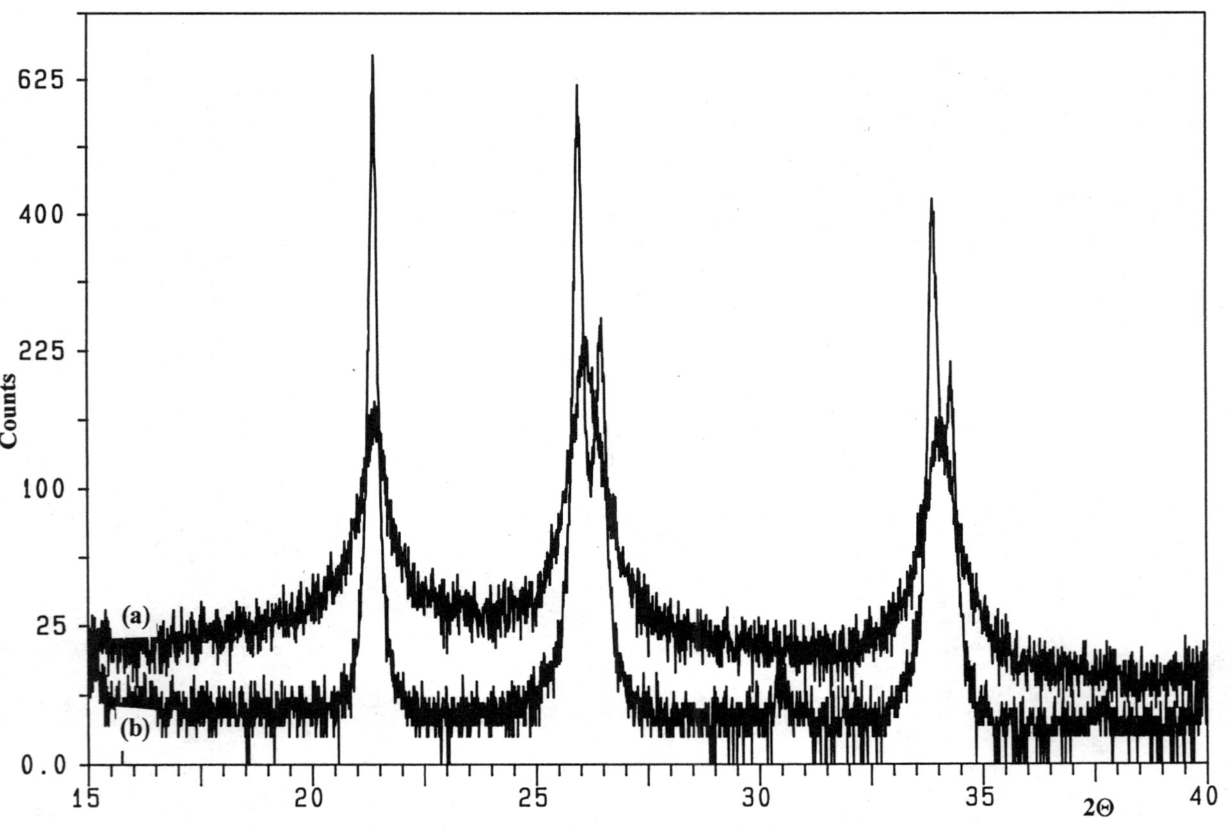

Figure 4. X-ray diffraction patterns; (a) silica supported catalyst pre-cursor calcined at 800°C, (b) pattern obtained from standard U_3O_8 reference sample.

The role of water is known to have an important affect on the oxidation of uranium [*15*] and since water was a major reaction product and potential VOC destruction catalysts may have to operate under wet conditions, U_3O_8 phase stability was investigated in a flowing air stream saturated with water at ambient temperature and pressure. The diffraction patterns from these studies are shown in figure 5. The results indicated that no change in the initial U_3O_8 phase occurred, even up to 600°C. Some thermal broadening and loss of peak resolution were observed but even after approximately 7 hours at elevated temperature patterns were superimposable with those obtained under ambient conditions.

Performance of Uranium Oxide Catalysts for VOC Destruction.

Benzene Destruction. The destruction of benzene was primarily investigated over U_3O_8, initial catalytic activity was observed at 380°C with trace benzene conversion. The conversion increased markedly as the temperature was raised, reaching 100% at 400°C. The sole reaction products were the carbon oxides CO and CO_2, produced with selectivities of 27% and 73% respectively. The sharp rise in conversion can be attributed to the increase in the temperature of the catalyst bed during the highly exothermic benzene combustion reaction (ΔH_{298} = -3302 kJ mol^{-1}). The temperature of the gas stream close to the exit from the catalyst bed was consistently in the region of 30°C higher than the reactor furnace temperature when the catalyst showed high conversion.

A hysteresis effect was evident for the catalytic activity when the reaction temperature was decreased from 400°C. Decreasing the temperature to 350°C resulted in a slight drop in benzene conversion to 95% whilst the CO and CO_2 product selectivities remained approximately constant. When the temperature was decreased further to 300°C the catalyst was inactive.

A comparison of the U_3O_8 combustion activity was made with Co_3O_4, a known highly active combustion catalyst for the destruction of benzene and other organic substrates [*16*]. Co_3O_4 was active at 350°C, 50°C lower than U_3O_8, benzene conversion increased with temperature but even at 450°C it was only 90%. Although Co_3O_4 was less active than U_3O_8 the former catalyst showed the advantage of 100% selectivity towards CO_2, the preferred product.

A new class of uranium oxide catalysts based on a uranium oxide silica supported system has been developed. The benzene conversion over U/SiO_2 was 100% at 400°C, CO and CO_2 were produced with selectivities of 27% and 73% respectively, and similar to those observed over U_3O_8. The activity of the U/SiO_2 catalyst was similar to U_3O_8, however a large hysteresis effect in the benzene conversion was particularly evident over the U/SiO_2 catalyst (figure 6).

A gradual decrease in the benzene conversion was observed on decreasing the reaction temperature from 400°C. The conversion decreased to 93% at 200°C. This hysteresis effect can be explained by local heating within the catalyst bed, due to the exothermic nature of the combustion reaction. Once the ignition temperature for the reaction has been exceeded this local heating effect is significant, maintaining the catalyst bed temperature above the ignition temperature, even when the temperature of the external heating source was considerably lowered. Although the actual bed temperature was not measured directly, the temperature of the exit gas stream from the

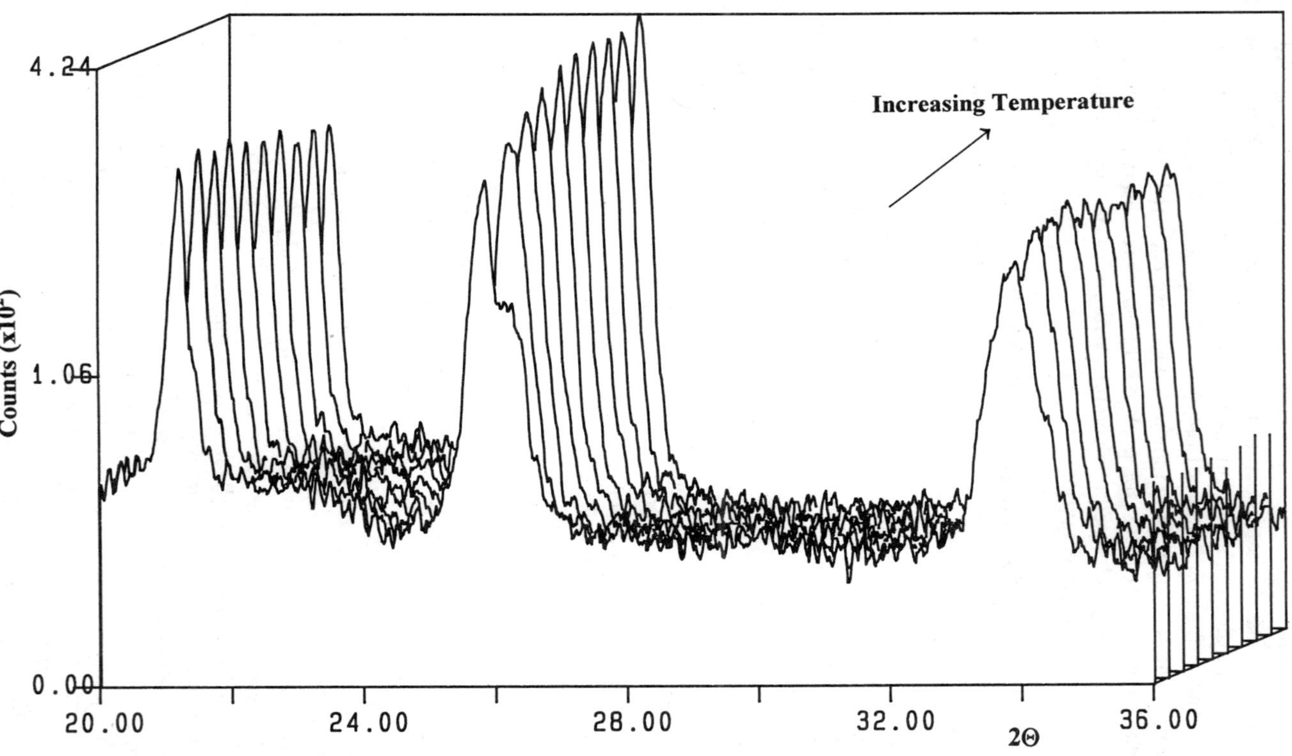

Figure 5, *In-situ* X-ray diffraction patterns of U_3O_8 from ambient to 600°C in a wet (4% water) flowing air atmosphere.

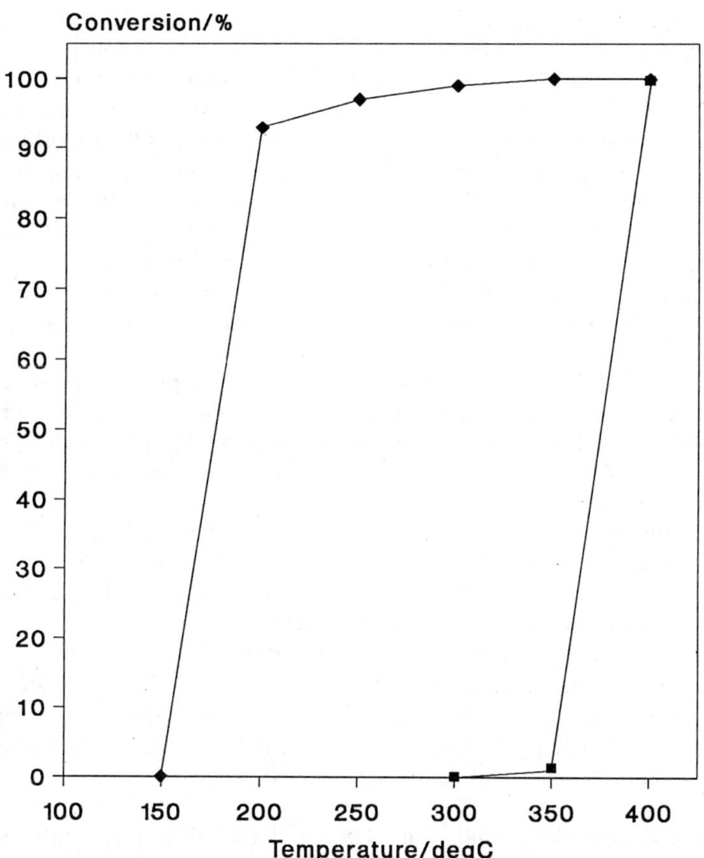

Figure 6. Benzene conversion hysteresis over U/SiO_2 at GHSV = 70,000h^{-1}; ■ increasing temperature, ◆ decreasing temperature.

catalyts bed was 325°C when the external furnace temperature was only 200°C, providing an indication as to the extent of the exothermic heating effect. Once initated the reaction was not autothermal, as decreasing the furnace temperature to 150°C resulted in a total loss of activity as the bed temperature decreased below the ignition temperature. The temperaure of the exit gas stream also deccreased to within ±2°C of the furnace temperature.

The uranium oxide supported catalyst has been developed further by the addition of cobalt, copper, chromium and iron dopant ions. The addition of cobalt to the U/SiO_2 system depressed the benzene conversion relative to the undoped catalyst. The conversion over Co/U/SiO_2 at 400°C was 91%, increasing steadily with temperature and reaching 100% at 500°C. The CO and CO_2 selectivities were not significantly affected by the addition of the cobalt component. The activity of the Cr/U/SiO_2 catalyst was similar to that observed with Co/U/ SiO_2 as. benzene conversion was slightly suppressed. At 500°C the Cr/U/SiO_2 catalyst did not completely destroy benzene, as a trace quantity was still present in the reactor effluent. The CO and CO_2 selectivities were not affected by the addition of chromium. Doping U/SiO_2 with iron suppressed the benzene conversion significantly as the conversion was only 4% at 400°C compared to 100% for the undoped material. However, the benzene conversion did increase significantly to 100% at 500°C. CO and CO_2 selectivities were similar to those observed for Co/U/SiO_2 and Cr/U/SiO_2. The addition of copper did not decrease the conversion of benzene when compared to the U/SiO_2 system as the conversion was 100% at 400°C. The results of benzene conversion at 400°C and GHSV = 70,000 h^{-1} are summarised in figure 7. Comparison is made at high conversion as this is the regime that VOC destruction must operate under.

The addition of copper had an extremely beneficial effect on the product distribution, as the selectivity towards CO_2 was increased across the entire temperature range. The CO_2 selectivity over the Cu/U/SiO_2 catalyst was at least greater than 99%, with only trace quantities of CO detected at 400°C and 450°C.

The activity of the empty reactor tube and the SiO_2 support at high space velocity were negligible, showing only 1% conversion over SiO_2 at 500°C. The low activity in these blank reactions indicated that the destruction of benzene was a heterogeneously initiated process, it also showed that the uranium oxide phase was the catalytically active component of the silica supported system.

Butane Destruction. Studies at GHSV = 70,000 h^{-1} to investigate the activity of U_3O_8 showed that the temperature for initial activity was 450°C. The maximum butane conversion was 80% at 600°C. CO and CO_2 were produced simultaneously at all catalytically active temperatures. CO was a major reaction product with a selectivity of 20% at 550°C decreasing to 14% at 600°C, the remaining product was CO_2.

A comparison of the destruction of butane has also been made with Co_3O_4 at equivalent gas hourly space velocity. The Co_3O_4 catalyst was active 150°C lower than U_3O_8, as initial activity was observed at 350°C. However, maximum butane conversion over Co_3O_4 was 75% compared to 80% obtained over U_3O_8. The maximum butane and oxygen conversions over Co_3O_4 were at 500°C, increasing the temperature to 600°C resulted in a decrease of both conversions. These observations suggest that whilst increasing the reaction temperature above 600°C would increase the butane conversion over U_3O_8 it would have an adverse effect on Co_3O_4, which may deactivate at high reaction temperatures.

The U/SiO_2 catalyst previously tested for benzene destruction has also been examined for the destruction of butane. At 500°C over U/SiO_2 butane conversion was 100%, this activity was significantly higher than that shown by U_3O_8 which was only 80% conversion at 600°C. The hysteresis in conversion which was characteristic for benzene combustion was also evident for butane combustion. Decreasing the temperature below 500°C caused the butane conversion to drop below 100%, however conversion remained above 70% down to 400°C but was inactive at 350°C. The hysteresis effect with butane was less pronounced than that observed with benzene, this effect may be related to the local heating in the catalyst bed during the combustion process. This heating effect would be greater for benzene as the combustion reaction is more exothermic, $\Delta H = -3302$ kJ mol^{-1}, compared to butane, $\Delta H = -2878$ kJ mol^{-1}.

The addition of cobalt to U/SiO_2 considerably suppressed butane conversion compared to the undoped material. At 500°C the $Co/U/SiO_2$ catalyst only showed 66% conversion against 100% over U/SiO_2. The conversion did increase as the temperature was raised but it was still only 82% at 600°C. Little change in the CO and CO_2 selectivities were observed relative to U/SiO_2. The $Cr/U/SiO_2$ catalyst showed considerably enhanced activity compared to all the other catalysts tested. This enhancement was significant at 400°C as butane conversion over $Cr/U/SiO_2$ was 93% whilst the U/SiO_2 catalyst was inactive. Butane conversion over $Cr/U/SiO_2$ increased steadily above 400°C reaching 100% at 550°C. The CO selectivity produced over $Cr/U/SiO_2$ was greater than the U/SiO_2 system. Doping with iron depressed butane conversion in a similar manner to the $Co/U/SiO_2$ catalyst although the effect was less pronounced. The CO and CO_2 selectivities remained relatively unchanged by the addition of iron. The $Cu/U/SiO_2$ catalyst showed a significant butane conversion of 95% at 450°C compared to 7% for U/SiO_2, this level of conversion over $Cu/U/SiO_2$ was equivalent to the $Cr/U/SiO_2$ catalyst (94%), although at the lower temperature of 400°C conversion over $Cu/U/SiO_2$ was only 3% compared to 93% over $Cr/U/SiO_2$. A comparison of butane conversion at 400°C and GHSV = 70,000 h^{-1} is summarised in figure 8.

The incorporation of the components cobalt, copper, chromium and iron into the U/SiO_2 system decreased the temperature for initial butane combustion activity by at least 50°C, although ultimately the cobalt and iron doped catalysts were less active.

As was observed for the destruction of benzene over the copper containing system the same enhancement in the CO_2 selectivity was also evident for butane destruction. CO_2 selectivity over $Cu/U/SiO_2$ was greater than 97% at all temperatures, and was 100% at low conversion and at 550°C. CO oxidation studies using U_3O_8 and the silica supported catalysts showed that oxidation rates over $Cu/U/SiO_2$ were considerably higher than any of the other uranium oxide based catalysts. The most active catalyst for CO oxidation was Co_3O_4. Therefore the two most active CO oxidation catalysts also produced the highest CO_2 selectivities during VOC destruction, suggesting that sequential CO oxidation was an important factor for the increased CO_2 production.

The activity for butane destruction in an empty tube and over SiO_2 was negligible and similar conclusions to those for benzene destruction can also be drawn, that is the

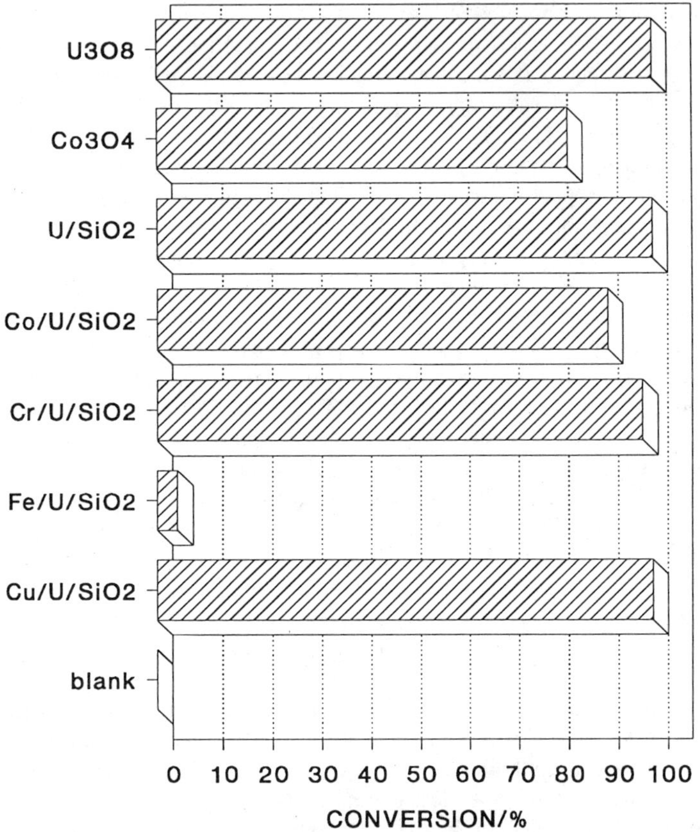

Figure 7. Benzene conversion summary at 400°C and GHSV = 70,000 h^{-1}.

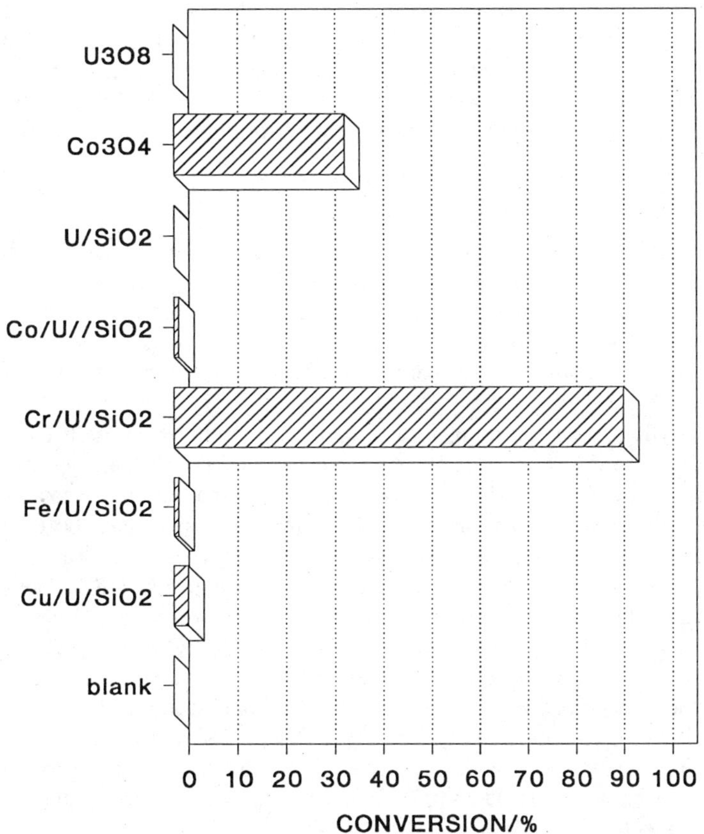

Figure 8. Butane conversion summary at 400°C and GHSV = 70,000 h^{-1}.

destruction of the substrate was initiated by the catalyst and the uranium oxide phase was the active component for butane destruction.

The destruction of benzene over U_3O_8 is a more facile process than for butane. This indicates that the interaction of benzene with the sites active for combustion on the oxide surface are different from those of butane, possible due to the electron rich character of the aromatic compound. However, similarities exit between the two processes as combustion is dominant with no partial oxidation products formed.

The addition of Cr and Cu clearly have an important enhancing effect on the catalytic activity of the supported catalyst for butane destruction. The same enhancement in activity was not observed for benzene destruction. The reasons for this enhancement are not clear, but electronic dopant effects modifying the combustion sites of U_3O_8 may play an important role for alkane destruction. Further studies using XPS techniques are now required to elucidate more fully the effects of the dopants.

Conclusions

U_3O_8 is an active catalyst for the destruction of benzene and butane, and generally showed higher activity than Co_3O_4. The high activity of U_3O_8 detailed in these studies is significant particularly when compared to the combustion activity of Co_3O_4, which is known to be highly active for many substrates [*16*]. *In-situ* XRD studies of U_3O_8 under typical operating conditions of dry and wet flowing air showed that U_3O_8 was stable up to 600°C.

A catalytic system based on silica supported uranium oxide has been synthesised and tested for the destruction of benzene and butane. Thermal analytical techniques showed that supporting uranyl nitrate on silica destabilised the nitrate compared to the pure starting material. Calcination of the silica supported catalyst at 800°C produced a dispersed U_3O_8 phase, which showed an increase in crystallite size on use. The conversion of benzene over U/SiO_2 matches the performance of U_3O_8, showing 100% conversion at 400°C. A significant hysteresis in benzene conversion over U/SiO_2 has also been observed, decreasing the temperature from 400°C maintained a conversion greater than 93% down to 200°C. This effect is explained by considering the local heating of the catalyst bed by the highly exothermic combustion reaction.

Relative to U_3O_8, the activity for butane destruction was enhanced considerably by supporting the uranium oxide on silica, U_3O_8 was still the active phase and the increased activity was most likely due to the higher surface area of the catalytically active component.

Doping U/SiO_2 with cobalt, chromium and iron suppressed the conversion of benzene and did not significantly alter the selectivity towards CO and CO_2. Similar effects were also observed for butane destruction when the U/SiO_2 system was doped with cobalt and iron. This suppression of conversion may be due to the formation of distorted U_3O_8 supported phases, as indicated by XRD characterisation studies, although it appears that such a phase produced by copper doping is at least as active as the undoped supported U_3O_8 catalyst. The addition of chromium to U/SiO_2 enhanced the butane conversion considerably, which was 93% at 400°C. Doping the U/SiO_2 catalyst with copper also had a beneficial effect for butane destruction and was the only dual component supported catalyst which did not suppress benzene conversion. The major effect associated with copper addition was to increase the selectivity towards CO_2 which was >99% for benzene and >97% for butane destruction. The increase in CO_2 selectivity during VOC destruction was due to secondary oxidation of CO.

Literature Cited

[*1*] Mukhopadhyay, N.; Moretti, E.C., Current and Potential Future Industrial Practices For Controlling Volatile Organic Compounds; American Institute of Chemical Engineers. Center For Waste Reduction Technology: New York, NY, 1993.
[2] Tichenor, B.A.; Palazzolo, M.A., Environ. Prog., **1987**, 6, 174.
[*3*] Zieba, A.A.;. Banaszak, T.;. Miller, R., Appl. Catal., **1995**, 124, 47.
[*4*] Spivey, J.J., Ind. Eng. Chem. Res., **1987**, 26, 2180.
[*5*]. Spivey, J.J.; Butt, J.B., Catal. Today, **1992**, 11, 465.
[*6*]. Heyes, Jr., C.J.; Irwin, G.; Johnson, H.A.; Moss, R.L., J. Chem. Tech. Biotechnol., **1982**, 32, 1025.
[*7*] Yao, Y.F.Y., J. Catal., **1973**, 28, 139..
[*8*] Yao, Y.F.Y.; Kummer, J.T., J. Catal., **1973**, 28, 124.
[*9*] Nozaki, F.; Ohki, K., Bull, Chem. Soc. Japan, **1972**, 45, 3473.
[*10*] Grasselli, R.K.; Suresh, D.D., J. Catal., **1972**, 25, 273.
[*11*] Keulks, G.W.; Yu, Z.; Krenzke, D.L., J, Catal., **1993**, 84, 38.
[*12*] Steenhof de Jong, J.G.; Guffens, C.H.E.; Van der Baan, H.S., J. Catal., **1972**, 26, 401.
[*13*] Collette, H.; Maroie, S.; Riga, J.; Verbist, J.J.; Gabelica, Z.; Nagy, J.B.; Derouane, E.G., J. Catal., **1986**, 98, 326.
[*14*] Allen, G.C.; Tempest, P., J. Chem. Soc., Dalton Trans., **1982**, 2169.
[*15*] Colmenares, C.A, Prog. Solid State Chem., **1984**, 12, 257.
[*16*] Golodets, G.I., Stud. Surf. Sci. Catal., **1983**, 15, 652.

METHANE OXIDATION

Chapter 6

Partial Oxidation of Methane on Low-Surface-Area SiO_2–Si-Supported Vanadia Catalysts

Miguel A. Bañares[1], Luis J. Alemany[1], Francisco Martin-Jiménez[2], J. Miguel Blasco[2], Manuel López Granados[1], Miguel A. Peña[1], and José L. G. Fierro[1,3]

[1]Instituto de Catálisis y Petroleoquimica, Consejo Superior de Investigaciones Cientificas, Campus Universitario de Cantoblanco, 28049 Madrid, Spain

[2]Departamento de Ingenieria Quimica, Universidad de Málaga, 29071 Málaga, Spain

Partial oxidation of methane (POM) has been carried out on a very low surface area and non-porous V_2O_5/SiO_2-Si catalyst at atmospheric pressure. The influence of a reactor dead volume before the catalyst bed as well as the role of radical initiators (NO) in the gas phase are discussed for the performance of the POM reaction. Mixing volume and/or addition of NO promote production of C_1 oxygenates and C_{2+} hydrocarbons at temperatures below 950 K. The catalyst provides a heterogeneous route for the production of C_1 oxygenates (methanol and formaldehyde) and C_{2+} hydrocarbons from methane. The extent of the latter reaction increases with increasing reaction temperature, CH_4 + O_2 mixing volume, and/or NO added. When NO addition increases the reactivity of methane and the yield of all the products, the reaction equilibrium is shifted to more oxidized products.

During the last decade the catalytic partial oxidation of methane (POM) to formaldehyde and methanol attracted a great deal of attention *(see, e. g., 1-3)*. Many kinetic studies have been carried out on high surface area silica-supported redox oxides of MoO_3 *(4-6)* and V_2O_5 *(7-9)*, oxometallate precursors *(10)*, and bare silica *(11,12)* using either oxygen or nitrous oxide as the oxidant, although very few of these attempted to correlate catalytic activity with catalyst structure and texture. In general, activity and formaldehyde yields reported by different research groups appear to be controversial and not particularly encouraging. Moreover, several authors have questioned the catalytic nature of the POM reaction *(11,13,14)*. Indeed, it has been stated that the yields for formaldehyde obtained using catalysts are lower than those found from the purely gas-phase reaction *(13)*. In addition, a heterogeneous-homogeneous reaction scheme that predicts the formation of formaldehyde on catalyst surface has been proposed *(14)*. The production of C_1 oxygenates is limited and the degradation to CO dominates, due both to

[3]Corresponding author

0097–6156/96/0638–0078$15.00/0

catalyst surface reactions *(13,15-17)* and to homogenous degradation in the hot zone downstream from the catalyst bed *(6,18)*. The chemical formulations for a coupling catalyst to yield C_{2+} hydrocarbons and a catalyst for the conversion of methane to C_1 oxygenates are, in general, quite different. However, we have recently observed a close link between methane coupling and conversion to C_1 oxygenates on vanadium oxide based catalysts *(15)*. From the literature it is evident that most catalysts require a high surface area silica support to promote methane activation, but this is detrimental to the selectivity of the reaction due to the relative ease with which C_1 oxygenates and C_{2+} hydrocarbons are oxidized by the high surface area of the silica carrier. The use of a catalytic system combining a large dead volume upstream from the catalyst bed with a very low surface area, such as that for non-porous silica-silicon supported vanadia catalysts, has proved to be active and selective for the production of C_1 oxygenates (methanol and formaldehyde) and C_{2+} hydrocarbons, decreasing the amount of non-selective CO_x products by ca. 50% compared to high surface area V_2O_5/SiO_2 catalysts *(15)*. A close relationship is observed between CO_x and C_{2+} hydrocarbon production. The appearance of CO_x (predominantly CO) is largely due to the participation of a deep side oxidation reactions of C_{2+} hydrocarbon precursors in the gas phase and on the silica surface. C_{2+} hydrocarbons appear to be produced on large surface area catalysts but are readily oxidized to CO as a consequence of the large surface area. In the present paper, the role of NO, a well known chemical radical initiator, and the thermal homogeneous activation *vs.* heterogeneous activation on non-porous V_2O_5/SiO_2-Si catalysts are studied and their effects on activity and selectivity are discussed.

Experimental

Materials. A non-porous silicon wafer with small amounts of boron (0.04 ppm B) was used as a carrier precursor. This carrier was calcined in air at 923 K in order to develop a layer of silica. The resulting silica-silicon material was impregnated with a vanadyl acetylacetonate solution in methanol, with the concentration selected to obtain a vanadium loading of 21 ppm (equivalent to 11% surface coverage as determined by XPS). Owing to the non-porous texture of the B-doped silicon material, vanadium oxide is deposited on the outer surface of the crystals, and its concentration can only be revealed by surface sensitive techniques, such as photoelectron spectroscopy. The impregnated V-precursor was then dried at 383 K for 12 h and finally calcined at 823 K for 2 h. The catalyst is referred to as 21V, where 21 denotes the vanadium loading expressed in ppm. The catalyst was crushed and sieved in the particle size range of 0.250-0.125 mm.

Experimental Methods. The catalysts were characterized by a variety of physical techniques. Powder X-ray diffraction patterns were recorded with a Philips PW 1010 vertical diffractometer using nickel filtered CuKα radiation ($\lambda = 0.1538$ nm), under constant instrumental parameters. For each sample, Bragg angles between 5 and 80° were scanned at a rate of 2°/min. Fourier transform IR spectra were recorded with a Nicolet 5ZDX spectrophotometer working with a resolution of 4 cm^{-1} from self-supporting wafers of 3% sample in KBr. Raman spectra were obtained using a Bruker FT Raman instrument with a 1064 nm exciting source. The samples were placed in a stationary sample holder and kept at room temperature and exposed to air. The power was fixed in the range of 20-40 mW in order to avoid sample vaporization.

The morphological study and chemical identification were carried out with a scanning electron microscope (SEM) Jeol model JSM-840 coupled to a silicon/lithium detector and a Kevex processor for energy dispersive X-ray analysis. Photoelectron spectra were acquired with a Fisons ESCALAB 200R spectrometer equipped with a hemispherical electron analyzer and a MgKα X-ray source (h.ν = 1253.6 eV) powered at 120 watts. A PDP 11/04 computer from DEC was used for collecting and analyzing the spectra. Finely divided samples were pressed into small copper holders and then placed in the pretreatment chamber. The samples were outgassed or calcined in the pretreatment chamber of the spectrometer at 873 K prior to being moved into the ion-pumped analysis chamber. The base pressure in the analysis chamber during data acquisition was maintained below $2x10^{-9}$ Torr (1 Torr = 133.33 Pa). The spectra were collected for 30 to 100 min, depending on the peak intensities, at a pass energy of 10 eV (1 eV = 1.602×10^{-19} J), which is typical of high resolution conditions. The intensities were estimated by calculating the integral of each peak after smoothing and subtraction of the "S-shaped" background and fitting of the experimental peak to a combination of Lorentzian and Gaussian curves. All binding energies (BE) were referenced to the Si 2p peak at 103.4 eV. This reference gave BE values within an accuracy of $\pm$ 0.2 eV.

Activity Measurements. Activity experiments were performed in a flow quartz microreactor (9 mm i.d.) working at 1 bar (10^5 Pa) total pressure. Samples of 200 mg and a methane pseudo-residence time of 4 gh/mol (CH_4 flow = 50.0 mmol/h) were used in all of the experiments. The reactor was designed so that there was a very low dead volume present downstream of the catalyst bed to prevent further decomposition of partial oxidation products *(1,6,18)*. A variable mixing volume reactor was used *(19)* so that methane and oxygen could be fed separately. The oxygen feed entrance could be regulated in order to control the volume of the mixture of methane and oxygen, thus providing a variable mixing volume reactor. The temperature profile inside the reactor dead volume is essentially isothermal upstream of the catalyst bed except near the ends of the furnace where heat exchange with the environment occurs. Blank reactions for dead volume and no dead volume configurations on SiO_2-Si substrate, inert packing SiC and on the empty reactor have been reported *(15)*. No significant differences in results could be observed between these and the empty reactor *(15)*.

The radical initiator NO (SEO-Air Liquide, 900 ppm in nitrogen) was cofed by means of mass flow controllers. The NO concentration in the reaction feed was 0.03% molar. Reaction feed consisted of $CH_4 + O_2$ + (900 ppm NO in N_2) = 50.0 + 25.0 + 35 mmol/h. The concentration of oxygen and methane was the same for experiments in the absence of the NO additive. In this case, the reaction feed was: $CH_4 + O_2 + He = 50.0 + 25.0 + 35.0$ mmol/h. The pressure was measured upstream of the catalyst bed and did not increase more than 0.15 atm above atmospheric pressure during the catalytic experiments (total pressure below ca. 1.1 atm). Under these conditions homogeneous conversion of methane to methanol and formaldehyde is negligible (13).

Safety considerations. The feed consisted mainly of a mixture of CH_4 and O_2. This mixture is flammable at methane concentrations below 46% CH_4 (53% O_2) and above 20% CH_4 (80% O_2). This was taken into consideration when the reaction mixture was changed. Once the catalytic experiment was completed, the reaction feed was switched to helium.

Toxic or dangerous reaction products were treated accordingly. CO causes haemoglobin disfunction with a threshold limiting value of 500 ppm in air, and is lethal at a concentration above 0.2 % vol. The threshold limiting value for methanol is higher (200 ppm), but long–term inhalation causes severe headache, vomiting and visual and digestive disorders. Finally, formaldehyde is also toxic and a carcinogenic agent (threshold limit value of 1 ppm), and inhalation and contact must be avoided. Consequently, a cold trap was placed at the exit of the on–line analytical system, intended for the removal of formaldehyde and methanol. After the cold trap, a catalytic cartridge (Hopcalite) converted the CO in the out–stream to CO_2. The resulting purified gases were emitted outside of the laboratory. The residue was properly disposed of after the experiments.

Results

Structure of the materials. XRD patterns of the V_2O_5/SiO_2-Si catalyst and the SiO_2-Si substrate show sharp diffraction peaks at theta 28.51 (I = 100), 47.34 (I = 52), and 56.2 (I = 23), characteristic of the silicon space group FD3m crystallized in a cubic system (JCPDS file # 27-1402). No other diffraction peaks than those arising from the silicon substrate could be observed on the V_2O_5/SiO_2-Si catalyst. Therefore, the silica layer that developed on top of the silicon crystal is either amorphous or not large enough to produce a diffraction pattern. The absence of any diffraction pattern of vanadium oxide excludes the formation of large V_2O_5 crystals. SEM micrographs display the existence of Si (111) developed from Si (100) as a consequence of high temperature treatments. EDX analysis revealed the presence of vanadium on the catalyst but no crystalline vanadia is observed. At very low surface loadings on silica, highly dispersed surface vanadium oxide species are observed *(20, 21)*.

An infrared spectrum from framework vibration modes of reference silica (Degussa Aerosil 200) shows bands at 1085, 795, and 460 cm^{-1}, which are characteristic of Si-O-Si modes *(22, 23)*. The band at 1085 cm^{-1} is the most intense, and that at ca. 460 cm^{-1} is broad and weak. Similar bands are observed on the silica-silicon substrate but they are shifted to lower energies (1060, 770, and 469 cm^{-1}). The V_2O_5/SiO_2-Si catalyst presents an intermediate situation (1080, 785, 460 cm^{-1}). The IR bands are least sharp on the SiO_2-Si substrate, sharper on the V_2O_5/SiO_2-Si catalyst, and sharpest on the reference silica. The precise frequencies of Si-O-Si modes on silica-silicon materials are a function of the thickness of the silica film *(22)*. A linear function between this shift and silica film thickness *(22)* was used in the present work to estimate a thickness for the silica layer of 500 nm. Bands of the V_2O_5/SiO_2-Si catalyst are better defined, as expected from the second calcination of the silica-silicon substrate after impregnation with the vanadium oxide precursor. We have also observed that the silica layer is further developed during methane oxidation experiments *(19)*.

Further information on the silica layer can be obtained using the surface sensitive photoelectron spectroscopy technique where the Si2p core level displays a single component at 103.4 eV, typical of SiO_2, and no features were recorded at 99.3 eV where metallic silicon is expected to appear. Thus, XPS measurements indicate that the silicon substrate must be covered by a layer of silica thicker than 2 nm, which is in agreement with the infrared estimate. It is important to emphasize in this respect that the catalyst still has a metallic appearance, indicating that the particles consist of a silicon core covered by a silica layer several nm thick.

The V 2p core level spectra of the catalysts show the characteristic spin-orbit splitting in the 520-516 eV binding energy (BE) range. Their profile is complex above ca. 519 eV as a consequence of overlapping of the less intense V $2p_{1/2}$ peak and the O 1s satellite arising from the non-monochromatized $MgK\alpha_{3,4}$ radiation used as the exciting source (Figure 1). The BE of a V $2p_{3/2}$ core level spectra of a fresh and a used V_2O_5/SiO_2-Si catalyst is at 517.2 eV, which indicates that $V^{(V)}$ is the dominant species *(24)*. The high V/Si atomic ratio determined by XPS should also be noted , despite the low bulk loading of vanadium oxide (21 ppm V_2O_5) (Table I). Due to the high surface sensitivity of the XPS technique, the observation of a rather intense vanadium XPS signal clearly suggests that vanadium oxide species remain preferentially on the surface. The very low surface area of the silica-silicon substrate also accounts for the rather high V/Si atomic ratio determined by XPS. The further development of the silica layer during reaction, as evidenced by infrared spectroscopy, may be responsible for the significant decrease in the V/Si atomic ratio after catalytic experiments. Vanadium oxide evaporation during reaction, due to the high temperatures used or diffusion into the silica layer, might also occur. It is also worth noting that the surface area is expected to

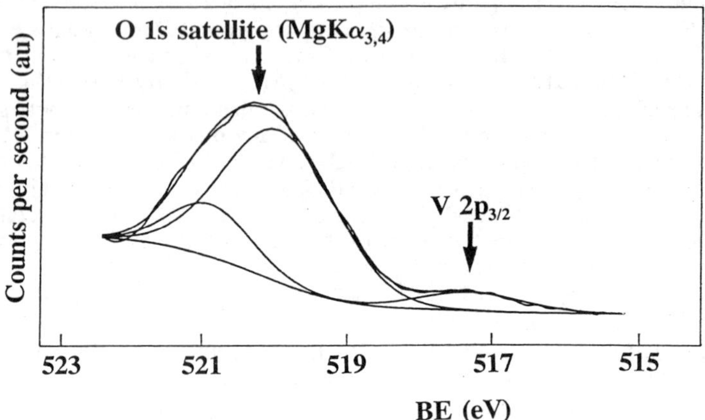

Figure 1. V2p core level spectrum for V_2O_5/SiO_2-Si catalyst.

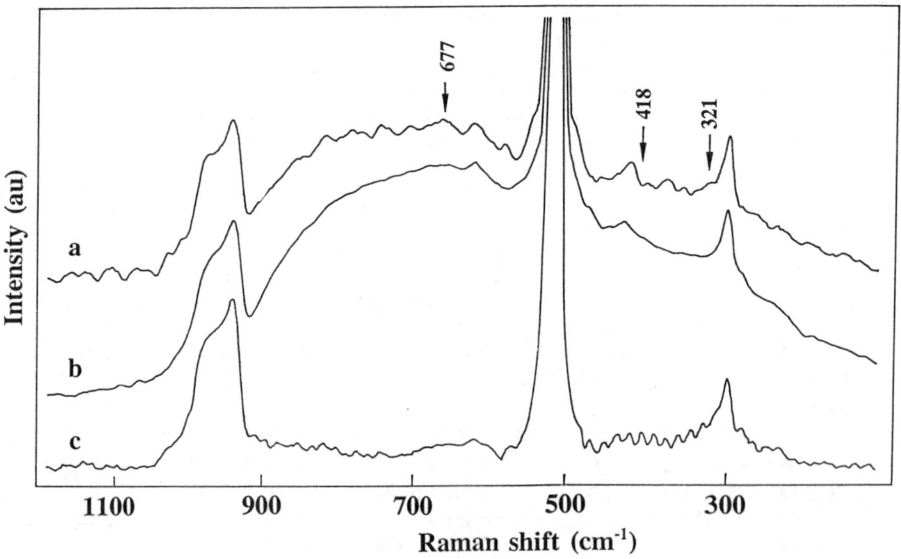

Figure 2. FT-Raman spectra of V_2O_5/SiO_2-Si catalyst under ambient conditions. (a), SiO_2-Si; (b), V_2O_5/SiO_2-Si catalyst fresh; and (c), V_2O_5/SiO_2-Si used in a reaction.

increase during silica layer development, with the subsequent increase in the number of silicon sites "visible" by XPS, thus decreasing the V/Si XPS atomic ratio.

Table I. Photoelectron spectroscopic characterization of the materials

Catalyst	Condition	I_V/I_{Si}	BE V $2p_{3/2}$ (eV)
V_2O_5/SiO_2-Si	Fresh	0.119	517.3
V_2O_5/SiO_2-Si	Used	0.022	517.2

Raman spectra of V_2O_5/SiO_2-Si and silica-silicon substrate are presented in Figure 2. The sharp band at 520 cm^{-1} corresponds to the first order Raman band of silicon, and those at ca. 430, and 300 cm^{-1} are characteristic of second order Raman modes *(25, 26)*. The Raman bands at ca. 970-960, 800, and 430 cm^{-1} can be assigned to the silica layer *(20)*. The incorporation of vanadium leads to very weak Raman features at ca. 677, 418, and 321 cm^{-1}, which are characteristic of hydrated, surface-dispersed vanadium oxide species on silica *(20)*. These also show a Raman band at ca. 990 cm^{-1}, which cannot be observed due to interferences with the Raman bands of the silicon substrate. After catalytic experiments, no differences are observed in the Raman bands. The change observed in baseline curvature is probably associated with further development of the silica layer.

Catalytic performance. To effectively assess the role of gas-phase reaction initiation, experiments were performed with different mixing volumes and temperatures. Since methane does not react in the absence of oxygen at temperatures used in our experimental conditions *(27)*, regulation of the $CH_4 + O_2$ mixing volume is a good means of controlling the extent of gas-phase activation of methane. Consequently, methane and oxygen were fed into the reactor through two independent concentric quartz tubes so that the position of the inner quartz tube determines the extent of homogeneous methane activation. NO was also used as an additive in the reaction feed to generate radicals in order to better evaluate their role in methane activation. Since the effect of variable $CH_4 + O_2$ mixing volume was already determined *(19)*, we only use two extreme configurations for this study: no-mixing volume and maximum mixing volume (0 and 9 mL, respectively). Additional generation of radicals by NO was carried out on both reactor configurations. During catalytic experiments formaldehyde and methanol (C_1 oxygenates), CO and CO_2 (CO_x products), and C_2H_n, C_3H_n , and trace amounts of C_4H_n (C_{2+} hydrocarbons), along with water were the only products observed. The effect of these variables on methane conversion and yield of the three groups of products: C_1-oxygenates, C_{2+} hydrocarbons, and non-selective oxidation products CO_x is presented in Figure 3.

As expected, any increase of catalyst temperature increases methane conversion. In agreement with previous studies *(15,19)* an increase of the $CH_4 + O_2$ mixing volume increases methane conversion. The use of NO also increases methane conversion and the combined use of NO and the $CH_4 + O_2$ mixing volume results in the largest increase in methane conversion. However, oxygen limiting conversion conditions prevent a further increase in methane conversion above ca. 900 K for this reaction configuration. The conversion of methane rendered by the combination of NO additive and $CH_4 + O_2$ mixing volume is larger than the simple sum of methane conversions recorded for NO promoted and $CH_4 + O_2$ mixing volume promoted reactions, over the reaction temperature range (Table II). This must be indicative of a synergetic effect originating from the radical character of methane activation and subsequent chain reactions. The

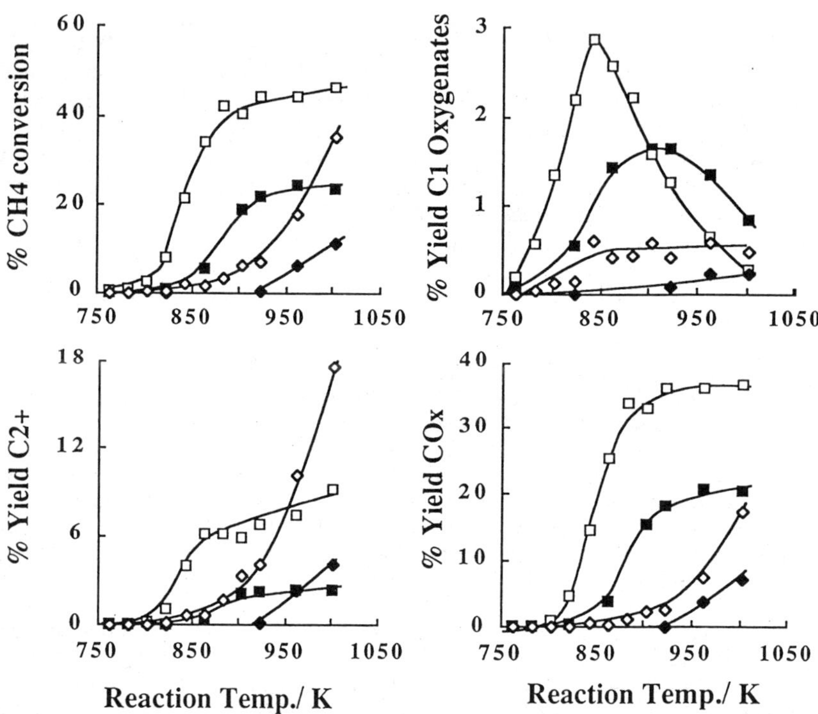

Figure 3. Methane conversion and yield to different groups of product on V_2O_5/SiO_2-Si catalyst. (□), reaction with NO; (◊), reaction mixture diluted in helium. Open symbols stand for configuration with mixing volume and solid symbols for the configuration with no mixing volume. Reaction conditions: Methane residence time: 4 g.h/Mol; CH_4/O_2=2 molar; mixing volume = 9 mL.

same trend is observed for C_1-oxygenate products. Maximum yield of C_1 oxygenates is shifted to lower temperatures in the presence of both the $CH_4 + O_2$ mixing volume and the NO additive. At medium reaction temperatures (ca. 900 K), production of C_{2+} hydrocarbon is equivalent for the NO promoted reaction and for the mixing volume promoted reaction. Combined radical generation, by NO initiator and $CH_4 + O_2$ mixing volume, increases production of C_{2+} hydrocarbons by ca. an order of magnitude. However, as reaction temperatures increase, production of C_{2+} hydrocarbon products in the presence of NO, for any $CH_4 + O_2$ mixing volume configuration, increases at a lower rate. The origin of the higher activity for NO promoted reactions can easily be understood from the yields of non-selective oxidation products, which dramatically increase at high reaction temperatures. As stated above, the conversion of methane is promoted by either $CH_4 + O_2$ mixing volume and/or NO additives, but the generation of radicals by homogenous reaction of $CH_4 + O_2$ yields smaller quantities of non-selective CO_x oxidation products. It is interesting to note, however, that NO additives promote and increase C_1 oxygenates production by a factor of ca. 3 (NO additive only) and a factor of ca. 6 (combined NO additive and $CH_4 + O_2$ mixing volume). The fact that the yield of C_1 oxygenates reaches a maximum in the presence of NO suggests that the degradation of C_1-oxygenates is also promoted by the NO additive, particularly at higher reaction temperatures.

Table II. Methane conversion for different conditions

Reaction Temp. (K)	*% CH4 Molar Conversion*			*Added CH_4[a] Conversion*
	NO	*MV[b]*	*(NO and MV[b])*	*(NO) + (MV[b])*
803	0.0	0.0	2.4	0.0
823	0.8	0.3	7.9	1.1
863	5.6	1.5	33.9	7.1
903	18.9	6.2	40.4	25.0
923	21.9	7.1	44.0	29.1
963	24.2	18.2	44.1	42.4

[a]Addition of methane molar conversion values from 2nd and 3rd column.
[b]MV: $CH_4 + O_2$ mixing volume

To further understand the effect of the NO additive on methane conversion, selectivity *vs.* conversion plots are presented in Figure 4 for all of the experimental conditions. For the sake of clarity selectivities to C_3 and C_4 hydrocarbons have been eliminated due to their very low values as a consequence of dilution. Reaction in the absence of all promoters (Figure 4A) does not result in a high conversion of methane (Figure 3). The high temperatures required to convert methane result in the elevated production of non-selective CO. Further oxidation to CO_2 can be observed at higher methane conversion (higher reaction temperature). Ethylene selectivity is always lower than that of ethane, which is oxidized to CO. A very small amount of methanol (ca. 1% molar selectivity) is produced. When $CH_4 + O_2$ mixing volume (9 mL) is present (Figure 4B) production of ethane is significantly higher and, as methane conversion increases, its oxidative dehydrogenation to ethylene becomes dominant. The selectivity to methanol at low reaction temperature is also higher than in the absence of mixing volume. The selectivity to the complete oxidation product CO_2 is always low, and CO

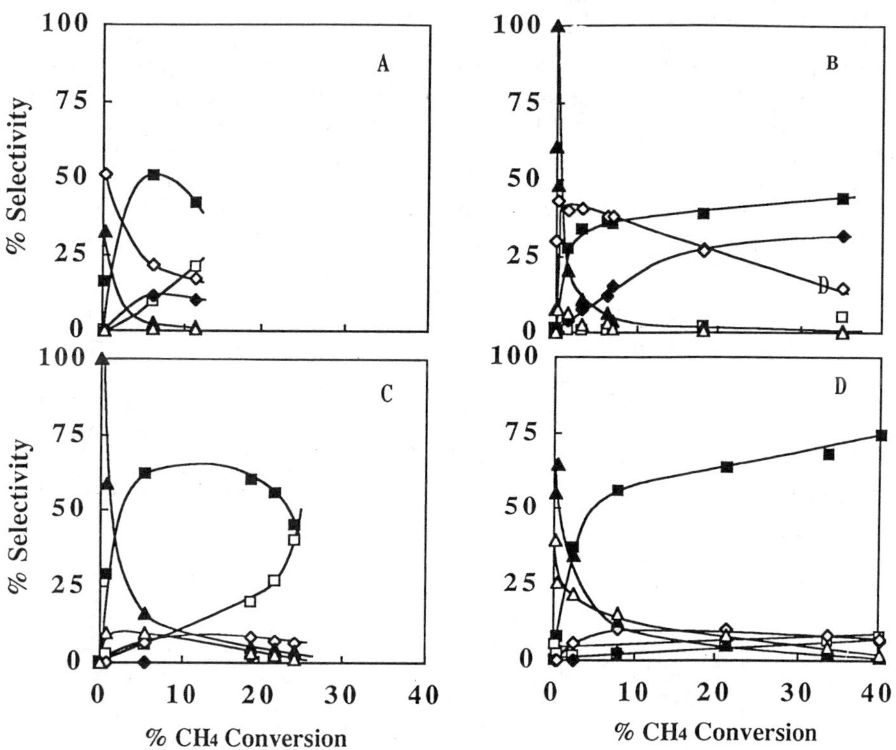

Figure 4. Selectivity vs. methane conversion with the following reaction configurations: (A), He, (B), He-Vol, (C), NO, and (D), NO-Vol. Reaction conditions as in Figure 3. (■), CO; (□), CO_2; (◆), C_2H_4; (◇), C_2H_6; (▲), HCHO; (△), CH_3OH.

selectivity flattens off, with no further degradation of CO being observed at increasing reaction temperature. When the reaction is promoted by NO additive only (Figure 4C) methane conversion increases but selectivity is essentially dominated by the non-selective CO product. C_{2+} hydrocarbon selectivity is largely decreased compared to the reaction promoted by $CH_4 + O_2$ mixing volume. At low methane conversion formaldehyde is the main oxidation product and methanol production is also significant. The selectivity to CO_2 increases to values near 50% molar selectivity, as methane conversion increases. Finally, when methane conversion is promoted by both NO and $CH_4 + O_2$ mixing volume (Figure 4D), the selectivity to C_1 oxygenates, formaldehyde and methanol, is dominant at low methane conversion. Selectivity to C_2H_n is moderate at any methane conversion for this configuration and CO is the main product at medium and high conversion of methane. Methanol and formaldehyde are promoted in the presence of NO at low conversion. The non-selective oxidation product CO is dominant at the expense of C_{2+} hydrocarbons at methane conversions higher than ca. 5 % compared to promotion by the $CH_4 + O_2$ mixing volume.

Figure 5 illustrates the relative conversion of oxygen *vs.* that of methane as a function of reaction temperature. Only ratios calculated at methane molar conversions above 1 % are presented. In the absence of any promotion of the reactivity, the O_2/CH_4 conversion ratio is ca. 2. When reaction is promoted by gas-phase activation of methane by oxygen, the O_2/CH_4 conversion ratio decreases to ca. 1.3 and does not increase at higher reaction temperatures. On the contrary, promotion of the reactivity by the NO additive increases the O_2/CH_4 conversion ratio to almost 3. Combined promotion of methane activation by the NO additive and the $CH_4 + O_2$ mixing volume results in an intermediate situation (ca. 2). If the activity originates from the catalyst alone or by promotion with NO additive, higher oxygen consumption is observed, corresponding to a decrease in selectivity and a higher production of C_1-oxygenates at low methane conversion (Figure 4). However, as reaction temperature increases, non-selective CO_x products are dominant (Figures 3 and 4). The participation of the $CH_4 + O_2$ mixing volume gas-phase activation of methane decreases oxygen consumption to ca. half of that observed for the presence of the NO additive alone. Combined activation yields an intermediate situation, reflected in a selectivities to oxidative coupling products.

Discussion

The catalyst used in the present work has a silicon core, which is completely covered by a thin layer of silicon dioxide developed by calcination treatments. This silica layer is of an amorphous nature as no diffraction patterns of silica crystalline phases were observed. Vanadium oxide deposited on top of this silica layer must be present as essentially dispersed surface vanadium oxide species since the diffraction pattern and Raman features of crystalline vanadia were absent. The V/Si XPS atomic ratios recorded are equivalent to those observed for silica-supported vanadium oxide catalysts with vanadium oxide loadings below the monolayer coverage *(28)*. Although the vanadium oxide loading is in the range of ppm, significantly lower than that of traditional high surface area silica-supported vanadium oxide catalysts *(5,28,29)*, the surface area is also lower (ca. 1 m^2/g *vs.* typically 100-200 m^2/g). Surface vanadium oxide loading is essentially identical to that of high surface area silica based catalysts. Consequently, vanadium sites on the silica-silicon substrate equivalent to those on silica supports are expected to be formed. Although the surface of V_2O_5/SiO_2-Si is essentially identical to classical silica-supported vanadium oxide catalysts, surface area differentiates both types of catalysts. Traditionally, catalytic reactions are promoted by high surface area supports, which increase the exposure of the active sites and promote

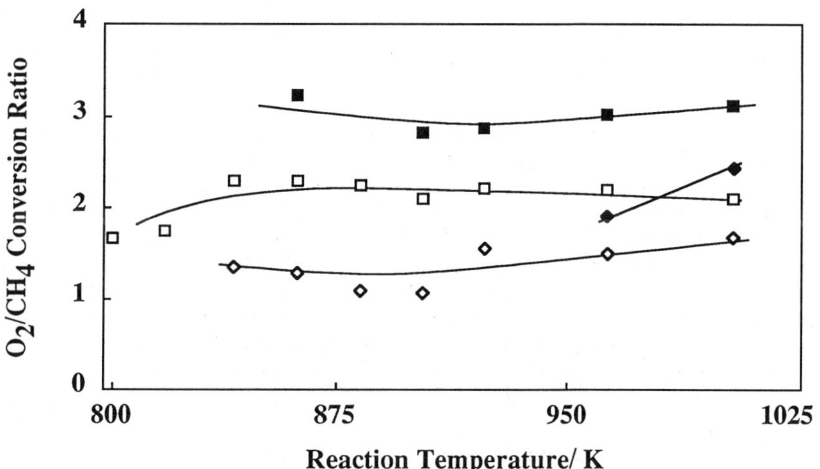

Figure 5. Molar conversion of O_2 *vs*. that of CH_4 plotted against reaction temperature. Reaction conditions as in Figure 3. (□), reaction with NO; (◊), reaction mixture diluted in helium. Open symbols stand for configuration with mixing volume and solid symbols for the configuration with no mixing volume.

activation of the reactants. The same method has also been used in methane conversion. However, it is now evident that the catalyst not only promotes methane activation, but also the degradation of reaction intermediates *(13,15-19,30,35,36)*. The use of different supported oxides has shown that better selectivites are achieved on less reactive systems *(31)*. It does appear that the most important handicap for selectivity for methane conversion is the surface area itself *(15,19,32,35,36)*, which seems to imply that there is no possibility to improving this reaction. However, the heterogeneous-homogeneous nature of the methane activation *(15,19,30,35-37)* permits us to avoid the problem of surface area. Radicals should be generated by means other than surface area and this additional generation of radicals proves to be a very interesting option *(16,30,33,35)*. It can be done in a heterogeneous way by a double-bed catalyst design *(33)* or by gas-phase production of radicals by addition of a radical initiator, like NO *(16)*, Since methane is converted to radicals by its reaction with oxygen in the gas phase, a simpler and inexpensive method to activate methane is to provide void volume before catalyst bed (dead-volume reactor configuration), so that a radical rich reaction mixture reaches the catalyst. We have already shown that gas-phase activation of methane by means of a $CH_4 + O_2$ mixing volume upstream the bed of the very low surface area V_2O_5/SiO_2-Si catalysts converts methane to a similar extent as on high surface area silica-supported vanadium oxide catalysts *(15, 19)*. However, a key difference has been observed: the decrease in surface area from ca. 200 m^2/g to ca. 1 m^2/g not only decreases the yield of CO to ca. one half, but also increases the yield of C_{2+} hydrocarbons in a complementary way, and doubles C_1 oxygenate production, as we have previously reported *(15)*. Results were identical using an empty reactor and a reactor SiO_2/Si support and correspond to gas phase reactions *(15)*. Methane conversion at 863 K increases from 2.0% using a reactor with 9 mL void volume before the bed of the SiO_2-Si substrate to 21.6% in the same reactor with 9 mL of void volume before the bed of the V_2O_5/SiO_2-Si *(15)*, showing the relevance of vanadium oxide on the SiO_2-Si substrate for the conversion of methane. Supported vanadium may provide local activation of oxygen thus increasing its reactivity and its ability to homolytically split peroxy radicals providing a route to methyl radicals *(15,19)*. The most remarkable feature of the V_2O_5/SiO_2-Si catalyst in a reactor with void volume before its bed was the very high yield of C_{2+} hydrocarbons and the presence of methanol. The absence of significant amounts of C_{2+} hydrocarbons in high surface area catalysts seems to be due to their further oxidation to CO on the high surface area catalysts *(15,36)*. A decrease in the oxidizing capacity of supported oxides on high surface area silica-supported oxides has proved to have a slight beneficial effect on C_2H_n products *vs.* CO *(31)*. This beneficial effect suggests that non-oxygenated intermediates are formed on the silica-supported catalysts during methane conversion. Clearly, a very important factor determining selectivity for methane conversion to C_1 oxygenates is the surface area of the support *(15,19,32,35,36)*. Surface area promotes the degradation of reaction intermediates by further oxidation resulting from collisions with the catalyst surface.

The effect of additional activation by adding NO provides further insight into the reaction mechanism. Figure 3 clearly shows the increase of methane conversion by generation of radicals. However, different methods of generating radicals do not result in the same catalytic performance. Promotion by gas-phase activation of methane by oxygen requires higher temperatures than is required by the presence of NO. A similar level of methane conversion is recorded at ca. 60 K higher in the absence of NO. Gas-phase activation of methane by $CH_4 + O_2$ mixing volume requires high reaction temperatures and enough mixing volume to be effective on V_2O_5/SiO_2-Si catalysts *(19)*. The incorporation of NO into the reaction feed with mixing volume provides an additional source of radicals that, as observed, increases methane conversion to a large extent. Due to the radical character of the reaction and chain reaction propagation, the combined production of radicals by gas-phase $CH_4 + O_2$ mixing volume and added NO is larger than the addition of each individual initiator (Table II). Undoubtedly, additional generation of radicals increases methane conversion, but the effect on the different groups of products (C_1 oxygenates, C_{2+} hydrocarbons, and non-selective CO_x) requires some additional comment.

C_1 oxygenates are promoted to a larger extent by NO than by $CH_4 + O_2$ mixing volume. In addition, the combined generation of radicals by NO and $CH_4 + O_2$ mixing volume yields larger amounts of C_1 oxygenates. However, the maximum amount is reached at temperatures lower than those for the reaction promoted by NO. The loss of C_1 oxygenates as reaction temperature increases must be due to further oxidation promoted by the radical rich reaction medium present. Profiles for C_{2+} hydrocarbons show different patterns, depending on whether NO is present or not. If the reaction is promoted by both NO and $CH_4 + O_2$ mixing volume there is a significant increase in the C_{2+} hydrocarbons yield. As reaction temperature increases the yield of C_{2+} hydrocarbons tends to increase at a more moderate rate in the presence of NO. In the absence of the $CH_4 + O_2$ mixing volume, the NO additive yields the same C_{2+} hydrocarbon product profile as the reaction promoted by combined NO and $CH_4 + O_2$ mixing volume, but values are significantly lower. In both cases, oxygen limiting conversion occurs at high temperatures to complicate catalytic analysis. More activation occurs from the $CH_4 + O_2$ mixing volume than from added NO, but more C_1-oxygenates are observed from added NO. In addition, as reaction temperature increases, the yield of C_{2+} hydrocarbons increases at a higher rate in the absence of NO. The yield of non-selective CO_x oxidation products is maximum for the reaction initiated by both NO and the $CH_4 + O_2$ mixing volume. The presence of a $CH_4 + O_2$ mixing volume shows a moderate increase for CO_x products (2.6% molar yield to CO_x at 923 K), which is more important for the NO promoted reaction (18.1 % CO_x molar yield at 923 K). In the presence of both NO and a $CH_4 + O_2$ mixing volume the molar yield goes up to 36.0 % CO_x at the same temperature.

There is a very important difference between promotion of the reaction by NO and a $CH_4 + O_2$ mixing volume. The increase of reactivity induced by NO is associated with a shift of the selectivity trends to oxygen containing products. The production of C_{2+} hydrocarbons requires an increase of $CH_4 + O_2$ mixing volume. The increased reactivity by $CH_4 + O_2$ mixing volume is largely associated with an increase of partial oxidation products. CO is not the main oxidation product for the $CH_4 + O_2$ mixing volume promoted reaction at medium reaction temperatures, but becomes important, although not the principal oxidation product, at very high reaction temperatures. Selectivity plots clearly highlight the remarkable increase of CO in the presence of NO (Figure 4) while the added mixing volume produces a general increase in all of the products, especially the selective C_{2+} hydrocarbon products.

There is an intermediate product distribution when NO and mixing volume are both present. Oxygen insertion is promoted by NO but C_{2+} hydrocarbons also increase. As a whole, the system is less selective, but yields of the C_1 oxygenates and C_{2+} hydrocarbons partial oxidation products also increase. The shift to further oxidation promoted by NO is also reflected by the distribution of C_2H_n hydrocarbons where a larger fraction of ethane is dehydrogenated to ethylene. For instance, at 923 K the reaction promoted by a $CH_4 + O_2$ mixing volume yields 1.11% C_2H_4, 2.73% C_2H_6, and 0.24% C_3H_n, whereas yields increase to 3.28% for C_2H_4, 2.94% for C_2H_6, and 0.48 % for C_3H_n when NO is present. Consequently, NO not only promotes C_{2+} hydrocarbons at intermediate reaction temperatures, but also increases oxidative dehydrogenation of ethane to ethylene to such an extent that ethylene becomes the main coupling product. Promotion of oxygen insertion by NO is also indicated by the increased presence of methanol, equivalent to that of formaldehyde. Unfortunately, as the reaction temperature increases, CO production becomes dominant in the presence of NO.

In agreement to previous results with variable mixing volumes and different surface areas, a general reaction scheme has been proposed (Figure 6). Methane can be activated via gas-phase or heterogenous means, which produces methyl radicals. Methyl radicals are in equilibrium with methyl peroxy radicals *(18,35)* and this equilibrium is shifted to CH_3O_2• below 950 K at atmospheric pressure (18). The presence of NO shifts selectivity trends towards oxygen insertion. Little activity is recorded in the absence of vanadium sites. The presence of vanadium on a SiO_2-Si substrate yields a remarkable increase in the production of C_1 oxygenates and C_{2+} hydrocarbons *(15)*. Most likely, interaction of CH_3O• with vanadium sites yields methyl radicals, which may desorb and then couple in the gas phase. This step will compete with further oxidation of the radicals on the surface. As surface area increases, further oxidation to CO_x becomes dominant at the expense of C_{2+} hydrocarbons and C_1 oxygenates. The reaction scheme for C_1 oxygenates is oversimplified since methanol and formaldehyde originate from different steps. As a whole, gas-phase activation of methane is desirable *vs.* heterogeneous. The use of a catalytic system combining gas-phase activation with a very low surface area catalyst lowers heterogeneous activation for methane, but at the same time, further oxidation of the reaction intermediates is minimized. Results reported here demonstrate that the presence of NO certainly increases gas-phase activation of methane, but also shifts the equilibrium to non-selective oxidation products thus decreasing the selectivity of the system. The larger conversion of oxygen for a given conversion of methane attained in the presence of NO seriously limits the yields of partial oxidation products at high conversion levels of methane.

Conclusions

Partial oxidation of methane on a very low surface area V_2O_5/SiO_2-Si catalyst at 780-950 K can be promoted by increasing gas-phase reactions, which demonstrates the gas-phase-heterogeneous nature of catalytic methane oxidation. Radicals are formed and can either be oxidized to C_1 oxygenates or coupled to C_{2+} hydrocarbons. The yield of C_1 oxygenates consists of formaldehyde and methanol in similar amounts. Although literature studies do not report production of methanol at atmospheric pressure, the observation of methanol in the present study must be due to the very low surface area of the catalyst employed. The reactivity increases with increasing reaction temperature, $CH_4 + O_2$ mixing volume, and/or a NO additive, i. e. with the generation of more radicals. On the whole, activation by NO promotes reactivity at much lower temperatures, but it also promotes the formation of deep-oxidation products, thus decreasing overall selectivity. This limits the maximum methane conversion due to oxygen limiting conversion. In any case, the combined promotion by added NO and mixing volume generates a remarkable increase in C_{2+} hydrocarbons at intermediate reaction temperatures, and a larger fraction of ethane produced is oxidized to ethylene due to the oxidative character of NO. With respect to the yields, production of ethylene is significantly increased.

Acknowledgements

This research was partially supported by the Commission of the European Union under grant JOU2-CT92-0040 and Comisión Interministerial de Ciencia y Tecnología, Spain, under projects MAT94-1307CE, and AMB93-1426CE.

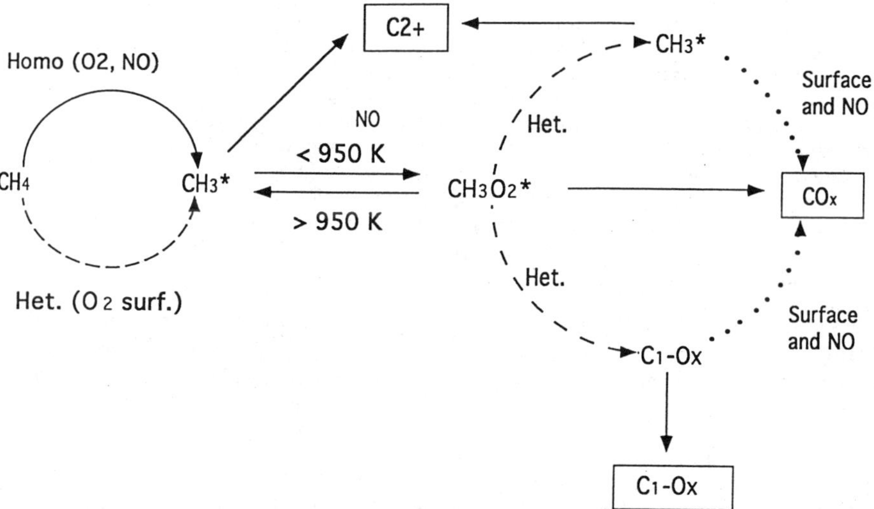

Figure 6. Reaction scheme for methane conversion. Solid line stands for gas-phase processes, dashed lines for heterogeneous ones and dotted lines for surface and/or NO promoted processes.

Literature Cited

(1) Pitchai, R., and Klier, K. *Catal. Rev.-Sci. Eng.* **1986**, *28*, 13
(2) Foster, N. R. *Appl. Catal.* **1985**, *19*, 1
(3) Brown, M. J., and Parkyns, N. D., *Catal. Today* **1991,** *8*, 305
(4) Khan, M. M., and Somorjai, G. A. *J. Catal.* **1985,** *91*, 263
(5) Spencer, N. D. *J. Catal.* **1988,** *109*, 187
(6) Bañares, M. A, and Fierro, J. L. G., In *Catalytic Selective Oxidation*; Oyama, S. T., and Hightower, J. W.; ACS Symposium Series 523, Washington, D. C., 1993; pp. 354-367
(7) Zhen, K. J., Khan, M. M., Mark, C. H., Lewis, K. B., and Somorjai, G. A. *J. Catal.* **1985,** *94*, 501
(8) Spencer, N. D., and Pereira, C. J. *J. Catal.* **1989,** *116*, 319
(9) Kennedy, M., Sexton, A., Kartheuser, B., Mac Giolla Coda, E., McMonagle, J. B., and Hodnett, K. B. *Catal. Today* **1992**, *13*, 447
(10) Kasztelan, S., and Moffat, J. B. *J. Catal.* **1987,** *106*, 512
(11) Kastanas, G. N., Tsigdinos, G. A., and Schwank, J. *Appl. Catal.* **1988,** *44*, 33
(12) Parmaliana, A., Frusteri, F, Miceli, D., Mezzapica, A., Scurrell, M. S., and Giordano, N. *Appl. Catal.* **1991**, *78*, L7
(13) Baldwin, T. R., Burch, R., Squire, G. D., and Tsang, S. C. *Appl. Catal.* **1991,** *74*, 137
(14) Garibyan, T. A., and Margolis, L. YA. *Catal. Rev. -Sci. Eng.* **1989**, *31*, 355
(15) Martín Jiménez, F., Blasco, J. M., Alemany, L. J., Bañares, M. A., Faraldos, M., Peña, M. A., and Fierro, J. L. G. *Cat. Lett.* **1995**, *33*, 279
(16) Irusta, S., Lombardo, E. A. and Miró, E. E. *Catal. Lett* **1994,** *29*, 339
(17) Walker, C. S., Lapszewicz, J. A., and Foulds, G. A. *Catal. Today* **1994,** *21*, 519
(18) Mackie, J. C., *Catal. Rev.-Sci. Eng.* **1991**, *33*, 169
(19) Alemany, L. J., Bañares, M. A., Martín Jiménez, F., Blasco, J. M., López Granados, M., Peña, M. A., and Fierro, J. L. G., submitted
(20) Das, N., Eckerdt, H., Hu, H., Wachs, I. E., Walzer, J. F., Jeher, F. J. *J. Phys. Chem.* **1993,** *97*, 8240
(21) Deo, G., Wachs, I. E., and Haber, J. *Critical Rev. in Surf. Chem.* **1994,** *4*, 141
(22) Thiry, P. A., Liehr, M., Pireaux, J. J., Sporken, R., Caudano, R., Vigneron, J. P., and Lucas, A. A. *J. Vac. Sci. Technol.* **1985,** *B3*, 1118
(23) Schaefer, J. A., and Göpel, W. *Surf. Sci.*, **1985,** *155*, 535
(24) See, for instance, "Practical Surface Analysis. Auger and X-Ray Photoelectron Spectroscopy", D. Briggs and M. P. Seah, Eds., Wiley, Chichester and New York (N. Y.), 1991
(25) Dietrich, B., Osten, H. J., Rucker H., Methfessel, M., and Zaumseil, P. *Phys. Rev. B* **1994,** *49*, 17185
(26) Tsang, J. C., Thompson, W. A., and Oehrlein, G. S. *J. Vac. Sci. Technol.* **1985,** *B3*, 1129
(27) Amenomiya, Y., Birss, V. I., Golezinowski, M., Galuszka, and Sanger, A. R. *Catal. Rev.-Sci. Eng.* **1990,** *32*, 163
(28) Faraldos, M. S., Bañares, M. A., Anderson, J. A., Wachs, I. E., and Fierro, J. L. G., *J. Catal.*, **1996**, in press
(29) Parmaliana, A., Fruster, I, F., Mezzapica, A., Miceli, D., Scurrell, M. S., and Giordano, N. *J. Catal.* **1993,** *148*, 514
(30) Vedeneev, V. I., Krylov, O. V., Arutyunov, V. S., Basevich, V Ya., Goldenberg, M. YA., and Teitel Boim,. M. A., Appl. *Catal. A* **1995,** *127*, 51
(31) Bañares, M. A., Alemany, L. J., López Granados, M., Faraldos, M., and Fierro, J. L. G. *Catal. Today*, submitted
(32) Cardoso, J. H., Bañares, M. A., and Fierro, J. L. G., in preparation

(33) Sun, Q., Di Cosimo, J. I. , Herman, R. G, Klier, K. and Bhasin, M. M. *Catal. Lett.* **1992,** *15*, 371
(34) Kartheuser, B., Hodnett, K. B., Zanthoff, H., and Baerns, M. *Catal. Letters* **1993**, *21*, 209
(35) Krylov, O. V., *Catal. Today*, **1993**, *18*, 209
(36) Aika, K., Fujimoto, N., Kobayashi, M., and Iwamatsu, E., *J. Catal.* **1991,** *127*, 1
(37) Lane, G. S., and Wolf, E. E., *J. Catal.*, **1988,** *113*, 144

Chapter 7

Transient Study of the Function of Oxygen in Methane Coupling over the Conductive Ceramics $Li_{0.9}Ni_{0.5}Co_{0.5}O_{2-x}$

D. Qin, A. Ovenston, A. Villar, and J. R. Walls

Department of Chemical Engineering, University of Bradford, West Yorkshire, BD7 1DP, United Kingdom

Partial oxidation of CH_4 to C_2H_4 and C_2H_6 (CH_4 coupling) over a non-stoichiometric catalyst $Li_{0.9}Ni_{0.5}Co_{0.5}O_{2-x}$, in which Li^+ cations and lattice oxygen anions are mobile under certain conditions, was investigated. The reaction spectrum was recorded dynamically and illustrated a considerable quantity of C_2H_4 and C_2H_6 formation above 700°C. Activation energies up to 800°C were estimated based on the Arrhenius empirical model. The results of both TPRS (temperature programmed reaction spectrum) and TPSR (temperature programmed surface reaction) indicate that active surface oxygen leads to the production of C_2H_4 and C_2H_6, while CO_2 is produced from both gaseous and surface oxygen. The spectra also suggest that at least two types of lattice oxygen are responsible for the formation of C_2s. TPR (temperature programmed reduction) showed two H_2 consumption peaks. However, TPO (temperature programmed oxidation) of pre-reduced samples showed five oxidative peaks, none of them representing metallic Ni or Co. The active site is proposed to contain lattice oxygen.

Since the pioneering work by Keller and Bhasin (*1*), the oxidative coupling of methane (OCM) to ethane, ethylene and higher hydrocarbons has attracted the attention of many scientists from academia and industries, as shown in a number of review papers (*2-4*). The kinetics and mechanism of OCM on a wide variety of catalysts have been the subject of many publications in recent years (*5-8*). It is generally agreed that C_2H_6 formation occurs via dehydrogenation of the CH_4 molecule on the catalytic active site, followed by gas phase coupling of the resulting methyl radicals. The types of active sites involved in the reaction vary with the catalysts and determine the reaction pathway for the formation of the undesired deep oxidation products CO and CO_2 (*9-11*). Furthermore, since the CH_4 molecule can be activated via adsorbed oxygen, lattice

0097–6156/96/0638–0095$15.00/0

oxygen, or gas-phase oxygen, different catalytic models are expected to be applicable to different catalytic systems.

Li/MgO catalysts have been extensively investigated and the work of Lunsford and co-workers has been particularly influential with regard to identification of catalytically active surface/lattice oxygen species and the overall reaction mechanism (*5,9,12-13*). Another useful class of materials includes transition metal oxides in which the metal ions exhibit variable valence and its doped-ions are reversibly removable under certain conditions. $Li_xNi_{1-y}Co_yO_{2-\delta}$, studied originally for use as an anode in solid state lithium batteries, is one such material; its Li ions can move with relative ease under an applied DC voltage (*14*). It is possible that the catalytic chemistry in such systems is significantly different from that of non-transition metal oxide catalysts (*15-18*).

Previous work in our laboratory has focused on the thermal stability, crystallographic structure and AC conductivity of $Li_xNi_{1-y}Co_yO_{2-\delta}$ as a function of stoichiometry (*19*). $Li_xNi_{1-y}Co_yO_{2-\delta}$ has the same crystallographic structural framework as that of $LiCoO_2$, which can be described as a close-face-centred-cubic packing of oxygen ions, where the octahedral sites are alternately occupied by Li^+ and Co^{3+}, making up alternate (111) layers of Li-O and Co-O. The substitution of Ni for Co does not change the structural framework of $LiCoO_2$. This fact suggests that Ni will occupy the same octahedral sites as Co. The substitution of Ni for Co also increases the thermal stability of the structural framework of $LiCoO_2$. Some of the Li in the structure may be easily removed or exchanged with other ions and hence result in non-stoichiometric proportions of the alkali component. The Li content will affect the volume of the lattice cell and the conductance but not the structural framework. In this report the compound previously found to be the most stable and possess optimum conductivity, $Li_{0.9}Ni_{0.5}Co_{0.5}O_{2-\delta}$, is considered (*19*).

Li is well known as a promoter for OCM and for its ability to create active sites (*9-13, 20*). Ni and Co are active catalysts for hydrocarbon activation. Therefore, $Li_{0.9}Ni_{0.5}Co_{0.5}O_{2-\delta}$ was expected to be a catalyst for OCM. Others have previously shown that $LiNiO_2$ is an active catalyst for OCM, but with a poor yield (*16-18, 21*). The non-stoichiometric material $Li_{0.9}Ni_{0.5}Co_{0.5}O_{2-\delta}$ was expected to have better performance as a catalyst for OCM. The kinetics of reaction and the AC conductivity changes during reaction will be presented elsewhere. Transient techniques in catalytic research are often used to gain insight into the reaction mechanism and kinetics of complex reactions (*22-25*). Frequently used methods include temperature programmed reaction spectra (TPRS), temperature programmed reduction/oxidation (TPR/TPO), temperature programmed desorption (TPD) and temperature programmed surface reaction (TPSR). OCM over $Li_xNi_{1-y}Co_yO_{2-\delta}$ was investigated using transient techniques, and the roles of surface and gaseous oxygen are discussed in this paper.

Experimental

$Li_{0.9}Ni_{0.5}Co_{0.5}O_{2-\delta}$ was prepared by mixing Li_2CO_3, NiO and Co_3O_4 in stoichiometric proportions, firing for two hours at 773 K and fusing for 48 hours under O_2 at 1273 K (*19*). X-ray diffraction (XRD) measurements were carried out using a Siemens

Diffractometer 5000 with Cu-Kα radiation at room temperature. The surface area was measured by Micromeritics ASAP 2000 equipment using N_2 as adsorbate. The results are summarised in Table I

Table I Properties of $Li_{0.9}Ni_{0.5}Co_{0.5}O_{2-\delta}$

Surface area 3.8 m^2/g

XRD crystallographic data

hkl	003	101	102	006	104	105	009
d	4.654	2.393	2.335	2.297	1.999	1.839	1.548

Experiments were carried out in a quartz micro-reactor within a Carbolite CST-12/50 furnace controlled by a Eurotherm programmed temperature controller. The flow rates of the gases were controlled by Brooks digital mass flow meters. The total space velocity was $1x10^5$ h^{-1} and 250 mg of catalyst were used for each experiment except where indicated. CH_4 and O_2, with Ar as inert gas, were mixed at a pre-set ratio (between 1 and 4) before being introduced to the micro-reactor. Any steam in the effluent was condensed in a cold-trap, part of the remaining effluent was directed to a VG-SXP 800 quadrupole mass spectrometer (QMS) on the multi-channel ion-monitoring mode for on-line monitoring and any remaining gaseous species were burnt off. Ar, O_2, CH_4, CO_2, H_2, CO, C_2H_4 and C_2H_6 were monitored, using the mass numbers 40, 32, 15, 44, 2, 28, 27 and 30 respectively. Mass number 28 arises from an overlap of signals from CO, C_2H_4, C_2H_6 and a trace of CO_2 and mass number 27 represents a mixture of both C_2H_4 and C_2H_6, the relative partial pressures of C_2H_4 and CO were determined by subtracting the total intensity from the contributions of other species, respectively. There was no evidence for the production of C_2s. A software package was used to analyse the data using the procedure described by Bao et al (26). The relative intensities of the fragments of the components for each species were taken into account and the sensitivity for each gas was estimated by using a calibration gas mixture containing Ar, CO_2, CO, CH_4, C_2H_4, and C_2H_6. Hence concentrations for each species could be estimated and conversions and selectivities determined. The following formulae were used:

$$CH_4 conversion(\%) = \frac{(2P_{C_2s}^{out} + P_{CO_2}^{out} + P_{CO}^{out})}{P_{CH_4}^{out} + (2P_{C_2s}^{out} + P_{CO_2}^{out} + P_{CO}^{out})} * 100\% \text{ and}$$

$$C_2s..selectivity(\%) = \frac{2P_{C_2s}^{out}}{(2P_{C_2s}^{out} + P_{CO_2}^{out} + P_{CO}^{out})} * 100\%$$

Since the measurement of ion currents were made under real reaction conditions, the background of each of the components was higher than that expected in

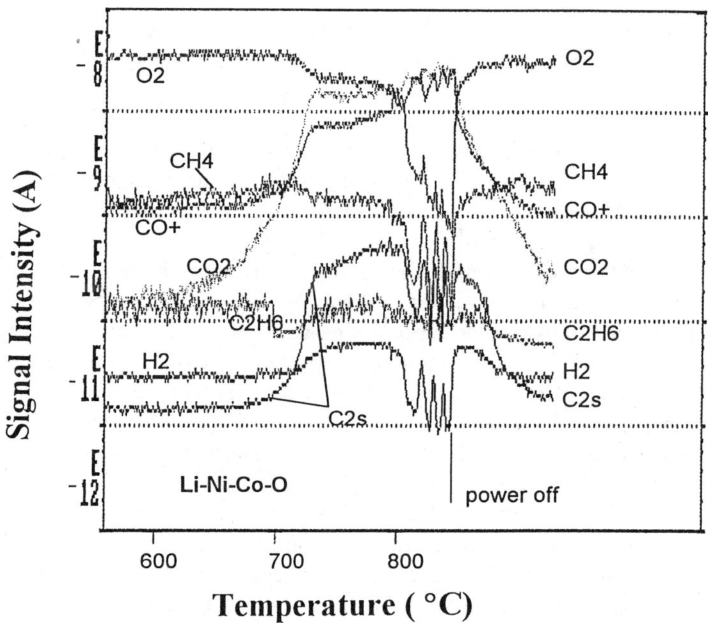

Figure 1. Temperature programmed reaction spectrum (TPRS) for the reaction of CH_4-O_2 over Li-Ni-Co-O catalyst. Ar:O_2:CH_4 = 1:1:1 at a space velocity of 10^5 h^{-1}. Temperature ramp of 30 °C/min.

a high vacuum chamber. In the absence of a catalyst, there was no reaction between CH_4 and O_2 until combustion occurred at about 800 °C. Care was taken to avoid the danger of combustion which could occur between CH_4 and O_2 within specific limits depending on the CH_4/O_2 ratio chosen.

On-line computer controlled Shimadzu DTA-50 equipment was used for differential thermal analysis (DTA). Experiments were conducted on powdered samples (10 mg) using a platinum crucible with α-Al_2O_3 as a reference. A temperature range of 20 °C-900 °C with a rate of increase of 20 °C/min. was employed.

Results

TPRS and Activation Energies for the Reaction of CH_4 and O_2. For a CH_4/O_2 ratio 1:1, the temperature programmed reaction spectrum (TPRS) of CH_4 and O_2 over Li-Ni-Co-O shows some production of C_2H_4 and C_2H_6 between 700-800°C (Figure 1). Initially, up to 700 °C, some complete oxidation of CH_4 to CO_2 and CO took place. Thus, above 700 °C, another mechanism apparently provided for partial oxidation with a higher CH_4 conversion. C_2H_4 and C_2H_6 exhibited significant increases but with an almost simultaneous increase in CO. CO_2 showed a rapid increase from about 650 °C, below that for the C_2H_4 and C_2H_6 increases. Above 800 °C, homogeneous combustion occurred with oscillations.

In contrast, a NiO catalyst co-precipitated with MgO shows negligible activity towards OCM (Figure 2). CO and CO_2 were the dominant products throughout the temperature range investigated. There were no evidence for the formation of C_2H_4 and only a slight indication of C_2H_6. CO_2 formed above 380 °C and its formation rate increased significantly at 420 °C. CO increased sharply at 420 °C. Below 420°C, CO_2 is formed from adsorbed oxygen in small amounts due to the very low surface area. During this stage, CH_4 conversion is negligible so its signal remained almost constant. The observations indicate that a different mechanism is involved in the reaction below and above 420°C (*27*). There was no evidence for the production of C_2s at temperatures up to 800 C. Thus CH_4 coupling is negligibly small over the NiO-MgO catalyst.

Since the reaction is highly exothermic, some of the heat of reaction caused the true catalyst temperature to be higher than that of the programmed furnace temperature. To avoid this problem, a stepwise temperature program (5°C/min ramp, holding temperature for 5 minutes every 20°C) was used to determine the CH_4 conversion and C_2s selectivities as functions of temperature for different CH_4/O_2 ratios. As shown in Figure 3, the highest yield of C_2s was achieved between 780 and 840°C using a CH_4/O_2 ratio of 4:1 (about 13% O_2 in feedstock). The concentrations of the product species depended on reciprocal temperature in Arrhenius fashion, with activation energies as shown in Table II. Further discussion about the effects of temperature and CH_4/O_2 ratio on CH_4 conversion, selectivities and activation energies, along with reaction kinetics, will be presented elsewhere. The main purpose of this paper is to consider the function of oxygen in the reaction.

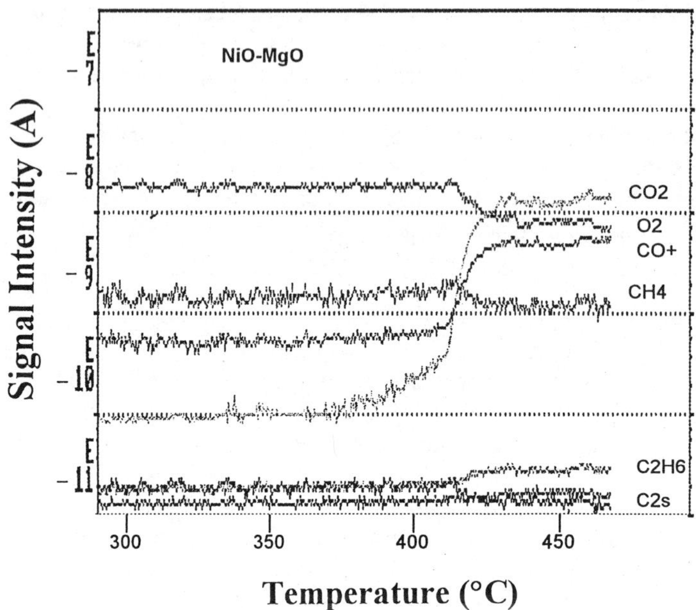

Figure 2. Temperature programmed reaction spectrum (TPRS) for the reaction of CH_4-O_2 over NiO-MgO catalyst. Ar:O_2:CH_4 = 1:1:1 at a space velocity of 10^5 h^{-1}. Temperature ramp of 30 °C/min.

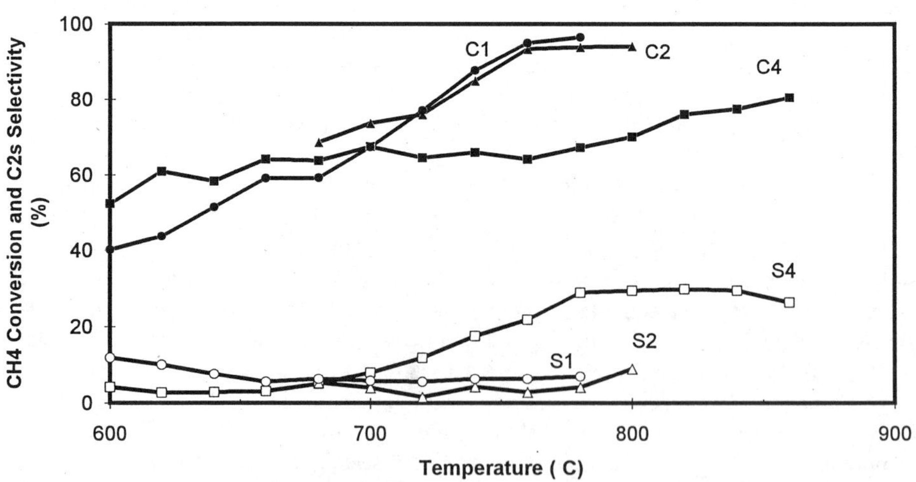

Figure 3. CH_4 conversion and C_2s selectivity as functions of temperature and CH_4/O_2 ratio. Space velocity of 10^5 h^{-1}. S: selectivity to C_2s; C: conversion of CH_4. The number following S or C represent the CH_4/O_2 ratio in the feedstock.

Table II Activation energies of products between 680 and 780 °C *

Products	C_2H_4	C_2H_6	CO	CO_2
kJ/mol	343 ± 5	105 ± 3	197 ± 4	254 ± 3

*: Space velocity 1×10^5 h^{-1}. The error was calculated to a 95% confidence limit

Comparison between Coupling and Combustion of CH_4. By maintaining the reactor temperature at 680 °C, significant increases in the formation of C_2s were observed (Figure 4a, using a CH_4/O_2 ratio of 1:1). CO_2 showed two significant peaks. One corresponded to the formation of C_2H_4, and another occurred before this peak. A temperature jump of the catalyst bed to 726 °C was recorded at the peak point due to the release of exothermic heat (i.e. $2CH_4 + O_2 \rightleftharpoons C_2H_4 + 2\,H_2O$ $\Delta H°_{25} = -282$ kJ/mol). This temperature increase was not detected during reaction at higher temperatures between 700-800 °C when a stepwise temperature program was used.

By setting the reactor temperature to 795 °C, combustion of CH_4 was observed for a CH_4/O_2 ratio of 1:1. During combustion, the propagation of radicals converts all hydrocarbons into CO_x and the TPRS spectrum shows oscillations (*28*) as were observed in Figure 1. One cycle of such an oscillation is demonstrated in Figure 4b. Consequently, CH_4 and O_2 decreased significantly while CO_x became dominant. No new C_2s peaks were observed during or prior to the combustion process. Combustion was terminated by switching off the power to the furnace.

Temperature Programmed Surface Reaction (TPSR). TPSR was carried out in order to investigate the role of lattice oxygen. The experiments were divided into two parts, a) reaction of surface lattice oxygen with gaseous CH_4 and b) reaction of adsorbed CH_4 with gaseous O_2. The first part consisted of two steps, i.e. pre-treatment and surface reaction. Ar (at 50 ml/min) was passed over a fresh sample of catalyst (1.6 g) for 30 minutes at 200 °C. The Ar was then replaced by O_2 at a flow rate of 50 ml/min. The sample was then heated in this atmosphere up 750 °C, and maintained at this temperature for 30 minutes in order to achieve a saturated oxidized state. Finally, in order to remove any physical and weakly-bonded chemically adsorbed oxygen, Ar was allowed to replace the O_2 for 30 minutes at 750 °C. It was believed that most of the molecular oxygen had been removed by this pre-treatment (Step I in Figure 5). The backgrounds of O_2 were showed in Step I and II of Figure 5, which represented 0.5 and 0.1 volume %, respectively. The high background, compared with that of a high vacuum chamber, was due to the desorption of O_2 from the inside-wall of the rig and was impossible to reduce since the measurement was carried out under real reaction conditions.

The surface reaction started at 750 °C after introducing CH_4 at 20 ml/min with Ar at 30 ml/min to replace Ar alone (Step II in Figure 5, CO signal is not shown in Figure 5). The formation of CO_2 ,CO and C_2s was immediately observed. Since the gaseous O_2 concentration was very low, it does not interfere with the observation of

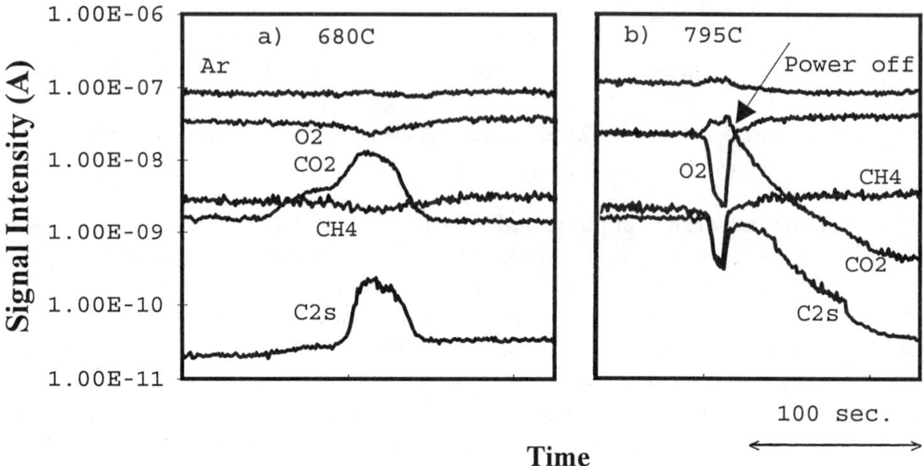

Figure 4. Temperature programmed reaction spectrum (TPRS) for comparison of CH_4 coupling reaction and combustion of over Li-Ni-Co-O. Ar:O_2:CH_4 = 1:1:1 at a space velocity of 10^5 h^{-1}. Temperature set at a) 680 °C and b) 795 °C, respectively.

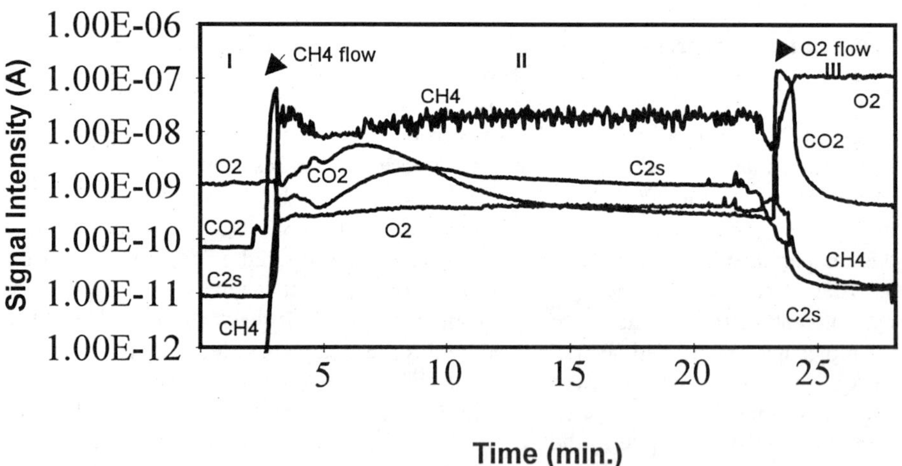

Figure 5. Temperature programmed surface reaction (TPSR) at 750 °C over Li-Ni-Co-O. Stage I, Ar flow (50 ml/min); Stage II, Ar and CH_4 flow (30 and 20 ml/min); and stage III, O_2 flow (50 ml/min).

surface oxygen of the catalyst. C_2s exhibited two peaks with respect of time. CO_2 exhibited three, two corresponding to the formation of C_2s and one preceding these peaks. The observation probably represents different lattice oxygen sites associating with different activities. The decrease of the formation of CO_2 and C_2s indicates the depletion of active surface oxygen during the process.

The second part of the experiment was for the reaction of adsorbed CH_4 with O_2. Following step II, O_2 was allowed to replace the CH_4 and Ar at a flow rate of 50 ml/min (Step III). O_2 would react with adsorbed CH_4 and any remaining gaseous CH_4. CO_2 formation was again significant but C_2s formation was negligible. That in the observation indicated that gaseous O_2 produced only CO_2 (and CO, not shown in Figure 5) singnificantly. The total extent of reaction was far less than Step II, indicating that there was an insignificant amount of CH_4 remaining on the catalyst.

Temperature Programmed Reduction (TPR). TPR and TPO techniques were used to investigate the lattice oxygen species in the sample. Before the TPR experiment, 50 ml/min Ar was flushed through a fresh sample for 30 minutes at 200 °C in order to clean adsorbed moisture and gases from the surface. The sample temperature was then decreased to room temperature whilst maintaining the Ar flow. The TPR experiment was operated at a temperature ramp of 30°C/ min up to 800°C using 25% H_2 in Ar with a total flow rate of 50 ml/min.

The H_2 consumption was recorded during the temperature ramp (Figure 6). H_2 consumption did not change below 400°C and exhibited two significant peaks at 480°C and 635°C. For the TPR experiment, the peaks indicated that some reduction of the sample had occurred. Since Ni and Co hydrides are not stable under these conditions, only lattice oxygen in the sample could react with H_2. The two peaks in Figure 6 indicated that there were at least two types of lattice oxygen. The first sharper peak illustrates a more rapid reaction with H_2. The area between the baseline and the curve represents the amount of H_2 consumed for each specific type of oxygen. The area ratio of the first peak to that of the second was about 1, thus the amount of each type of lattice oxygen is probably of similar magnitude.

Temperature Programmed Oxidation (TPO). TPO experiments were conducted using separate samples pre-reduced at 650 °C at different H_2 partial pressures for 4 hours. Since the sensitivity of the micro reactor-QMS was not sufficiently high to distinguish several reduced species, DTA (Differential Thermal Analysis) was used. Air with a flow rate of 15 ml/min was used as the oxidant and a temperature ramp of 20 °C/min was employed.

Several exothermic peaks were observed between 180°C and 480°C (Figure 7). The higher H_2 partial pressures resulted in more pronounced peaks. Two significant peaks appeared between 350 °C and 400 °C. The partial pressure of H_2 during pre-reduction strongly affected the appearance of some peaks, e.g. peaks 2 and 5, or the height of peak 3. It is significant to note that there was no clear correlation of these peaks with that expected for metallic Ni (about 300 °C) or Co (The first peak for Co is at about 500 °C, not shown in Figure 7.). Hence the samples were not easily reduced to a metallic state under these pre-reducing conditions.

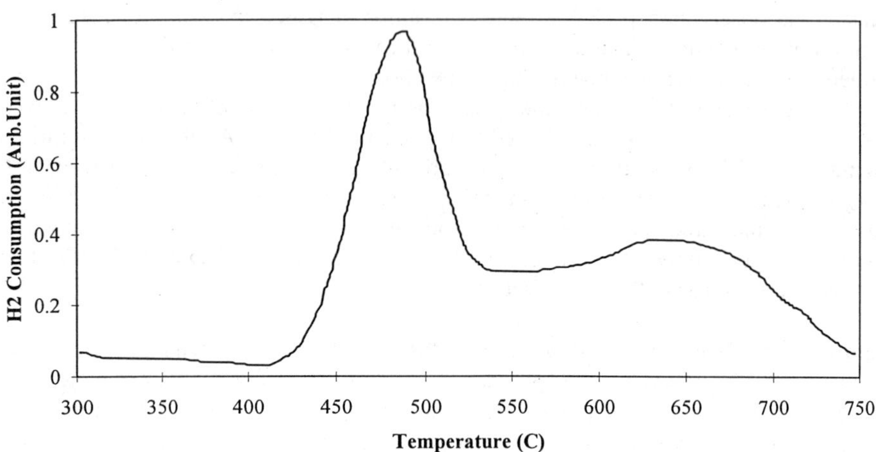

Figure 6. Temperature programmed reduction (TPR) over Li-Ni-Co-O. Sample weight 1.6 g. 25% H_2 in Ar at total flow rate of 50 ml/min. Temperature ramp of 30 °C/min.

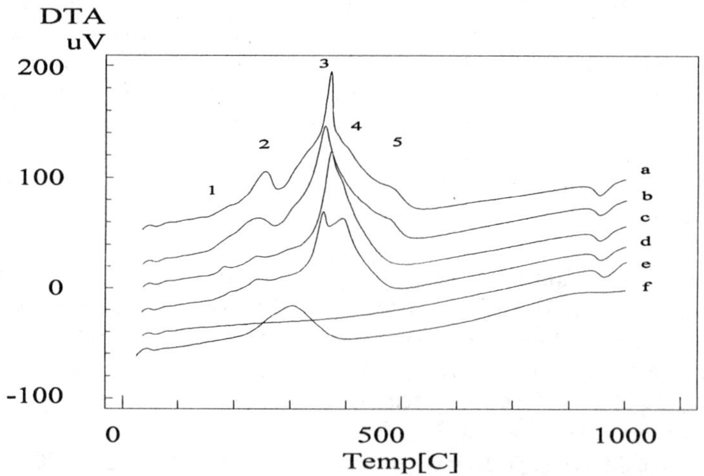

Figure 7. Temperature programmed oxidation (TPO) over pre-reduced Li-Ni-Co-O using DTA. Sample weight 10 mg, air as active gas (15 ml/min). Temperature ramp of 20 °C/min. Samples were pre-reduced in H_2 a) 58%, b) 53%, c) 33% and d) 13% (balanced by Ar). e) fresh Li-Ni-Co-O, f) Ni metal

The endothermic peak at 860°C arises from the collapse of the $Li_xNi_{1-y}Co_yO_{2-\delta}$ structure and a release of O_2 (*19*). The appearance of the peak for each case indicated that the structure was not destroyed by the reduction and oxidation processes. Thus catalyst could be safely used below 860 °C without decomposition.

Discussion

Active centre for OCM. $Li_xNi_{1-y}Co_yO_{2-\delta}$ consists of alternate (111) layers of Li-O and Ni/Co-O. Li, Ni and Co can freely form Li-O-Co, Li-O-Ni, Ni-O-Ni, Co-O-Co or Ni-O-Co superstructures. The oxidation state of Co could be either +3 or +4, and that of Ni could be either +2 or +3 (*18-19*). The variable oxidation state of these two transition metals enables easy exchange or transfer of electrons. The internal redox reaction is also able to supply oxygen for catalytic oxidation. This process can be represented thus,

$$12\ Li_{0.9}Co_{0.5}^{3+}Ni_{0.5}^{3+}O_{1.95} \rightleftharpoons 6\ NiO + 2\ Co_3O_4 + 2\ O_2 + 5.4\ Li_2O$$

In order to conserve the electronic and mass balances, the oxidation state Co^{3+} or Ni^{3+} is used in the formula. The reverse reaction enables regeneration of the original structure or replenishment of the lattice oxygen. The lattice oxygen in $Li_xNi_{1-y}Co_yO_{2-\delta}$ may be exchanged with gaseous O_2 and is active for OCM (*21, 25*).

Otsuka et al (*21*) suggested that lattice oxygen would promote the CH_4 dissociative adsorption on a Ni^{3+}-O^{2-} centre for a $LiNiO_2$ catalyst and OCM would proceed via a redox mechanism. The process could be described as follows (*21, 24, 11*),

$$2\ \overset{+}{Li}\text{-}\overset{2-}{O}\text{--}\overset{3+}{Ni}\text{-}O \cdots (H\text{---}CH_3) \longrightarrow 2\ \overset{+}{Li}\text{-}\overset{-}{O}(H)\text{--}\overset{2+}{Ni}(CH_3)\text{-}O \longrightarrow H_2O + C_2H_6 + Li_2O + 2NiO$$

The Ni^{3+}-O^{2-} centre activates CH_4 and produces C_2H_6, which is then dehydrogenated to C_2H_4. Li_2O and NiO will be regenerated by gaseous O_2. The $Li_xNi_{1-y}Co_yO_{2-\delta}$ catalyst has Ni^{3+}-O^{2-} centres and produces C_2H_4 and C_2H_6 in addition to CO_x, while the NiO-MgO catalyst has Ni^{2+}-O^{2-} centres and produces only CO and CO_2. However, since NiO is one of the intermediate products, it may catalyze CO_x production. This is one of the reasons why CO_x is simultaneously observed during the production of C_2H_4 and C_2H_6. Co could also form a similar structure to Ni. Hence, there would be two types of active centres as observed by the two peaks for C_2H_4 and C_2H_6 formation in Figure 5.

CO_2 Formation and Oxygen. The undesirable production of CO_2 is a key factor controlling the selectivity and yield of C_2H_4 and C_2H_6. Since there are two peaks for CO_2 formation simultaneous to those of C_2H_4 and C_2H_6, it is possible that there is a) deep oxidation of C_2H_4 and C_2H_6 and/or b) NiO formed during the oxidative coupling process catalyses CH_4 deep oxidation. It was also found that CO_2 is formed well before any production of C_2H_4 and C_2H_6 (Figure 1) and the formation of CO_2 precedes a peak before those for C_2H_4 and C_2H_6 (Figure 4a & 5). These observations indicate that the formation of CO_2 can be either independent of or dependent upon C_2 production. The experiments also illustrated that CO_2 was formed from both surface and gaseous reactions, which agrees with the report of Wolf et al (*25*) that $C^{16}O^{16}O$, $C^{18}O^{16}O$ and $C^{18}O^{18}O$ were observed when using a $^{18}O^{18}O$ reaction with CH_4 over a $LiNiO_2$ catalyst. Thus, the excess of gaseous O_2 should be avoided in order to suppress CO_2 formation.

Surface lattice oxygen is an active species responsible for the formation of C_2H_4 and C_2H_6. During the reaction, it is gradually consumed and needs to be replenished from gaseous O_2 or bulk lattice oxygen. The latter is only a limited source and replenishment from gaseous O_2 remains necessary. If there is not enough active lattice oxygen, the reaction will cease, as was observed in TPSR (Figure 5). An increase of the proportion of oxygen will increase the chance of replenishment of lattice oxygen but also increase the risk of complete oxidation. As shown in Figure 3, experiments with lower CH_4/O_2 ratios exhibited reduced selectivity to C_2s. According to the isotope exchange experiments on $LiNiO_2$ (*21*), the rate of replenishment of lattice oxygen is not particularly fast; the exchange rate of oxygen between gaseous O_2 and lattice oxygen did not instantly reach equilibrium. This implies that gaseous O_2 always exists in the reactor no matter what the rate of formation of C_2H_4 and C_2H_6 is. This is harmful to the coupling reaction since the reaction of gaseous O_2 with surface species apparently makes no significant contribution to the formation of C_2s.

Even at the optimum situation, in which all gaseous O_2 is transformed into lattice oxygen, the selectivities to C_2H_4 and C_2H_6 will, according to the result of TPSR in Figure 5, still be limited by the formation of CO_2.

Some feasible ways of increasing the selectivities to C_2 hydrocarbons are by separating CH_4/O_2 feeds using a membrane reactor (*29*) and by modifying the chemistry and the surface structure of the catalysts.

TPR and TPO. The use of active lattice oxygen is responsible for the production of C_2H_4 and C_2H_6, as well as CO_2 and CO; the reaction of gaseous oxygen with surface species results in deep oxidation products alone. Among the types of active lattice oxygens, two were identified to be active for OCM. It is important when designing a good catalyst for high OCM activity to distinguish the types of lattice oxygen and their functions.

The TPR profile showed two H_2 consumption peaks, which implied there were at least two types of lattice oxygen. Coincidentally, the surface oxidation profile (Figure 5) also showed two peaks for C_2H_4 and C_2H_6 formation. However, The TPO profile showed more than five peaks for samples which had been subjected to different pre-reduction conditions. The reduction is thought to occur as follows,

$$12\ Li_{0.9}Co_{0.5}{}^{3+}Ni_{0.5}{}^{3+}O_{1.95} + 4H_2 \rightleftharpoons 6\ NiO + 2\ Co_3O_4 + 4\ H_2O + 5.4\ Li_2O$$

However, it seems the $Li_{0.9}Co_{0.5}Ni_{0.5}O_{2-x}$ structural framework is stable and not destroyed by pre-reduction since the typical exothermic peak at 860°C was observed for each sample. The peaks in TPO may represent different superstructures such as Ni-O-Ni, Co-O-Co, Co-O-Ni, Li-O-Ni, Li-O-Co, etc. Ni and Co can have different valences which result in more complicated superstructures such as Ni^{3+}-O^{2-}. The major peaks were located between 300 and 500 °C. This range is lower than that for the H_2 consumption peaks observed in the TPR profile. This is due to high thermodynamic stability of Ni or Co oxides. The Gibbs free energies at 25 °C (298 K) (*30*) of Ni or Co oxidation may be compared to those of NiO or Co_2O_3 reduction thus:

$$Ni + O_2 \rightleftharpoons NiO \qquad \Delta G^\circ_{298} = 211 \quad (kJ/mol)$$

$$Co + 2/3O_2 \rightleftharpoons Co_3O_4 \qquad \Delta G^\circ_{298} = 258 \quad (kJ/mol)$$

$$NiO + H_2 \rightleftharpoons Ni + H_2O \qquad \Delta G^\circ_{298} = 17 \quad (kJ/mol)$$

$$Co_3O_4 + 4\,H_2 \rightleftharpoons 3Co + 4\,H_2O \qquad \Delta G^\circ_{298} = 47 \quad (kJ/mole)$$

The ΔG°_{298} data above can only be used as references for oxidation and reduction, since thermodynamic data for the more complex superstructures are not available. Further work is necessary to identify each peak with a specific superstructure and correlate it with OCM reactivity.

Conclusions

A non-stoichiometric ceramic $Li_{0.9}Co_{0.5}Ni_{0.5}O_{2-x}$, in which Li^+ cations and lattice oxygen anions are mobile under certain conditions, was investigated for the partial oxidation of CH_4 to C_2H_4 and C_2H_6 (CH_4 coupling) using transient techniques. TPRS was recorded and showed that a considerable quantity of C_2H_4 and C_2H_6 was formed at temperatures above 700°C. Yields of C_2s above 20% were obtained for a CH_4/O_2 feedstock ratio of four. In-situ transient techniques indicated that surface centres like Ni^{3+}-O^{2-} might be responsible for CH_4 activation and C_2H_4 and C_2H_6 production. CO_2 is formed from reaction of both surface and gaseous oxygen, and could be either independent of or dependent upon C_2H_4 and C_2H_6 production. TPR and TPO suggest many superstructures such as Ni-O-Ni, Co-O-Co, Co-O-Ni, Li-O-Ni, Li-O-Co, etc. with variable valences of Ni and Co could exist during reduction. Lattice oxygens in specific environments may be active for partial oxidation.

Acknowledgements

The authors thank the British Engineering and Physical Sciences Research Council (EPSRC) for financial support for this research.

References

1. Keller, G.E., Bhasin, M.M., *J.Catal.*, **1982,** *73*, 9

2. Ross, J.A., Bakker, A.G., Bosh, H., Ommen, J.G.van, Ross, J.R.H., *Catal.Today*, **1987**, *1*, 133
3. Lee, J.S., Oyama, S.T., *Catal.Rev.Sci.Eng.*, **1988** , *30(2)*, 249
4. Amenomiya, Y., Birss, V.I., Goledzinowski, M.,Galuszka, J., Sanger, A.R., *Catal.Rev. Sci. Eng,* **1990**, *32*, 163
5. Ito, T.,Wang, J.X., Lunsford, J.H., *J.Amer.Chem.Soc.*, **1985**, *107*, 5062
6. Asami, K., Shikada, T., Fujimoto, K., Tominaga, H., *Ind.Eng.Chem.Res.*,**1987**, *26*, 2348
7. Ross, J.A., Korf, S.J., Veehof, R.H.J., Ommen, J.G.van, Ross, J.R.H., *Appl.Catal.Today,* **1989**, *52(1-2)*, 147
8. Miro, E.E., Santamaria, J.M., Wolf, E.E., *J.Catal.*, **1990**, *124*, 465
9. Driscoll, D.J., Martir, W., Wang, J.X.,Lunsford, J.H., *J.Amer.Chem.Soc.*, **1985**, *107*, 58
10. Iwamatsu, E., Aika, K.I., *J.Catal.*, **1989**, *117*, 416
11. Lane, G.S., Wolf, E.E., *J.Catal.*, **1988,** *113*, 144
12. Lin, C.H., Wang, J.X., Lunsford, J.H., *J.Amer.Chem.Soc.*, **1987,** *109*, 4808
13. Ito, T., Lunsford, J.H., *Nature,* **1985,** *314*, 721
14. Delmas, C., Saadoune, I., *Solid State Ionics,* **1992**, *53-56*, 370
15. Hinsen, W., Baerns, M., *Chem.-Ztg.*, **1983**, *107*, 56
16. Otsuka, K., Liu, Q., Hatano, M., Morikawa, A., *Chem.Lett.*, **1986**, 903
17. Otsuka, K., Liu, Q., Hatano, M., Morikawa, A., *Inorg.Chim.Acta,* **1986**, *118*, L23
18. Hatano, M., Otsuka, K., *Inorg.Chim.Acta,* **1988**, *146*, 243
19. Ovenston, A., Qin, D., Walls, J.R., *J.Mat.Sci.*, **1995**, *30*, 2496
20. Qin, D., Chang, W., Chen, Y., Zhou, J., Gong, M., *J.Catal.*, **1993**, *142(2)*, 719
21. Hatano, M., Otsuka, K., *J.Chem.Soc.Faraday Trans.I,* **1989**, *85(2)*, 199
22. Waugh, K., *Appl.Catal.,* **1988**, *43*, 315
23. Creten, G., David, J.L., Froment, G.L., *J.Catal.*, **1995**, *154*, 151
24. Moggridge, G.D., Badyal, J.P.S., Lambert, R.M., *J.Catal.*, **1991**, *32*, 92
25. Miro, E., Santamaria, J., Wolf, E.E., *J.Catal.*, **1990**, *124*, 451
26. Bao, X., Muhler,M., Schlogl, R., and Ertl, G., *Catal. Letters.*, **1995**, *32*, 185
27. Ovenston, A., Qin, D., Walls, J.R., *to be published*
28. Putnam, A.A., *Combustion-Driven Oscillations in Industry, Fuel and Energy Science Series*, editor Beér,J.M.; Publisher: American Elsevier, New York, 1971
29. Balachandran, U., Dusek, J.T., Sweeney, S.M., Poeppel, R.B., Mieville, R.L., Maiya, P.S., Kleefish, M.S., Pei, S., Kobylinski, T.P., Udovich, C.A., Bose, A.C., *American Ceramic Society Bulletin*, **1995**, *70(1)*, 71
30. *Lange's Handbook of Chemistry (13th)*, Editor Dean, J.A.; Publisher: McGraw-Hill Book Company, 1985.

Chapter 8

Oxidative Coupling of Methane by Adsorbed Oxygen Species on $SrTi_{1-x}Mg_xO_{3-\delta}$ Catalysts

Keiichi Tomishige, Xiao-hong Li, and Kaoru Fujimoto

Department of Applied Chemistry, Graduate School of Engineering, University of Tokyo, Hongo, Bunkyo-ku, Tokyo 113, Japan

It was found that ethane was formed by the stoichiometric reaction between methane and an adsorbed oxygen species on $SrTi_{1-x}Mg_xO_{3-\delta}$ with high selectivity (> 80 %) at much lower temperatures (550-750 K) than those typically used for the catalytic reaction. The properties of the adsorbed oxygen species were investigated by means of temperature programmed desorption, and the role of Mg^{2+} was studied using XRD, BET, and the measurements of the exchange rate between lattice oxide ions and gas phase oxygen. The added Mg^{2+} ions seem to be located at the surface and bulk Ti^{4+} site of Sr-Ti mixed oxides, where oxide ion defects are formed because of differences in ion charges. Surface oxide ion defects play an important role in oxygen adsorption, while bulk defects promote the mobility of oxide ions. $SrTi_{0.5}Mg_{0.5}O_{3-\delta}$ was initially active for oxidative coupling of methane at low temperature (873 K), but combustion of methane predominated under steady state conditions, due to a change in adsorbed oxygen species induced by the adsorption of CO_2.

Oxidative coupling of methane to produce C_2 hydrocarbons has been recognized as a promising route for the direct conversion of methane to C_2 hydrocarbons, and this reaction is very important in terms of the chemical utilization of natural gas. This is because methane is the main component of natural gas, whose reserve is comparable to that of petroleum. Numerous metal oxides have been known to be effective catalysts for the oxidative coupling of methane. Much research on the reaction mechanism and the active site for the oxidative coupling of methane has shown that the activation of methane molecule to the methyl radical plays an key role in this reaction, and that certain kinds of oxide ion species are responsible for the activation of methane to the methyl radical (*1, 2, 3*). These results indicate that activation of molecular oxygen is as important as activation of methane in the oxidative coupling of methane. On some alkali and alkaline earth metal oxides the surface O^- species has been reported to be responsible for the activation of methane, and oxide ion vacancies were the active sites for the activation of oxygen molecules to the O^- species (*4, 5, 6*). Superoxide ions (O_2^{2-}) were recognized to be active for the coupling reaction from the investigation on Na_2O_2, BaO_2, and SrO_2 (*7, 8*)

0097–6156/96/0638–0109$15.00/0

catalysts, and it has been suggested that on alkaline earth metal oxides, an oxygen molecule interacted with a coordinately unsaturated lattice oxide ion O^{2-} to form the O_2^{2-} species (*9*).

It is well known that in perovskite-type oxide (ABO_3) systems, the replacement of A and/or B site cations by other metal cations often brings about the formation of lattice defects (*10, 11*). Perovskite oxides with lattice oxide ion defects can adsorb and activate oxygen for the oxidative coupling of methane. It was reported that compounds related to perovskite with the formula $SrTi_{1-x}Mg_xO_{3-\delta}$ have the ability to adsorb oxygen, and that these can be utilized as lean-burning oxygen sensors (*12*). In addition, we recently found that the adsorbed oxygen species on $SrTi_{1-x}Mg_xO_{3-\delta}$ had the ability to activate methane at much lower temperatures than the usual conditions for this catalytic reaction, to form ethane with high selectivity (*13*).

In the current study the properties of the oxygen species adsorbed on $SrTi_{1-x}Mg_xO_{3-\delta}$ catalysts which are active for oxidative coupling were investigated. The formation of and role of the lattice oxide ion defects which occur by replacing Ti^{4+} with Mg^{2+} ions were studied by means of temperature programmed desorption of adsorbed oxygen species and the exchange reaction between lattice oxide ions and $^{18}O_2$ oxygen in the gas phase.

Experimental

$SrTiO_3$, Sr_2TiO_4 and $SrTi_{1-x}Mg_xO_{3-\delta}$ (X= 0.1, 0.3, 0.5) catalysts were prepared by calcining the proper stoichiometric mixture of powdered $SrCO_3$ (Wako Pure Chemical industries), TiO_2 (Aerosil), and MgO (Koso Chemical) at 1473 K in air for 2h. MgO and SrO were prepared by calcining MgO and $SrCO_3$, respectively, under the same conditions. Crystal structures were investigated by X-ray diffraction (Rigaku RINT2400, Cu K_α). Surface areas were measured by the BET method.

Temperature programmed desorption (TPD) of O_2 was obtained using a closed circulating system equipped with a quadrupole mass spectrometer. After the sample was evacuated at 1123 K for 0.5 h, 6.7 kPa of oxygen gas (Takachiho Trading, 99.9%) was exposed to the sample (sample weight: 0.65 g) at a particular temperature, and then the sample was cooled to room temperature under an O_2 atmosphere. Under the 6.7 kPa of oxygen used, the amount of O_2 adsorption reached a saturation level. TPD spectra were measured by heating the oxides at the heating rate of 10 K/min after evacuation at room temperature. The desorbed O_2 was analyzed by the mass signal intensity of m/e=32.

The amount of O_2 adsorption was measured by a volumetric method using the vacuum line (with a dead volume of 45 cm^3). The sample weight was 1.0 g in this experiment. The evacuation treatment was carried out at 1123 K for 0.5 h and oxygen adsorption conditions were the same as those used for the TPD experiment.

The exchange reaction rate between the lattice oxide ions and the $^{18}O_2$ oxygen (Isotec Inc, 98.5 atom% ^{18}O) in the gas phase was measured in a closed circulating system (with a dead volume of 190 cm^3) equipped with a quadrupole mass spectrometer. The sample (sample weight: 0.40 g) was evacuated at 1123 K for 0.5 h, cooled down to the reaction temperature, then the $^{18}O_2$ gas was introduced into the sample. The isotopic distribution of gas phase oxygen was estimated by the mass signal intensities of m/e=32, 34, and 36, and the exchange reaction rate was calculated from the concentration of ^{16}O ($2^{16}O_2$ + $^{16}O^{18}O$) in the gas phase. The reaction pressure was 2.0 kPa, and the reaction temperature was 473−673 K.

Temperature programmed reaction (TPR) of methane with adsorbed oxygen species on the catalysts was carried out in a fixed bed flow reaction system equipped with an FID gas chromatograph. The O_2 adsorption was done at 1123 K and then the sample was cooled to 373 K under flowing air. Next, CH_4 (Takachiho Trading, 99.99%) diluted with Ar (Takachiho Trading, 99.995%) was passed through the reactor with a partial pressure of methane of 20 kPa. The sample weight was 0.50 g. The TPR of methane was measured in the range of 373 — 773 K at first, then TPD spectra were recorded by heating in an Ar flow after the reactor was purged once with Ar in order to analyze the adsorbed product in the range of 773 — 1200 K. The amounts of the products were analyzed by using FID gas chromatograph equipped with methanator which converted CO and CO_2 into methane. We used the Porapak QS was used as the separating column. Selectivity was calculated using by the TPR result combined with the TPD result. For the series of these experiments the heating rate was 6.3 K/min.

Catalytic oxidative coupling of methane was carried out in the same apparatus as that used for TPR measurement. Reaction conditions were a total pressure 0.1 MPa, CH_4: O_2: N_2 : Ar= 5: 1: 4: 20, a total flow rate 30 ml/min, a sample weight 0.5 g, and a reaction temperature of 873 K.

Results and Discussion

Crystal Structure and Surface Area. Figure 1 shows the XRD patterns of $SrTiO_3$, $SrTi_{1-x}Mg_xO_{3-\delta}$ (X=0.1, 0.3, 0.5), and Sr_2TiO_4. The structure of $SrTiO_3$ was found to be the perovskite type, whereas that of Sr_2TiO_4 was K_2NiF_4 type. For these oxides no impurities were observed. In contrast, $SrTi_{1-x}Mg_xO_{3-\delta}$ (X=0.1, 0.3, 0.5) oxides seem to consist of more than two phases. For $SrTi_{1-x}Mg_xO_{3-\delta}$ (X= 0.3, 0.5), the peak at the diffraction angle 2θ =62.3° which was attributed to MgO was especially clearly observed. This indicates that all the Mg^{2+} ions were not incorporated into the Sr-Ti mixed oxides. In the diffraction angle range of 2θ = 31.0°-33.0°, Sr-Ti mixed oxides have characteristic strong peaks with diffraction angles for the samples in Figure 1 listed in Table 1. The positions of the diffraction peaks are highly dependent on the amount of the additive Mg^{2+} ions and the atomic ratio of Sr^{2+} to Ti^{4+}. If all of the additive Mg^{2+} ions replaced the B sites of $SrTiO_3$, only the peaks at the same positions as those of $SrTiO_3$ would be observed. But fact new peaks not observed in the XRD pattern of $SrTiO_3$ were observed in the XRD of $SrTi_{1-x}Mg_xO_{3-\delta}$ (X=0.1, 0.3, 0.5). This is the consistent with the presence of MgO which was not incorporated. According to the JCPDS data, the peaks at 31.39° and 32.41° are from $Sr_4Ti_3O_{10}$, and the peaks at 31.59° and 32.53° are correspond to $Sr_3Ti_2O_7$. From the comparison between these standard data and the results in Table 1, $SrTi_{0.9}Mg_{0.1}O_{3-\delta}$ phase probably consists of $SrTiO_3$ as the major phase and $Sr_4Ti_3O_{10}$, as the minor phase, and $SrTi_{0.5}Mg_{0.5}O_{3-\delta}$ consists of Sr_2TiO_4 as the major phase and $Sr_3Ti_2O_7$ as the minor one. This interpretation can explain the position of the peaks smaller than the characteristic peaks in the diffraction angle range 2θ = 31.0°-33.0°.

Table 1. The diffraction angle in the XRD patterns of $SrTiO_3$, $SrTi_{1-x}Mg_xO_{3-\delta}$, Sr_2TiO_4 oxides.

Samples	Diffraction angle 2θ / degree*		
$SrTiO_3$			32.42
$SrTi_{0.9}Mg_{0.1}O_{3-\delta}$		31.79	32.39
$SrTi_{0.7}Mg_{0.3}O_{3-\delta}$	31.38	31.74	32.41
$SrTi_{0.5}Mg_{0.5}O_{3-\delta}$	31.28	31.67	32.44
Sr_2TiO_4	31.31		32.53

* Cu Kα XRD.

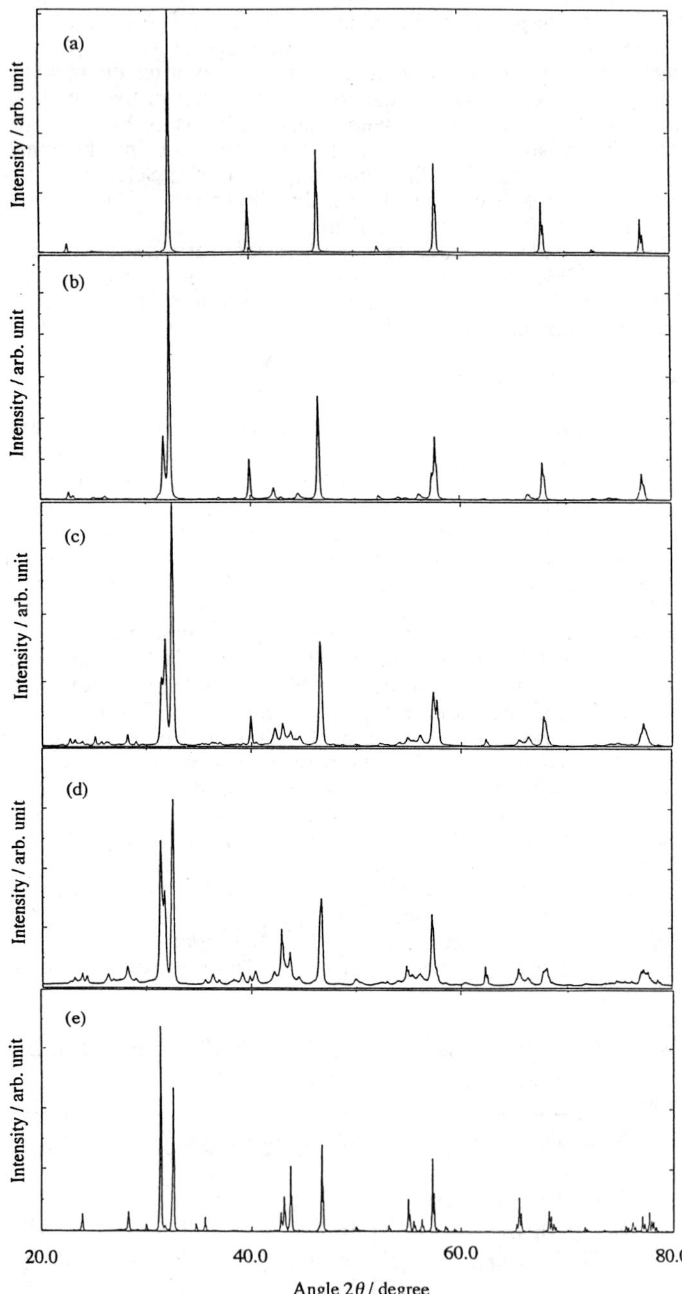

Figure 1. Powder X-ray diffraction patterns of (a) $SrTiO_3$, (b, c, d) $SrTi_{1-x}Mg_xO_{3-\delta}$ (x= 0.1, 0.3, 0.5), and (e)Sr_2TiO_4.

In addition, each peak width observed on XRD patterns of $SrTi_{0.5}Mg_{0.5}O_{3-\delta}$ was much broader than those for $SrTiO_3$ and Sr_2TiO_4. The BET surface area of these three oxides was so small (< 3.3 m^2 g^{-1}) and was not so different from each oxide listed in Table 2. Therefore this peak broadening does not seem to be due to the size of crystallites. We think the partial substitution of Mg^{2+} (with an ionic radius: 0.066 nm) for Ti^{4+} (with an ionic radius: 0.068 nm) in Sr-Ti mixed oxides distorts the crystal structure. We do not understand clearly why the relative peak intensity of the first and second maximum peaks on $SrTi_{0.5}Mg_{0.5}O_{3-\delta}$ and Sr_2TiO_4 is different.

Table 2 shows the BET surface area of the oxides. The oxides except for MgO have very low surface areas. The surface area of the $SrTi_{1-x}Mg_xO_{3-\delta}$ (X=0.1, 0.3, 0.5) oxides was a little higher than that of $SrTiO_3$ and Sr_2TiO_4. This phenomenon is probably due to the MgO impurity in the $SrTi_{1-x}Mg_xO_{3-\delta}$ (X=0.1, 0.3, 0.5) oxides, each major phase seems to have a lower surface area than that listed in Table 2.

Table 2. Surface areas and the amounts of oxygen adsorption.

Oxides	Surface area / m^2 g^{-1} a)	The amount of oxygen adsorption / μmol g^{-1} b)
$SrTiO_3$	2.0	not determined
Sr_2TiO_4	0.7	3.5
$SrTi_{0.9}Mg_{0.1}O_{3-\delta}$	3.3	1.6
$SrTi_{0.7}Mg_{0.3}O_{3-\delta}$	3.1	5.8
$SrTi_{0.5}Mg_{0.5}O_{3-\delta}$	2.5	8.2
MgO	16.5	not determined
SrO	0.4	5.6

a) The surface area was estimated by the BET method.

b) The amount of oxygen adsorption was determined by the volumetrical method. The temperature at which 6.7 kPa of oxygen gas was exposed to the 1.0 g sample was 1123 K.

Oxygen Adsorption and Temperature Programmed Desorption. Table 2 shows the amount of oxygen adsorption for each oxide. The amount of oxygen adsorption on the $SrTi_{1-x}Mg_xO_{3-\delta}$ oxides increased with increased amounts of Mg^{2+}. While the amount of oxygen adsorption on MgO is too small to be determined by the volumetric method even though MgO had rather a high surface area, SrO adsorbed a large amount of oxygen. Oxygen adsorbed on Sr_2TiO_4, but it did not adsorb on $SrTiO_3$.

Figure 2 shows temperature programmed desorption of adsorbed oxygen on $SrTi_{1-x}Mg_xO_{3-\delta}$ (x=0.1, 0.3, 0.5), $SrTiO_3$ (Figure 2a), Sr_2TiO_4, MgO, and SrO (Figure 2b). TPD profiles on $SrTi_{1-x}Mg_xO_{3-\delta}$ (x=0.1, 0.3, 0.5) were very different from those on Sr_2TiO_4 and SrO. Oxygen species on $SrTi_{1-x}Mg_xO_{3-\delta}$ (x=0.1, 0.3, 0.5) desorbed at temperatures of 400-850 K with a maximum at about 620 K, while oxygen on Sr_2TiO_4 desorbed at 550-850 K with a maximum at about 720 K. $SrTi_{0.5}Mg_{0.5}O_{3-\delta}$ was found to contain the Sr_2TiO_4 phase from XRD results, but the oxygen desorption temperature on $SrTi_{0.5}Mg_{0.5}O_{3-\delta}$ was about 100 K lower than that on Sr_2TiO_4. This indicates that the surface structure of $SrTi_{0.5}Mg_{0.5}O_{3-\delta}$ was different from that of Sr_2TiO_4 though they have similar bulk structures. This difference of the surface structure is caused by the addition of Mg^{2+}. The amount of oxygen adsorption on $SrTi_{1-x}Mg_xO_{3-\delta}$ (x=0.1, 0.3, 0.5) is highly dependent on the amount of the additive Mg^{2+} ions, but the oxygen desorption temperature is independent of the amount of Mg^{2+}. These results strongly suggest that similar oxygen adsorption sites were formed on $SrTi_{1-x}Mg_xO_{3-\delta}$ (x=0.1, 0.3, 0.5) oxide surfaces, though they have different bulk structures as indicated by their XRD

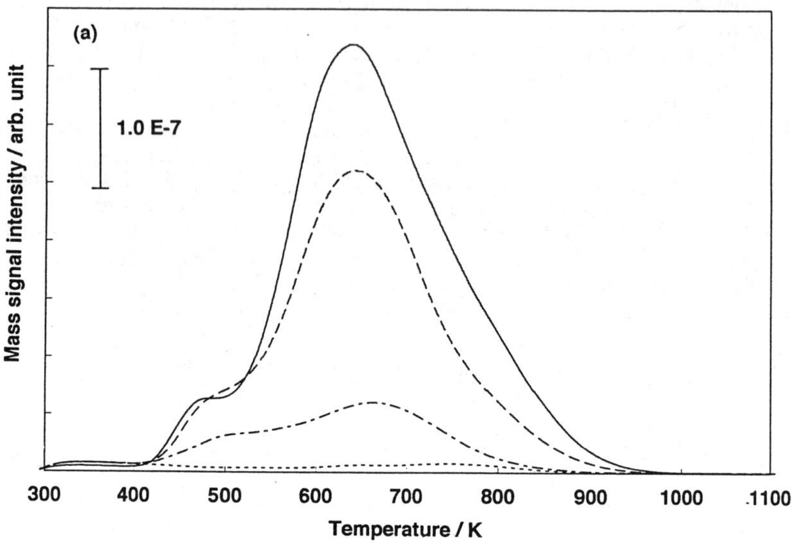

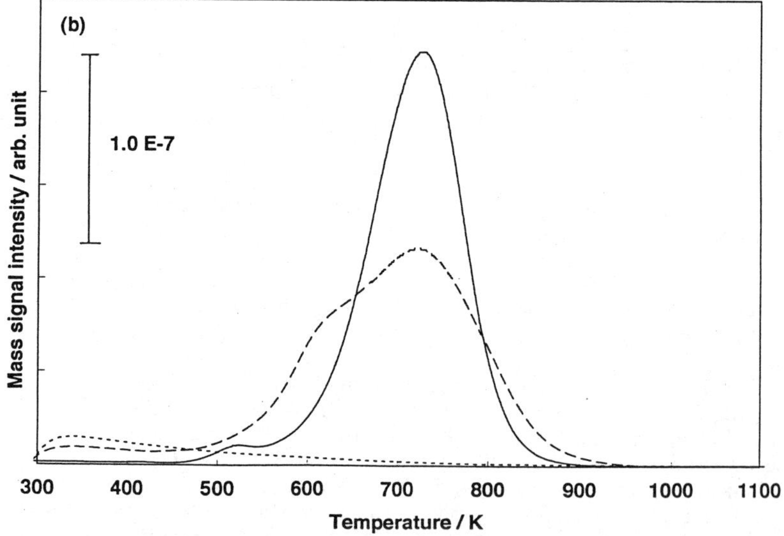

Figure 2. Temperature programmed desorption profiles of adsorbed oxygen on samples. (a) the dependence on the amount of Mg^{2+} on $SrTi_{1-x}Mg_xO_{3-\delta}$ x=0.5 (————), 0.3 (– – – – – –), 0.1(—·—·—) and $SrTiO_3$ (- - - - - - -). (b) Sr_2TiO_4 (————), SrO (– – – – –) and MgO (- - - - - - -).Sample weights: 0.65 g except for SrO and MgO (0.30 g), heating rate: 10 K/min, temperature samples were exposed to oxygen gas : 1123 K, pressure: 6.7 kPa.

patterns. We think that the replacement of surface Ti^{4+} sites of Sr-Ti mixed oxides of various compositions by Mg^{2+} forms the oxygen adsorption site. In this case, oxide ion defects are formed because of the ion charge difference between Mg^{2+} and Ti^{4+}, and this oxide ion defects seem to be involved in oxygen adsorption. The total of mass signal intensities for each oxides is almost proportional to the amount of oxygen adsorption for each, as determined by the volumetric method.

Figure 3 shows the TPD profiles of adsorbed oxygen on $SrTi_{0.5}Mg_{0.5}O_{3-\delta}$, and its dependence on the temperature at which oxygen gas is exposed to the sample. The TPD profiles for samples exposed at temperatures 573- 1123 K are almost the same. When the exposure temperature was 523 K or 473 K, the amount of oxygen adsorption was about 65% and 33% of the saturation amount, respectively. From these results it appears that most adsorbed oxygen species are formed in the temperature range of 473 - 573 K, indicating that oxygen adsorption has as activation energy. This activation energy is probably due to electron transfer from the lattice oxide ion (O^{2-}) to the oxygen molecule in order to form an adsorbed oxygen ion species. According to Figure 2a, the temperature range of the desorption peaks were found to be very wide, indicating that some kinds of oxide species were present on the surfaces on these oxides. The peak can be divided into three parts, the first low temperature peak (peak top: 450 K), the second large peak (peak top: 620 K), and the third part which is at temperatures above 650 K. The existence of the third part seems to make this TPD profile broad. The largest peak is attributed to the oxygen species adsorbed on the oxide ion defect. As the temperature range of the third peak apparently agrees with the TPD profile of adsorbed oxygen of the Sr_2TiO_4, this is attributed to the oxygen species adsorbed on the Sr_2TiO_4-like surface of the $SrTi_{0.5}Mg_{0.5}O_{3-\delta}$ oxide. In addition, this peak temperature range agrees with that of oxygen desorption from SrO. It has been reported that oxygen interacts with coordinatively unsaturated oxide ions of alkaline earth oxide surfaces, or surface basic sites (*9*). We think that oxygen adsorption occurs on a surface basic site of Sr_2TiO_4. We do not clearly understand the cause of the lowest temperature peak on $SrTi_{1-x}Mg_xO_{3-\delta}$ (x=0.1, 0.3, 0.5).

Exchange Reaction between $^{18}O_2$in the Gas Phase and Lattice Oxide Ions. Figure 4 shows the Arrhenius plot of the exchange reaction between $^{18}O_2$in the gas phase and lattice oxide ions on $SrTi_{0.5}Mg_{0.5}O_{3-\delta}$, Sr_2TiO_4 , and $SrTiO_3$. It was found that the activity of the oxygen exchange reaction on $SrTi_{0.5}Mg_{0.5}O_{3-\delta}$ is much higher than that on Sr_2TiO_4 and $SrTiO_3$. The activation energy of this reaction was 41, 54, and 60 kJ mol^{-1} on $SrTi_{0.5}Mg_{0.5}O_{3-\delta}$, Sr_2TiO_4, and $SrTiO_3$, respectively. The addition of Mg^{2+} decreased the activation energy for the exchange reaction. The amount of surface oxide ion can be estimated from the lattice constant (a_0) and the BET surface area (*S*), assuming that the surface is the (001) face. The calculated amount of the surface oxide ion is 2.7 x 10^{-5} mol g-cat^{-1} on $SrTi_{0.5}Mg_{0.5}O_{3-\delta}$ (*S*=2.5 m^2 g^{-1}, a_0=0.388 nm, this lattice constant is thought to be same as that of Sr_2TiO_4), 7.7 x 10^{-6} mol g-cat^{-1} on Sr_2TiO_4 (*S*=0.7 m^2 g^{-1}, a_0=0.388 nm), and 2.2 x 10^{-5} mol g-cat^{-1} on $SrTiO_3$ (*S*=2.0 m^2 g^{-1}, a_0=0.390 nm). In each experiment, the initial amount of ^{18}O was 3.0 x 10^{-4} mol per 0.4 g-cat. The atomic ratio of ^{16}O in the gas phase oxygen after reaction at 623 K for 0.5 h was 61% (4.6 x 10^{-4} mol g-cat^{-1}) on $SrTi_{0.5}Mg_{0.5}O_{3-\delta}$, and 12% (9.0 x 10^{-5} mol g-cat^{-1}) on Sr_2TiO_4. As calculated based on this value, the lattice oxides ion of 17 and 11 layers in $SrTi_{0.5}Mg_{0.5}O_{3-\delta}$ and Sr_2TiO_4, respectively, were exchanged with gas phase oxygen. This fact suggests that the addition of Mg^{2+} promotes diffusion of the bulk oxide ion by formation of oxide ion defects in the bulk. This is supported by reported decrease of resistivity with the addition of Mg^{2+} (*12*). The exchange reaction between $^{18}O_2$ in the gas phase and lattice oxide ions proceeded even at low temperature. This indicates that the dissociation of the oxygen molecule easily proceeds on these oxides, and it is suggested that the oxygen is adsorbed dissociatively on the surface.

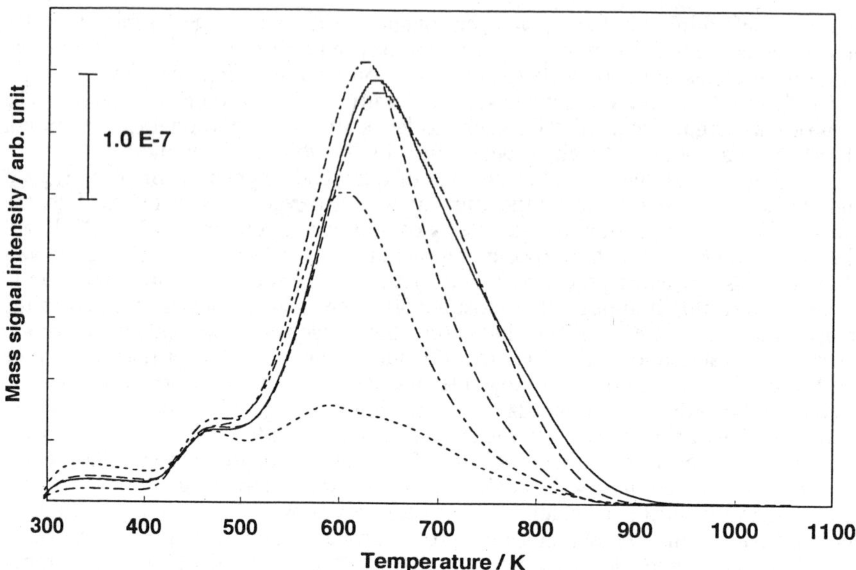

Figure 3. The dependence of TPD profiles of adsorbed oxygen on $SrTi_{0.5}Mg_{0.5}O_{3-\delta}$ on the sample exposure temperature to oxygen gas. The exposing temperature of oxygen : 1123 K (————), 673 K (- - - - -), 573 K (—·—·—), 523 K(—··—··—), 473 K (- - - - - -).Sample weight: 0.65 g, heating rate: 10 K/min, oxygen pressure: 6.7 kPa.

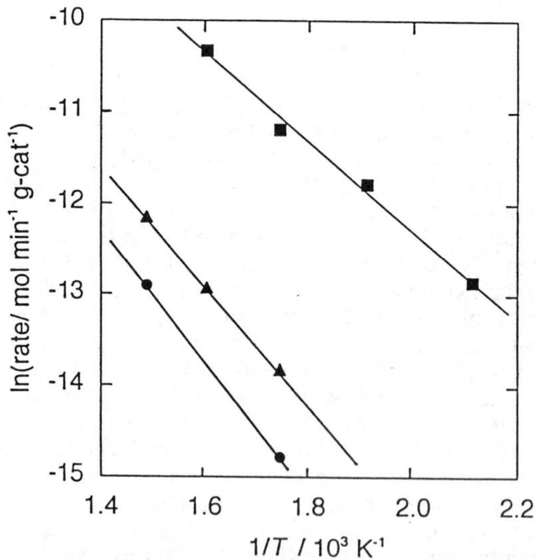

Figure 4. The Arrhenius plot of the exchange reaction between $^{18}O_2$ in the gas phase and the lattice oxide ions on $SrTi_{0.5}Mg_{0.5}O_{3-\delta}$ (■), Sr_2TiO_4 (▲), and $SrTiO_3$ (●).The initial pressure of $^{18}O_2$: 2.0 kPa, reaction temperature: 473 - 673 K, Sample weight: 0.4 g.

Reaction between Methane and Adsorbed Oxygen Species. Figure 5 shows the results of the temperature programmed reaction of methane with adsorbed oxygen species on $SrTi_{0.5}Mg_{0.5}O_{3-\delta}$ and Sr_2TiO_4, followed by temperature programmed desorption. The TPR product was only C_2H_6, and the amount of other products (C_2H_4, CO, CO_2) were below the detection limit. Only CO_2, which was probably produced by the oxidation of methyl radicals and the combustion of methane or ethane, was observed by TPD. The formation of C_2H_6 on $SrTi_{0.5}Mg_{0.5}O_{3-\delta}$ proceeded at temperatures higher than 500 K and its maximum was at about 650 K, while for Sr_2TiO_4 it proceeded at temperatures higher than 600 K and its maximum was at 700 K. This result indicated that methane was activated to methyl radical at much lower temperatures than those typically used for the catalytic reaction, and the activity on $SrTi_{0.5}Mg_{0.5}O_{3-\delta}$ was much higher than that on Sr_2TiO_4. In addition, $SrTiO_3$ showed no activity for C_2H_6 formation in the TPR experiment, because oxygen can not be adsorbed on $SrTiO_3$. The temperature range of C_2H_6 formation by TPR was similar to that of oxygen desorption. This agreement suggests that the ethane was formed by the oxidative coupling of methane by adsorbed oxygen species. When methane was exposed to $SrTi_{0.5}Mg_{0.5}O_{3-\delta}$ without adsorbing oxygen species, no products were observed in this low temperature range. This supports the proposal that the methane was activated at this low temperature by adsorbed oxygen species on $SrTi_{0.5}Mg_{0.5}O_{3-\delta}$.

The total amounts of products and the selectivities of C_2 hydrocarbon formation on $SrTi_{0.5}Mg_{0.5}O_{3-\delta}$ and Sr_2TiO_4 are listed in Table 3. The amounts of O_2 adsorption and the BET surface areas are shown together. From these results, C_2 hydrocarbon formation by the reaction between methane and the adsorbed oxygen species on $SrTi_{0.5}Mg_{0.5}O_{3-\delta}$ and Sr_2TiO_4 was found to be highly selective. The amount of reactive oxygen estimated by the TPR products corresponded to that by the volumetric measurement of adsorbed oxygen very well. This supports the proposal that methane was oxidized by adsorbed oxygen species and that the reactions proceeded stoichiometrically. It has reported that weakly adsorbed oxygen species on perovskite oxides are involved in the combustion reaction, and oxygen species strongly bonded to the surface are effective for the formation of C_2 hydrocarbons (*14, 15, 16*). This is not true for the adsorbed oxygen species on $SrTi_{0.5}Mg_{0.5}O_{3-\delta}$, most likely because the kind of adsorbed oxygen species was different. Judging from the results that surface oxygen species on $SrTi_{1-x}Mg_xO_{3-\delta}$ have the ability to activate methane at about 600 K, and that gas phase oxygen can exchange with lattice oxygen ions, combined with the suggestion that the oxygen adsorption site is the oxide ion defect, the oxygen ion species may be the O^- species. Further spectroscopic studies are necessary to elucidate the chemical state of this adsorbed oxygen species.

Table 3. The total amounts and the selectivities to products by TPR of methane on $SrTi_{0.5}Mg_{0.5}O_{3-\delta}$ and Sr_2TiO_4.

Catalysts	total amount of TPR products[a)] C_2H_6	CO_2[b)]	sel. of C_2H_6 formation	amount of O_2 adsorption[c)]	sel. of O_2 for C_2 formation[d)]
	C-μmol/g-cat		%	μmol/g-cat	%
$SrTi_{0.5}Mg_{0.5}O_{3-\delta}$	11.2	2.5	82	8.2 (7.8)[e]	34
Sr_2TiO_4	4.0	1.2	77	3.5 (3.4)[e)]	29

a) The formation of C_2H_4 and CO was below the detection limit. b) The amount of CO_2 was estimated by TPD after TPR. c) Measurement by the volumetrical method. d) The ratio of the O_2 for C_2 formation to total adsorbed O_2. e) The O_2 consumption was estimated by the amount of products.

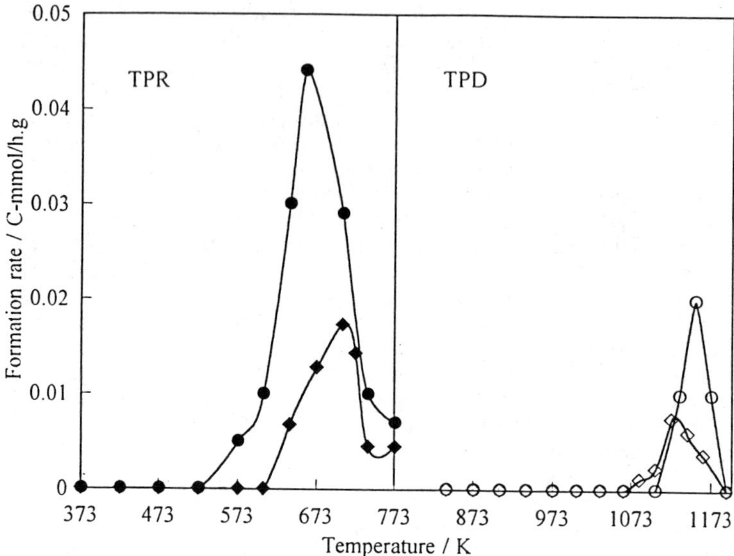

Figure 5. The profiles of the TPR of methane with adsorbed oxygen species and TPD on $SrTi_{0.5}Mg_{0.5}O_{3-\delta}$ and Sr_2TiO_4. Sample weight: 0.50 g, heating rate: 6.3 K/min, sample was exposed to O_2 was at 1123 K. ●: C_2H_6 formation by TPR of $SrTi_{0.5}Mg_{0.5}O_{3-\delta}$, ○: CO_2 desorption from $SrTi_{0.5}Mg_{0.5}O_{3-\delta}$, ◆: C_2H_6 formation by TPR of Sr_2TiO_4, ◇: CO_2 desorption from Sr_2TiO_4.

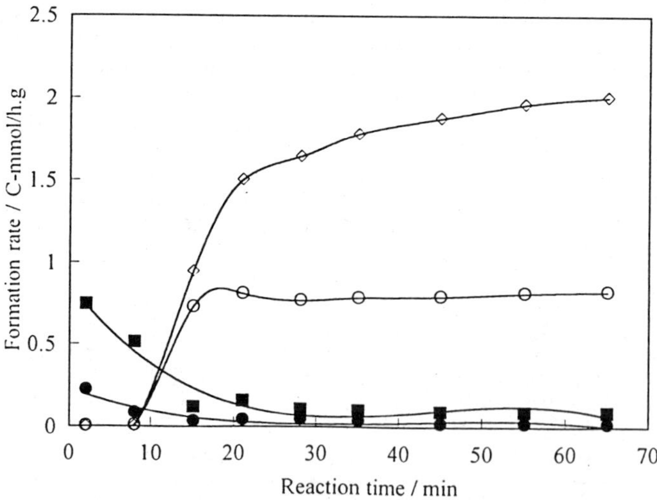

Figure 6. The dependence of the formation rate of C_2H_6 (■), C_2H_4 (●), CO (○) and CO_2 (◇) for the oxidative coupling of methane over a $SrTi_{0.5}Mg_{0.5}O_{3-\delta}$ catalyst on the reaction time. The reaction conditions: the total pressure 0.1 MPa, CH_4: O_2: N_2 : Ar=5: 1: 4: 20, reaction temperature 873 K, total flow rate 30 ml/min, sample weight 0.5 g.

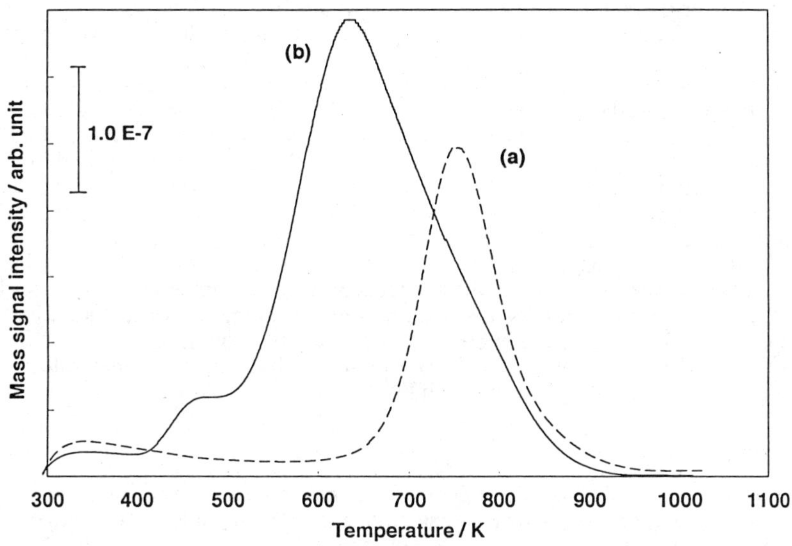

Figure 7. TPD profiles of adsorbed oxygen on $SrTi_{0.5}Mg_{0.5}O_{3-\delta}$ preadsorbed with CO_2 .(a) $SrTi_{0.5}Mg_{0.5}O_{3-\delta}$ preadsorbed with CO_2 , (b) $SrTi_{0.5}Mg_{0.5}O_{3-\delta}$ after evacuation at 1123 K. Sample weight: 0.65 g, heating rate: 10 K/min, oxygen pressure: 6.7 kPa, O_2 exposure temperature: 873 K, CO_2 pressure: 1.3 kPa, CO_2 adsorption temperature: 873 K.

Catalytic Reaction and Deactivation. The results of the TPR indicated that methane was activated by adsorbed oxygen species on $SrTi_{1-x}Mg_xO_{3-\delta}$ at higher temperatures than 800 K, and TPD results indicated that oxygen was activated and adsorbed at higher temperatures than 673 K. Based on these data, the catalytic oxidative coupling of methane on $SrTi_{0.5}Mg_{0.5}O_{3-\delta}$ catalyst at 873 K was studied. Figure 6 shows the reaction time dependence of the formation rate of the products. Initially the main products were C_2 hydrocarbons, and as CO was not observed, the selectivity of C_2 hydrocarbon formation seemed to be considerably high. But the rate of formation of C_2 hydrocarbons decreased gradually and that of CO and CO_2 increased drastically with the reaction time. In this experiment, the total amounts of C_2H_6 and C_2H_4 produced were 75 and 20 μmol $g\text{-}cat^{-1}$, respectively. In this case, 58 μmol O_2 per 1.0 g of the catalyst was used for the oxidative coupling of methane. The turn over number per oxygen adsorbed was found to be 7.0. It is certain that the oxidative coupling of methane by adsorbed oxygen species proceeded catalytically. The change of catalytic properties with the reaction time was probably due to the adsorption of CO_2 on the catalyst surface. To check this, TPD profiles of oxygen adsorbed on $SrTi_{0.5}Mg_{0.5}O_{3-\delta}$ on which CO_2 was preadsorbed at 873 K were obtained and these are shown in Figure 7. The TPD profiles indicated that the properties of the adsorbed oxygen changed drastically with the CO_2 adsorption, and that oxygen species on the surface of $SrTi_{0.5}Mg_{0.5}O_{3-\delta}$ onto which CO_2 had been adsorbed was probably the active species for the combustion of methane. In this case, after CO_2 was desorbed from $SrTi_{0.5}Mg_{0.5}O_{3-\delta}$ by evacuating at 1123 K, the initial catalytic activity was reproduced. Therefore the structure of $SrTi_{0.5}Mg_{0.5}O_{3-\delta}$ did not irreversibly change by CO_2 adsorption. At the present stage, it is certain that oxidative coupling of methane proceeded at lower temperature (873 K) than usual, and CO_2 adsorption seemed to promote the formation of adsorbed oxygen species active for methane combustion, not oxidative coupling. In order for the selective oxidative coupling of methane to succeed at this low temperature, CO_2 adsorption on these catalysts must be inhibited.

Conclusion

1) Ethane was formed by the reaction between methane and adsorbed oxygen species on $SrTi_{0.5}Mg_{0.5}O_{3-\delta}$ with high selectivities (>80 %) at much lower temperatures (550 K-750 K) than those usually used for the catalytic reaction.
2) The oxygen adsorption sites on $SrTi_{1-x}Mg_xO_{3-\delta}$ seem to be the surface oxide ion defects formed by the substitution of Mg^{2+} for surface Ti^{4+} in Sr-Ti mixed oxides.
3) The oxide ion mobility in $SrTi_{1-x}Mg_xO_{3-\delta}$ was much higher than that in Sr_2TiO_4 and $SrTiO_3$, due to bulk oxide ion defects.
4) $SrTi_{0.5}Mg_{0.5}O_{3-\delta}$ catalysts are initially active for the oxidative coupling of methane at low temperature (873 K), but the adsorption of CO_2 on the surface of $SrTi_{0.5}Mg_{0.5}O_{3-\delta}$ during catalytic reaction resulted predominantly in combustion of methane under steady state conditions. This is due to the change of the adsorbed oxygen species induced by the adsorption of CO_2.

Literature Cited

(1) Lee, J. S.; Oyama, S. T.; *Catal. Rev.-Sci. Eng.*, **1988**, 30, 249.
(2) Amenomiya, Y.; Birss, V. I.; Goledzinowski, M.; Galuszka, J.; Sanger, A. R.; *Catal. Rev.-Sci. Eng.*, **1990**, 32, 163.
(3) Voskresenskaya, E. N.; Rouguleva, V. G.; Anshits, A. G.; *Catal. Rev.-Sci. Eng.*, **1995**, 37, 101.
(4) Ito, T.; Wang, J.; Lin, C.; Lunsford, J. H.; *J. Am. Chem. Soc.*, **1985**, 107, 5062.

(5) Lin, C. H.; Wang, J. X.; Lunsford, J. H.; *J. Am. Chem. Soc.*, **1985**, 109, 4808.
(6) Lin, C. H.; Wang, J. X.; and Lunsford, J. H.; *J. Catal.*, **1988**, 111, 302.
(7) Otsuka, K.; Said, A. A.; Jinno, K.; Komatsu, T.; *Chem. Lett.*, **1987**, 77.
(8) Sinev, M. Y.; Korchak, V. N.; Krylov, O. V.; *Kinet. Catal.*, **1986**, 27, 1274.
(9) Indovina, V.;Cordisci, D.; *J. Chem. Soc., Faraday Trans. I*, **1982**, 78, 1705.
(10) Tejuca L. G.; Fierro, L. G.;Tascon, J. M. D.; *Adv. Catal.*, **1989**, 36, 237.
(11) Shimizu, T.; *Catal. Rev.-Sci. Eng.*, **1988**, 34, 355.
(12) Yu, C.; Shimizu, Y.; Arai, H.; *Chem. Lett.*, **1986**, 563.
(13) Li, X.; Fujimoto, K.; *Chem. Lett.*, **1994**, 1581.
(14) Tagawa, T.; Imai, H.; *J. Chem. Soc., Faraday Trans. I*, **1988**, 84, 923.
(15) France, J. E.; Shamsi, A.; Ahsan, M. Q.; Energy and Fuels, **1988**, 2, 235.
(16) Nitadori, T.; Kurihara, S.; Misono, M.; *J. Catal.*, **1986**, 98, 221.

Higher Alkane Oxidation

Chapter 9

Partial Oxidation of Butane at Microsecond Contact Times

D. A. Goetsch, P. M. Witt, and L. D. Schmidt

Department of Chemical Engineering and Materials Science, University of Minnesota, Minneapolis, MN 55455

Using a single layer of woven Pt/Rh gauze as a chemical reactor for the oxidation of n-butane, we find it possible to achieve greater than 90% oxygen conversion with high selectivity to olefins and oxygenated hydrocarbons at contact times on the order of 100 μsec at atmospheric pressure. This reactor configuration operates by rapidly heating the reactants to temperatures greater than 800°C in about 5 μsec followed by rapid quenching to avoid homogeneous chain branching reactions between butane and oxygen. The highly transparent gauze allows most of the reactants to pass between the wires of the gauze without undergoing reaction or being appreciably heated. This gas then mixes with hot products produced on the wire surface or within the boundary layer about the wires to rapidly quench and preserve intermediate products. It is critical to quench the products to temperature around 400°C within less than 1 msec to avoid successive homogeneous decomposition reactions of intermediates such as olefins and aldehydes. This process is capable of directly producing oxygenates and olefins at carbon selectivities of greater than 70%.

Catalytic partial oxidation offers great potential converting abundant and inexpensive light alkanes into value added chemicals such as olefins and oxygenates. A crucial feature of such processes is the ability to attain high surface reaction rates to minimize contributions from non-selective homogeneous reactions and thus allow selective catalytic partial oxidation processes to dominate. Previously, we showed that porous α-alumina foams coated with Rh [1,2] or with Pt [3] give essentially complete conversions of both fuel and O_2 with residence times on the order of milliseconds. An essential factor of the reactor used is that gases remain at nearly ambient temperature until they enter the catalyst and are heated to temperatures in excess of 1000°C by the

0097–6156/96/0638–0124$15.00/0

chemical reactions occurring on the catalytic surface. Autothermal operation combined with large axial Peclet number results in the gas being heated from 20 to 1000°C in less than 0.1 msec. High mass transfer rates within the monolith favor radical termination via surface interactions at the expense of the non-selective homogeneous propagation and chain branching reactions, thus minimizing the contribution of non-selective free radical homogeneous chemistry.

RESULTS AND DISCUSSION

In the results reported here, we have replaced the metal coated 1 cm thick porous ceramic monolith with a single layer of woven gauze made of Pt-10%Rh wire (nominally 76 μm wire diameter and 81% transparency) such as that used for NH_3 oxidation or HCN synthesis. [4] We find that the single gauze catalyst gives very different results compared to monoliths or multiple gauze layers. The superficial contact time (wire diameter divided by gas velocity at inlet conditions) was varied from 10 to 500 μsec.

Reactions between n-butane and oxygen were carried out in a 15 mm diameter and 38 cm long quartz tube with a single layer of Pt-10%Rh gauze at 1.4 atm. The reactor was constructed by joining 15 mm ID and 18 mm ID quartz tubes together. The resulting seat was used in conjunction with a 15 mm ID quartz insert to hold the gauze in place. This arrangement avoids a sudden contraction or enlargement that might disturb flow over the gauze. A layer of alumina cloth was used to seal the space between the insert and reactor wall. Two layers of external insulation where used to minimize radial temperature gradients. The inner layer consisted of a 1.25 cm of Kaolin wool, while the outer layer consisted of 3.8 cm of fiberglass covered with a aluminum foil backed paper. The reactor was configured vertically and the feed gas fed from the bottom of the reactor. The feed gas could be preheated by using a heating tape wrapped around the first 10 cm of reactor length. Figure 1 shows a schematic of the reactor configuration.

The n-butane to oxygen ratio was varied between 2:1 to 4.8:1 and the system was operated nearly adiabatically. As a result, changes in feed composition or feed rate result in different wire temperatures and mixed product gas temperatures. The wire temperature was determined by an optical pyrometer while the product gas temperature was measured by a chromel-alumel thermocouple sheathed in quartz. The pyrometer was calibrated by against a resistively heated sample of silicon under ultra high vacuum. The uncertainty in temperature measurement with the pyrometer was estimated to be ±20°C over the observed temperature range. The thermocouple probe was capable of axially traversing reactor to measure gas temperature as a function of distance downstream of the catalytic gauze.

The partial oxidation of butane involves several reactions between butane and oxygen. Depending on the stoichiometry of the feed, conditions vary from complete combustion (1), to syngas formation (2), to olefin formation (3), and to oxygenate formation (4).

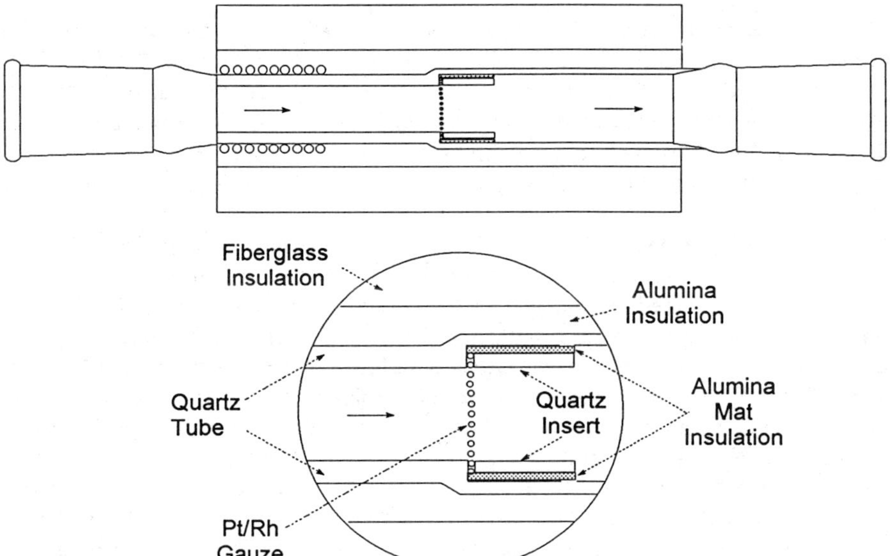

FIGURE 1. Schematic of the reactor. A single layer of Pt/Rh gauze is held in place by a quartz sleeve that has the same internal diameter as the inlet section of the reactor (15 mm). A layer of fibrous alumina is used to seal the space between the sleeve and reactor wall. A heating tape wrapped around the inlet section of the reactor can be used to preheat the feed.

$$C_4H_{10} + 6.5O_2 \rightarrow 4CO_2 + 5H_2O \tag{1}$$

$$C_4H_{10} + 2O_2 \rightarrow 4CO + 5H_2 \tag{2}$$

$$C_4H_{10} + 0.5O_2 \rightarrow C_4H_8 + H_2O \tag{3}$$

$$C_4H_{10} + O_2 \rightarrow C_4H_8O + H_2O \tag{4}$$

The water-gas-shift reaction (5) also takes place at these conditions, but is kinetically rather than equilibrium limited.

$$CO + H_2O \rightarrow CO_2 + H_2 \tag{5}$$

In addition, at temperatures greater than 400°C, the ketones and particularly the aldehydes produced will undergo decarbonylation to give lower oxygenates such as formaldehyde and acetaldehyde which can also undergo decarbonylation to give CO. These reactions can be written as

$$C_4H_8O \rightarrow \bullet CHO + \bullet C_3H_7 \tag{6}$$

$$\bullet CHO + H\bullet \rightarrow HCHO \tag{7}$$

$$\bullet CHO \rightarrow CO + H\bullet \tag{8}$$

$$\bullet C_3H_7 \rightarrow C_3H_6 + H\bullet \tag{9}$$

Similar reactions to equations 6-9 can be written for decomposition of the ketones produced from the partial oxidation of butane to give acetaldehyde, ethylene, and CO. Thus, the partial oxidation of butane produces parent oxygenates and olefins in addition to lower oxygenates and olefins.

Figure 2 shows the conversion of butane and oxygen as function of the butane to oxygen ratio in the feed, and the corresponding downstream gas temperature for operation at 265 μsec and 1.4 atm. At a butane to oxygen feed ratio of two, greater than 90% oxygen conversion and 40% butane conversion are achieved and the product gas exits the reactor at 530 °C. Increasing the butane to oxygen ratio from about 2:1 to 3.5:1 results in little reduction in oxygen conversion but a significant decrease in butane conversion from 40% to about 10% and a corresponding decrease in exit gas temperature. Further increases in butane to oxygen ratio have little effect on the butane conversion but the oxygen conversion decreases from 90% to less than 70%. When the butane to oxygen ratio reaches 4.8, the reaction on the catalytic surface extinguishes.

Product carbon selectivities are plotted as a function of butane to oxygen ratio in Figure 3. Olefins are selectively produced at butane to oxygen ratios of about 2:1. Increasing the butane to oxygen ratio decreased the olefin selectivity from 60% to less than 20% over the feed composition range studied. Altering the feed composition

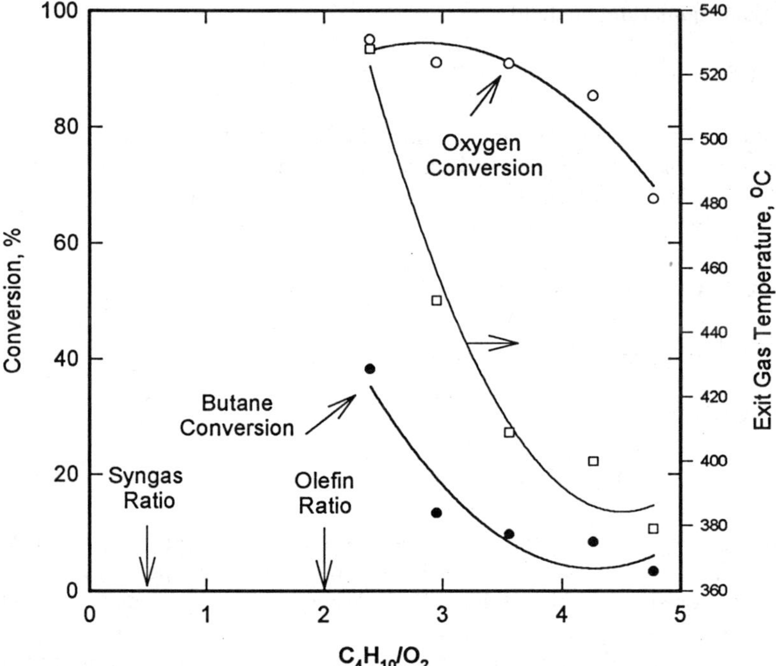

FIGURE 2. Butane conversion, oxygen conversion, and exit gas temperature as a function of butane to oxygen ratio in the feed at a contact time of 265 μsec and 1.4 atm total pressure over a single layer of Pt-10%Rh gauze.

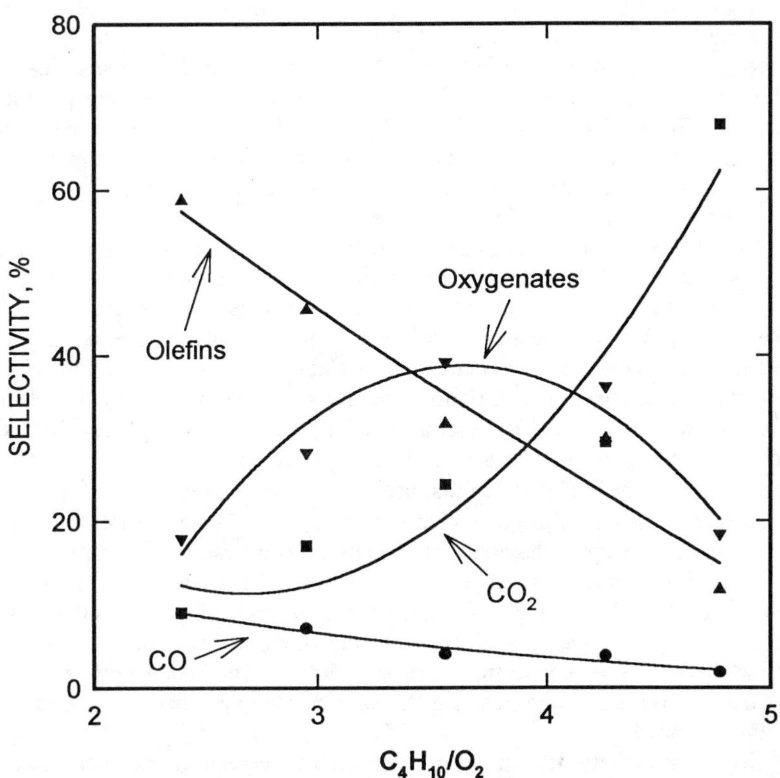

FIGURE 3. Carbon selectivities for total olefins, total oxygenates, CO, and CO_2 as a function of butane to oxygen ratio in the feed at a contact time of 265 μsec and 1.4 atm total pressure over a single layer of Pt-10%Rh gauze.

from 2:1 to 3.5:1 decreased the olefin composition and increased the oxygenate selectivity from less than 20% to more than 40%. Further increases in butane to oxygen ratio resulted in decreased selectivity to oxygenates. Figure 3 shows that the CO_2 to CO selectivity ratio increases with increasing butane to oxygen ratio. This is not surprising since the exit gas temperature correspondingly decreases from 530°C to about 370°C. However, optical pyrometry shows that the wire temperature decreases from 850°C to 800°C which corresponds to approximately 20% less radiative heat loss. The CO_2 yield decreased from 3.4% to 2.2% over the same range of conditions. Although the CO selectively decreases slightly with increasing butane to oxygen ratio in the feed, the large decrease in butane conversion corresponds to a significant yield reduction in CO. Alternatively, the CO yield increases with increasing product gas temperature. These data suggest that CO_2 is primarily produced on the catalytic surface while CO is made by decarbonylation of oxygenate intermediates in the gas phase. This serves to emphasize the importance of rapid quenching in order to produce oxygenates selectively and in high yield.

The olefins produced by oxidative dehydrogenation of butane consist of 1-butene, cis and trans 2-butene, and lower olefins such as ethylene and propylene. Figure 4 shows that as the butane to oxygen ratio increases, the selectivity to lower olefins decreases. We observe a corresponding decrease in alkane selectivity, primarily the methane selectivity. Since the cracking of alkane C-C bonds is endothermic, the decrease in temperature results in a corresponding decrease in lower olefin selectivity. However, the decrease in cracked olefin selectivity can also be explained by reduced oxygen concentration at the wire surface to catalyze the oxidative dehydrogenation of butane and subsequent β-elimination that produces cracked olefins.

Experiments were also conducted with a fixed feed composition (3.5:1 butane to oxygen ratio) to examine the effect of velocity on conversion, selectivity, and yields at 1.4 atm and a product gas temperature of 410°C. The exit gas temperature was maintained at 410°C by adjusting the amount of preheat. Figure 5 shows that the butane conversion decreased slightly with increasing contact time or decreasing velocity while the oxygen conversion increased slightly. The butane conversion results suggest that the reaction is initiated on the wire surface and that the reaction rate is mass transfer limited.

Carbon selectivity data in Figure 6 show that oxygenate and olefin selectivity increases with decreasing contact time. Also, the ratio of olefin to oxygenate selectivity remains relatively constant over the range of velocities or contact times studied. Figure 7 shows the carbon selectivity among lower olefins and the butylenes. Unlike Pt coated monoliths which are selective to the lower olefins, a single layer of Pt-10%Rh gauze primarily produces butylene. [5] Comparison of the butylene selectivity to the oxygenate selectivity gives a constant ratio which suggests that the butylenes and the oxygenates are produced by a parallel reaction pathway.

The oxygenates produced in this system consist of formaldehyde, acetaldehyde, methanol, ethanol, acetone, acrolein, methyl ethyl ketone, methyl vinyl ketone, butanal, and crotonaldehyde. Carbon selectivities for the various oxygenate products are shown in Figure 8. The predominate oxygenates are primarily formaldehyde and acetaldehyde.

The high transparency (81%) of the gauze combined with short contact times

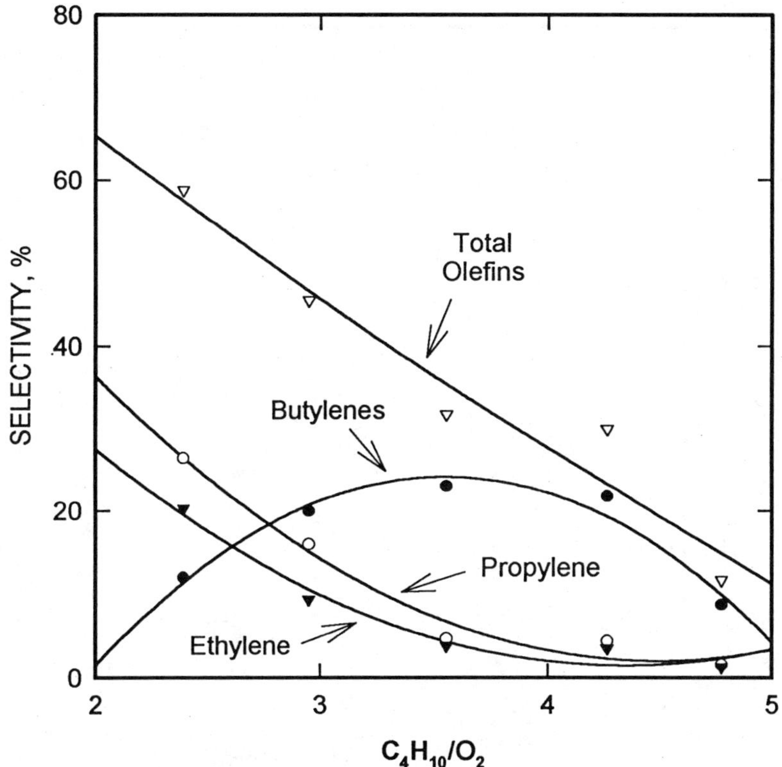

FIGURE 4. Butylene, Propylene, and Ethylene carbon selectivities as function of butane to oxygen ratio in the feed at a contact time of 265 μsec and 1.4 atm total pressure over a single layer of Pt-10%Rh gauze.

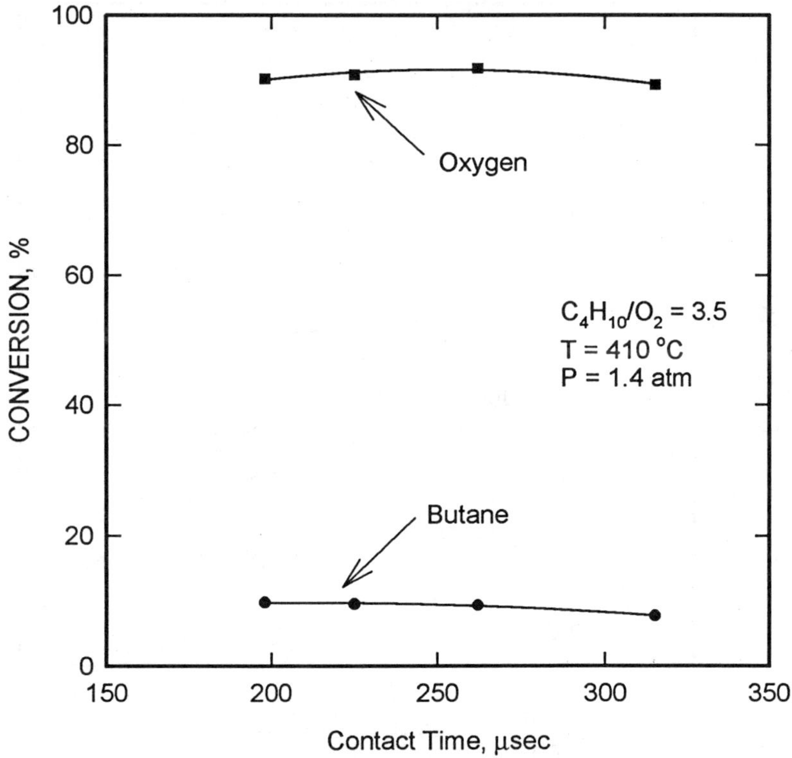

FIGURE 5. Butane and oxygen conversion as a function of superficial contact time over a single Pt-10%Rh gauze at 1.4 atm, 410°C, and C_4H_{10}:O_2 ratio of 3.4.

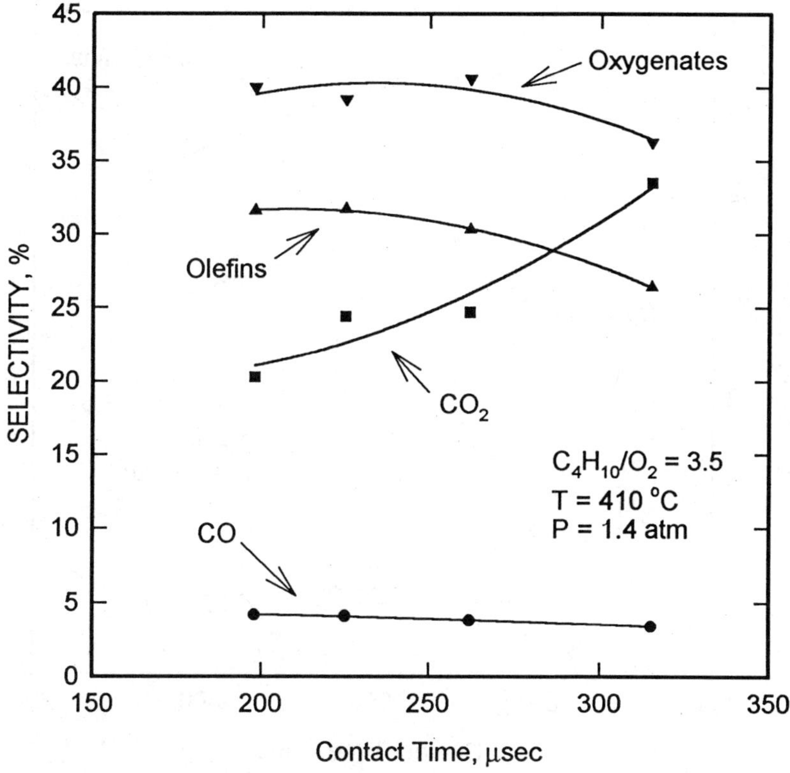

FIGURE 6. Carbon selectivities for total olefins, total oxygenates, CO, and CO_2 as a function of superficial contact time over a single Pt-10%Rh gauze at 1.4 atm, 410°C, and C_4H_{10}:O_2 ratio of 3.4.

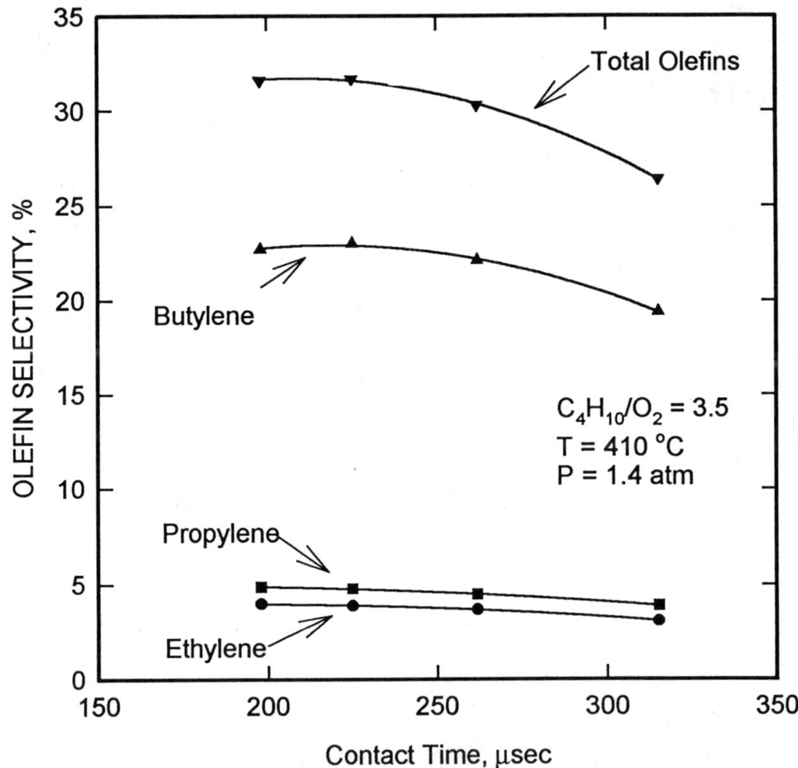

FIGURE 7. Butylene, Propylene, and Ethylene carbon selectivities as a function of superficial contact time over a single Pt-10%Rh gauze at 1.4 atm, 410°C, and C_4H_{10}:O_2 ratio of 3.4.

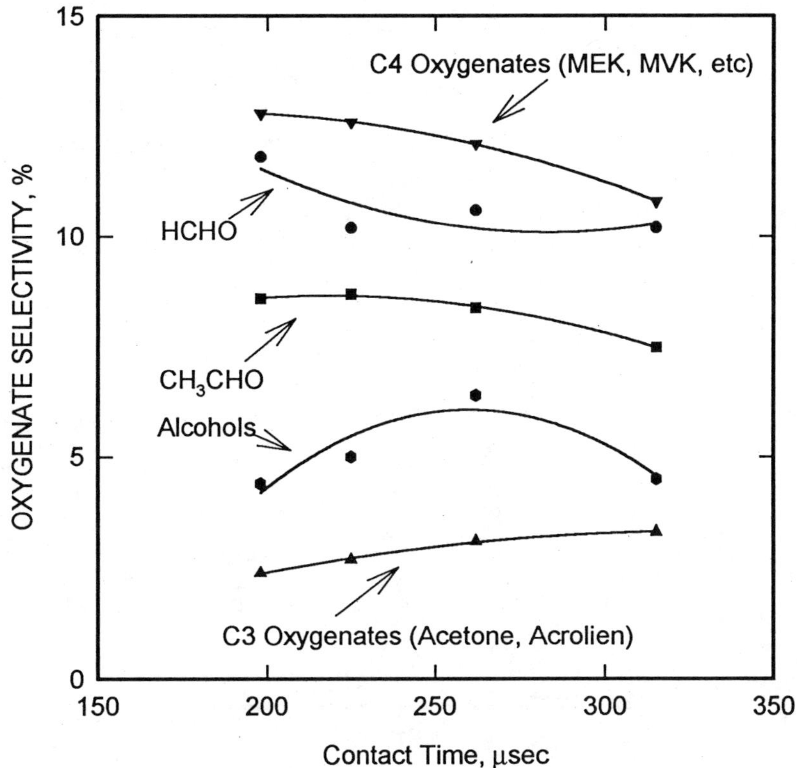

FIGURE 8. Carbon selectivities for oxygenates produced from the partial oxidation of butane with a single Pt-10%Rh gauze at 1.4 atm, 410°C, and C_4H_{10}:O_2 ratio of 3.4.

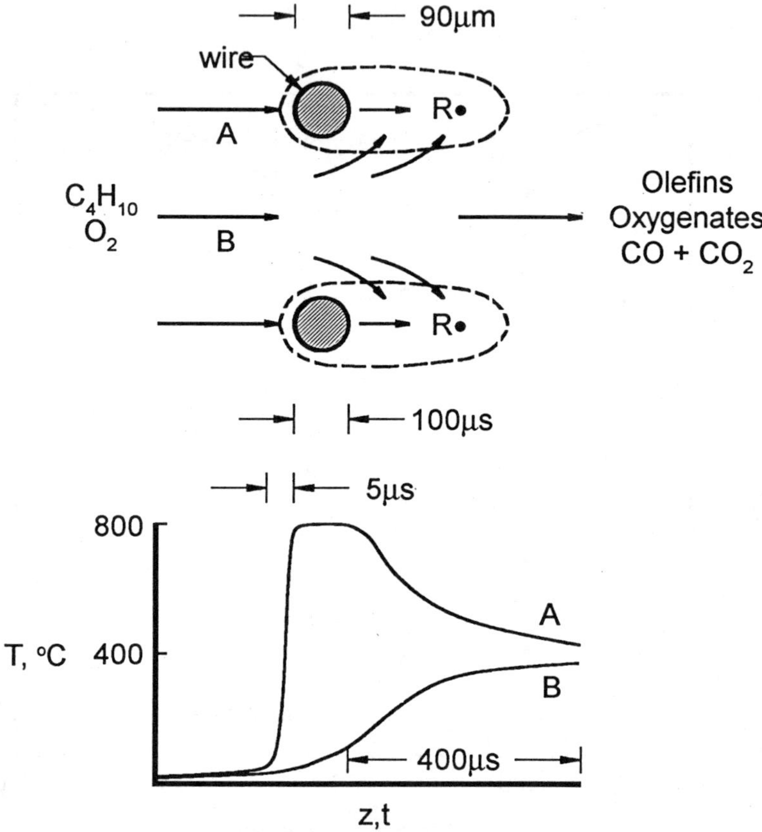

FIGURE 9. Sketch of temperature profiles for a single layer of gauze for gas flowing around single gauze wires and between gauze wires. Gas flowing close to the wire is rapidly heated (~5 μsec) to the wire surface temperature (~800°C), remains at this temperature for the contact time with the wire (~100 μsec), and is rapidly quenched to 400°C by mixing with 25°C gas flowing between the wires (~400 μsec). (Adapted from ref. 8)

leads to a significant amount of reactants passing between the catalytic wires. The high oxygen conversion achieved with a single gauze strongly suggests that significant homogeneous reaction takes place in the wake region behind the gauze wires. However, reaction products in the wake are rapidly quenched by the cold feed gas that by-passes the catalyst.

We believe that butylene and aldehydes survive in the presence of greater than 800°C catalyst surfaces because the products are rapidly quenched by the cold feed gas that by-passes the wire surface. The mixing of the hot reacted gases (contacting the wires) and the cold unreacted gases (passing between the wires) occurs within a few wire diameters downstream of the gauze, which is at >800°C, and cools the products to about 400°C within 200 μsec. This allows selective production and survival of oxygenates and is shown in Figure 9.

The Reynolds number varies between 1 to 3 over the range of experimental conditions. Under these conditions, boundary layer separation from a single wire in would not be expected. [6] However, the pressure gradient induced by the opening, the large temperature gradient between the wire and the gas, and the reduction in molecular weight all contribute to destabilize the boundary layer. [7] In addition, roughening of the wire surface with exposure time to reaction conditions introduces form drag further contributing to premature boundary layer separation. [8] The temperature of the gas within the boundary layer and the wake region behind the wire will be nearly the same as the wire surface while the gas passing between the wires will be significantly cooler. The difference in viscosity between the hot and cold gas results in shear boundary causing mixing of the hot and cool streams much in the same manner as a jet introduced into another medium.

Increasing the number of gauze layers from one to five decreases the transparency from 81 to 13% and results in an increase in olefin selectivity from 36 to 63% and a decrease in oxygenate selectivity from 40 to 14%, Table I. The exit gas temperature increased to 830°C and the inability to rapidly quench reaction products results in increased amounts of olefins by cracking reactions and decarbonylation of oxygenates.

TABLE I

Conversions and Selectivity in Single and Multiple Gauze Reactors For Butane Oxidation

	Conversion, %		Carbon Selectivity, %					Temperature, °C	
	C_4H_{10}	O_2	CO	CO_2	$C_4^=$	$<C_4^=$	Oxygenates	Catalyst	Gas
1 layer	10	90	4	20	27	9	40	800	400
5 layers	38	>99	9	9	15	48	14	900	830

The data suggest that oxygenates are produced by a heterogeneously assisted homogeneous reaction mechanism. [9,10] The first step must take place on the catalyst surface since the feed gas enters at temperatures too low for significant homogeneous radical initiation to occur. However, within the wake region the gas is at temperatures high enough to allow homogeneous reaction. Reaction will continue to propagate in the wake region until enough cold feed gas has been entrained to reduce the temperature enough to stabilize the alkyl peroxides responsible for chain branching.

The constant selectivity ratio between formaldehyde and acetaldehyde can be explained by an intramolecular reaction mechanism. For butane, the alkyl peroxy radical can attack at a contiguous secondary carbon forming a cyclic transition state. Following electron rearrangement the activated complex decomposes to acetaldehyde and an ethoxide radical. At temperatures greater than 360°C, the ethoxide radical can spontaneously decompose to formaldehyde and a methyl radical. [11] The methyl radical undergoes further reaction to give either formaldehyde, formic acid, CO, or CO_2. Thus, the intramolecular pathway would produce formaldehyde to acetaldehyde in ratios between 2:1 and 1:1 depending on the reaction temperature. The reaction pathway can be represented as follows:

+ ·CH3
-H
+O2
-H
- ·OOH
+ + ·CH3

(10)

The alkyl peroxy radical can also abstract an internal hydrogen and then decompose to butylene and hydroperoxide radical. This is consistent with the constant selectivity ratio observed between oxygenates and butylene.

SUMMARY

Rapid chemical preheat coupled with rapid quench by unreacted gases by-passing the catalyst make single gauze catalysts uniquely capable of converting alkanes into useful chemicals with high selectivity and conversions at very high temperatures where reaction rates are extremely fast. These laboratory reactors produce about 20 kg/day of aldehydes and butylene from butane, and a one foot diameter reactor should be capable of producing approximately 3.5 metric tons/day of these chemicals. The relative ease with which formaldehyde can be removed from the product stream by water washing should allow even higher yields by recycling unreacted products. The ability to directly produce valuable chemicals such as formaldehyde from natural gas could have significant impact on the practice of commercial partial oxidation reactions.

ACKNOWLEDGEMENT

We would like to acknowledge the NSF under grant CTS-9311295 and by the Department of Energy under grant DE-FG0288ER 13878-A02 for their support of this project.

REFERENCES

1. Hickman, D. A., Haupfear, E. A., Schmidt, L. D.; *Catal. Lett.*, **17**, 223 (1993).
2. Hickman, D. A., Schmidt, L. D.; *AIChE J.*, **39**, 1164 (1993).
3. Huff, M., Schmidt, L. D.; *J. Phys. Chem.*, **97**, 11815 (1993).
4. Satterfield, C. N.; *Heterogeneous Catalysis in Industrial Practice*; McGraw-Hill, Inc.: New York, New York (1991).
5. Huff, M., Schmidt, L. D.; *J. Catal.*, **155**, 82 (1994).
6. Coutanceau, M., Bouard, R.; *JFM*, **79**, 231 (1977).
7. Schlichting, H.; *Boundary Layer Theory, 7th Ed.*; McGraw-Hill, Inc., New York, New York (1979).
8. Goetsch, D. A., Schmidt, L.D.; *Science*, **271**, 1560 (1996).
9. Pfefferle, L. D., Pfefferle, W. C.; *Catal. Rev. Sci. Eng.*, **29**, 219 (1987).
10. Lunsford, J. H., Morales, E., Dissanayake, D., Shi, C.; *Int. J. Chem Kinetics*, **26**, 921 (1994).
11. Benson, S. W.; *Thermochemical Kinetics: Methods for the Estimation of Thermochemical Data and Rate Parameters*, 2nd Ed., John Wiley & Sons, New York, 1976.

Chapter 10

The Catalytic Activity of Wells–Dawson and Keggin Heteropolyoxotungstates in the Selective Oxidation of Isobutane to Isobutene

Fabrizio Cavani[1], Clara Comuzzi[2], Giuliano Dolcetti[2], Richard G. Finke[3], Arianna Lucchi[1], Ferruccio Trifirò[1], and Alessandro Trovarelli[2]

[1]Dipartimento di Chimica Industriale e dei Materiali, Viale Risorgimento 4, 40136 Bologna, Italy
[2]Dipartimento di Scienze e Tecnologie Chimiche, Via Cotonificio 108, 33100 Udine, Italy
[3]Department of Chemistry, Colorado State University, Fort Collins, CO 80523

The catalytic performance of the Wells-Dawson-type polyoxotungstate, $K_6P_2W_{18}O_{62}$, in the oxydehydrogenation of isobutane to isobutene has been studied, and compared with the activity of the corresponding Keggin compound, $K_3PW_{12}O_{40}$. The Wells-Dawson compound is characterized by a lower activity than the Keggin salt, but by a remarkably higher selectivity to the olefin. The activity of the Wells-Dawson catalyst is strongly dependent on the hydrocarbon content in the feed, and increases with increased partial pressure of isobutane. This unexpected autocatalytic behavior is explained by a proposed mechanism where the olefin product induces a modification of the catalyst surface, with the creation of more active sites. In addition, at the highest hydrocarbon content in the feed, the contribution of heterogeneously-initiated, homogeneous gas-phase reactions become important, thus favoring high selectivity to isobutene.

Interest in heteropolyanions as catalysts for the liquid-phase and gas-phase oxidation of hydrocarbons has been growing rapidly in the last years (*1,2*). The unique features of these materials, their high acidity and tunable redox properties, make them ideal for a variety of applications. Heteropolycompounds have found industrial applications for many years as catalysts for the oxidation of methacrolein to methacrylic acid, and are well studied as catalysts for acid-catalyzed reactions (alkylation of isobutane, etherification) and as catalysts for the homogeneous and heterogeneous oxidation of olefins and paraffins (*1,2*). Their reactivity and selectivity can be fine-tuned, for example by modifying their composition and structure through incorporation of transition metal ions or alkali metal counterions.

Among the different classes of heteropolyanions (Keggin, Anderson, Wells-Dawson and Waugh), Keggin-type complexes account for the majority of studies and applications in homogeneous and heterogeneous reactions, probably because of their

0097–6156/96/0638–0140$15.00/0

higher thermal stability under reaction conditions, their easy preparation, and the fact that the Keggin-anion structure has been known since 1934. However, the use of Wells-Dawson type polyoxoanions as oxidation catalysts has been limited to homogeneous, liquid-phase oxidation of olefins (*3,4*). More specifically, no example or application of a gas-phase, heterogeneous catalyst using a Wells-Dawson type polyoxoanion has previously been reported in the scientific literature.

Among the organic substrates which are studied as raw materials for oxifunctionalization reactions, paraffins deserve particular attention. The oxidative activation of paraffins is a current challenge in catalysis: new low cost, widely available, raw materials for the petrochemical industry is the goal here (*5*). Among light paraffins, isobutane is one of the currently most interesting targets due to the increasing market for isobutene, which in turn is a raw material for the synthesis of oxidation products such as methacrylic acid (currently produced via the environmentally unfriendly acetone-cyanohydrin route), or for the synthesis of methyl t-butyl ether, an octane booster for reformulated gasolines. Heteropolyoxomolybdates have, for example, been investigated as catalysts for the one-step transformation of isobutane to mixtures of methacrolein and methacrylic acid (*6,7*).

Herein we report the catalytic reactivity of Wells-Dawson-type and Keggin-type heteropolyoxotungstates for the oxidation of isobutane. Potassium salts of $(PW_{12}O_{40})^{3-}$ and of $(P_2W_{18}O_{62})^{6-}$ were prepared and characterized, since it is known that potassium salts exhibit a considerably enhanced structural stability with respect to the parent acid form, or in comparison to salts containing smaller cations (*1*).

Preparation of Catalysts

The Wells-Dawson compound of composition $K_6P_2W_{18}O_{62}\cdot 10H_2O$ was prepared as detailed elsewhere (*8*); its purity and composition were checked by FT-IR, solution ^{31}P NMR, and elemental analysis. $K_3PW_{12}O_{40}$ was prepared by precipitation from an aqueous solution containing Na_2WO_4 and H_3PO_4. HNO_3 was added to the solution until a pH lower than 1.0 was reached, and then KNO_3 was added for precipitation of the Keggin salt. The resultant $K_3PW_{12}O_{40}$ purity was checked by XRD and IR. Before catalytic tests, all compounds were calcined at 450°C for 3 h in air. Surface area analyses were carried out by measuring N_2 adsorption isotherms with a Carlo Erba Sorptomatic 1900 Instrument interfaced with a computer. The infrared spectra were collected as KBr pellets with a Digilab FTS-40 instrument.

Thermal Evolution of $K_6P_2W_{18}O_{62}.10H_2O$ and $K_3PW_{12}O_{40}$

The surface areas of the "as synthesized" powders were respectively 1.5 and 15 m^2/g for $K_6P_2W_{18}O_{62}\cdot 10H_2O$ and for $K_3PW_{12}O_{40}$. Treatment in air at 450°C does not affect the textural properties; in the case of the Wells-Dawson compound, $K_6P_2W_{18}O_{62}\cdot 10H_2O$, a slight modification of surface area (from 1.5 to 3 m^2/g) is observed, which is probably related to the loss of crystallization water from the material, which occurs at around 150°C, as detected by thermogravimetric analysis. Figure 1 shows for comparison the FT-IR spectra of the catalysts before and after thermal treatment in air at 450°C for 2h. The vibrational bands typical of P-O (ca.

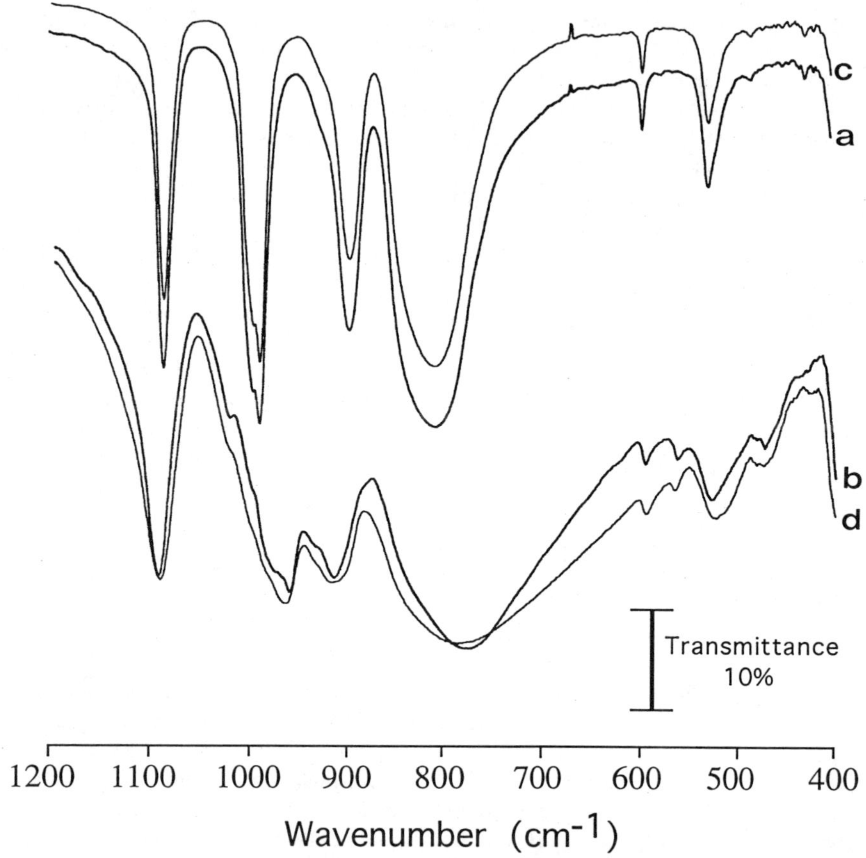

Figure 1. FT-IR spectra of the $K_3PW_{12}O_{40}$ and of the $K_6P_2W_{18}O_{62}$ catalysts as synthesized (*a* and *b*, respectively), and after calcination at 450°C (*c* and *d*, respectively).

1090 and 1020 cm^{-1} in the Wells-Dawson and in the Keggin compound, respectively), W=O (ca. 780-800 cm^{-1}) and W-O-W (ca. 910 and 950 cm^{-1} for the Wells-Dawson compound, and ca. 980-1000 cm^{-1} for the Keggin) are observed. No major modification of the IR pattern is induced by treatment in an oxidizing atmosphere up to 500°C, indicating that the polyoxoanion primary structure remains intact. After reaction (at least 300 h time-on-stream), the IR patterns of the unloaded catalysts did not differ from those of the starting materials, indicating that the catalysts were stable in the reaction environment. This was also confirmed by checking the catalytic activity which was constant with time on stream.

Catalytic Activity of the Wells-Dawson and Keggin Compounds in Isobutane Oxidation

The catalytic activity for the gas-phase oxidative dehydrogenation of isobutane was examined using a stainless steel flow reactor operating at atmospheric pressure, all as previously described (*6*). Three grams of granulated catalyst (particle size 0.3-0-5 mm) were loaded for tests. Results were collected after approximately 50 h time-on-stream, and no effect of catalyst deactivation was observed even after 300 h time-on-stream. Analysis of the products was done as described previously (*6*). Carbon balance was always in the range 95-110%.

The effect of the reaction temperature on the catalytic performance of $K_3PW_{12}O_{40}$ and $K_6P_2W_{18}O_{62}$, both calcined at 450°C, is shown in Figures 2 and 3, respectively. The main products in both cases were isobutene and carbon oxides; minor amounts of propylene also formed, as well as acetic acid, methacrolein and methacrylic acid. Significantly, the main products obtained were different from those obtained under the same conditions with Keggin polyoxomolybdates; in this case, in fact, the $K_3PMo_{12}O_{40}$ catalyst yielded methacrolein and methacrylic acid as the main products of selective oxidation, and isobutene was obtained only in minor amounts (*6*). This fact points out the different reactivity of molybdenum and tungsten in polyoxometalates; Mo^{VI} is able to i) abstract hydrogen and ii) insert oxygen onto the activated hydrocarbon, while W^{VI} can only perform the first type of oxidative reaction. The Mo^{VI} vs. W^{VI} greater electronegativity and lower M-O bond strength account for the different reducibility and reoxidizability (*9*) of the two metal ions.

Results show that the Keggin, $K_3PW_{12}O_{40}$, and the Wells-Dawson, $K_6P_2W_{18}O_{62}$, compounds are characterized by a different catalytic performance. The Keggin compound was clearly more active, activating isobutane already at 350°C (total oxygen conversion was reached at approximately 420-430°C), while the Wells-Dawson type catalyst was almost inactive at this temperature. This difference might be explained by the different values of surface area of the two compounds. However, the different apparent activation energies for isobutene formation in the two cases (30 kcal/mol for the Wells-Dawson compound, 15 kcal/mol for the Keggin compound) also indicates a different nature or reactivity of the active site. The Wells-Dawson compound was, however, much more selective to isobutene; in addition, remarkably, this selectivity to isobutene was roughly constant over the entire examined range of temperature, at a residence time of 3.6 s. This might indicate that isobutene is kinetically stable under the reaction conditions, and does not undergo consecutive oxidative degradation reactions. The Wells-Dawson compound is more selective even

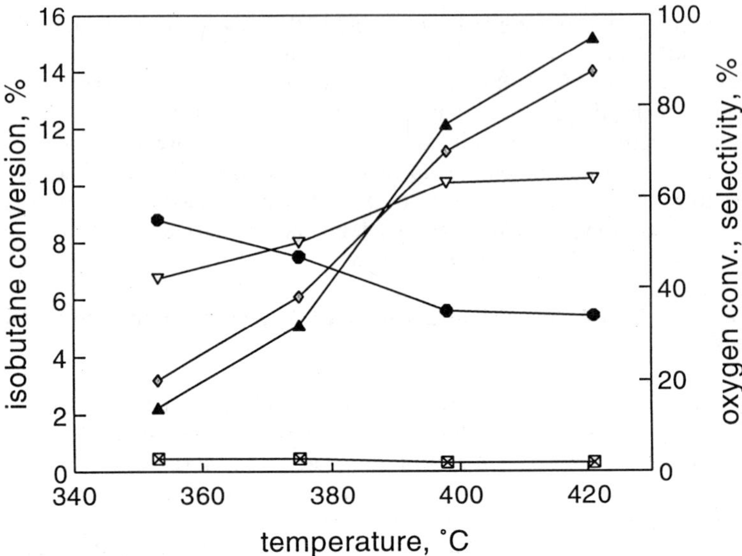

Figure 2. Isobutane and oxygen conversions, and selectivity to the products as a function of the reaction temperature for the $K_3PW_{12}O_{40}$ catalyst. Symbols: isobutane conversion (◇); oxygen conversion (▲), selectivity to isobutene (●), selectivity to carbon oxides (▽), selectivity to the other products (⊠). Other conditions: residence time 3.6 s, feedstock composition: 26% isobutane; 13% oxygen; 12% water; balance helium.

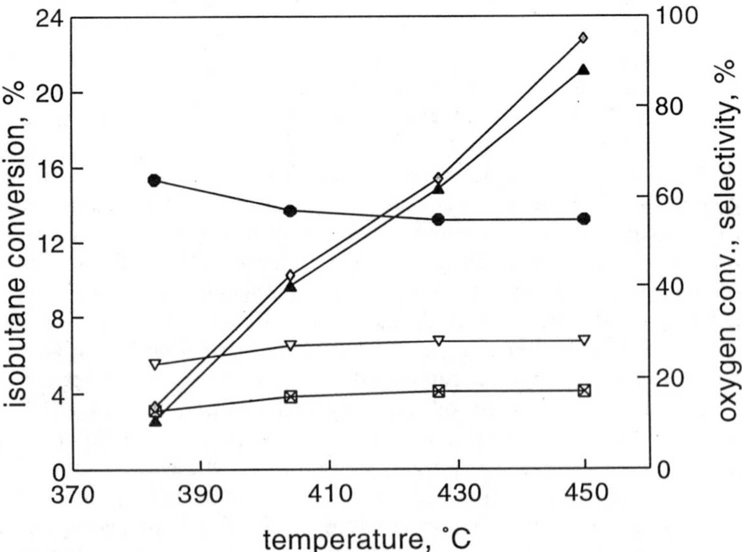

Figure 3. Isobutane and oxygen conversions, and selectivity to the products as a function of the temperature for the $K_6P_2W_{18}O_{62}$. Symbols and conditions as in Figure 2.

if we compare the results obtained on the two catalysts at comparable isobutane conversion, despite the higher temperature at which the Wells-Dawson catalyst is active; the higher selectivity accounts for a lower consumption of dioxygen, thus allowing a higher maximum isobutane conversion in comparison to the Keggin compound.

The autocatalytic effect at high isobutane concentration

The effect of an increasing concentration of isobutane in the feed on the Keggin and Wells-Dawson catalysts are shown in Figures 4 and 5, respectively. The higher activity of the Keggin compound is confirmed; the isobutane conversion remains approximately constant (8-9%) over the complete range of isobutane concentrations examined, as expected for the case of a reaction order close to 1. The oxygen conversion correspondingly increased when the hydrocarbon concentration was increased, because of the higher hydrocarbon-to-oxygen ratio in the feedstock, and the selectivity to isobutene was significantly improved.

On the contrary, the behavior of the Wells-Dawson compound proved rather unexpected. Under the conditions examined, we can distinguish two regions: a first region, for hydrocarbon concentrations lower than approximately 15-20%, in which the catalyst exhibited a very low activity, and the conversion was approximately constant, implying a reaction order close to 1 (i.e., a behavior similar to that observed with the Keggin catalyst, apart from the lower activity), and a second region, for isobutane concentrations higher than 20%, where the conversion steadily increased, a trend which clearly indicates a reaction order higher than 1. The effect was completely reversible, and therefore it cannot be explained as due to a structural irreversible transformation, but rather might be due to a temporary and reversible surface modification. This behavior clearly points to an autocatalytic effect at high isobutane partial pressure, possibly due either to i) an *in-situ* modification of the active centers when the gas-phase composition is varied, or to ii) one of the products acting itself as a reactant or as catalyst for the reaction, or acting as a modifier of the catalyst surface.

Several different possibilities could explain this effect, so additional tests were carried out in order to further define the phenomenon:

1) A first possibility is that an increase in the isobutane concentration might favour the reduction of the catalytic surface (for instance, of the W^{VI} ions), with formation of more active $W^{V\bullet}$ centers. In favour of this hypothesis is the observation of an induction period when changing the feed composition, a phenomenon which might well be explained by a progressive modification of the catalyst surface. This induction period is shown in Figure 6, which illustrates the progressive variation in catalytic performance after varying the feed composition from 42% to 8% isobutane content in the feed (i.e., from a gas phase composition under which the catalyst is active to a situation under which it is inactive), and vice versa from 26% to 42% isobutane.

The reduction of tungsten should be disfavoured under more oxidizing conditions, that is at higher oxygen concentration in the feedstock. On the contrary, Figure 7 reports data showing a remarkable increase in the catalytic performance when the oxygen content in the feed is increased. The oxygen effect is similar to that obtained by increasing the isobutane concentration. Since an increase in either

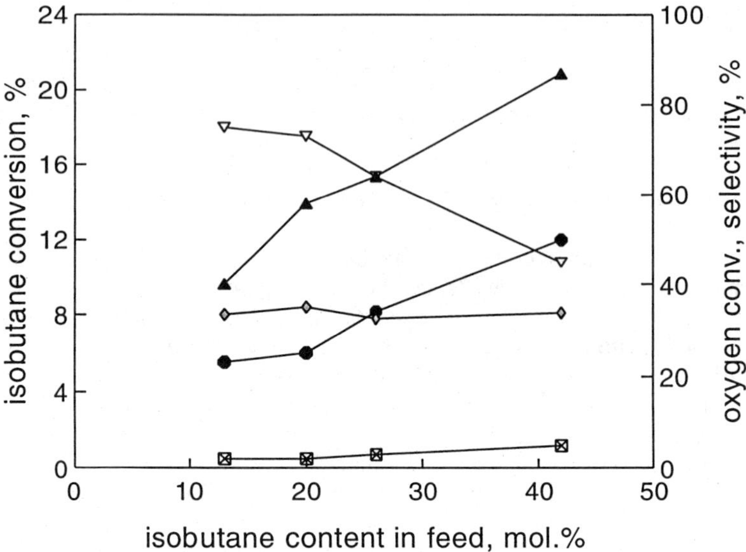

Figure 4. Isobutane and oxygen conversions, and selectivity to the products as a function of the isobutane content in the feed for the $K_3PW_{12}O_{40}$ catalyst. Symbols are the same as in Figure 2. Other conditions: T 470°C, residence time 1 s, feedstock composition: 13% oxygen; 12% water; balance helium.

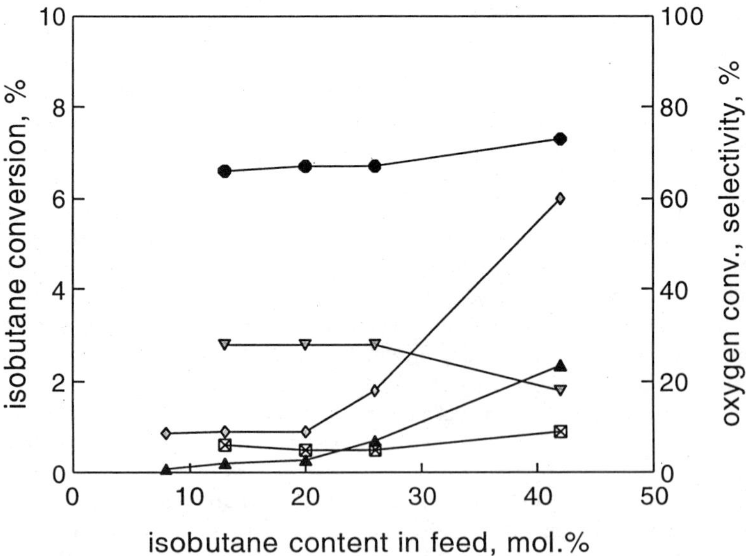

Figure 5. Isobutane and oxygen conversions, and selectivity to the products as a function of the isobutane content in the feed for the $K_6P_2W_{18}O_{62}$ catalyst. Symbols and conditions are the same as in Figure 4.

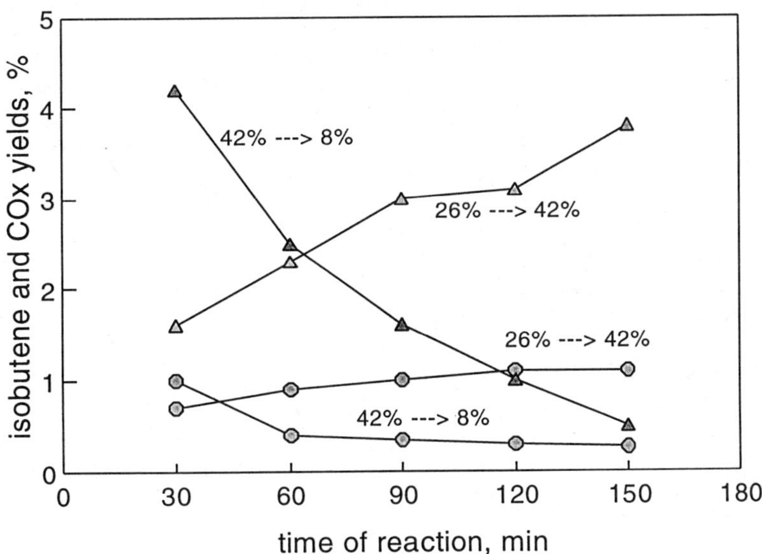

Figure 6. Isobutene (△) and carbon oxides (O) yields as a function of the time of reaction after changing the isobutane content in the feedstock from 26% to 42%, and from 42% to 8%. Time = 0 min corresponds to the moment at which the change is made. Other conditions: T 470°C, residence time 1 s, feedstock composition: 13% oxygen; 12% water; balance helium.

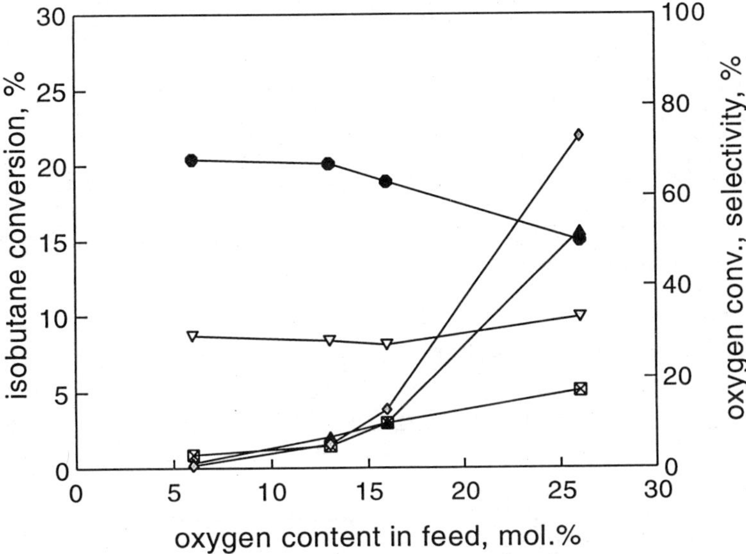

Figure 7. Isobutane and oxygen conversions and selectivities to the products as functions of the oxygen content in the feed for the $K_6P_2W_{18}O_{62}$ catalyst. Reaction conditions: T 470°C, residence time 1 s, feedstock composition: 26% isobutane; 12% water; balance helium. Symbols are the same as in Figure 2.

isobutane or oxygen clearly increases the concentration of the products, it is possible that the products (rather than isobutane) may change the catalyst surface to make it more active, thus yielding the observed autocatalytic effect. Water, a product of the reaction, was shown to not be playing this role, since water was also fed at low isobutane concentration, but a low activity was seen under these conditions.

In order to study the possible role played by the isobutene product, tests were done by co-feeding 8% isobutane and 2% isobutene, together with oxygen and water, at 470°C and a residence time 1 s. The results reported in Table I indicate that, under these conditions and in the presence of isobutene, the isobutane was more reactive than it was in the absence of added isobutene, but that primarily carbon oxides were formed (for comparison, the results obtained when oxidizing only 8% isobutane and only 2% isobutene are also given). This effect might be due either to a modification of the surface induced by the presence of the gas-phase olefin, with the formation of sites active in isobutane oxidation, or to the ignition of gas-phase homogeneous reactions due to the presence of the more labile, allylic C-H bond in isobutene. An analogous activation effect was obtained when co-feeding 2% propylene and 8% isobutane; in this latter test isobutane was more reactive, but with the formation of only carbon oxides, and no isobutene.

Table I. Results obtained by co-feeding isobutane and isobutene.

Hydrocarbon in feed, mol.%	*CO_x in the outlet, mol.%*	*iC_4H_8 in the outlet, mol.%*
8% iC_4H_{10} + 2% iC_4H_8	2.4	1.8
8% iC_4H_{10}	0.02	0.05
2% iC_4H_8	0.6	1.85

Reaction conditions: T 470°C, residence time 1 s, O_2 13 mol.%, H_2O 12 mol.%, balance He.

In order to define more clearly the role of the catalyst redox level, the calcined catalyst was first reduced by treatment in a diluted H_2 flow at 400°C, and then subjected to the reaction environment. The effect of the isobutane content in feed on the catalytic activity is displayed in Table II.

Table II. Reactivity of the $K_6P_2W_{18}O_{62}$ catalyst prereduced in H_2/He

iC_4H_{10} in feed, mol.%	*iC_4H_{10} conv./ O_2 conv., mol.%*	*Sel.i-C_4H_8, mol.%*	*Sel.CO_x, mol.%*	*Sel.others, mol.%[a]*
13	1.9/10.0	0	88	13
26	2.3/15.2	39	57	4

Reaction conditions: T 470°C, residence time 1 s, O_2 13 mol.%, H_2O 12% mol.%, balance He.

[a]Others are propylene, and traces of methacrolein and methacrylic acid.

The reduced catalyst was more active than the untreated one at low hydrocarbon concentration, but was also characterized by a very low selectivity to the isobutene. Even though the pre-reduced catalyst soon re-equilibrates in the reaction environment, the results obtained point out the role of the surface degree of oxidation. This is also in agreement with the results obtained by co-feeding the isobutane and an olefin (either isobutene or propylene), which led to an increase in the conversion of isobutane, but with the exclusive formation of carbon oxides.

Figure 8 shows the UV-Vis Diffuse Reflectance spectra of: (*a*) the $K_6P_2W_{18}O_{62}$ sample after calcination, (*b*) the Keggin $K_3PW_{12}O_{40}$ sample after calcination, (*c*) the $K_6P_2W_{18}O_{62}$ sample after reaction under conditions at which the catalyst showed very poor activity (8% isobutane in feed), (*d*) the $K_6P_2W_{18}O_{62}$ sample after reaction under conditions at which the catalyst was active (42% isobutane in feed), and (*e*) the $K_6P_2W_{18}O_{62}$ sample after pre-reduction treatment in diluted hydrogen. Also shown is the UV-Vis spectra (*f*) of the Keggin catalyst after reaction. In all cases the catalyst was unloaded after cooling in helium, in order to avoid possible modifications in the surface average redox level during cooling. Samples (*a*) and (*c*) were pale yellow, while sample (*d*) was light grey.

The spectra indicate that the main difference between the spent "active" Wells-Dawson catalyst (*d*) and the "inactive" catalyst (*c*) is an increase of the absorption in the range 450-900 nm which is observed in the former sample, and which disappears by reoxidation of the catalyst. This absorption is very intense in the Keggin compound after reaction (*f*), which is more active than the Wells-Dawson catalyst.

Octahedral W^V is known to exhibit an absorption centered at around 700 nm (*9*); this absorption is seen, for instance, in spectrum (*e*) relative to the sample reduced in hydrogen, while it is not present in spent samples. In addition, catalysts after reaction did not exhibit the blue color typical of reduced heteropolyblues. However, when the oxidized catalyst was subjected to a reducing treatment in a diluted H_2 stream with increasing temperature, and the corresponding UV-Vis spectra were recorded, it was found that in correspondence with the progressive reduction of tungsten, the spectrum first exhibited an increase in the absorption at wavelengths higher than 450 nm, then only at temperatures higher than 350°C the absorption centered at 700 nm appeared. In addition, a change in the relative intensities of the bands in the 250 to 450 nm range occurred. During heating, the sample turned from the original pale yellow colour to a grey-yellow colour (after reduction at temperatures higher than 150°C but lower than 300°C), to a pale green and, only after prolonged reduction at 400°C, the compound was definitely blue. This behavior is likely due to the fact that the first electron(s) furnished to the compound are delocalized throughout the structure; therefore, W^V entities are only formed when an energetic barrier to electron delocalization is established (*9*), that is for high reduction degrees. These data support the hypothesis that one difference between the "active" and the "inactive" catalyst is the extent of surface reduction which develops under the different operating conditions.

Another aspect concerns the significant differences in the charge-transfer spectral region between the calcined Keggin and Wells-Dawson samples (samples (*a*) and (*b*)), which can be tentatively assigned to differences in the coordination sphere of tungsten at the surface of the crystallites.

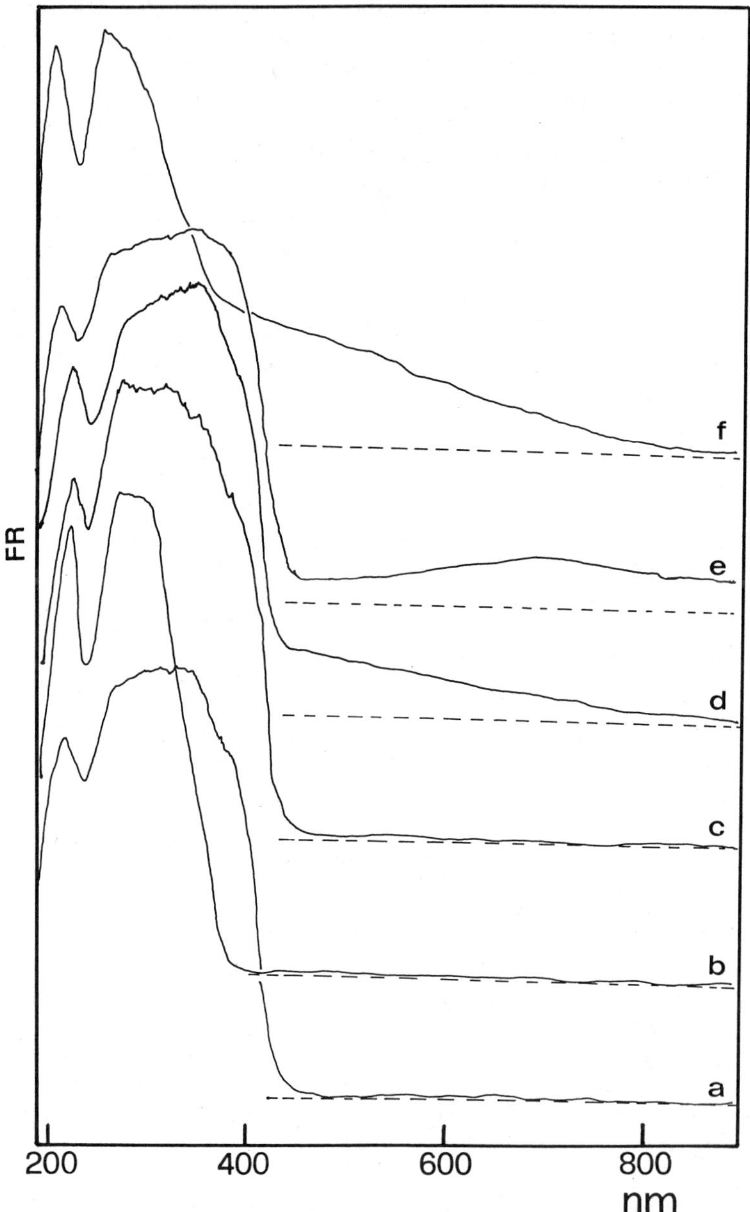

Figure 8. UV-Vis Diffuse Reflectance spectra of fresh and used catalysts. See the text for further explanation.

2) A second possibility to explain the observed autocatalytic effect is that an increase in isobutane concentration might favor the formation of carbonaceous substances, which may act themselves as active sites. However, this hypothesis can be ruled out due to the fact that an increase in activity is also observed on increasing the oxygen partial pressure (Figure 7), conditions which should greatly hinder formation of active surface carbon.

3) A third possible explanation is that surface-initiated, homogeneous, radical reactions may occur at high isobutane and high oxygen concentration, analogously to what has been reported in the case of methane oxidative coupling and of ethane and propane oxydehydrogenation to the corresponding olefins (*10-13*). In this hypothesis, after the activation of the hydrocarbon by the catalyst, the radical intermediate can be ejected into the gas phase and be converted to the olefin via a radical chain-reaction, following well-known mechanisms (*14*):

$$(CH_3)_3CH + cat \longrightarrow [(CH_3)_3C\bullet]_{ads} + cat\text{-}H$$
$$[(CH_3)_3C\bullet]_{ads} \longrightarrow (CH_3)_3C\bullet$$
$$(CH_3)_3C\bullet + O_2 \longrightarrow (CH_3)_2C{=}CH_2 + HO_2\bullet$$
$$HO_2\bullet + (CH_3)_3CH \longrightarrow HOOH + (CH_3)_3C\bullet$$
$$HOOH \longrightarrow 2\ OH\bullet$$
$$(CH_3)_3CH + OH\bullet \longrightarrow (CH_3)_3C\bullet + H_2O \text{ etc.}$$

These reactions sometimes exhibit an order higher than 1 with respect to the hydrocarbon. In this explanation for the observed autocatalysis, at low hydrocarbon concentration there is insufficient isobutane in the gas phase to sustain the homogeneous chain reaction, and the mechanism is mainly heterogeneous; under these particular conditions, the catalyst shows, therefore, a poor activity. Moreover, surface-initiated, homogeneous radical-type mechanisms in light paraffins oxydehydrogenation are known to be more selective than completely heterogeneous mechanisms (*10,13*). The autocatalytic effect observed might also be due to a contribution of isobutene in the chain-propagation, due to the enhanced reactivity of the weaker allylic C-H bond (86 kcal/mol) with respect to the stronger C-H bond in isobutane (91 kcal/mol).

In summary, the differences observed between the Keggin and the Wells-Dawson catalyst may be due not only to the different surface area, but also to differences in the reactivity of tungsten. The Keggin compound is characterized by an higher activity and lower selectivity, its catalytic performance is less sensitive to variations in the feedstock composition, and the operating mechanism is also less dependant upon the reaction conditions. On the other hand, different surface properties may lead to a more selective mechanism, as in the case of the Wells-Dawson compound.

The proposed working mechanism of isobutane oxidation on $K_6P_2W_{18}O_{62}$ and $K_3PW_{12}O_{40}$ catalysts

The data obtained indicate that the operating mechanism is a sensitive function of the reaction conditions; in addition, it is likely that more than one of the possibilities

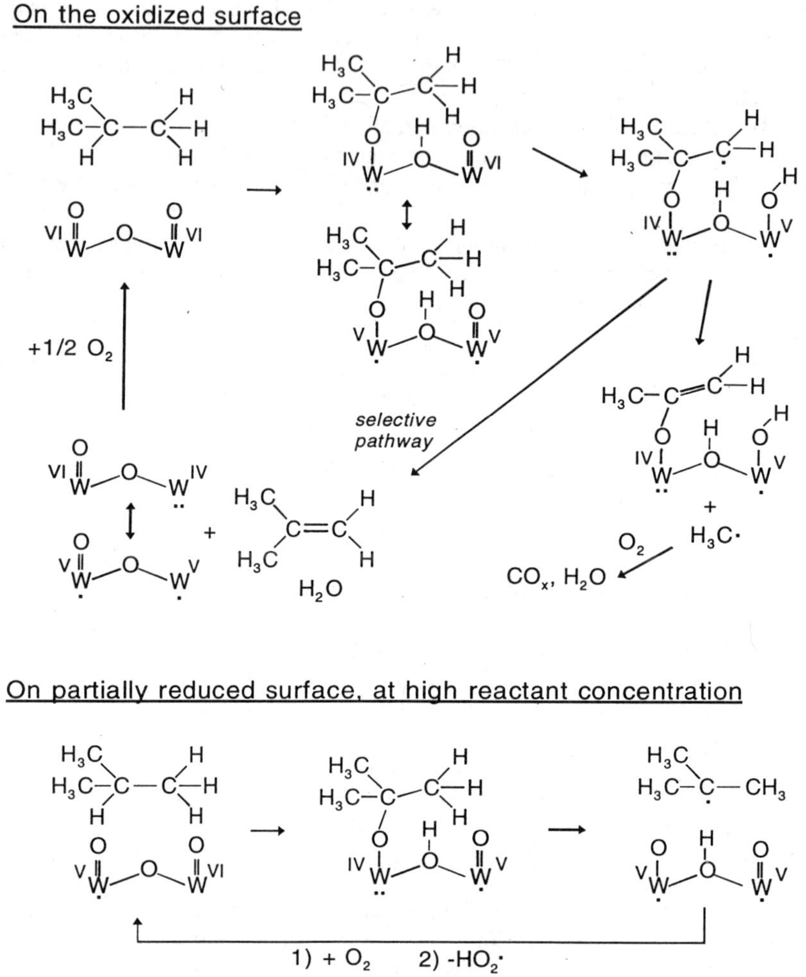

On the oxidized surface

+1/2 O_2

selective pathway

H_2O

O_2

CO_x, H_2O

$H_3C\cdot$

On partially reduced surface, at high reactant concentration

1) + O_2 2) -$HO_2\cdot$

$$(CH_3)_3C\cdot \xrightarrow{O_2} (CH_3)_3COO\cdot \longrightarrow (CH_3)_2C{=}CH_2 + HO_2\cdot$$

$$(CH_3)_3CH + HO_2\cdot \longrightarrow H_2O_2 + (CH_3)_3C\cdot$$

$$H_2O_2 \longrightarrow 2\ OH\cdot;\quad (CH_3)_3CH + HO\cdot \longrightarrow H_2O + (CH_3)_3C\cdot \quad \text{etc.}$$

Figure 9. Proposed working mechanism for the oxydehydrogenation of isobutane to isobutene.

above reported could contribute to the observed behavior. Figure 9 shows the proposed reaction mechanism, one that is summarized in the discussion which follows. This mechanism is proposed with insights from solution spectroscopic studies of reduced, oxygen-deficient forms of heteropolyoxomolybdates and tungstates (the so-called "heteropolyblues") (*15-17*), which generally show adjacent M-O-M sites (M=W, Mo, V and others) in close electronic communication (*15-19*).

We propose that isobutane is activated on the surface through the formation of an adsorbed species which, according to literature indications (*20*), may be an alkoxy species. The extent of catalyst reduction affects the activity of the surface sites responsible for isobutane activation: a partially reduced surface leads to more active sites, but the reaction is less selective, possibly due to a lower availability of surface oxygen, which in turn does not allow the selective transformation of the adsorbed intermediate into the olefin. Under these conditions the intermediate evolves towards the formation of carbon oxides, possibly by oxidative attack from gaseous oxygen. This is confirmed by the performance of the pre-reduced compound, as well as by the results obtained by co-feeding isobutane and an olefin: paraffin conversion was increased but with formation of carbon oxides only. On the contrary, an oxidized surface is less active, but favours the heterogeneous transformation of the adsorbed intermediate species to isobutene.

At isobutane concentration lower than the 15-20%, the Wells-Dawson compound shows poor activity, much lower than that of the Keggin catalyst. This is due to the lower surface area and, likely, also to the different reducibility of the tungsten ion. Under these conditions the surface of the Wells-Dawson catalyst is oxidized (as it was after the calcination treatment); this oxidized surface is less active, but it is selective.

When the paraffin concentration (or the oxygen concentration) in feed is increased, the concentration of isobutene in the gas phase is also increased. Under these conditions, two main effects are likely obtained: i) on one hand, the higher reducing power of the isobutene makes the tungsten more reduced, thus yielding a more active catalyst; ii) on the other hand, the high hydrocarbon concentration in the gas phase makes the contribution of the homogeneous selective mechanism of isobutene formation the prevailing one. A reduced catalyst may initiate more radicals, increasing the amount of gas-phase homogeneous reactions. In other words, under increased paraffin concentration the overall activity (i.e., the step of isobutane activation) may be governed by the higher catalyst reduction degree, while the selectivity is mainly governed by the homogeneous gas-phase mechanism. The fundamental role of isobutene (which may also participate in the gas phase mechanism, due to its more labile C-H bond) yields the observed autocatalytic effect.

Acknowledgements. MURST and CNR (Progetto Strategico Tecnologie Chimiche Innovative) are gratefully acknowledged for financial support.

Literature Cited

1 Misono, M. *Catal. Rev. Sci. Eng.* **1987**, *29*, 269.
2 Ono, Y. In *Perspectives in Catalysis*; Thomas, J.M., Zamaraev, K.I., Eds.; Blackwell Scientific Publ.: Oxford, 1992; p. 431.

3 Mizuno, N.; Lyon, D. K.; Finke, R. G. *J. Catal.* **1991**, *128,* **84**.
4 Mansuy, D.; Bartoli, J. F.; Battioni, P.; Lyon, D. K.; Finke, R. G. *J. Am. Chem. Soc.* **1991**, *113,* 7222.
5 Kung, H. H. *Adv. Catal.* **1994**, *40*, 1.
6 Cavani, F.; Etienne, E.; Favaro, M.; Galli, A.; Trifirò, F.; Hecquet, G. *Catal. Lett.* **1995**, *32*, 215.
7 Yamamatsu, S.; Yamaguchi, T. Eur. Patent 425,666 A1, 1989, assigned to Asahi Chem. Co.
8 Lyon, D. K.; Miller, W. K.; Novet, T.; Domaille, P. J.; Evitt, E.; Johnson, D. C.; Finke, R. G. *J. Am. Chem. Soc.* **1991**, *113,* 7209.
9 Pope, M. T. *Heteropoly and Isopoly Oxometalates*; Springer Verlag; New York; 1983.
10 Cavani, F.; Trifirò, F.; *Catal. Today* **1995**, *24*, 307.
11 Burch, R.; Crabb, E. M. *Appl. Catal. A: General* **1993**, *100*, 111.
12 Burch, R.; Chalker, S.; Loader, P.; Thomas, J.; Ueda, W. *Appl. Catal., A: General* **1992**, *82*, 77.
13 Kennedy, E. M.; Cant, N. W. *Appl. Catal.* **1991**, *75*, 321.
14 Sheldon, R. A.; Kochi, J. K. *Metal-Catalyzed Oxidations of Organic Compounds*; Academic Press; New York; 1981.
15 Pope, M. T., In *Mixed-valence Compounds*; Brown, D. B., Ed.; D. Reidel Publishing: Dordrecht, 1980; p. 365.
16 Prados, R. A.; Pope, M. T. *Inorg. Chem.* **1976**, *15*, 2547.
17 Kazansky, L. P.; Fedotov, M. A. *J. Chem. Soc. Chem. Commun.* **1980**, 644.
18 Piepgrass, K.; Pope, M.T. *J. Am. Chem. Soc.* **1987**, *109*, 1586.
19 Kawafune, I. *Chem. Lett.* **1989**, 185.
20 Finocchio, E.; Busca, G.; Lorenzelli, V.; Willey, R. J. *J. Catal.* **1995**, *151*, 204.

Chapter 11

Factors Determining the Selectivity in the Oxidative Dehydrogenation of Propane over Boria–Alumina Catalysts

O. V. Buyevskaya[1,3], M. Kubik[1], and M. Baerns[2]

[1]Lehrstuhl für Technische Chemie, Ruhr-Universität Bochum, Universitätstrasse 150, D–44780 Bochum, Germany
[2]Institut für Angewandte Chemie, Berlin-Adlershof e.V., Rudower Chaussee 5, D–12484 Berlin, Germany

Boria-alumina catalysts are active in the selective oxidative dehydrogenation of propane. For undoped Al_2O_3, carbon oxides were the main products while the addition of 30 wt% B_2O_3 led to a significant increase in propene selectivity; propene yields of 18 % were achieved at 823 K. Pulse experiments with C_3D_8 in the temporal-analysis-of-products reactor and spectroscopic studies showed that an addition of boria in the amount of less than a monolayer eliminates hydroxyl groups of γ-alumina involved in H-D exchange with propane. The suppression of the dissociative adsorption of propane resulted in a decrease of catalyst activity but did not improve propene selectivity. Oxygen isotopic exchange between gas-phase $^{18}O_2$ and surface ^{16}O was used to study the role of oxygen activation in the overall process; only for high boria loading no oxygen exchange products were detected. The suppression of both, dissociative adsorption of propane and oxygen, was found to be necessary for the improvement of propene selectivity in the oxidative dehydrogenation of propane over boria-alumina catalysts. It was assumed that higher propene selectivity can be achieved by secondary reactions of propyl radicals formed on the surface after their release into the gas phase due to diminished surface oxidation.

Boria-alumina catalysts were originally proposed by Murakami et al. (*1*) for the partial oxidation of ethane. Only low yields of acetaldehyde in the order of 1.0 % were obtained; the ethylene yield, however, amounted to 14.6 %. Recently, we have found

[3]Current address: Institut für Angewandte Chemie, Berlin-Adlershof e.V., Rudower Chaussee 5, D–12484 Berlin, Germany

0097–6156/96/0638–0155$15.00/0

(*2, 3*) that this type of catalyst is also active in the oxidative dehydrogenation of propane. Propene yields of up to 18 % achieved on B_2O_3(30 wt %)/Al_2O_3 are close to the best results for the oxidative dehydrogenation of propane summarized in the recent review of Kung (*4*). In this work, we present further results on catalytic activity as well as on surface and bulk characterization of boria-alumina catalysts. Primary steps of propane and oxygen interaction with catalytic surfaces were studied by pulsing C_3D_8 and $^{18}O_2$ in the temporal-analysis-of products (TAP) reactor. Finally, it is the aim of this study to elucidate the reaction pathways for selective and non-selective product formation.

Experimental

B_2O_3/Al_2O_3 catalysts with boria contents of 0.17, 15 and 30 wt % and KOH(0.27 wt %)/γ-Al_2O_3 were prepared by impregnation of γ-Al_2O_3 (83.8 m^2/g) with boric acid or KOH dissolved in warm water. After stirring for 1 h the excess of water was evaporated; the remainding solid was dried overnight at 383 K and calcined at 873 K for 14 hours. A B_2O_3 (25.5 wt %)/γ-Al_2O_3 catalyst was prepared following the procedure described above; the dried precursor was calcined at 1073 K according to (*5*) in order to obtain the 2 $Al_2O_3 \cdot B_2O_3$ phase; γ-Al_2O_3 was pretreated at 873 K for 14 hours before catalytic testing.

For the catalytic experiments a fixed-bed reactor equipped with on-line gas chromatography was used. It was operated at ambient pressure and T = 823 K using 1 g of catalyst and flow rates from 20 to 60 ml/min; p(C_3H_8) and p(O_2) were varied from 10 to 40 and 20 to 50 kPa, respectively.

Surface areas were measured by the 1-point B.E.T. method. Phase compositions were determined by X-ray diffraction using monochromatic copper K_α radiation. For IR measurements a Unicam RS FTIR spectrometer with a commercial diffuse reflectance accessory (Spectra Tech, model 003-102) was used. After the pretreatment of the powdered catalyst at 773 K in a continous flow of nitrogen (V_{STP}= 10 ml/min) the spectra of the catalysts were recorded at 823 K.

The interaction of C_3D_8, C_3H_8 and $^{18}O_2$ with the catalytic metal oxide surfaces (γ-Al_2O_3; B_2O_3/γ-Al_2O_3; KOH/γ-Al_2O_3) was studied by pulse experiments *in vacuo* (P = 10^{-4} Pa) in the TAP reactor (*6*) in the temperature range from 753 K to 843 K using 0.2 g of catalyst. The catalyst was first exposed to oxygen pulses at reaction temperature and kept under vacuum for 10 min before the pulse experiments. A $C_3D_8(C_3H_8)$/Ne = 1:1 or $^{18}O_2$/Ne = 1:1 mixture was pulsed over the catalyst; pulse sizes amounted to $(1 \div 7) \cdot 10^{15}$ molecules/pulse.

Results and Discussion

Characterization. Results on catalyst characterization are summarized in Table I. The addition of boria to alumina resulted in a decrease in surface area from 83.8 m^2/g (γ-Al_2O_3) to 74.9 and 29.4 m^2/g for boria contents of 15 and 30 wt %, respectively. In the sample with low content of boria (0.17 wt %) only the presence of a γ-Al_2O_3 (JCPDS 37-1462) phase was detected by means of X-ray diffraction. In the catalysts with boria loadings of 15 and 30 wt % the diffraction patterns identified B_2O_3 (JCPDS 6-297) in addition to γ-Al_2O_3. For the sample containing 30 wt % B_2O_3, the formation of the 2 $Al_2O_3 \cdot B_2O_3$ phase (JCPDS 29-10) was additionally observed. This phase was detected as a major phase in the B_2O_3 (25.5 wt %)/γ-Al_2O_3 catalyst calcined at 1073 K. XRD patterns of all samples are, however, diffuse in nature indicating that the presence of additional amorphous phases cannot be excluded.

Table I. Catalysts used and their characterization

Catalyst	S_{BET}, m^2/g	Phases identified by XRD
γ-Al_2O_3	83.8	γ-Al_2O_3
B_2O_3 (0.17 wt %)/γ-Al_2O_3 *)	83.9	γ-Al_2O_3
B_2O_3 (15 wt %)/γ-Al_2O_3	74.9	B_2O_3; γ-Al_2O_3
B_2O_3 (25.5 wt %) /γ-Al_2O_3	62.1	2 $Al_2O_3 \cdot B_2O_3$ (major); B_2O_3; γ-Al_2O_3
B_2O_3 (30 wt %) /γ-Al_2O_3	29.4	B_2O_3; γ-Al_2O_3; 2 $Al_2O_3 \cdot B_2O_3$
KOH(0.27 wt %)/γ-Al_2O_3 *)	81.0	not examined

*) The amount of B and K corresponds to about 3.5 % of a hypothetical monolayer when assuming 10^{19} atoms/m^2.

Catalytic Results. The effect of the addition of 0.17 wt % boria to alumina on the catalytic properties is illustrated by results given in Table II. A significant decrease in the propane conversion was observed on the doped catalyst compared to pure

Table II. Results on catalytic activity of B_2O_3(0.17 wt %)/γ-Al_2O_3 and of γ-Al_2O_3 (T = 823 K, τ = 2 g·s/ml, p(C_3H_8) = 10 kPa, p(O_2) = 20 kPa, N_2 - balance)

Catalyst	Conversion, %		Selectivity, %				
	C_3H_8	O_2	C_3H_6	C_2H_4	C_3H_4O	CO	CO_2
γ-Al_2O_3	42.2	76.5	11.4	4.6	0	47.3	36.6
B_2O_3(0.17wt %)/γ-Al_2O_3	24.8	41.1	18.8	3.4	0	44.0	33.9

alumina. It is difficult to compare product selectivities because of different degrees of propane conversion. A slight increase in propene selectivity on B_2O_3(0.17 wt %)/ γ-Al_2O_3 compared to undoped alumina can be due to lower propane conversion. In

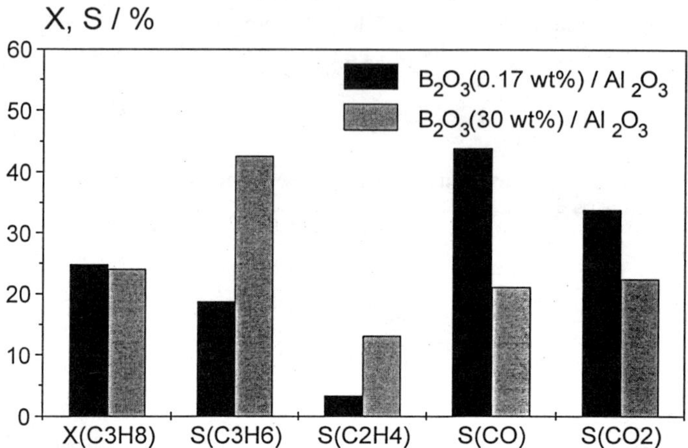

Figure 1. Propane conversions and product selectivities over B_2O_3 (0.17 wt %)/ γ-Al_2O_3 and B_2O_3 (30 wt %)/γ-Al_2O_3 catalysts (T = 823 K, $p(C_3H_8)$ = 10 kPa, $p(O_2)$ = 20 kPa).

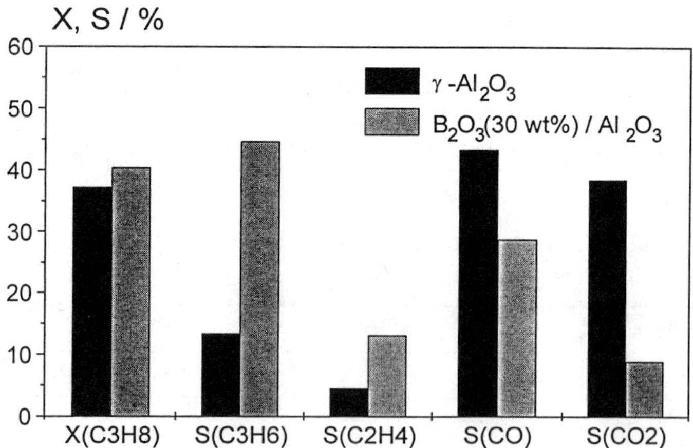

Figure 2. Propane conversions and product selectivities over γ-Al_2O_3 and B_2O_3 (30 wt %)/γ-Al_2O_3 catalysts (T = 823 K, $p(C_3H_8)$ = 25 kPa, $p(O_2)$ = 50 kPa).

any case no significant suppression of total oxidation was observed with an addition of a low amount of boria.

Product distribution on B_2O_3/γ-Al_2O_3 catalysts was significantly affected by boria loading. Selectivities towards C_3H_6, C_2H_4 and CO_x obtained over a catalyst with a high content of B_2O_3 (30 wt %) are given in Figure 1 in comparison with a sample containing 0.17 wt % of B_2O_3 at similar values of propane conversion (X = 24.1 - 24.8 %). With increasing boria content C_3H_6 selectivity increased from 18.8 to 42.6 % coinciding with a decrease in CO_x selectivity from 77.9 to 43.6 %.

For higher degrees of propane conversions (37.2 - 40.4 %) the comparison of product selectivities is given in Figure 2 for undoped γ-Al_2O_3 and B_2O_3(30 wt %)/γ-Al_2O_3. In accordance to the catalytic results presented in Figure 1 it is obvious that high amounts of boria on alumina had a positive effect on the formation of the selective products e.g. propene and ethylene, for a wide range of propane conversions. The total oxidation of propane and its dehydrogenation products to CO_x was significantly diminished.

As derived from XRD data, the formation of a new phase, i.e. 2 $Al_2O_3{\cdot}B_2O_3$, was detected in the B_2O_3(30 wt %)/Al_2O_3 catalyst. An attempt was made to synthesize this phase in its pure form and to examine its catalytic performance. Catalytic results ($X(C_3H_8)$ = 21.6 %; $S(C_3H_6)$ = 16.9 %; $S(C_2H_4)$ = 3 % at T = 823 K, τ = 4 g·s/ml; $p(C_3H_8)$ = 10 kPa, $p(O_2)$ = 20 kPa) on a B_2O_3 (25.5 wt %)/γ-Al_2O_3 catalyst consisting of 2 $Al_2O_3{\cdot}B_2O_3$ as the major crystalline phase showed that this phase is a non-selective one and therefore cannot be considered as an active one for the formation of propene and ethylene on B_2O_3 (30 wt %) /γ-Al_2O_3.

Active Sites. From the results on the oxidative transformation of ethane over B-Al-O catalysts only little is known about the nature of the active sites. Mainly it was observed that an addition of boria to alumina effectively suppresses the formation of CO_x and that the formation of ethylene increases with increasing the content of B_2O_3 (*1, 7*). Since the boria-alumina catalysts lack redox-properties, catalyst activity should be determined to a large extent by surface acidity/basicity. Although TPD-measurements of acidity/basicity of B_2O_3(0-30 wt %)/Al_2O_3 using NH_3, pyridine and SO_2 as probe molecules were performed by Colorio et al. (*6*) no definite correlation between the catalytic properties and the acidity in the selective transformation of ethane to ethylene was derived. In the oxidative dehydrogenation of propane, the abstraction of a H atom from the propane molecule with the formation of an adsorbed propyl radical and surface OH group was found to be the primary step on V-Mg-O catalysts (*8*). For the oxidation of low alkanes on V-containing catalysts, it was shown by Kung (*4*) that the selectivities can be explained by the probability of surface alkyl species reacting with a reactive surface oxygen. Additionally, Corma et al. (*9*) showed that on V-Mg-O catalysts selectivity is determined by the acid-base character of the lattice oxygen present on the surface: the lower the nucleophilicity of the oxygen species, the higher the propene selectivity.

It should be noted that studying the nature of the acidic sites on boria-alumina by means of IR-spectra of adsorbed pyridine and ammonia failed due to strong absorption of boria itself in the region from 1100 to 1700 cm^{-1} masking any adsorbate

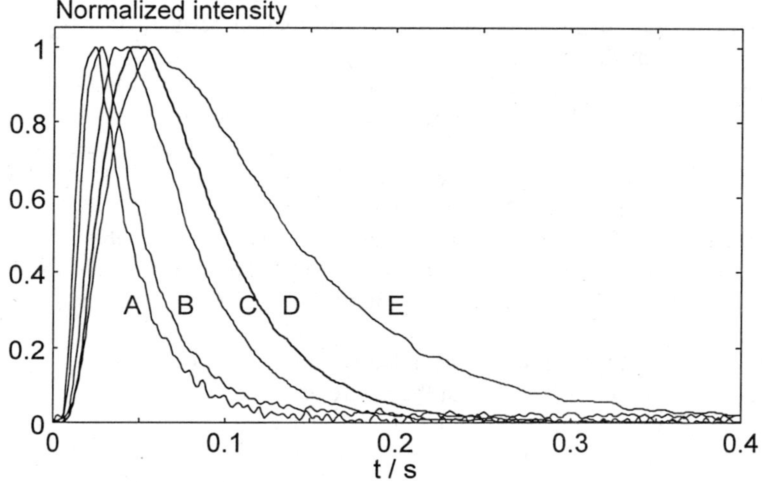

Figure 3. Normalized responses of propane isotopes (curve A - C_3D_8 ; curve B - C_3D_7H ; curve C - $C_3D_4H_4$; curve D -$C_3D_3H_5$; curve E - C_3H_8) when pulsing C_3D_8 over γ-Al_2O_3 at 803 K.

peaks. Thus, the changes in Lewis and Brønsted acidity caused by boria doping could not be elucidated by this conventional method. Therefore, H-D exchange reaction of propane (C_3D_8) with surface OH groups was used for elucidating the changes in surface properties of Al_2O_3 by addition of B_2O_3.

As shown in Figure 3, during pulsing of deuterated propane over γ-Al_2O_3 at 803 K, a sequence of the response signals for the various isotopes of propane was detected leading to the assumption of a consecutive pathway of formation: $C_3D_8 \rightarrow C_3D_7H \rightarrow ... \rightarrow C_3H_8$

It was proved by pulsing C_3H_8 that no conversion of propane to propene and CO_x occured over γ-Al_2O_3 at 803 K. Thus, conversion of deuterated propane ($X(C_3D_8)$ = 92 %; T = 803 K, $4 \cdot 10^{15}$ molecules/pulse) was only due to H-D exchange with surface hydroxyl groups and the measured m/e did not arise from propene or CO_x. When performing the same pulse experiments over B-doped catalysts no formation of isotopes of propane and other products, e.g., propene and CO_x, were detected when C_3D_8 was pulsed over B_2O_3(0.17 wt %)/γ-Al_2O_3 and B_2O_3(30 wt %)/γ-Al_2O_3 in the temperature range between 753 and 843 K. These results showed that addition of only 0.17 wt % of B_2O_3 corresponding to less than monolayer surface coverage of Al_2O_3 resulted in the complete elimination of hydroxyl groups involved in the reaction of H-D exchange with propane. The dissociative adsorption of propane which effectively occurs on the γ-alumina surface is suppressed by a very low B_2O_3 loading.

The changes in the surface composition of Al_2O_3 caused by addition of B_2O_3 can be monitored by means of IR spectroscopy. IR absorption bands between 3700 and 3800 cm^{-1} were assigned to different types of isolated hydroxyl groups bonded to the surface of alumina (*10,11*) while a band at 3200 cm^{-1} was reported for OH stretching vibrations in $B(OH)_3$ (*12*). IR spectra of γ-Al_2O_3 and B_2O_3/γ-Al_2O_3 catalysts at 823 K are shown in Figure 4. There was no significant change in the intensity of the main band of alumina at 3672 cm^{-1} on B_2O_3(0.17 wt%)/γ-Al_2O_3 while the shoulders above 3700 cm^{-1} which can be assigned to isolated OH groups of alumina (*10*) were not observed. Furthermore, no additional band at 3200 cm^{-1} corresponding to $B(OH)_3$ was detected in this catalyst. For B_2O_3(30 wt%)/γ-Al_2O_3, no absorption bands corresponding to OH groups of alumina (υ: 3500 - 3800 cm^{-1}) were observed; the only detected band (υ = 3204 cm^{-1}) was due to $B(OH)_3$.

From transient and spectroscopic studies it can be concluded that indeed the fraction of "active" OH groups taking part in the H-D exchange reaction is rather low; these results are in agreement with findings of Larson and Hall (*13*) who showed that isotopic exchange between CH_4 and CD_4 over alumina occured via hydroxyl groups although a small fraction of them were involved. The exchange behaviour of acyclic alkanes with deuterium over γ-alumina has also been investigated by Robertson et al. (*14*). The mechanism is believed to involve the heterolytic dissociation of the C-H bond with the formation of carbanionic intermediates (*14*).

The effect of elimination of „active" hydroxyl groups on the catalytic performance is illustrated by results on propane oxidation over γ-Al_2O_3 and B_2O_3(0.17 wt %)/γ-Al_2O_3 given in Table II. A significantly low propane conversion was obtained

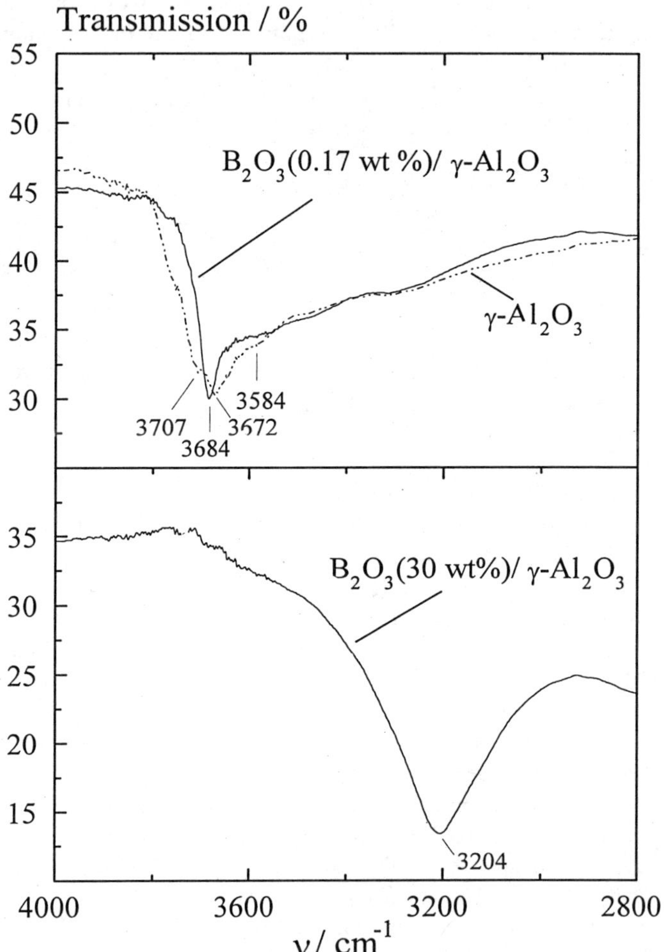

Figure 4. Transmission FT-IR spectra of γ-Al_2O_3, B_2O_3(0.17 wt%)/γ-Al_2O_3 and B_2O_3(30 wt%)/γ-Al_2O_3 at 823 K after calcination in N_2 flow for 1 h at 773 K.

over a doped catalyst compared to pure alumina under the same reaction conditions. Thus, the changes in the surface properties, e.g., surface acidity, due to elimination of part of the hydroxyl groups was accompanied by suppression of the dissociative adsorption of propane which led to a decrease in the activity towards propane conversion. Since the changes in the surface properties with the addition of a low boria amount was mainly related to the elimination of mobile hydrogen, as confirmed by pulse experiments with C_3D_8, a decrease in the Brønsted acidity can be assumed.

In order to clarify further the relationship between the catalyst activity in propane oxidation and reactivity of surface hydrogen in H-D exchange, a K-doped γ-Al_2O_3 catalyst was prepared and its activity was tested using both pulse experiments with C_3D_8 in the TAP reactor and a conventional mode of propane oxidation in a fixed bed reactor. Catalytic results on propane oxidation ($p(C_3H_8) = 10$ kPa, $p(O_2) = 20$ kPa, T = 823, $\tau = 2$ g·s/ml) showed some decrease in the propane conversion on the KOH/Al_2O_3 catalyst ($X(C_3H_8) = 36.8$ %) compared to undoped alumina ($X(C_3H_8)$ = 42.2 %); there was no significant change in the product selectivities. Pulsing C_3D_8 over KOH(0.27 wt %)/γ-Al_2O_3 at 803 K resulted in a low propane conversion of 4 %. The formation of C_3D_7H (see Figure 5) was detected as the only reaction product. Thus, a decrease in the reactivity of surface hydroxyl groups occured due to addition of K, but there was, however, no complete loss of activitity for the reaction of H-D exchange with C_3D_8. IR spectra presented in Figure 6 showed that a remarkable change in the range corresponding to OH stretching vibrations was observed on KOH/γ-Al_2O_3. For the K-doped catalyst, the main band at 3492 cm^{-1} can be assigned to KOH according to Buchanan (*15*) but the presence of isolated OH in alumina is shown by the peak at 3707 cm^{-1}.

From results described above it can be concluded that a catalyst shows high activity when propane activation occures via dissociative adsorption, but the formation of non-selective products are dominant in this case.

Oxygen Activation. For the further elucidation of surface properties which can determine the product selectivity, oxygen interaction with catalytic surfaces was studied by means of pulse experiments with $^{18}O_2$ in the TAP reactor. In this type of experiments the activity towards oxygen dissociative adsorption can be clarified using oxygen isotopic exchange between gas-phase $^{18}O_2$ and surface ^{16}O. Results on pulsing of $^{18}O_2$ over various catalysts are summarized in Table III. It should be noted that $^{18}O^{16}O$ was detected as the only product of the oxygen exchange reaction on all catalysts studied. Transient responses of $^{18}O_2$ pulsed and $^{18}O^{16}O$ formed are given in Figure 7 and 8 for γ-Al_2O_3 and B_2O_3(0.17 wt %)/γ-Al_2O_3, respectively.

As shown in Table III, there was no difference in the yield of exhange product $^{18}O^{16}O$ for the reaction of $^{18}O_2$ with surface oxygen of γ-Al_2O_3 and B_2O_3(0.17 wt%)/γ-Al_2O_3. Transient responses of isotopic oxygens on both catalysts (see Figure 7 and 8) were similar. Thus, the dissociative adsorption of oxygen was not suppressed by a low boria loading. As was mentioned above no improvement in selectivity to propene and ethylene was observed with an addition of 0.17 wt % B_2O_3. Contrary to results from the B-containing catalyst a significant increase in the

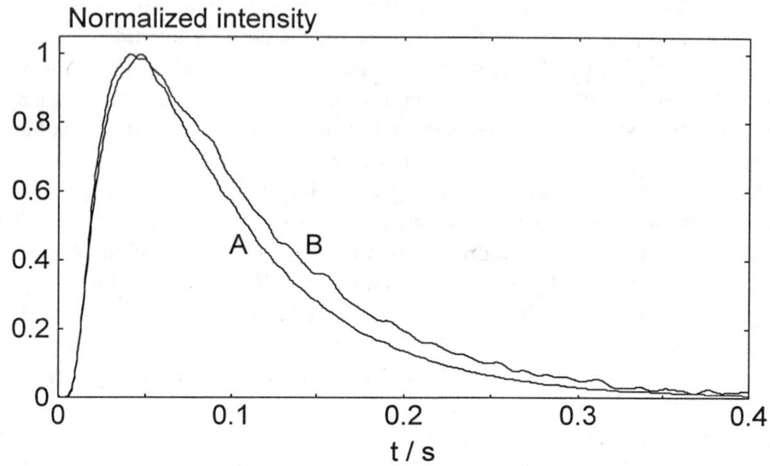

Figure 5. Normalized responses of propane isotopes (curve A - C_3D_8; curve B - C_3D_7H) when pulsing C_3D_8 over KOH(0.27 wt %)/γ-Al_2O_3 at 803 K.

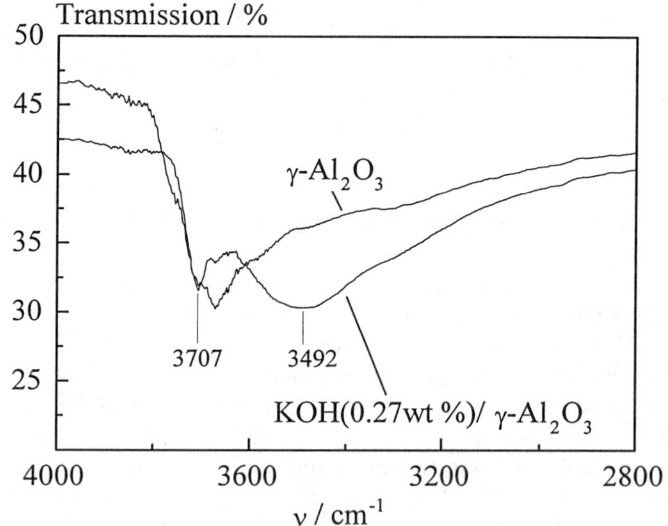

Figure 6. Transmission FT-IR spectra of γ-Al_2O_3, KOH(0.27 wt%)/γ-Al_2O_3 at 823 K after calcination in N_2 flow for 1 h at 773 K.

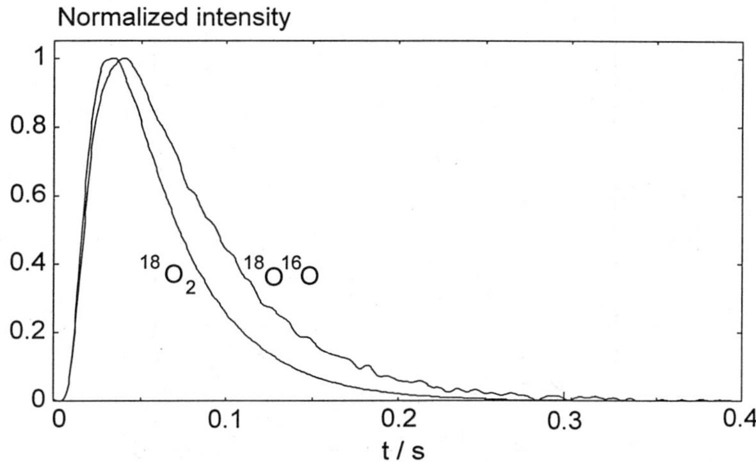

Figure 7. Normalized responses of oxygen isotopes while pulsing $^{18}O_2$ over γ-Al_2O_3 at 803 K.

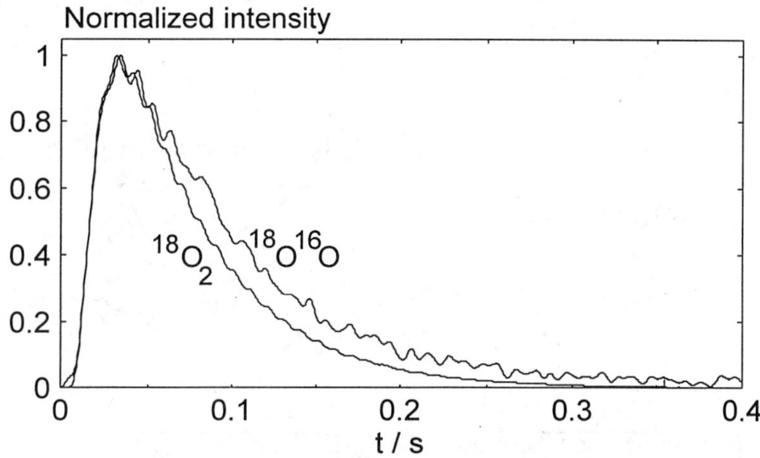

Figure 8. Normalized responses of oxygen isotopes while pulsing $^{18}O_2$ over B_2O_3(0.17 wt %)/γ-Al_2O_3 at 803 K

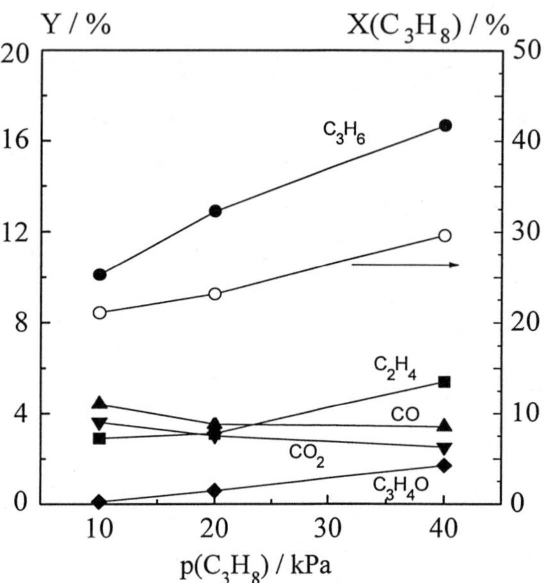

Figure 9. Effect of propane partial pressure on propane conversion and yields of products over B_2O_3(30 wt%/)Al_2O_3 (T= 823 K, τ=4 g·s/ml, $p(O_2)$ = 20 kPa).

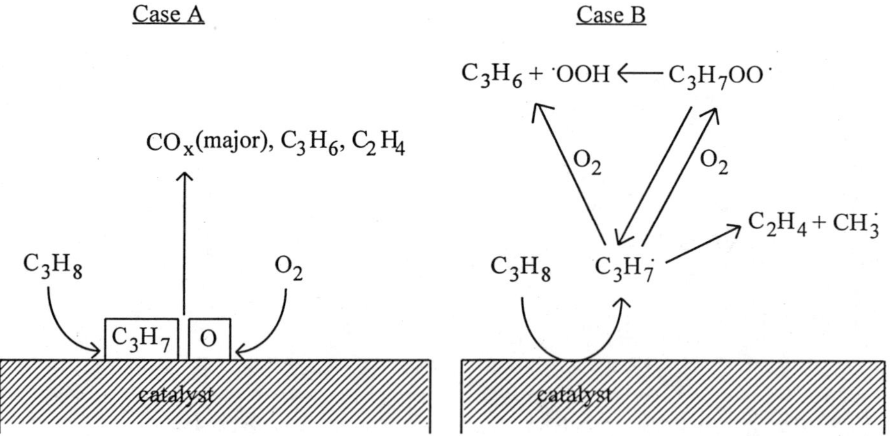

Figure 10. Probable reaction scheme for the oxidative dehydrogenation of propane.

Table III. Results on the oxygen exhange between gas-phase $^{18}O_2$ and surface oxygen ($^{16}O_s$) at different temperatures in the pulse experiments.

Catalyst	Yield of $^{18}O^{16}O$, %		
	753 K	803 K	843 K
γ-Al_2O_3	0.2	0.8	1.1
B_2O_3(0.17 wt %)/γ-Al_2O_3	0.2	1.0	1.2
B_2O_3 (30 wt %) /γ-Al_2O_3	0	0	0
KOH(0.27 wt %)/γ-Al_2O_3	1.8	11.7	15.0

activity towards oxygen exchange was observed on the KOH/γ-Al_2O_3 catalyst, which, however, did not showed high selectivity to propene. For the catalyst containing 30 wt % B_2O_3 with higher propene selectivity, no oxygen exchange was detected after pulsing $^{18}O_2$ over the whole temperature range, i.e. the dissociative adsorption of oxygen does not occur effectively enough to be detected under the conditions applied. As mentioned above, only hydroxyl groups corresponding to $B(OH)_3$ were detected on this catalyst. It is not clear whether the relationship between the nature of hydroxyl groups and catalyst activity for the oxygen exchange exists but present results showed that complete elimination of the OH groups on alumina coincided with a loss of activity towards oxygen dissociative adsorption.

Effect of p(C_3H_8). The effect of propane partial pressure was studied over the B_2O_3(30 wt %)/γ-Al_2O_3 catalyst at 823 K and p(O_2) = 20 kPa. Results showing propane conversion and yields of products are presented in Figure 9. An increase in p(C_3H_8) from 10 to 40 kPa resulted in a rise of C_3H_8 conversion from 21.1 to 29.6 % coinciding with an increase in the yields of propene (10.1 % $\rightarrow$ 16.7 %), ethene (2.9 % $\rightarrow$ 5.4 %) and acrolein (0.1 % $\rightarrow$ 1.7 %); this corresponds to an increase in selectivity of propene, ethene and acrolein and to a suppression of total oxidation. The yield of CO_x decreased from 8.0 % to 5.7 %. For the oxidation of propane in the absence of the catalyst under the same reaction conditions, propane conversion did not exceed 1.5 %. Thus, the effect of propane partial pressure which promotes both propane conversion and the formation of selective products cannot be ascribed to homogeneous gas-phase reactions only. Detailed kinetic analysis is required for the elucidation of the reaction pathways but from the present results it may be assumed that propane activation occurs at least partly via the formation of propyl radicals; a bimolecular reaction between a $C_3H_7O_2\cdot$and C_3H_8 with regeneration of the chain carrier $C_3H_7\cdot$suggested for the B-P-O catalyst (*16*) cannot be excluded for propane oxidation over B_2O_3(30 wt %)/γ-Al_2O_3 catalyst.

Reaction Mechanism. The above findings can be summarized in the reaction scheme given in Figure 10: the interaction of propane with catalytic surface can occur via the dissociative adsoption of propane (case A) which can lead, however, to the formation of mainly non-selective products. Most probably, dissociatively adsorbed oxygen takes part in the formation of CO_x while the nature of the oxygen species involved in the abstraction of hydrogen from propane with the formation of a propyl radical is an open question. The suppression of both dissociative adsorption of propane and oxygen was found to be necessary for the improvement of selectivity in the oxidative dehydrogenation of propane (case B). It can be assumed that higher propene selectivity can be achieved by secondary reactions of propyl radicals formed on the surface after their release into gas phase. Thus, the formation of selective products can be significantely improved by the suppression of surface oxidation.

Conclusions

The addition of B_2O_3 to γ-Al_2O_3 led to significant changes in the surface properties which affected both, catalyst activity and selectivity in the oxidatve dehydrogenation of propane. Pulse experiments with C_3D_8 and $^{18}O_2$ in the TAP reactor supported by spectroscopic studies were found to be suitable for elucidation of the reaction pathways which depended on boria loading. A propene yield of 18 % obtained on a 30 wt% B_2O_3/γ-Al_2O_3 catalyst was due to elimination of dissociative adsorption of both propane and oxygen. It can be tentatively assumed that a change in the surface acidity caused by the absence of isolated hydroxyl groups is one of the reasons for decreased catalyst activity observed in experiments on the catalyst with a low boria loading. An increase in selectivity towards propene with increasing boria content coincided with diminished surface oxidation due to suppression of oxygen dissociative adsorption.

Acknowledgments

This work was financially supported by German Federal Ministry for Education, Science, Research and Technology, BMFT Contract No. 03D0001A7.

Literature Cited

1. Murakami, Y.; Otsuka, K.; Wada, Y.; Morikawa, A. *Chem. Lett.* **1989**, 535.
2. Baerns, M.; Buyevskaya, O. V.; Kubik, M.; Maiti, G.; Ovsitser, O.; Seel, O. *210 ACS National Meeting,* August 20-24, **1995,** Chicago, oral presentation; to be published in Catal.Today.
3. Baerns, M; Bouevskaia, O.; Kubik, M.; Maiti, G. *German Pat. Appl. 19530454.3,* **1995**.
4. Kung, H. H. *Adv. Catal.* **1994**, *40*, 1.
5. Lehmann, H.-A.; Teske, K. *Z. Anorg. Allgem. Chem.* **1973**, *400*, 169.
6. Gleaves, J. T.; Ebner, J. R.; Kuechler, T.C. *Catal. Rev.-Sci. Eng.* **1988**, *30(1)*, 49.
7. Colorio, G. C.; Bonnetot, B.; Vedrine, J. C.; Auroux, A. *Stud. Surf. Sci. Catal.* **1994**, *82*, 143.
8. Chaar, M. A.; Patel, D.; Kung, M. C.; Kung, H. H. *J. Catal.* **1987**, *5*, 483.
9. Corma, A.; López Nieto, J. M.; Paredes, N. *J. Catal.* **1993**, *144*, 425.

10. Knözinger, H.; Ratnasamy, P. *Catal. Rev. -Sci. Eng.* **1978,** *17(1)*, 31.
11. Peri, J. B.; Hannan, R. B. *J. Phys.Chem.* **1960**, *64*, 1526.
12. Broadhead, P.; Newman, G. A. *Spectrochim. Acta A* **1972**, *28*, 1915.
13. Larson, J. G.; Hall, W. K. *J. Phys.Chem.* **1965,** *69*, 3080.
14. Robertson, P. J.; Scurrell, M. S.; Kemball, C. *J. Chem. Soc. Faraday Trans.1* **1975**, *71*, 903.
15. Buchanan, R. A., *J. Chem. Phys.* **1959**, *31*, 870.
16. Otsuka, K.; Uragami, Y.; Komatsu, T.; Hatano, M. *Stud. Surf. Sci. Catal.* **1991**, *61*, 15.

Chapter 12

The Oxidative Dehydrogenation of Propane over Molybdenum-Containing Catalysts

The Effect of Support Pretreatment on Catalytic Performance

Azra Yasmeen, Frederic C. Meunier, and Julian R. H. Ross[1]

Centre for Environmental Research, Department of Chemical and Environmental Science, University of Limerick, Limerick, Ireland

Previous work from this laboratory has shown that of a series of molybdena catalysts supported on various supports, molybdena supported on titania was the most promising catalyst for the oxidative dehydrogenation of propane and that the yields obtained with this material compared well with those of some of the best catalysts for this reaction reported in the literature. This paper reports work which showed that an even more effective catalyst was molybdena supported on an alumina which had been calcined at higher temperatures to reduce its total surface area. The effect of different calcination temperatures of the alumina support was examined in some detail and it was found that calcination at 1200°C gave the best results. It has been shown that a coverage at least equivalent to a monolayer is necessary for the optimum performance of the catalyst.

In a previous paper [1], we compared the behaviour of a series of catalysts consisting of molybdena supported on a variety of oxides for the oxidative dehydrogenation of propane to propene:

$$C_3H_8 + 1/2\ O_2 \rightarrow C_3H_6 + H_2O.$$

We found that of the materials made, the titania-supported catalyst was the most selective for comparable conversions of propane. We also found that the coverage of the support had to be more than a monolayer in order to obtain the optimum selectivity to propene. Addition of vanadium and niobium oxides to the titania-supported molybdenum oxide gave an increase in the activity of the resultant catalyst compared with that of the unpromoted material without any loss of selectivity.

The catalysts examined in the previous work referred to above were prepared using oxide supports which had a variety of different surface areas ranging from 12 to 211 m^2g^{-1} which had been calcined at either 450 or 650 °C. Hence, as the molybdena loading in all cases was maintained constant at approximately 5 wt%, the differences between the catalysts could have been due to different surface coverages by the molybdena rather than being an inherent property of the molybdena-support combination. The present paper therefore compares the results obtained with a series of molybdena-containing catalysts prepared using supports which had been sintered so that the surface areas were approximately constant. The data obtained show that of these materials, molybdena supported on alumina which had been calcined at 1200°C gave the highest yields of propene, the resultant yields being higher than those reported previously for molybdena on titania.

Further work has therefore been carried out to examine the use of alumina as a support for molybdena and this paper thus describes the effect of molybdena loading and of the effect of calcination temperature of the support, as well as the effect of the addition of vanadia and niobia to the

[1]Corresponding author

0097–6156/96/0638–0170$15.00/0

most selective Mo/Al_2O_3 catalyst. It is shown that a 5 wt% Mo/Al_2O_3 gives the optimum yield of propylene, slightly above 10%, under the conditions used for the measurements. This result compares very favourably with results obtained under the same conditions for a $NiMoO_4$ catalyst prepared according to the method described by Mazzochia et al. [2].

Experimental

Table I lists some of the catalysts used in this work. The supports used were titania (anatase, BDH), niobia (prepared by calcining hydrated niobia (HCST GmbH)), γ-alumina (Ketjen), silica (BDH), magnesia (BDH) and zirconia (monoclinic, Gimex).

Table I. Details of Mo-containing catalysts prepared with different supports

Catalyst	Temperature of calcination of support/°C	BET area of support $/m^2g^{-1}$	BET area of catalyst $/m^2g^{-1}$	Molybdenum content /wt% Mo
Mo/TiO_2	uncalcined	41	31.3	5.04
Mo/Nb_2O_5	630	12.9	9.2	5.22
Mo/Al_2O_3	1200	47	29.3	5.15
Mo/SiO_2	1150	44.5	22.4	4.42
Mo/MgO	uncalcined	58	88.0	5.4
Mo/ZrO_2	650	46.4	42.0	5.4

All the catalysts, except for Mo/TiO_2, were calcined at 650°C; Mo/TiO_2 was calcined at 450°C.

The alumina, silica and zirconia were calcined at the temperatures shown so that their surface areas were of the same order as those of the titania and magnesia. The niobia, for which the area was much lower than the other samples, was calcined at 630°C to ensure that the T modification was obtained. (When hydrated niobia is heated, it first forms the TT phase at above about 537°C and then transforms to the T phase at about 600°C which is stable up to about 1000°C at which temperature it forms the H-phase [3]; the T-phase is characterised by a doublet in the XRD pattern at d= 3.13Å.) The molybdena-containing materials were prepared by the incipient wetness technique using an aqueous solution of ammonium heptamolybdate (Rhone Poulenc, Normapur AR) maintained at 70°C, followed by drying overnight at 70°C and calcination at 650°C for two hours; an exception was the Mo/TiO_2 sample which was calcined at 450°C to avoid further sintering of the support. In all the cases shown in Table I, the Mo content was approximately 5 wt% (corresponding to about 7.5wt% MoO_3); in several cases, it was necessary to use a series of impregnation steps to attain this loading, chosen to be just above monolayer coverage [4].

Table II. Details of a series of 5 wt% $Mo\text{-}Al_2O_3$ catalysts prepared using a series of alumina supports calcined at different temperatures.

Catalyst	Temperature of calcination of support /°C	BET area of support $/m^2g^{-1}$	BET area of catalyst $/m^2g^{-1}$
Mo/Al_2O_3	900	173	110.2
Mo/Al_2O_3	1000	125.27	94.8
Mo/Al_2O_3	1200	47	29.3
Mo/Al_2O_3	uncalcined	211	189.0

All the catalysts were calcined at 650°C.

Table II gives details of a further series of Mo/Al_2O_3 catalysts which was prepared using alumina calcined at various temperatures; the Mo contents were adjusted so that the surface coverages were approximately the same as those of the samples shown in Table I, just above the monolayer

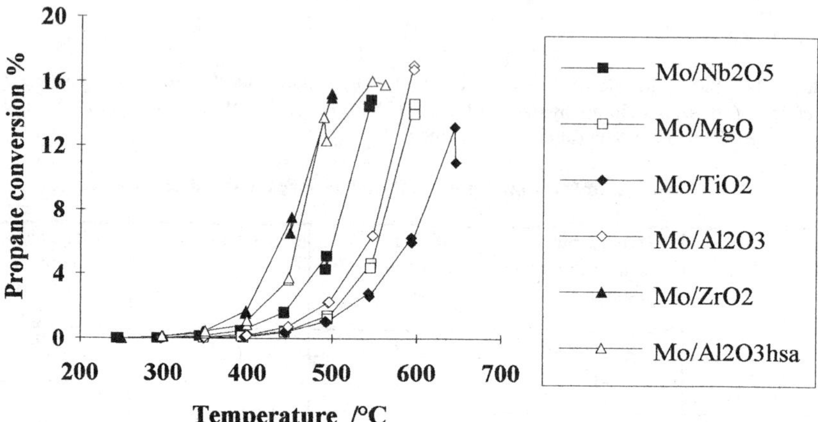

Figure 1. Plots of propane conversion versus reaction temperature for a series of molybdena catalysts (see Table I) supported on various oxides of low surface area.

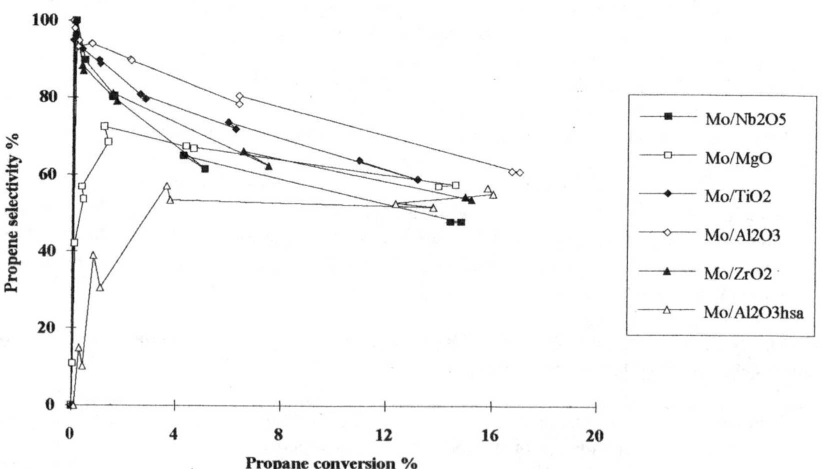

Figure 2: Propene selectivity versus conversion over molybdena supported on different oxides for the experiments shown in Figure 1.

coverage. Finally, an alumina-supported sample containing MoO_3 and V_2O_5 on alumina was prepared by impregnation using a solution of ammonium and ammonium heptamolybdate and another containing MoO_3, V_2O_5 and Nb_2O_5 was prepared from an oxalic acid solution of the same two salts together with niobium oxalate. The total surface areas of the supports and the Mo-containing catalysts were obtained using a Micromeretics Gemini system (BET method, N_2 adsorption at 78 K).

The catalytic behaviours were obtained using a quartz atmospheric pressure flow reactor (4 mm internal diameter) in which particles of the catalyst samples (600 mg, 0.3 - 0.6 mm) were held by quartz wool plugs. The temperature of the reactor was measured by an external thermocouple placed just after the catalyst. The standard reaction mixture used for the tests presented here consisted of a flow containing 29.4% propane (Air Products, 99.9%) and 9.6% oxygen (BOC, 99.9%) with the balance being helium (BOC, 99.9%). Analysis of the main products (propene and carbon oxides) was carried out using an on-line Varian 3300 gas chromatograph (TCD detector, HayesSep Q column). For each sample, the reaction temperature was raised in steps of 50°C over the range 250 to 650°C, each step being maintained for 1 h.

Results and Discussion

Figure 1 shows the propane conversion as a function of reaction temperature measured for the series of MoO3 catalysts prepared on different supports which had been calcined at different temperatures so that their areas were of the order of 40-50 m^2g^{-1} (Table I). The results shown for the MgO and TiO_2 supports are the same as those which were presented previously [1]. The silica-supported material had no measurable activity under the conditions used for these experiments. The order of activities of the catalysts based on the various supports was: $ZrO_2 > Nb_2O_5 > Al_2O_3 > MgO > TiO_2 >> SiO_2$. This differs slightly from the order found previously for supports of widely differing areas [1]: $ZrO_2 > Al_2O_3 > Nb_2O_5 > TiO_2 > SiO_2$. The most significant difference between these results and those reported previously was a substantial decrease in the activity (i.e. an increase in the temperature required for a given conversion by some 50°C) of the alumina-supported material compared with that of the material prepared on a higher surface area alumina (area ca. 230 m^2/g) for which the data are also shown (denoted as hsa) for comparison purposes and the total inactivity of the silica-supported material.

Figure 2 shows a series of plots of the selectivity towards propene versus the conversion of propane, these plots corresponding to the data of Figure 1; the balance of the products was made up predominantly of CO_2 in all cases although there was also a small quantity of oxygenated products formed. At any conversion of propane, the selectivity towards propene was highest for the alumina-supported material and the results for the other supports were in the order: $TiO_2 > MgO > ZrO_2 > Nb_2O_5$. Compared with the results reported previously for the high-surface area alumina calcined at lower temperatures (also shown), the selectivities obtained with the low surface area alumina are significantly higher, particularly at lower conversions: the highest yield of propene (conversion x selectivity) obtained for the highest propane conversion (corresponding to approximately 100% conversion of the oxygen) with the low surface area support was ca. 10.5% . The value obtained with the low surface area support compared favorably with a value of (13.15%) obtained with a $NiMoO_4$ material prepared according to the method given in ref. [2]. There was also a slight improvement in the yield for the Mo/ZrO_2 sample over what had previously been reported [1] for the same support calcined at a lower temperature but this yield was not as high as that obtained with the alumina-based material. In consequence, it was decided to examine in more detail the effect of the temperature of calcination of the support and of the MoO_3 loading on the catalytic properties of Mo/Al_2O_3.

Figure 3 shows the propane conversion as a function of reaction temperature for a series of samples prepared with Al_2O_3 supports calcined at different temperatures (Table II); also shown for comparison purposes are data obtained for the support calcined at 1200°C but without any MoO_3. The most active sample was that which was calcined at the lowest temperature and the least active was the support alone. Corresponding selectivity vs. conversion plots are given in Figure 4 which shows that the highest selectivity for any conversion was obtained with the support calcined at the highest temperature, a gradual improvement being obtained with increasing calcination temperature. The selectivity of the support alone was much lower than those of the molybdena materials, this being a

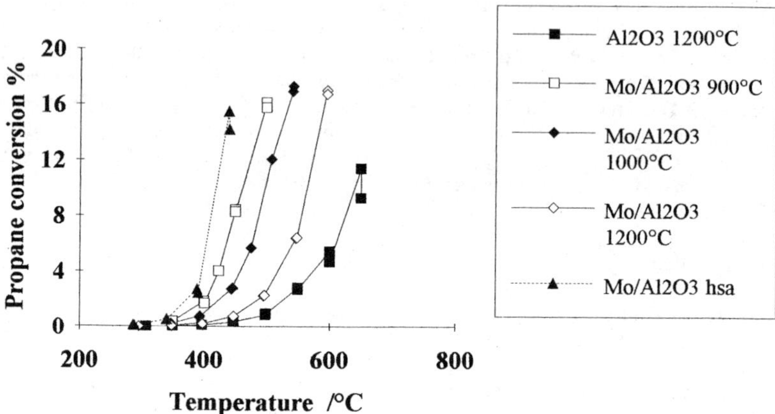

Figure 3. Plots of propane conversion versus reaction temperature for molybdena catalysts (Table II) supported on alumina calcined at various temperatures.

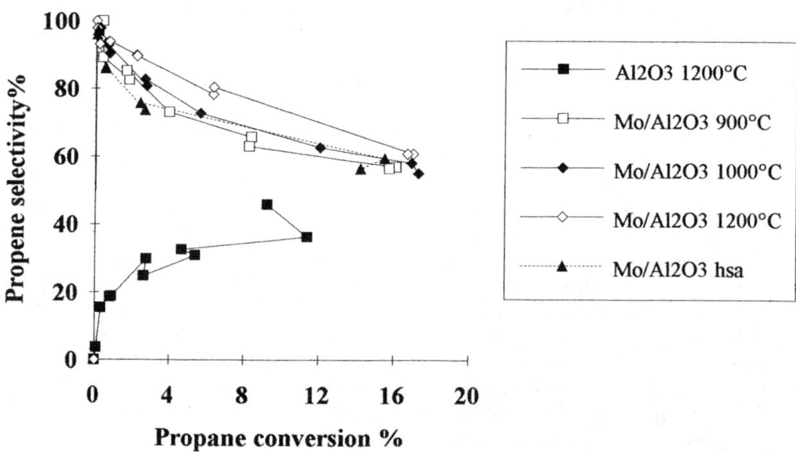

Figure 4. Propene selectivity versus conversion over molybdena supported on alumina calcined at various temperatures for the experiments shown in Fig. 3.

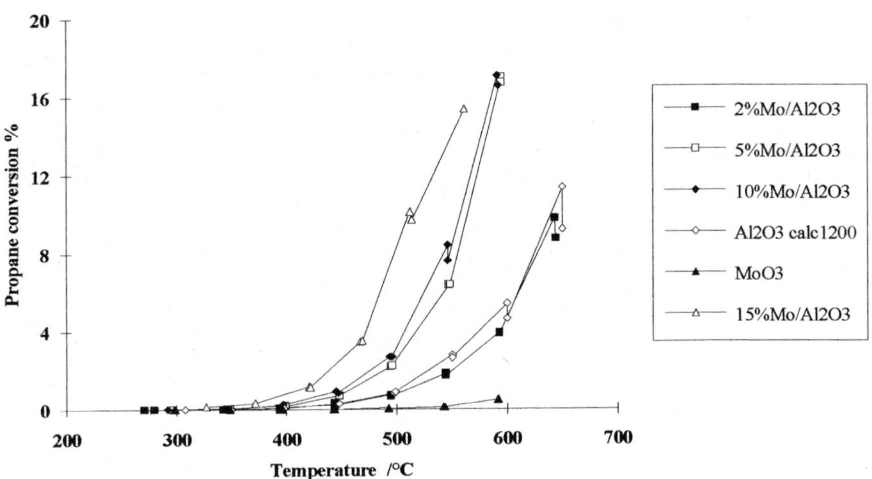

Figure 5. Plots of propane conversion versus reaction temperature for catalysts with different loadings of molybdena supported on alumina calcined at 1200°C.

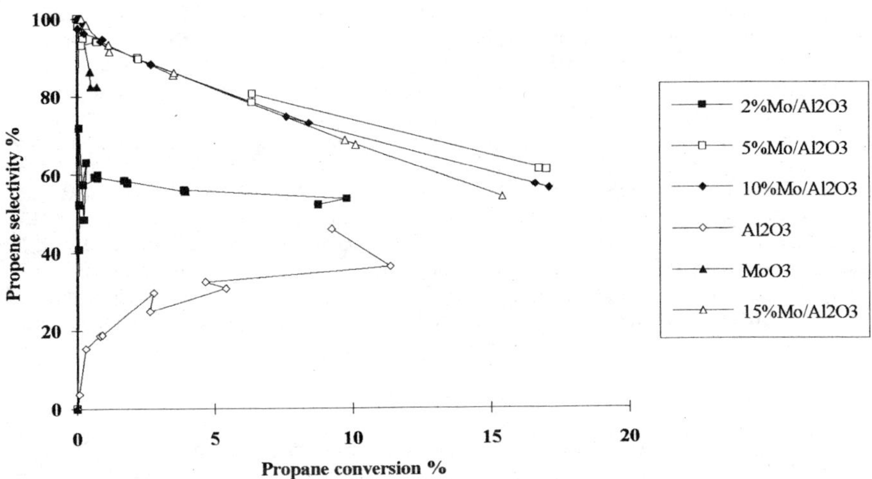

Figure 6. Propene selectivity versus propane conversion for catalysts with different loadings of molybdena supported on alumina calcined at 1200°C.

result of the total oxidation of the propane to give CO_2; the total oxidation probably occured on acidic sites on the alumina which were covered by molybdena species with the supported molybdena catalyst.

We now present equivalent data for the effect of MoO_3 loading on the catalytic behaviour, these having been obtained with the support calcined at 1200°C. Fig. 5 shows the propane conversion as a function of reaction temperature for various different MoO_3 loadings and also for pure MoO_3 and for the Al_2O_3 alone. The activity of the MoO_3 was very low and the activities of the 2 wt% MoO_3 sample and the alumina support were somewhat higher but very similar; the activities of the other materials were significantly higher and increased with increasing MoO_3 contents. Figure 6 shows the corresponding selectivity vs. conversion plots. The selectivities of the catalyst with 2 wt% MoO_3 were higher than those of the support alone but were considerably lower than those of the materials with higher MoO_3 contents. For the latter, the activity and selectivity data were significantly different only at higher conversions; above conversions of ca. 6%, the selectivities of the material with 5 wt% MoO_3 were the highest. This was the sample discussed above which gave a yield of 10.5% of propene. It is interesting to note that the samples with greater than 5 wt% of MoO_3 also gave up to ca. 5.2 % selectivity to oxygenates, this contrasting with the material with 2 wt% which gave ca. 2% oxygenates or with the support alone which gave none. We can conclude from these results that it is necessary to cover the support by molybdena species so that no uncovered alumina sites are available for total oxidation. Similar conclusions have been reached by other authors for different reactions over supported molybdena catalysts; see for example reference [5].

The yields obtained with the alumina supports calcined at high temperatures were significantly better than those obtained with the same supports calcined at lower temperatures (Figures 3 and 4 and Table II) which contained more than a monolayer of molybdena, the optimum yield for the 19.5wt% MoO_3/Al_2O_3 sample being 8% (with a corresponding yield to oxygenates of 1.9%). It would therefore appear that the improved behavior of the catalysts for which the support had been calcined at higher temperatures is not due alone to a better coverage of the support by the active phase, as discussed above, but also to a more ideal interaction between the active phase and the support, possibly due to a change in the degree of dehydration of the support or in its crystal structure with increase in calcination temperature. It is also possible that the change in pore structure of the material calcined following calcination at high temperature is responsible for improved selectivities. X-ray diffraction did not show any difference in the structures of the molybdena which waslargely amorphous. Further work is being carried out to investigate the nature of the oxide-support interaction and the effect of pore structure on the performance of the catalysts and to try to increase further their selectivities.

It was shown previously [1] that the activity of a MoO_3/TiO_2 catalyst could be improved significantly without any significant change in the selectivity by adding either vanadia or a mixture of vanadia and Nb_2O_5 to the formulation. Equivalent MoO_3 + V_2O_5 and MoO_3 + V_2O_5 + Nb_2O_5 materials supported on low-area alumina were also prepared and tested when it was found that the propene selectivities and yields were lowered significantly by the presence of these promoters. Under the conditions of the experiments of Figures 1 and 2, the maximum conversions obtained with these materials were 15 and 14 % respectively and the corresponding selectivities to propene were 52 and 50 %. Further details of these results will be presented elsewhere [6]. We must conclude here that the interaction between the molybdena and the alumina discussed above is influenced in a different way by the incorporation of these two elements compared to the case of titania.

Acknowledgements

Part of this work was funded by the European Community (Human Capital and Mobility Programme, Grant NoCHRX CT92 0065.) which also provided a fellowship for F.C.M. We thank Drs. R.H.H. Smits and K. Seshan for useful discussions.

Literature Cited

1. Meunier, F.C., Yasmeen, A. and Ross, J.R.H. paper presented at the 13th Meeting of the *North American Catalysis Society*, Snowbird, Utah, June **1995**; submitted for publication in *Catal. Today.*

2. Mazzochia, C., Aboumrad, C., Diagne, C., Tempesti, E., Herrman J.M. and Thomas, G., *Catal. Lett.*, 10 **(1991)** 181-192.
3. Smits, R.H.H., Ph.D. thesis, University of Twente, The Netherlands **(1994).**
4. Grzybowska, B., Kess, P., .Grzybowski, R., Wcislo,K., Barbaux, Y., and Gengember, L., *Stud. Surf. Sci. Catal.*, 82 **(1994)** 151-158.
5. Fransen, T., Ph.D. thesis, University of Twente, The Netherlands **(1977).**
6. Yasmeen, A., Meunier, F.C., and Ross, J.R.H., to be published.

Chapter 13

Partial Oxidation of C_5 Hydrocarbons to Maleic and Phthalic Anhydrides over Molybdate-Based Catalysts

Umit S. Ozkan, Rachel E. Gooding, and Brian T. Schilf

Department of Chemical Engineering, Ohio State University, 140 West 19th Avenue, Columbus, OH 43210

The study presented in this paper demonstrates the use of molybdena-based catalysts in the partial oxidation of n-pentane and 1-pentene to phthalic and maleic anhydrides. The catalysts used are two-phase materials consisting of molybdenum oxide and a simple molybdate. The catalysts have been characterized by BET surface area measurement, laser Raman spectroscopy, X-ray diffraction, X-ray photoelectron spectroscopy and temperature-programmed desorption techniques. The reaction studies have been performed using a fixed-bed flow reactor. The phenomenon of contact synergy and the effect of reaction parameters have been examined. The reaction and TPD experiments suggest that MoO_3 has the sites for the activation of pentane. At lower temperatures, it is also possible to perform the oxygen insertion steps and/or C-C bond formation steps. However, at higher temperatures, the cracking and complete oxidation reactions become more dominant, changing the product distribution in favor of lower alkanes and alkenes and carbon oxides. The presence of two phases (MoO_3 and $MnMoO_4$) in close proximity appears to control the complete oxidation step while allowing the desorption of partially oxidized and/or coupled products.

There is a growing interest in converting C5 fraction from naptha steam crackers to value added products. While C5 hydrocarbons have not yet found a market in which they can be sold, recent legislation has reduced their direct use even further by limiting the amount of C5 alkanes that can be added to gasoline. Therefore, research that can lead to a process which can convert C5 hydrocarbons into useful intermediates or products is quite relevant. Furthermore, the activation and partial oxidation of lower alkanes continues to pose a major challenge to catalysis researchers. Especially, in the case of pentane oxidation to form phthalic anhydride, the challenge is even greater since the catalyst needs not only activate the alkane molecule, but also promote the formation of C-C bonds in an oxidizing environment.

One of the earliest studies on the partial oxidation of C5 hydrocarbons was reported by Butt and Fish (*1 - 3*) using vanadia catalysts in the reaction of 1-pentene and several branched chain pentenes. One of the major reaction products observed was maleic anhydride. More recent studies have focused on the partial oxidation of

0097–6156/96/0638–0178$15.00/0

pentane to phthalic and maleic anhydride over V-P-O catalysts (*4-13*). The number of studies using other catalytic systems is smaller (*14, 15*).

The study presented in this paper demonstrates the use of molybdena-based catalysts in the partial oxidation of n-pentane and 1-pentene to phthalic and maleic anhydrides. The catalysts used are two-phase materials consisting of molybdenum oxide and a simple molybdate. Our previous work on C4 hydrocarbon oxidation (*16-19*) have shown significant synergy effects over these catalysts. Present study makes use of some of the earlier findings from our previous work while extending the investigation to C5 hydrocarbons. The catalysts have been characterized by BET surface area measurement, laser Raman spectroscopy, X-ray diffraction, X-ray photoelectron spectroscopy and temperature-programmed desorption techniques. The reaction studies have been performed using a fixed-bed flow reactor. The phenomenon of contact synergy and the effect of reaction parameters have been examined.

Experimental

Catalysts were prepared using precipitation and wet impregnation methods described earlier (*19*). All catalysts were calcined in oxygen at 500 °C. The surface areas, determined by BET technique using krypton were in the range of 0.5-1 m^2/g.

The reactor was made out of quartz tubing with an inner diameter of 4.0 mm. The amount of catalyst used was adjusted to provide a constant total surface area of 0.56 m^2 in the reactor. The feed consisted of 2.5 % hydrocarbon (n-pentane or 1-pentene), 10% oxygen and balance nitrogen. The flow rates were controlled by Tylan mass flow controllers. Two different retention time (τ) values were used at 0.47 and 0.35 seconds. The temperature was varied in the range of 375 to 475 °C. The reactor was connected to a "u-tube" condenser using Cajon quick connect vacuum fittings. The condenser which was maintained at room temperature was used to trap the large organic molecules for the HPLC analysis.

The analytical system consisted of two parts, Gas chromatography (GC) and High Performance Liquid Chromatography (HPLC). The on-line gas chromotograph (Hewlett-Packard 5890A) was equipped with both thermal conductivity and flame ionization detectors and was used to analyze species that did not condense at room temperature. The columns used were Chromosorb PAW, HayeSep D and Molecular Sieve 5A. The HPLC system consisted of a Spectra Physics Degasser (model SCM 400), a Spectra System gradient pump (model 4000) and a Spectra System model MUV2000 detector. The column used was a Spherisorb 5 ODS-2 C-18 reverse phase column. At the beginning of the analysis, the solvent consisted of 50% acetonitrile and 50% water and at the end of 5 minutes, the solvent consisted of 60/40 acetonitrile/water.

The temperature-programmed desorption experiments were performed using a system described previously (*20*). The catalysts were calcined *in-situ* prior to any adsorption. The adsorbates, pentane and pentene (5%), were passed over the catalyst at room temperature for 2 hours at a flow rate of 30 cm^3(STP)/min. After flushing the catalyst for two hours with helium, the temperature program was started, with a heating rate of 15 °C/min up to 600 °C, followed by an isothermal step at 600 °C for 20 minutes. The species desorbing from the surface were analyzed by a mass spectrometer.

Results and Discussion

The catalysts used in this study were a pure molybdenum trioxide phase and a two-phase catalyst bringing molybdenum trioxide in close contact with a manganese molybdate phase. The characterization studies did not show any detectable changes in either of the two phases when they were brought into contact with each other.

The observed production rates of maleic anhydride, phthalic anhydride, and carbon oxides in pentane and pentene oxidation have been summarized in Figure 1. In both pentane and pentene oxidation, the single-phase catalyst was seen to have a higher activity than the two-phase catalyst. However, the higher activity was mostly for complete oxidation products. In pentane oxidation, MoO_3 was found to give a higher formation rate for maleic and phthalic anhydrides at the lowest temperature. However, as the temperature was raised, the formation rate of both of the anhydrides became higher over the $MnMoO_4/MoO_3$ catalyst. Another point to note is that the observed rate of formation of both maleic and phthalic anhydride increased by an order of magnitude as the feed was changed from pentane to pentene. The carbon oxide formation rate was consistently higher over the MoO_3 catalysts than over the two-phase catalyst. Other products observed included maleic acid, phthalic acid, acetaldehyde, formaldehyde, acetone, cracking products, dehydrogenation and isomerization products.

Figure 2 shows the observed production rates for maleic acid, phthalic acid, acetaldehyde, formaldehyde, acetone, and cracking products in pentene oxidation. Interestingly, the trends observed for maleic anhydride and phthalic anhydride are seen to hold for the maleic and phthalic acids as well. While there is no maleic or phthalic acid formation observed at 375 °C, both observed rates become significant at the high temperature end. At 425 °C, the single phase catalyst gives higher yields of the two acids. At 475 °C, however, the two-phase catalyst has the higher yield in both products. The reaction sequence between the acids and the anhydrides is not clear at this point. However, the possibility of the anhydrides hydrating in the presence of water, which is present in larger quantities at higher conversion levels cannot be ruled out. For the other partial oxidation products, namely, acetaldehyde, formaldehyde and acetone, the $MnMoO_4/MoO_3$ performs significantly better than the single phase catalyst at higher temperatures. Other species observed in these reactions are C4 and C3 cracking products with MoO_3 consistently giving higher cracking product yields. Also observed in pentene oxidation experiments is pentane, indicating the reversibility of the dehydrogenation step.

A comparison of selectivities for the two catalysts is presented in Figure 3. The selectivity values represent data obtained at 475 °C and at a constant 1-pentene conversion level of 90%. The corresponding oxygen conversion values for $MnMoO_4/MoO_3$ and MoO_3 were 76% and 99 % respectively. As is apparent from oxygen conversion levels, when the pentene conversion is kept constant, the single-phase catalyst shows a higher complete oxidation activity. This is especially pronounced in CO_2 selectivity. The selectivities for the partial oxidation products, especially maleic and phthalic anhydrides and maleic and phthalic acids were consistently higher over the $MnMoO_4/MoO_3$ catalysts than over the MoO_3 catalysts. The largest difference was observed in the selectivities to maleic anhydride and phthalic anhydride.

Experiments were also performed by varying the space time and keeping other parameters the same. While the major trends did not change when the space time was increased, the most significant difference was observed in the cracking products over MoO_3 (Figure 4). The yield for cracking products was more than doubled when the space time was increased from 0.35 s to 0.475 s over MoO_3 catalysts at 425 and 475 °C, temperatures at which the reaction medium was severely oxygen deficient.

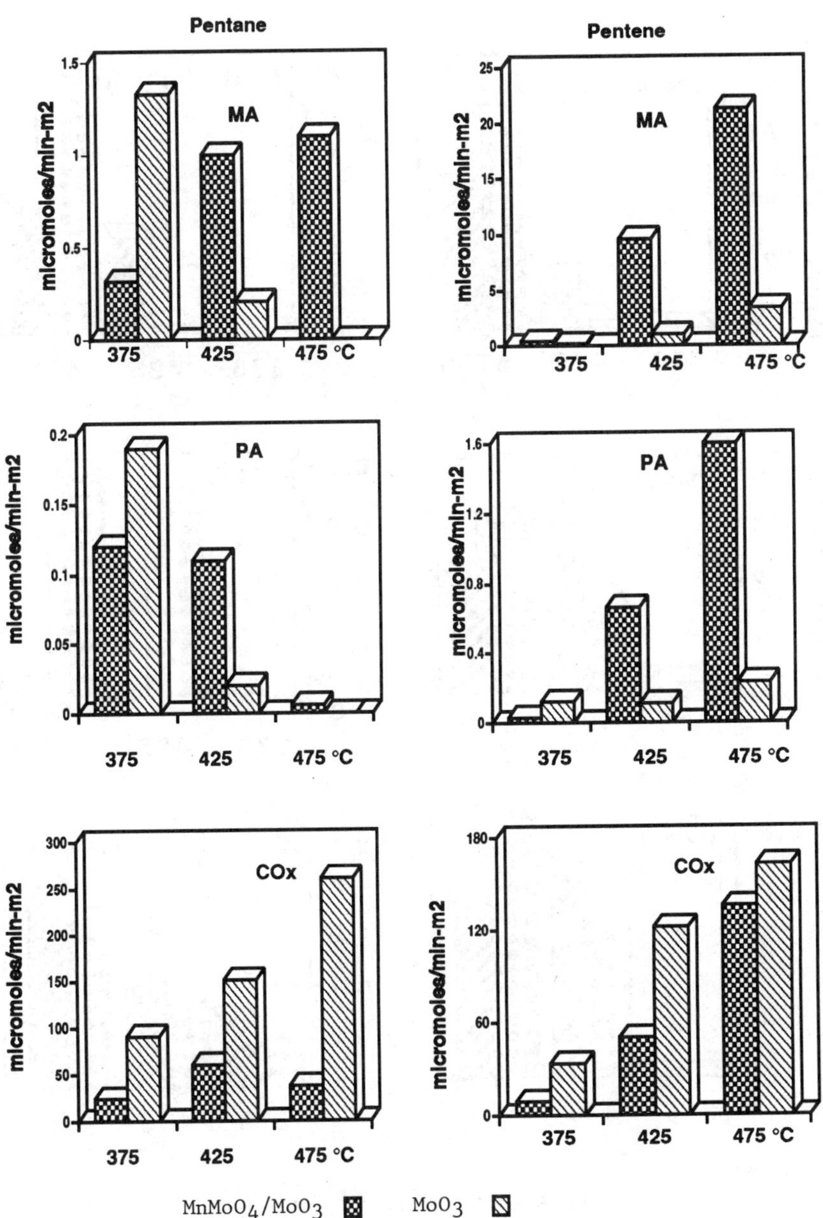

Figure 1. Formation rates for maleic anhydride, phthalic anhydride, and CO_x in pentane and pentene oxidation.

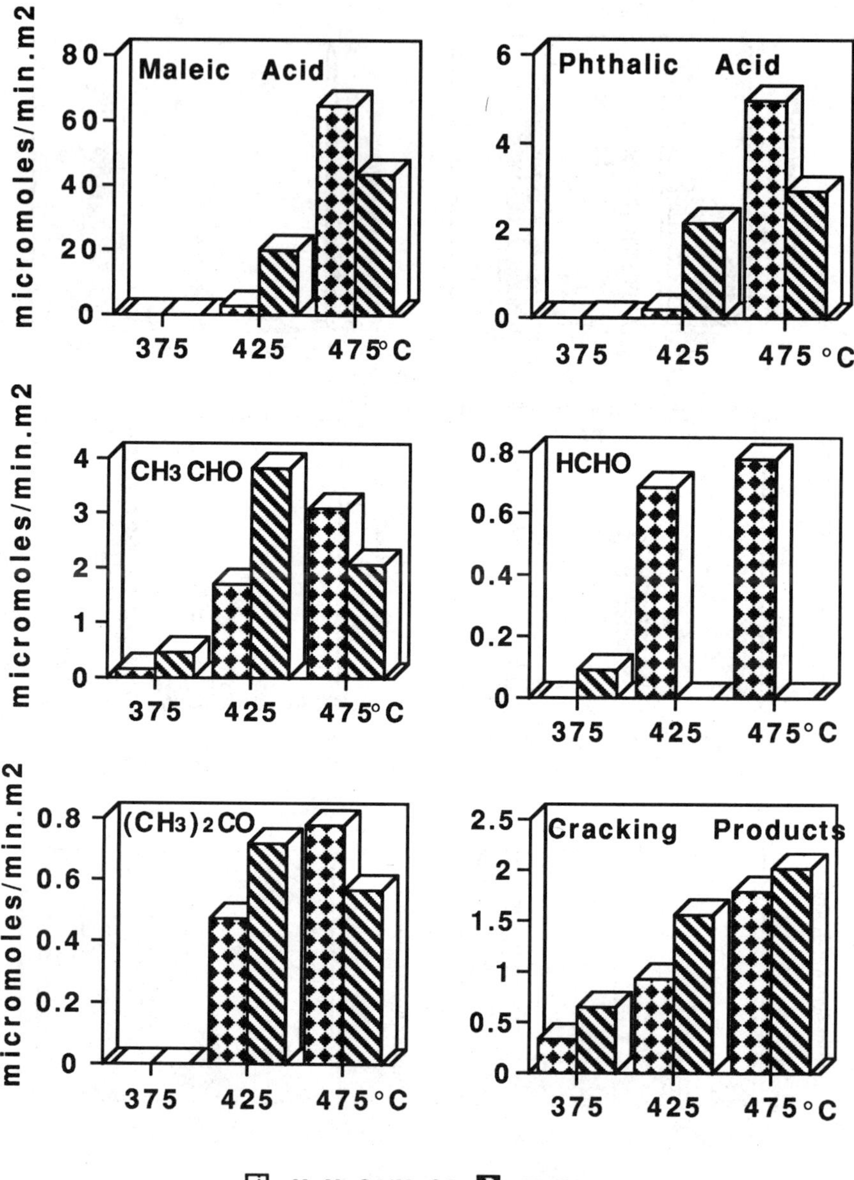

Figure 2. Formation rates of maleic acid, phthalic acid, acetaldehyde, formaldehyde, acetone and cracking products in 1-pentene oxidation.

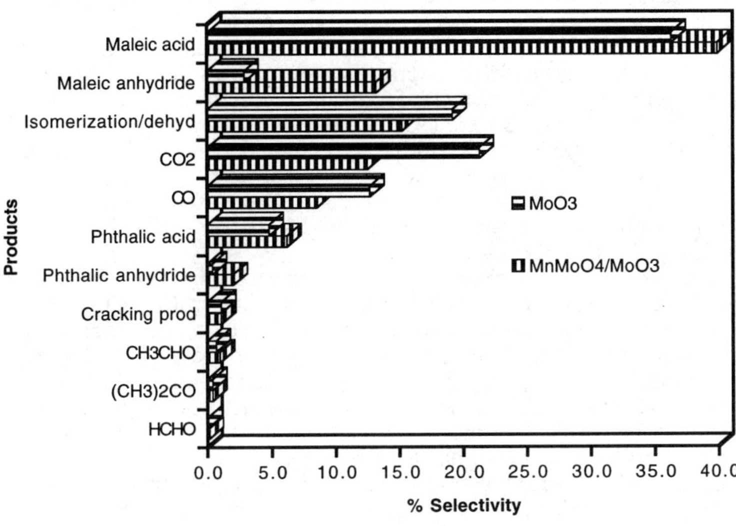

Figure 3. Comparison of selectivities at equal conversion of 1-pentene at 475 °C.

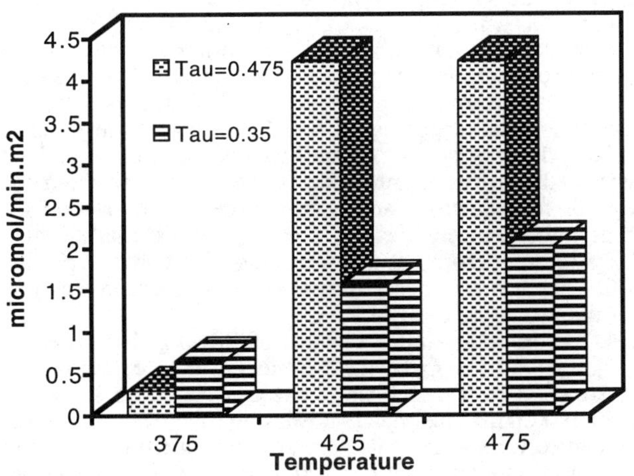

Figure 4. Formation rates of cracking products over MoO_3 catalyst at two different space velocities (1-pentene oxidation).

In addition to the reaction studies, the temperature-programmed desorption experiments also showed significant differences in the adsorption/desorption characteristics of pure MoO_3 and MoO_3 brought into contact with a simple molybdate phase. The desorbing species over both catalysts were pentane, pentene, pentadiene, methyl butane, acetone, acetaldehyde, formaldehyde, CO, and CO_2. The ions used to analyze the first three products were 72, 55, and 67 respectively, while mass to charge ratios of 27, 29, 30, 43, and 58 were used in order to discern between methyl butane, acetone, acetaldehyde and formaldehyde. Carbon monoxide and carbon dioxide were detected with mass-to-charge ratios of 28 and 44, respectively. From $MnMoO_4/MoO_3$ catalysts, maleic anhydride was also seen to desorb. The presence of maleic anhydride was confirmed with mass-to-charge ratios of 26, 54, and 98. There was no phthalic anhydride desorption on either catalyst.

Figure 5 shows the temperature-programmed desorption profiles of n-pentane, 1-pentene, 1,3-pentadiene and 2-methyl butane resulting from pentane adsorption. The two-phase catalyst exhibits desorption features for all C5 species at about 170°C. Among the four species observed, the lowest intensity belongs to n-pentane, the product of reversible adsorption, followed by 2-methyl butane. Pentane shows a second desorption peak at 355 °C over this catalyst, a fact that suggests that there are likely to be two different adsorption sites for pentane over this catalyst. The pure MoO_3 catalyst, on the other hand, has the largest desorption peaks for n-pentane and 2-methyl butane (210°C) and relatively smaller peaks for 1-pentene and 1,3-pentadiene (160°C), suggesting that hydrogen abstraction steps proceed more readily on the two-phase catalyst.

CO and CO_2 desorption profiles resulting from n-pentane TPD are presented in Figure 6. The CO desorption profile for $MnMoO_4/MoO_3$ catalyst is characterized by two small peaks at 200 and 400 °C and one very large peak at 550 °C. MoO_3, on the other hand, shows only one small desorption feature at a considerably higher temperature at 600 °C. The CO_2 desorption profiles show similar trends with the two-phase catalyst giving one strong peak at 470 and a shoulder peak at 400, while the single-phase catalyst has two very weak CO_2 signals at 500 and 600 °C.

The temperature-programmed desorption profiles for the oxygenated products resulting from pentane adsorption are presented in Figure 7. Partial oxidation products from the two-phase catalyst are maleic anhydride, which shows a triple maxima, with the most intense peak signal appearing at 460°C, formaldehyde desorbing at 390°C, acetaldehyde which desorbs at 550°C, and acetone desorbing at 355°C. As was previously mentioned, no maleic anhydride is seen to desorb from the pure MoO_3 surface, while both formaldehyde and acetaldehyde desorb at 600°C. These two species desorb at higher temperatures for the single-phase catalyst than for the two-phase catalyst. Acetone desorbs at a lower temperature than the other oxygenated products, 210°C. Acetone also desorbs at a lower temperature for the single-phase catalyst than for the two-phase catalyst.

The TPD desorption profiles obtained following 1-pentene adsorption are presented in Figures 8-10. Over the $MnMoO_4/MoO_3$ catalyst, 1-pentene, 1,3-pentadiene and 2-methyl butane each show two desorption peaks at about 120°C and 200°C. MoO_3 catalyst on the other hand shows only one desorption peak around 135 °C for each of the three C5 species. Also seen desorbing from both of the catalysts is n-pentane, which has a doublet in the 120-200 °C range and a single peak near 350 °C. The pure MoO_3 catalyst shows only one pentane desorption peak at 250°C (Figure 7).

Figure 8 shows the carbon oxide desorption profiles resulting from pentene adsorption. CO is seen to desorb at 540 °C from the two-phase catalyst while it desorbs at 600 °C from the MoO_3 catalyst. The CO_2 desorption profiles show a major maximum around 535 °C for the two-phase catalyst. The CO_2 desorption is negligibly small over the MoO_3 catalyst.

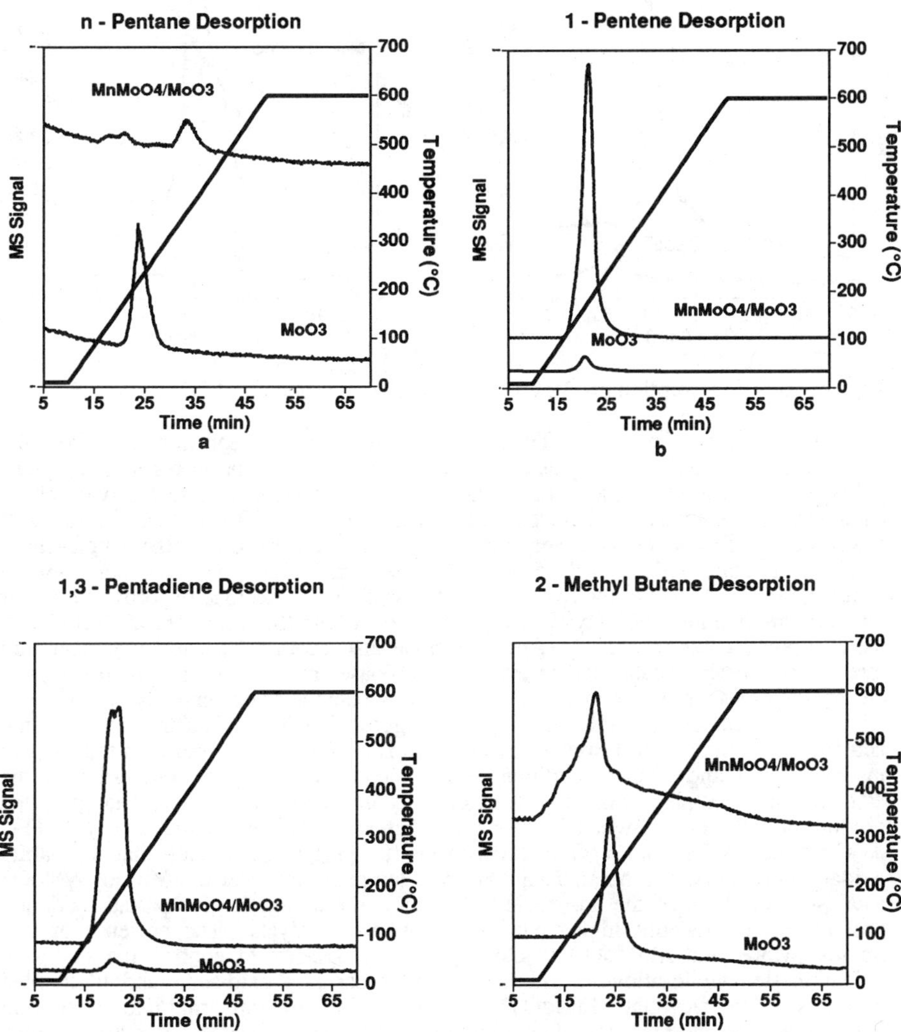

Figure 5. Desorption profiles for n-pentane, 1-pentene, 1,3-pentadiene, and 2-methyl butane in n-pentane TPD.

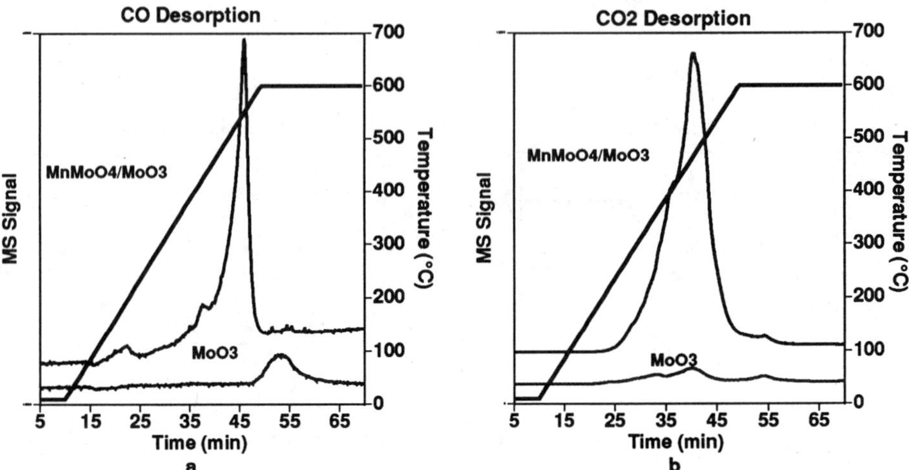

Figure 6. Desorption profiles for CO and CO_2 in n-pentane TPD

As seen in the case of n-pentane TPD, no maleic anhydride is seen to desorb from the pure MoO_3 surface. The two-phase catalyst shows three desorption features in maleic anhydride desorption profiles. Formaldehyde shows two peaks over the two- phase catalyst when pentene is the adsorbate gas. These peaks, at 410°C and 600°C, are both lower than the formaldehyde desorption peak of 600°C observed over the single-phase catalyst. Acetaldehyde also desorbs at a higher temperature over the single-phase catalyst, 600°C, than over the two-phase catalyst, 535°C. Acetone desorbs at a lower temperature than the other oxygenated products over both the single-phase, 240°C, and the two-phase catalyst, 315°C. Again, acetone also desorbs at a lower temperature over the single-phase catalyst than over the two-phase catalyst. There was no oxygen desorption observed over either catalyst following pentane or pentene adsorption.

The reaction and TPD experiments suggest that MoO_3 has the sites for the activation of pentane. At lower temperatures, it is also possible to perform the oxygen insertion steps and template addition steps. However, at higher temperatures, the cracking and complete oxidation reactions become more dominant, changing the product distribution in favor of lower alkanes/alkenes and carbon oxides. The TPD data and the results from reaction experiments are in agreement such that the partial oxidation products, in general, desorb at lower temperatures over the MnMoO4/MoO3 catalysts regardless of the adsorbate. Similarly, the selectivity for partial oxidation products are consistently higher over the two-phase catalysts. The presence of two phases (MoO_3 and $MnMoO_4$) in close proximity appears to control the complete oxidation step while allowing the desorption of partially oxidized and/or coupled products. The fact that $MnMoO_4/MoO_3$, in general, exhibits multiple desorption features suggest the presence of multiple adsorption/reaction sites over this catalyst. However, the detailed characterization work conducted over these catalysts in regard to their behavior in C4 oxidation reactions did not give any indication of a new phase being formed (*16-19*). It appears that the role of the molybdate phase is through modification of the sites on MoO_3 surfaces, by either limiting the oxidation activity of the sites or by allowing the rapid desorption of partial oxidation products. Although this study provides no direct evidence for the mechanism of such a site modification, an oxygen spillover effect proposed by Delmon and co-workers (*20-22*) and suggested by our earlier work (*17*) remains as a possibility.

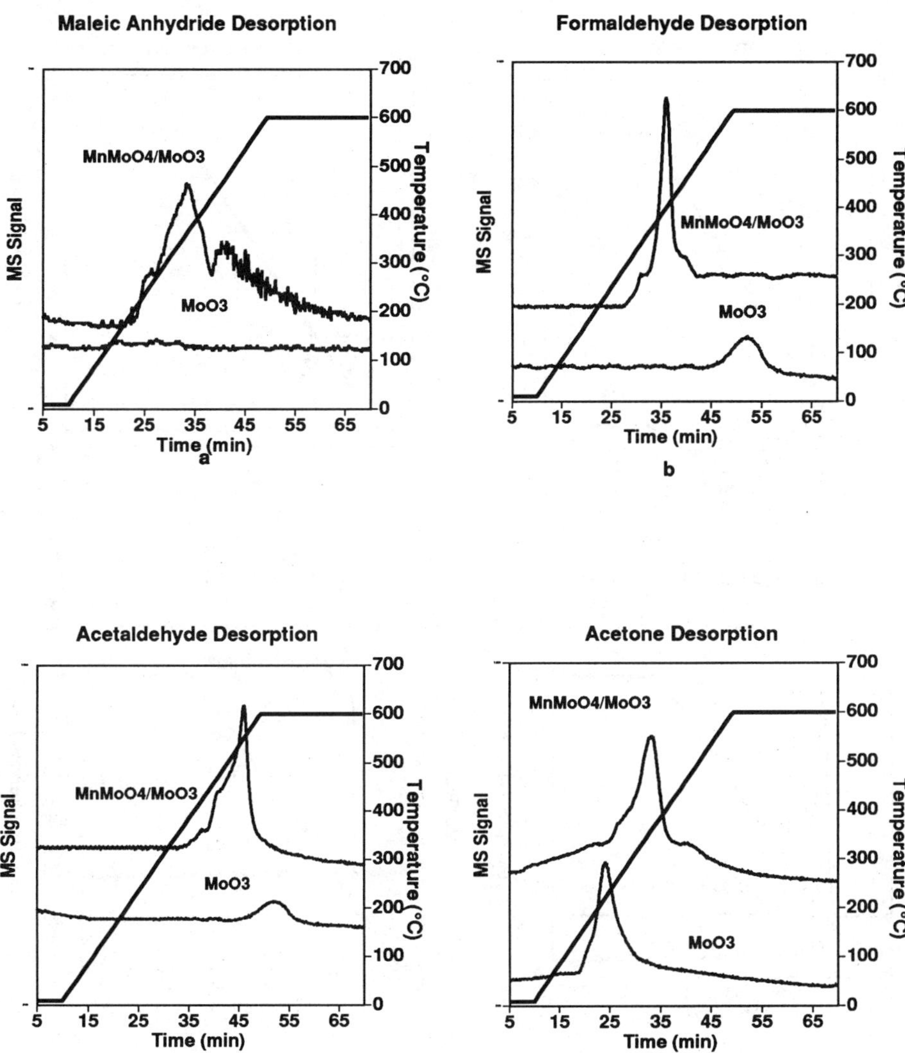

Figure 7. Desorption profiles for maleic anhydride, formaldehyde, acetaldehyde and acetone in n-pentane TPD.

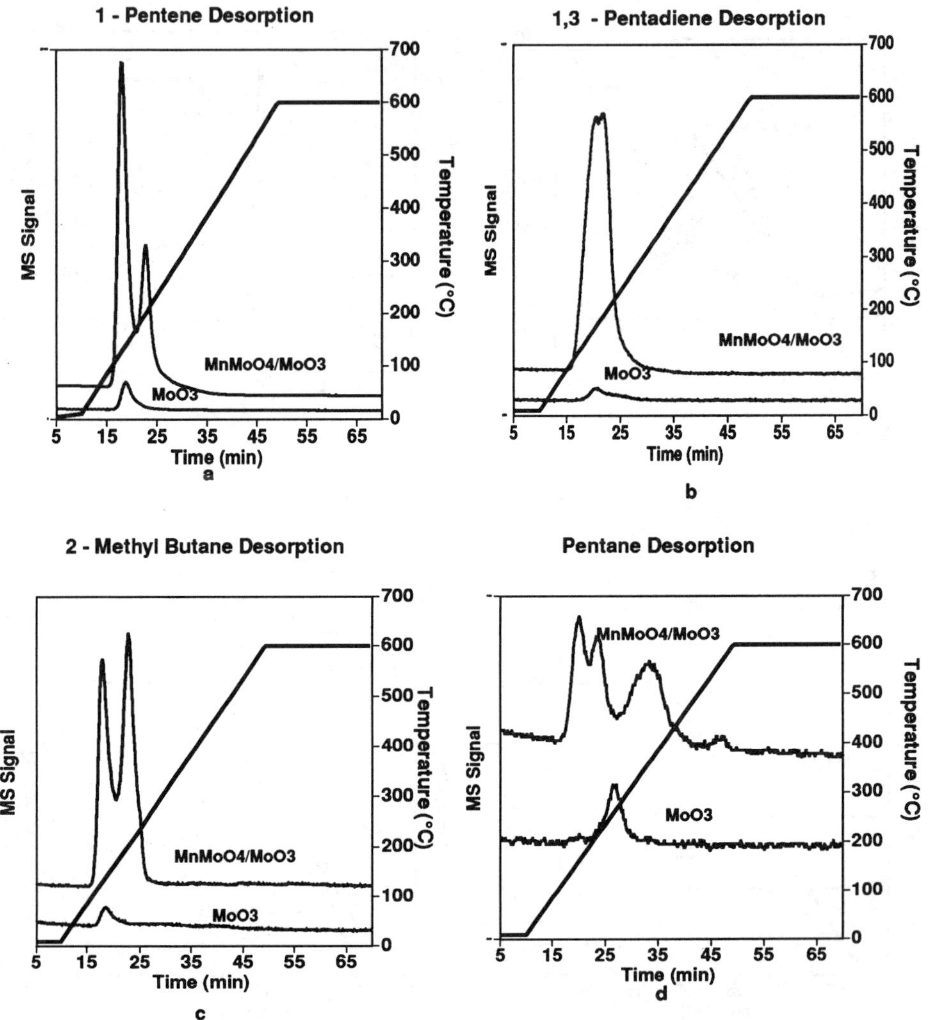

Figure 8. Desorption profiles for 1-pentene, 1,3-pentadiene, and 2-methyl butane, and n-pentane in 1-pentene TPD.

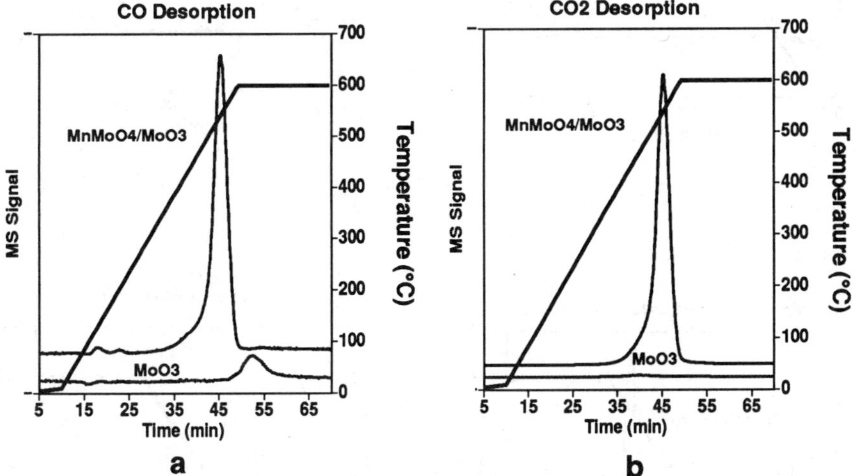

Figure 9. Desorption profiles for CO and CO_2 in 1-pentene TPD.

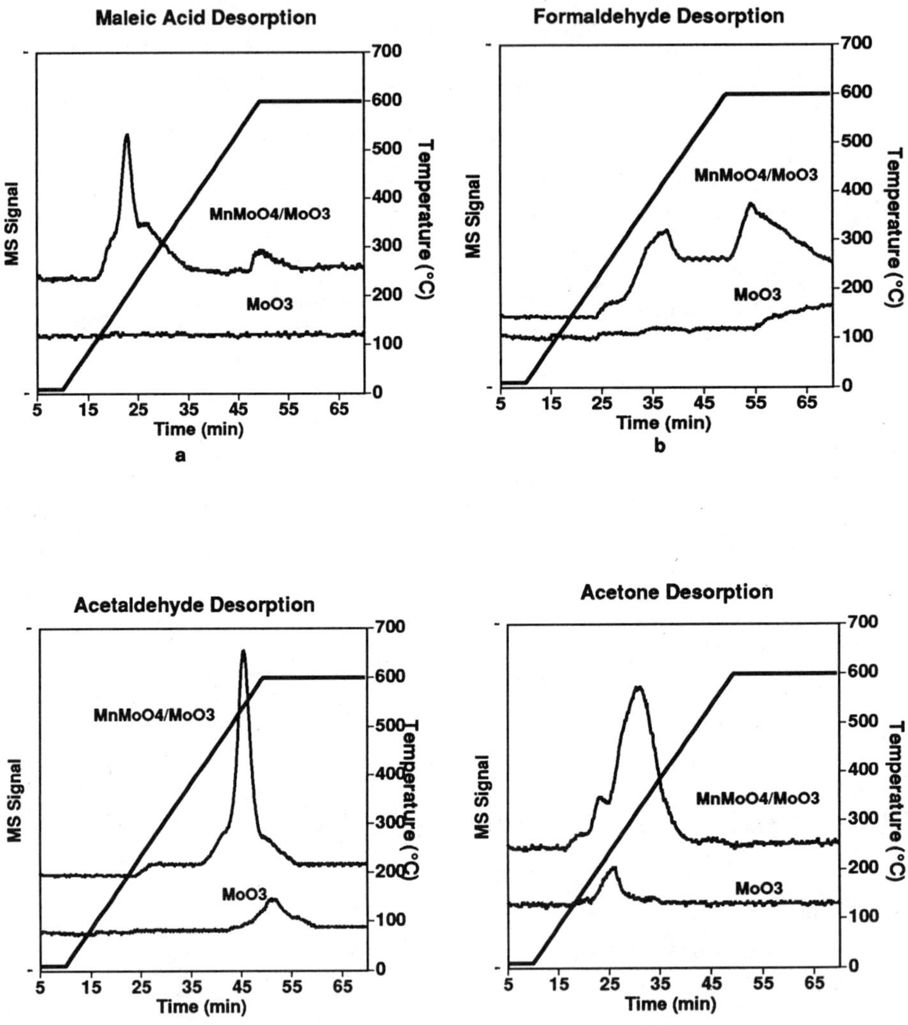

Figure 10. Desorption profiles for maleic anhydride, formaldehyde, acetaldehyde and acetone in 1-pentene TPD.

Acknowledgment

The financial support provided for this work by the National Science Foundation (CTS-9412544) and by the Matheson Corporation is gratefully acknowledged.

Literature Cited

1. Butt, N.S.; Fish, A. *J. Catal.*. **1966**, *5*, 205.
2. Butt, N.S.; Fish, A.,*J. Catal.*. **1966**, *5*, 494.
3. Butt, N.S.; Fish, A. *J. Catal.*. **1966**, *5*, 508.
4. Busca,G.; Centi,G. *J. Am. Chem. Soc.* **1989**, *111*, 46.
5. Busca, G.; Ramis,G.; Lorenzelli,V. *J. Mol. Catal.*. **1989**, *50*, 231.
6. Centi,G.; Burattini,M.; Trifiró,F. *Appl. Catal.*. **1987**, *32*, 353.
7. Centi, G.; Nieto, J.L.; Pinelli, D.; Trifiró, F.; Ungarelli, F. in *New Developments in Selective Oxidation;* Centi, G., Trifiró, F. Eds. Elsevier Science Publishers B.V.: Amsterdam, The Netherlands, 1990; pp 635.
8. Centi, G.; Trifiró, F. *Chem. Engr. Sci.* **1990**, *45,* 2589.
9. Centi, G.; Gleaves, J.T.; Golinelli, G.; Trifiró, F. in *New Developments in Selective Oxidation*; Delmon, B.; Ruiz, P. Eds.; Elsevier Science Publishers B.V.: Amsterdam, The Netherlands, 1992, pp 231.
10. Centi, G.; Gleaves, J.T.; Golinelli, G.; Perathoner, S.; Trifiró, F. in *Catalyst Deactivation*, Bartholomew, C.H.; Butt, J.B. Eds.; Elsevier Science Publishers B.V.: Amsterdam, The Netherlands, 1991, pp 449.
11. Gleaves, J.T.; Centi, G. *Catal. Today*. **1993**, *16*, 69.
12. Golinelli, G.; Gleaves, J.T. *J. Mol. Catal.* **1992**, *73*, 353.
13. Trifiró, F., *Catal. Today*. **1993**, *16*, 91.
14. Birkeland, K.E.; Harding,W.D.; Owend, L; Kung, H.H. *ACS Division of Petroleum Chemistry, Inc. Preprints*, **1993**, *38*(4)880.
15. Centi, G.; Nieto, J.L.; Iapalucci, C. *App. Catal.* **1989**, *46* 197.
16. Gill, R.C.; Ozkan, U.S. *J. Catal.*, **1990**, *122*, 452.
17. Ozkan, U.S.; Driscoll, S.A.; Zhang, L.; Ault, K.L. *J. Catal.*, **1990**, *124*, 183.
18. Ozkan, U.S.; Smith, M.R.; Driscoll, S.A.; *J. Catal.*, **1992**, *134*, 24.
19. Ozkan, U.S.; Smith, M.R.; Gill, R.C. *J. Catal.*, **1989**, *116*, 171.
20. Ozkan, U.S.; Cai, Y.; Kumthekar, W.M.; *J. Catal.*, **1993**, *142,* 182.
21. Weng, L.T.; Ruiz, P.; Delmon, B.; Duprez, D.; *J. Mol. Catal.*, **1989**, *52,* 349.
22. Delmon, B.; *Surface Review and Letters*, **1995**, *2*, 25.
23. Weng, L.T.; Delmon, B.; *Appl. Catal.*, **1992**, *81*, 141.

Chapter 14

Activation of *n*-Pentane on Magnesium–Vanadium Catalysts

S. A. Korili[1], P. Ruiz[2], and B. Delmon[2]

[1]Aristotle University and Chemical Process Engineering Research Institute, P.O. Box 1520, GR–54006 Thessaloniki, Greece
[2]Unité de Catalyse et Chimie des Matériaux Divisés, Université Catholique de Louvain, Place Croix du Sud 2/17, B–1348 Louvain-la-Neuve, Belgium

The activation of n-pentane in the presence of oxygen has been studied over two magnesium-vanadium oxides, magnesium pyrovanadate, $Mg_2V_2O_7$, and magnesium orthovanadate, $Mg_3V_2O_8$, prepared by the citrate method. The catalytic behavior of mixtures consisting of magnesium vanadate and antimony oxide, 50-50 % wt, was also investigated. The X-ray diffraction patterns of the vanadates have shown that they were of high purity, while XPS measurements revealed a slight surface enrichment in magnesium. At 350-500°C and atmospheric pressure, similar total n-pentane conversions were achieved with the two magnesium vanadate phases; the main products were linear unsaturated C_5 hydrocarbons, i.e. pentenes and pentadienes, and carbon oxides. Magnesium orthovanadate was more selective than the pyrovanadate towards the pure dehydrogenation reaction in comparison with the combustion one, and its selectivity was in general increased by the addition of Sb_2O_4. The latter had little effect on the behavior of the pyrovanadate phase. Decreasing the oxygen to alkane ratio in the feed led to a decrease in total conversion and an increase in selectivity, while the C_5 hydrocarbon product distributions were not influenced appreciably. The extent of the homogeneous reaction between pentane and oxygen was negligible for oxygen to alkane ratios in the feed equal to 0.5 and 1.0, but increased dramatically when the ratio was raised to 2.0.

Vanadium containing catalysts have attracted considerable research interest due to their performance in selective oxidation reactions. Among them, magnesium-vanadium compounds have been repeatedly reported as effective catalysts for the oxidative dehydrogenation of hydrocarbons, such as transformation of propane to propene *(1-3)*, butane to butene and butadiene *(2)*, and even ethylbenzene to styrene *(4)*, producing mainly unsaturated hydrocarbons with negligible formation of oxygenated organic products.

In spite of the great extent of research done, the conclusions drawn on the activity and selectivity of these materials are still controversial. Although it is generally accepted that the rate determining step is the breaking of the first C-H bond of the hydrocarbon to form an alkyl species on the catalyst surface, the way

0097–6156/96/0638–0192$15.00/0

that this alkyl species is transformed into selective oxidation products is still a matter of debate. Some researchers focus on reactivity and on the effect of the atomic arrangement of the active sites *(2, 5)*, proposing different explanations for various reactants. Redox properties of catalysts have been found to correlate well with selectivity *(1)*, while the environment of the active phase may also play a role through possible cooperation of phases *(1)* or contamination by residual elements *(2)*. It appears that there is a great number of factors influencing catalyst performance, and what is very important, the nature of the hydrocarbon being activated is probably one of them. It is therefore challenging to test the behavior of the magnesium vanadates in the oxidative dehydrogenation of other hydrocarbons as well, as for example pentane.

In fact, surprisingly little research has been done hitherto on C_5 - particularly n-pentane- oxidative dehydrogenation, regardless of the catalyst. Some results were reported in the 70's, most of them included in U.S. patents, where C_5 components were a part of a broad range of hydrocarbons investigated *(6-9)*. The investigations dealt mainly with the reactions of branched and unsaturated C_5 molecules, such as the production of isoprene from isopentane *(6)*, from isoamylenes *(7)*, and from a mixed feed consisting from isoamylenes, isopentane and n-pentane *(8)*, or the production of straight chain alkadienes from alkenes *(9)*. Manganese ferrites, bismuth molybdates and composite samples combining iron with nickel and antimony, or cobalt and phosphorous, were tested as potential catalysts in those studies. Recent publications on the subject are dealing mainly with the selective oxidation of pentane over vanadium-phosphorous catalysts for the formation of oxygenated products *(10, 11)*.

During the oxidation reactions, catalysts often undergo reduction and deactivation. On the other hand, it has been observed that some oxides, such as Sb_2O_4, although inactive directly in the catalytic reaction, can improve catalyst behavior by regeneration of active sites through the action of a surface mobile oxygen species. The phenomenon is known as the remote control mechanism, and the oxides providing oxygen are characterized as donors *(12)*.

The scope of the present work is the study of the oxidative dehydrogenation reaction of n-pentane over two magnesium-vanadium catalysts, namely magnesium pyrovanadate, $Mg_2V_2O_7$, and magnesium orthovanadate, $Mg_3V_2O_8$. Mixed catalysts of these phases with antimony oxide have also been prepared and tested, in order to test whether the combination of this oxide with the active phase leads to an improvement of the catalyst performance.

Experimental

Catalyst Preparation. The catalysts were prepared by the citrate method *(13)*. $Mg(NO_3)_2.6H_2O$ (Fluka purum p.a., >99%) was dissolved in distilled water at ambient temperature. The citric acid (Merck GR, >99.5%) first, and then a slurry of NH_4VO_3 (Merck GR, >99%), were added to this solution, always under agitation at room temperature. A few ml of HNO_3 65% (Merck GR) were also added to the resulting solution, in order to avoid any precipitation. The metal salts were in suitable proportions for the desired catalyst composition, and the quantity of citric acid was such that the anions were in 10% excess compared to those stoichiometrically required for the cations. The final transparent solution was evaporated at 30°C under reduced pressure in a rotavapor, up to the formation of a very dense homogeneous liquid, which was dried in a vacuum oven at 80°C for 24 h. The precursor formed this way, was decomposed in air by heating in an oven at 300°C for 16 h. The solids were subsequently calcined in the same oven in air at 550°C for 20 h.

The mechanical mixtures with Sb_2O_4 were prepared by mixing the catalysts with the oxide in fine powder form, dispersing the mixtures in n-pentane under vigorous agitation at room temperature, and then evaporating the solvent by overnight drying at 80°C. Sb_2O_4 was prepared by calcination of Sb_2O_3 (Merck GR, >99%) at 550°C for 20 h. The composition of each mixture was 50% catalyst - 50% Sb_2O_4, on a weight basis. The mechanical mixtures were used as prepared, without additional calcination or other treatment.

Catalyst Characterization. Characterization of the prepared catalysts and mechanical mixtures included determination of their elemental composition, identification of formed phases, measurement of their specific surface areas and evaluation of their surface composition.

Chemical Analysis. The bulk elemental composition of the catalysts was determined by atomic absorption on a Philips PC 8210 spectrometer.

Phase Identification. X-ray diffraction (XRD) patterns of the solids were obtained on a Siemens D5000 powder diffractometer with Ni-filtered CuKα radiation operated at 40 kV and 50 mA.

Surface Area. Specific surface areas were measured by Kr adsorption at -196°C on a Micromeritics ASAP 2000 static apparatus using the BET method. Krypton is considered a suitable adsorbate for the determination of relatively small surface areas (< 20 m^2/g) because of its low saturation vapor pressure which minimizes possible errors in dead volume corrections *(14)*.

Surface Composition. The surface composition of the catalysts was evaluated by X-ray photoelectron spectroscopy (XPS) measurements on a Fisons SSI X-probe spectrometer, model SSX 100/206, equipped with a monochromatized microfocus Al X-ray source (1486.6 eV). The sample powders were pressed into small stainless steel troughs of 4 mm diameter, introduced in the spectrometer at room temperature and outgassed to a pressure of 10^{-7} Torr. Analysis was made in high vacuum, ~$5x10^{-9}$ Torr. The spot size was 1.4 mm^2 and the pass energy was set at 50 eV, while a low energy flood gun set at 6 eV with a nickel grid placed 3 mm above the samples was used for compensation of charging effects.

The exact binding energies were calculated with respect to the C-(C, H) component of the 1*s* adventitious carbon peak which was fixed at 248.8 eV. The peaks recorded were C 1*s* (284.8 eV), O 1*s* (~530 eV), Mg 2*s* (~89 eV), V $2p_{3/2}$ (~517.5 eV) and Sb $3d_{3/2}$ (~540 eV). The recorded spectra were decomposed to 85/15 Gaussian/Lorentzian curves, after subtraction of the non-linear background, using a least squares fitting routine. Element atomic ratios on the surface of the samples were calculated from the relative intensities of the decomposed peaks, using the sensitivity factors supplied by the manufacturer, that is 1.00 for C 1*s* , 2.49 for O 1*s*, 0.64 for Mg 2*s* , 5.49 for V $2p_{3/2}$ and 9.62 for Sb $3d_{3/2}$.

Catalyst Testing. Catalytic tests were performed in a continuous flow microreactor operating at near atmospheric pressure. This microreactor was a Pyrex U-shape tube mounted vertically in a tubular furnace. The temperature of the catalytic bed was monitored by a thermocouple entering the bed through a side thermowell. Temperature control was performed with an independent thermocouple located outside the reactor and connected to the furnace, in order to avoid oscillation phenomena. n-Pentane was fed to the reactor from a certified high purity gas tank (Indugaz, special gas mixture) containing 4 % vol. of n-pentane in helium. Other gases used were oxygen and helium, both from Air Liquide and > 99.995% purity, which were used without any further purification. Mass flow controllers monitored and controlled the flow of gases to the reactor, and the flow in the outlet was continuously measured with a bubblemeter located at the exit of the system. The pressure drop in the reactor, which was always negligible, was monitored by a pressure gauge connected to the reactor inlet. The gases were mixed before entering the reactor by passing through a 35 cm stainless steel tube full of glass beads and were heated to 100°C before entering the reactor. The outlet of the reactor and the lines further on were also heated to 100°C to avoid any condensation of the products. Products and reactants emerging from the reactor were diverted through a four-way valve to the sampling valves of a gas chromatograph, equipped with a thermal conductivity detector and a double-column system, consisting of a Porapak Q and a Durapak n-Octane on Porasil C column. This GC system was used for the majority of the experiments reported in the present paper, but in the process of the work it was upgraded to a more powerful and flexible one, consisting of two detectors, a thermal conductivity and a flame ionization, and three columns permitting full analysis of all reactants and possible product isomers.

Fresh catalysts were ground, pelletized, then crushed and sieved, and only the fraction -800+500 μm was used in the tests. In a typical experiment, 500 mg of catalyst were spread on a Pyrex frit fitted in the reactor, between two layers of glass beeds to avoid any fluidization. The gas feed contained 3.5 % vol. n-pentane, oxygen, at an oxygen to alkane molar ratio equal to 0.5 or 1.0, and the balance helium. The total flow was 25 cm^3/min (ambient conditions), which corresponded to a space velocity of the order of 10^3 h^{-1}. The usual duration of a catalytic run was about 1h, during which several injections were made to the gas chromatograph. The results reported here correspond to a time-on-stream of 30-45 min. Existence of hot spots in the catalytic bed during reaction, was not evident in our experiments, since temperature fluctuations were kept to a minimum: catalyst temperature varied no more than ±2°C during runs.

Results

Catalyst Properties. The composition and specific surface area values of the fresh catalysts and the mechanical mixtures are shown in Table I. The magnesium to vanadium ratios corresponding to the bulk composition of the samples, as measured by atomic absorption spectroscopy, were very close to the theoretical ones (within experimental error). The BET surface areas of the catalysts were relatively large for these types of materials, due to the low temperatures applied during calcination. Usually, the surface areas of similar materials reported in the literature *(2, 4, 5)* are below 1 m^2/g, this being probably the result of calcination temperatures as high as 800°C. The BET surface area of the Sb_2O_4 used in the preparation of the mechanical mixtures was 0.5 m^2/g.

The X-ray patterns of the pure fresh catalysts revealed the existence of a single phase in each sample, the one corresponding to the stoichiometry used for preparation. The XRD pattern of the fresh magnesium orthovanadate catalyst is presented as an example in Figure 1a. The structure of the magnesium pyrovanadate catalyst

Table I. Catalyst Compositions and Specific Surface Areas

Catalyst	*Bulk Mg/V atom ratio*[a]	*Surface area (m^2/g)*
$Mg_2V_2O_7$	1.0 (1.0)	6
$Mg_3V_2O_8$	1.6 (1.5)	16
$Mg_2V_2O_7$ - Sb_2O_4	1.0 (1.0)	3
$Mg_3V_2O_8$ - Sb_2O_4	1.6 (1.5)	9

[a] Measured by atomic absorption spectroscopy. Numbers in parentheses denote the theoretical values.

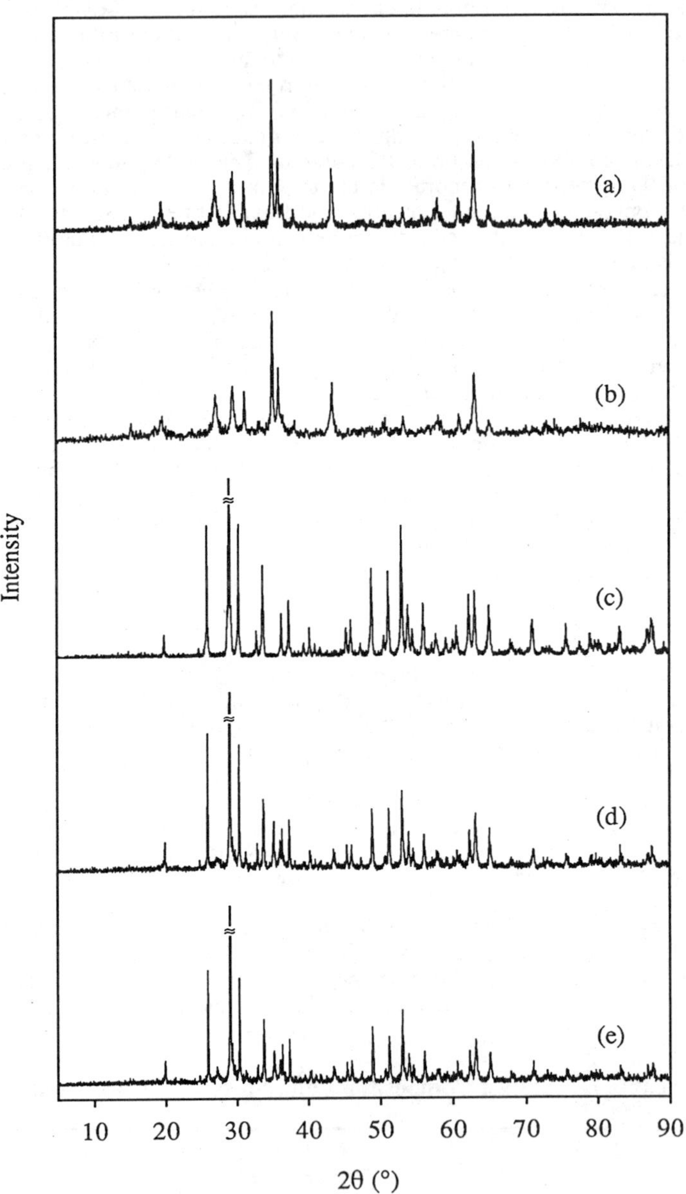

Figure 1. X-Ray diffraction patterns of catalysts.
(a) $Mg_3V_2O_8$ fresh (b) $Mg_3V_2O_8$ used (c) Sb_2O_4 fresh
(d) $Mg_3V_2O_8$ - Sb_2O_4 fresh (e) $Mg_3V_2O_8$ -Sb_2O_4 used.

corresponded to the more stable alpha phase. Also the structure of the antimony oxide corresponded to the alpha phase, containing only minor, unidentified impurities (Figure 1c). As expected, the patterns of the freshly prepared mechanical mixtures included only lines attributable to the particular catalyst used and Sb_2O_4; the XRD pattern of the mechanical mixture of magnesium orthovanadate with Sb_2O_4 is shown in Figure 1d.

The atom ratios of the main elements on the surface of the samples, calculated from XPS measurements, are shown in Table II. For both pure catalysts, the surface Mg/V ratios were higher than the bulk ones. Similar observations for increased magnesium content on the surface of magnesium vanadates, have been made by other researchers *(1, 15)*. No significant differences in the Mg/V surface ratios were observed between the pure catalysts and their mechanical mixtures. In the latter materials, the surface Sb/(V+Mg+Sb) ratios were lower than the ones corresponding to their bulk composition.

Table II. Atom Ratios on the Catalyst Surface[a]

Catalyst	*Mg/V atom ratio*		*Sb/(V+Mg+Sb) atom ratio*		
	fresh	*used*	*fresh*	*used*	
$Mg_2V_2O_7$	1.39	1.36	-	-	
$Mg_3V_2O_8$	2.06	1.91	-	-	
$Mg_2V_2O_7$ - Sb_2O_4	1.23	1.14	0.20	0.29	(0.30)
$Mg_3V_2O_8$ - Sb_2O_4	2.23	1.95	0.08	0.14	(0.28)

[a] Measured by XPS. Numbers in parentheses denote the theoretical values.

In general, no significant changes of physicochemical properties have been observed when comparing the samples before and after the catalytic tests. No formation of new phases was observed either for the pure catalysts, or for the mechanical mixtures. The XRD patterns of the magnesium orthovanadate catalyst and its mechanical mixture, both already used in reaction studies, are shown in Figure 1b and 1e, for comparison with the patterns of the fresh samples. Concerning the mechanical mixtures, the proportion of antimony on the surface increased for the samples after testing. When using pure catalysts in some cases where oxygen was totally consumed during the reaction, drastic changes of color from white-like to black, and an increased surface coverage with carbon, have been observed for the samples after testing.

Reaction Studies. We checked for the possible occurrence of gas phase homogeneous reaction by performing several series of tests in an empty reactor at temperatures 300-500°C and with oxygen to pentane ratios in the feed equal to 0.5, 1.0 and 2.0. When the ratio was 0.5 or 1.0, the empty reactor showed practically no activity in the temperature range examined. For example, for the oxygen to paraffin ratio equal to one and at 500°C, the total conversion of pentane was just 1.5%, and the products were pentenes and C_2 - C_3 light hydrocarbons. When the oxygen to paraffin ratio in the feed was changed to 2.0, the increase of the extent of the homogeneous reaction was impressive. Even at 350-400°C the n-pentane conversion was as high as 30%, and the main products were C_5 alkenes, C_2 and C_3 hydrocarbons, carbon oxides, and probably oxygenated organic compounds. As the temperature increased to 450°C, the conversion dropped abruptly to 2%, then increased slowly with temperature up to 5% at 500°C. In this temperature range, products were pentenes, C_2 and C_3, while oxygenated molecules disappeared entirely.

In the catalytic experiments, reaction products were mainly pentenes (1-pentene, trans- and cis-2-pentene), pentadienes (trans- and cis-1,3-pentadiene), carbon oxides and water. Oxygenated organic products have not been detected. In some cases, light hydrocarbons, mainly C_2 and C_3, were also produced in small amounts due to cracking reactions. With increasing temperature, total conversion of n-pentane, as calculated from the reactant concentration in the inlet and outlet of the reactor, increased for all samples tested. Conversions of n-pentane achieved at various reaction conditions are presented in Figures 2 and 3, along with the corresponding selectivities. It should be mentioned that at the early stages of this research, analytical system limitations did not permit full analysis of carbon oxides. Therefore, in order to have a common basis for comparing selectivities of all the samples presented here, we use in this paper the selectivity to C_5 dehydrogenation products relative to selectivity to CO_2 only, that is the ratio of moles of n-pentane converted to C_5 unsaturates to the moles of n-pentane converted to CO_2. Since the CO_2 proportion in carbon oxides varied with temperature and active phase present in the sample, comparison is better to be restricted to results obtained at one temperature with samples containing the same active phase.

As can be seen in Figures 2 and 3, the total n-pentane conversions achieved by the two pure magnesium vanadates under the same reaction conditions had very similar values. The extent of the dehydrogenation reaction and the extent of full oxidation to carbon oxides were different, with the orthovanadate phase being in all cases more selective to dehydrogenation compared to the pyrovanadate one. The same observations can be made when comparing the two mechanical mixtures. Total pentane conversion achieved with the mechanical mixtures was in all cases slightly higher than the one corresponding to the amount of active phase contained in the samples. When compared on selectivity terms, the mechanical mixture containing magnesium pyrovanadate had similar behavior to the pure catalyst, while the mixture containing orthovanadate was better than the pure catalyst.

Hydrocarbon product distributions obtained at reaction temperatures 400, 450 and 500°C, and oxygen to paraffin ratios in the feed equal to 1.0 and 0.5, are shown in Figures 4 and 5. The predominant dehydrogenation products were in most cases the 2-pentenes. As the temperature and the conversion increased, the pentadiene fraction also became significant, especially with the pure catalysts. Hydrocarbon distributions obtained with mechanical mixtures did not differ markedly from the ones with the corresponding catalysts for temperatures up to 450°C. The mechanical mixtures produced in general less alkadienes than the pure phases, the difference coming almost totally from the 1-pentene fraction. At 500°C, the selectivity patterns of the pure and the mixed catalysts were much more different, with the mixtures always producing less alkadienes than the corresponding pure samples. Oxygen to pentane ratio had no major influence on the hydrocarbon product distributions for all samples tested, apart from a decrease of the alkadiene product fraction with increasing oxygen content in the feed.

The possible existence of mass and energy transport limitations under our catalytic test conditions has been checked by application of standard literature criteria *(16)*. It was found this way that even the more probable limiting steps, such as intraparticle mass transfer, interparticle heat and mass transfer and interphase heat transfer, do not prevail in most cases, especially at low conversions. At higher conversions, the increased complexity of the reaction scheme makes the calculation of the several parameters included in these criteria rather ambiguous. Therefore, the results of the calculations under these conditions are less accurate, although being still on the safe side. Some individual experiments in the low conversion range, where pure magnesium orthovanadate at a particle size of -500+300 μm was used, confirmed that there were no transport limitations.

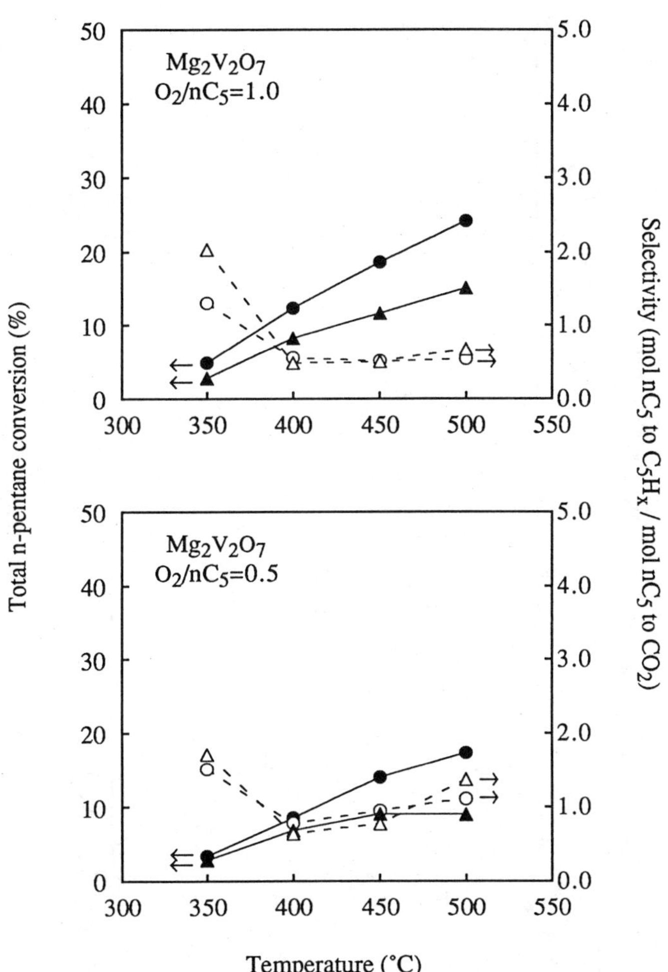

Figure 2. Conversion and selectivity results for the magnesium pyrovanadate containing catalysts.
● Total conversion, pure catalyst. ▲ Total conversion, mechanical mixture.
○ Selectivity, pure catalyst. Δ Selectivity, mechanical mixture.

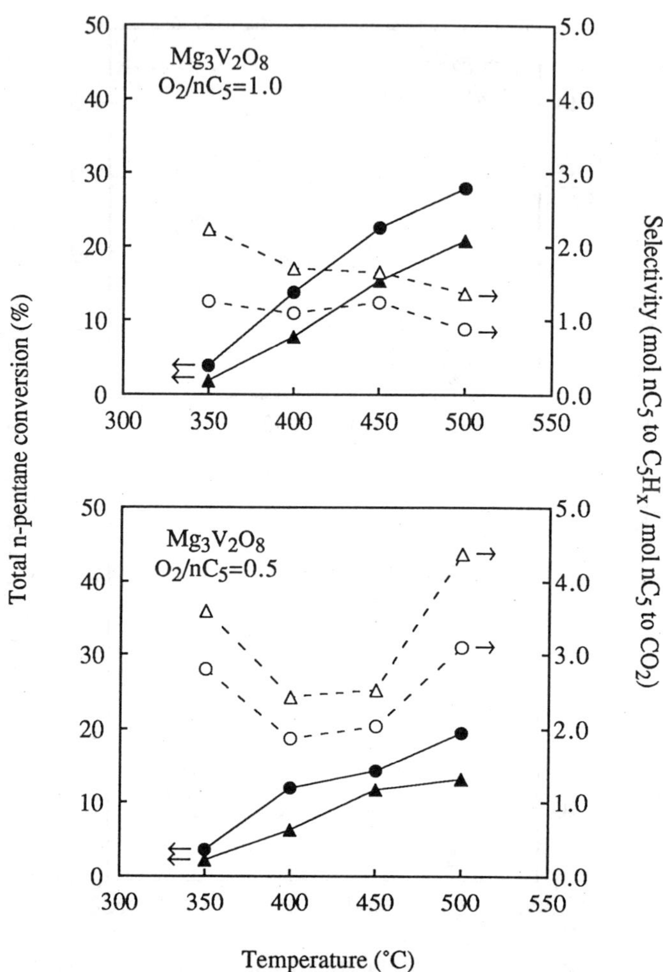

Figure 3. Conversion and selectivity results for the magnesium orthovanadate containing catalysts.
● Total conversion, pure catalyst. ▲ Total conversion, mechanical mixture.
○ Selectivity, pure catalyst. △ Selectivity, mechanical mixture.

Discussion

It should be noted that the magnesium vanadate samples were composed of a single phase, as revealed by their XRD patterns. This was a consequence of the preparation method used. The main advantage of the citrate method is that the combination of the existing elements in one phase, if chemically permitted, is favored. Phase impurities, if present, were in very low quantities, their XRD peaks being of the intensity of the background noise. There was an increased magnesium content on the surface though, observed for all samples by XPS, which did not lead to the formation of a phase detectable by X-ray diffraction. This probably reflects the existence of some surface positions which are deficient of vanadium ions or the formation of finely dispersed magnesium on the crystals of the vanadates. From the results to date on the oxidative dehydrogenation of n-pentane, there is no evidence of whether this increased Mg content on the surface affects catalyst activity or selectivity. In the case of propane, it has been reported *(1)* that the excess magnesium formed in magnesium - vanadium catalysts during their preparation, favors selectivity to propene.

For the mechanical mixtures, the apparent antimony content on the surface was found to be less than the bulk one. This can be attributed to differences in particle size between the two phases mixed, which resulted in an apparent surface enrichment of the more finely dispersed phase, in this case the magnesium vanadate, in a manner similar to what has already been reported for other two phase mixtures *(17)*. This particle size difference must be greater for the case of the mechanical mixture containing the orthovanadate phase than for the one containing the pyrovanadate, since the difference between the specific surface areas is much higher in the first case (Table I). This is also in accordance with the observation that the deviation of the surface Sb/(V+Mg+Sb) ratios from the theoretical values was larger for the $Mg_3V_2O_8+Sb_2O_4$ sample than for the $Mg_2V_2O_7+Sb_2O_4$ one (Table II).

The apparent antimony proportion on the surface of the mechanical mixtures increased after testing. The formation of a new phase on the sample surface during the catalytic reaction is not very probable, because the binding energies of elemental lines were the same for the samples prior to and after testing. A more plausible explanation is that carbon, deposited preferably on the catalyst active sites, shielded the XPS emissions of the corresponding elements, this resulting in a relative increase of the Sb signal.

Both pure magnesium vanadates proved to be effective for the activation of n-pentane, with the orthovanadate being the most selective for dehydrogenation products, i.e. alkenes and alkadienes. The enhanced selectivity of the ortho phase is in accordance with what has been observed for n-butane *(2)* and heavier molecules, like ethylbenzene *(4)*, while in the case of propane the selectivity of the pyro phase to propene is higher than that of the ortho phase *(1, 5)*. In the magnesium orthovanadate catalyst, active sites consist of isolated VO_4 tetrahedra, while in the pyrovanadate the active sites are V_2O_7, which are in fact corner-sharing VO_4 units. It is logical to suppose that an adsorbed alkyl intermediate, formed by the initial activation of the n-pentane molecule, can take up more oxygen from V_2O_7 than from VO_4. This could result in the pyrovanadate phase being less selective to dehydrogenated molecules than the orthovanadate one, as observed in the results of the present study.

The selectivity towards dehydrogenation products, generally appeared to decrease with increasing temperature (Figures 2 and 3). In fact, this decrease was due to the increase in the conversion. This inverse variation of selectivity with conversion, a common trend for catalytic oxidation reactions, is usually not very much influenced by temperature itself. An enhanced dehydrogenation selectivity was observed at 500°C for an oxygen to pentane ratio equal to 0.5, especially for the orthovanadate containing samples (Figures 2 and 3). This was probably a result of the fact that oxygen in the gas phase was almost totally consumed in these cases and only the pentane was in contact with the catalyst surface at the exit part of the catalyst bed. This could alter the development of selectivity, due to reaction of surface oxygen only, but no experiments were carried out with n-pentane alone fed to the reactor, in order to check the reactivity at these conditions. The relative amount of carbon monoxide produced could also play a role. If the CO_x produced under these conditions contained a greater proportion of CO, then the selectivity, as presented in Figures 2 and 3, would appear to be enhanced. However, in experiments where both carbon oxides were analyzed, the proportion of monoxide in carbon oxides was not higher in all cases of total oxygen consumption *(18)*.

The fact that the 2-pentenes were the predominant hydrocarbon products was probably the result of the greater probability of breaking first a methylene C-H bond instead of a methyl C-H bond. The methylene bonds are easier to break because they are usually weaker than the methyl ones *(19)*. This suggestion is consistent with the fact that the light hydrocarbons produced were usually C_2 and C_3, and is further supported by results of the literature concerning oxidation reactions of hydrocarbons of various chain lengths *(20-22)*. Under similar reaction conditions, a tendency of increase of the dehydrogenation rate is observed with increasing chain length of the hydrocarbon, that is with increasing number of methylene C-H bonds.

The two pure magnesium vanadates examined exhibited similar pentane conversions when tested under the same conditions. It should be pointed out that the specific surface area of $Mg_2V_2O_7$ was significantly lower than that of $Mg_3V_2O_8$. Therefore, it can be assumed that for the same surface area, the pyrovanadate catalyst could give higher conversions than the orthovanadate one, thus exhibiting higher specific activity (i.e. per unit surface area). This decrease of conversion per surface area with increasing magnesium content, indicates that the vanadium ions are responsible for developing the catalyst activity for n-pentane oxidative dehydrogenation. In our case, this can only be considered as a trend and cannot yet be extended to any quantitative activity conclusions, since the conversion levels were generally rather high. The conversion increased with temperature for all samples (Figures 2 and 3). The conversion levels at the higher temperatures examined were sometimes lower than expected, due to total oxygen consumption.

The addition of Sb_2O_4 to the catalysts had a general positive effect on the conversions achieved. At each temperature, total n-pentane conversion with each one of the mechanical mixtures was somewhat higher than the value corresponding to the amount of active phase in the sample. This observation is true even at the low temperature - low conversion points. Under the assumption of differential reactor operation, this could be extended to relative sample activities, thus reflecting an improvement of the catalysts through a cooperation between magnesium vanadates and antimony oxide. The latter material is inactive for the reaction when tested pure. However, the reaction studied here is in fact a very complicated scheme and its kinetics are unknown for the moment, so any conclusions on catalyst activities based on conversions should be addressed with caution. It is better that the results are interpreted in terms of conversion only.

Antimony oxide had little effect on catalyst selectivity for the pyrovanadate phase, but in the case of the orthovanadate, selectivity to dehydrogenation was generally improved with the mechanical mixtures. This improvement cannot be attributed solely to the lower conversions obtained with the mechanical mixtures, since in many cases the increase in selectivity was greater than the one resulting from the decrease in conversion alone. Further investigation is necessary to clarify the role of Sb_2O_4 as creator of selective or inhibitor of non-selective reaction sites and to confirm the existence of a remote control mechanism, as it has been observed in the case of other light alkanes such as butane *(23)*.

The effect of changing the oxygen to alkane ratio in the feed on the catalyst performance was in general the same for the two phases examined. Decreasing this

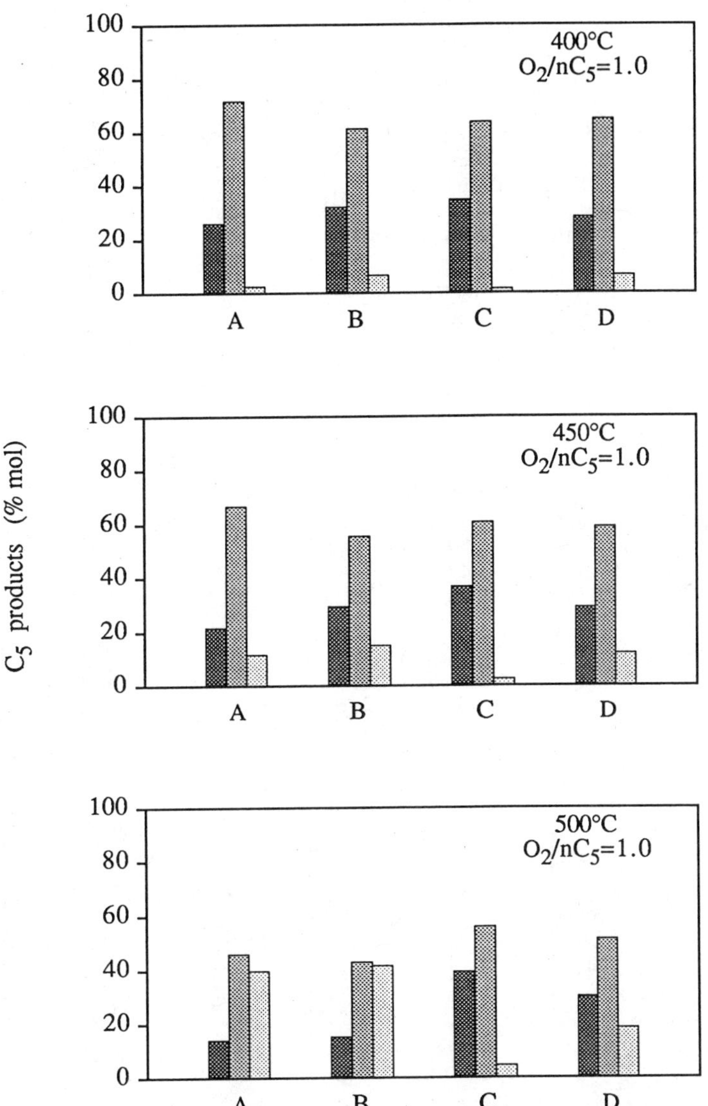

Figure 4. C_5 hydrocarbon product distribution for oxygen to n-pentane ratio in the feed equal to 1.0. Total n-pentane conversions are shown in Figures 2 and 3.
A: $Mg_2V_2O_7$ B: $Mg_3V_2O_8$ C: $Mg_2V_2O_7$-Sb_2O_4 D: $Mg_2V_2O_7$-Sb_2O_4
▩ 1-pentene ▩ 2-pentenes ▢ pentadienes

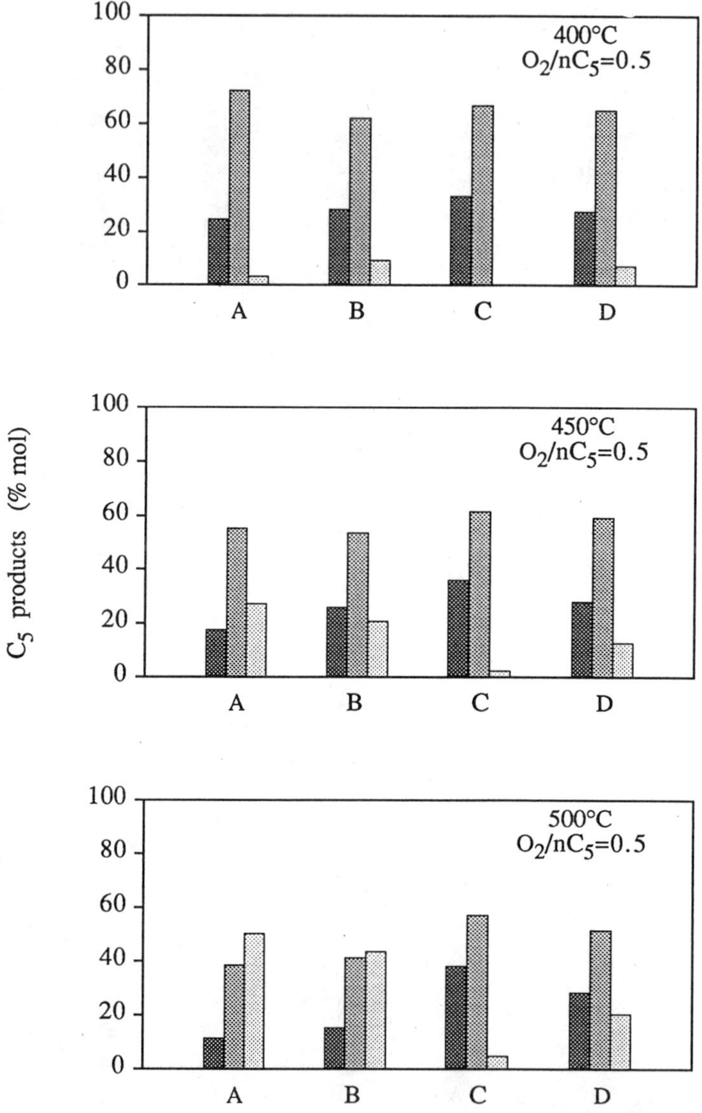

Figure 5. C_5 hydrocarbon product distribution for oxygen to n-pentane ratio in the feed equal to 0.5. Total n-pentane conversions are shown in Figures 2 and 3.
A: $Mg_2V_2O_7$ B: $Mg_3V_2O_8$ C: $Mg_2V_2O_7$-Sb_2O_4 D: $Mg_2V_2O_7$-Sb_2O_4
■ 1-pentene ■ 2-pentenes □ pentadienes

ratio from 1.0 to 0.5 resulted in a decrease of the total n-pentane conversion and an increase of the dehydrogenation selectivity. In the case of the samples containing the magnesium pyrovanadate phase, both pure and in the mechanical mixture, the increase in selectivity could be attributed almost totally to the decrease in conversion. In the case of the samples containing the ortho phase and at low oxygen to alkane ratio, the dehydrogenation selectivity was higher than the one accounted for solely by the decrease of n-pentane conversion. Analogous trends of decreasing conversion and increasing selectivity with decreasing oxygen content in the feed have been reported for similar oxidation reactions of ethane *(24)*, propane and butane *(20)* and ethylbenzene *(4)* on vanadium containing catalysts. In most of these cases, the effect on selectivity was more pronounced than the one on conversion.

The distribution of the dehydrogenation products, pentenes and pentadienes, was not much affected by the oxygen to n-pentane ratio in the feed (Figures 4 and 5). This supports the claim that the dehydrogenation and the combustion reactions usually proceed via parallel paths *(21, 22)*. This is especially true with short residence times, while at longer contact times, where a parallel-consecutive scheme of reaction is more probable, it is possible that the dehydrogenation reaction steps depend in the same way on the oxygen partial pressure.

The oxygen to n-pentane ratio in the feed affected appreciably the extent of the homogeneous reaction in the empty reactor. At low ratios, the oxygen was insufficient to activate the alkane molecule in the absence of catalyst. At higher oxygen content in the feed, homogeneous reaction took place, but then it seemed that there were two possible ways to proceed, depending on the temperature. Low temperature homogeneous reaction must be an exothermic process, possibly autocatalytic and leading to formation of both dehydrogenated and oxygenated products. High temperature homogeneous reaction has such a mechanism that the conversion rises very slowly with temperature and the products are only hydrocarbons. These results are in accordance with previous observations on hydrocarbon oxidation and combustion reactions *(25)* and the turnover temperature, in our case around 400°C, depends on reaction conditions and the reactor type.

Conclusions

Using the citrate method, high purity magnesium vanadates can be prepared, their actual bulk composition corresponding to the element stoichiometry used for preparation. Magnesium ortho and pyrovanadate prepared in this way proved to be effective for the activation of n-pentane, with the ortho phase being the most selective towards dehydrogenation products. Addition of antimony oxide to the ortho phase had in general a positive effect on catalyst selectivity. No major changes in catalyst performance were observed when antimony oxide was combined with the pyro phase. The homogeneous reaction at the temperature range examined was strongly influenced by the oxygen to alkane ratio, being favored by higher oxygen content in the reactor feed. Dehydrogenation product distributions varied mainly with the active phase present and with conversion and temperature.

Acknowledgments. S. A. Korili gratefully acknowledges the financial support of the European Union (Human Capital and Mobility Project ERBCHBI-CT94-1102). She also wishes to thank M. Genet for his help in XPS analyses and Dr A. Gil for valuable discussions.

Literature Cited

1. Gao, X.; Ruiz, P.; Xin, Q.; Guo, X.; Delmon, B. *J. Catal.* **1994,** *148,* 56.
2. Kung, M. C.; Kung, H. H. *J. Catal.* **1992,** *134,* 668.
3. Burch, R.; Crabb, E. M. *Appl. Catal. A* **1993,** *100,* 111.

4. Chang, W. S.; Chen, Y. Z.; Yang, B. L. *Appl. Catal. A* **1995,** *124,* 221.
5. Siew Hew Sam, D.; Soenen, V.;Volta, J. C. *J. Catal.* **1990,** *123,* 417.
6. Cichowski, R. S. *U.S. Patent No. 3,758,609* **1973**.
7. Cichowski, R. S. *U.S. Patent No. 3,862,910* **1975**.
8. Manning, H. E. *U.S. Patent No. 4,026,920* **1977**.
9. Watanabe, T.: Echigoya, E. *Bull. Jap. Petr. Inst.* **1972**, *14*, 100.
10. Centi,G.; López Nieto, J. M.; Iapalucci, C.; Brückman, K.; Serwicka, E. M. *Appl. Catal.* **1989,** *46,* 197.
11. Michalakos, P.M.; Kung, M.C.; Jahan, I.; Kung, H.H. *J. Catal.* **1993,** *140,* 226.
12. Weng, L.T.; Ruiz, P.; Delmon, B. In *New Developments in Selective Oxidation by Heterogeneous Catalysis;* Ruiz, P.; Delmon, B., Eds.; Elsevier: Amsterdam, 1991; 399.
13. Courty, P.; Ajot, H.; Marcilly, C.; Delmon, B. *Powder Tech.*, **1973**, *7*, 21.
14. Gregg, S. J.; Sing, K. S. W. *Adsorption, Surface Area and Porosity, 2nd edition*; Academic Press Inc.: London, 1982.
15. Guerrero-Ruiz, A.; Rodriguez-Ramos I.; Fierro, J. L. G.; Soenen, V.; Herrmann, J. M.; Volta, J. C. In *New Developments in Selective Oxidation by Heterogeneous Catalysis;* Ruiz, P.; Delmon, B., Eds.; Elsevier: Amsterdam, 1991; 203.
16. Mears, D. E. *Ind. Eng. Chem. Process Des. Develop.* **1971**, *10(4)*, 541.
17. Weng, L. T.; Spitaels, N.; Yasse, B.; Ladriere, J.; Ruiz, P.; Delmon, B. *J. Catal.* **1991,** *132,* 319.
18. Korili, S. A.; Ruiz, P.; Delmon, B. submitted for publication.
19. *Handbook of Chemistry and Physics, 56th edition*; Weast, R. C., Ed.; CRC Press Inc.: Cleveland, OH, 1976.
20. Chaar, M. A.; Patel, D.; Kung, H. H. *J. Catal.* **1988,** *109* , 463.
21. Kung, H. H.; Michalakos, P.; Owens, L.; Kung, M.; Andersen, P.; Owen, O.; Jahan, I. in *Catalytic Selective Oxidation*; Oyama, S. T.; Hightower, J. W., Eds.; ACS Symposium Series 523, American Chemical Society: Washington, DC, 1993; 389.
22. Mamedov, E.A.; Cortés Corberán, V. *Appl. Catal. A* **1995,** *127,* 1.
23. Ruiz, P.; Bastians, P.; Caussin, L.; Reder, R.; Daza, L.; Acosta, D.; Delmon, B. *Catalysis Today* **1993,** *16,* 99.
24. Juárez López, R.; Godjayeva, N. S.; Cortés Corberán, V.; Fierro, J. L. G.; Mamedov, E. A. *Appl. Catal. A* **1995,** *124*, 281.
25. Benson, S. W.; Nangia, P. S. *Account Chem. Res.* **1979**, *12 (7)*, 223.

Chapter 15

Identification of Active Sites and Structure Sensitivity of the Oxidative Dehydrogenation of Propane over Vanadium Magnesium Oxide Catalysts

Amalia Pantazidis and Claude Mirodatos[1]

Institut de Recherches sur la Catalyse, Centre National de la Recherche Scientifique, 2 avenue Albert Einstein, F–69626 Villeurbanne Cédex, France

The oxidative dehydrogenation of propane was studied over a series of VMgO catalysts with a V content varying from 5 to 45 wt.%. Selectivity was shown to be little dependent on V content, while a high dependency of the intrinsic activity on the surface composition and structure was observed. Thorough catalyst characterization under reaction conditions demonstrated that this sensitivity was related to the combination of surface V^{5+} units with ionic vacancies to activate propane molecules. Thus, the highest propene yields were obtained when V content and phase sintering were tuned. This corresponded to an optimized redox potential, a key parameter in the Mars - Van Krevelen mechanism.

INTRODUCTION

VMgO catalysts are known to be active and selective for the oxidative dehydrogenation of propane (ODHP) (1-4). Initially, most of the authors focused on the catalytic performances of the pure VMgO phases. Chaar et al. (1) claimed that magnesium orthovanadate ($Mg_3V_2O_8$) was responsible for catalytic activity and selectivity, while Siew Hew Sam et al. (2) concluded that magnesium pyrovanadate (α-$Mg_2V_2O_7$) was the most selective phase. Recently, other researchers stated that the biphasic VMgO catalysts exhibited quite different catalytic performances from the pure phases (3-4). Gao et al. (3) stated that the selectivity of the orthovanadate was promoted by being in intimate contact with a coexisting pyrovanadate or excess MgO phase. They suggested a synergetic effect, which was attributed to reconstruction or recontamination phenomena. Corma et al. (4) proposed that the better catalytic performance observed on samples with low V/Mg surface atomic ratios was related to the presence of isolated VO_4 tetrahedra. They also suggested a correlation between catalytic activity and the redox properties of the catalysts, as well as with the acid-base character of the oxygen present on the catalyst surface, based on TGA and XPS results respectively. It can be stressed however that no detailed characterization of the VMgO catalysts under conditions close or similar to the ODHP conditions was

[1]Corresponding author

0097–6156/96/0638–0207$15.00/0

attempted, although major chemical and structural changes of the solids might be expected under reaction conditions.

The present work aims at understanding the ODHP performance of VMgO catalysts, optimized by varying vanadium content and calcination temperature. A combined approach of *ex situ* and *in situ* characterization methods (XRD, XPS, ESR, ^{51}V NMR, FT-IR, UV-vis, electron microscopy, electrical conductivity, microcalorimetry and transient kinetics) led to a description of the active sites under the reaction conditions.

EXPERIMENTAL

Preparation Method. $Mg(OH)_2$ was first precipitated from a magnesium nitrate solution (Prolabo) by a potassium hydroxide solution (Prolabo). After centrifugation, washing and drying, $Mg(OH)_2$ was added to a hot aqueous solution of NH_4VO_3 (Aldrich) containing 1 vv. % of NH_4OH. The suspension was evaporated to dryness and then dried overnight at 120°C. The resulting solid was crushed and calcined at 550°C for 6 h under an O_2/He flow. The vanadium content was varied from 4.6 to 44.7 wt.%. The number placed before the formula V/VMgO corresponded to the weight percent of vanadium in the calcined sample. The reference VMgO phases of ortho-, pyro- and metavanadate were prepared from the 32V, 36V and 44V/VMgO catalysts, respectively, after calcination at 800°C.

Surface Area. Surface areas were measured by nitrogen adsorption, according to the BET method.

X-ray Diffraction. Samples were analyzed on a Phillips PW 1710 diffractometer with the Cu Kα radiation under standard acquisition conditions. JCPDS-ICDD standard spectra software was used to determine the phases.

Nuclear Magnetic Resonance of ^{51}V. NMR spectra were obtained on a Bruker MSL-300 spectrometer operating at 78.86 Hz under static or rotating conditions. The spectra were recorded at room temperature and were referenced to $VOCl_3$. The VMgO spectra were compared to the ones obtained for the reference VMgO phases i.e. ortho-, pyro- and metavanadate.

Electron Spin Resonance. ESR spectra were recorded with a Varian E9 spectrometer using the 9.3 GHz (X band mode) microwave frequencies. A dual cavity was used and the g values were measured by comparison with a DPPH sample (g=2.0036). VMgO catalysts were tested after calcination at 550°C and after 50 h of reaction at 500°C. The samples were outgased for 12 h in a special cell allowing ESR measurements without further contact with air and humidity. The measurements were recorded at 77K. The spectra were referenced to $VOSO_4$. The concentration of V^{4+} was determined and compared to the total amount of V ions given by chemical analysis. Assuming that the V ions could only be V^{4+} or V^{5+}, a bulk ratio V^{4+}/V^{5+} was determined.

X-ray Photoelectron Spectroscopy. XPS measurements were carried out on a VG-model ESCA III spectrometer. The exciting radiation was Mg*Ka* (1253.6 eV). The binding energies were calibrated with respect to the C1s energy at 284.9 eV corresponding to the carbon impurities. Atomic concentration ratios were calculated accounting for their cross section and mean free path, after correcting the intensity ratios with theoretical sensitivity factors proposed by the manufacturer. The samples were pre-treated directly on the sample holder in a heated reaction chamber under flowing gases before being transferred into the XPS chamber. The samples were analyzed (i) after pretreatment under a 1% O_2/He mixture with increasing temperature (5°C/min) up to 560°C and (ii) after the same pretreatment followed by reaction at 560°C for 30 min under the reacting mixture C_3H_8/O_2/He = 7.1/3.7/89.2%.

Electrical Conductivity. The catalyst powder (0.1<d<0.2 mm) was slightly compressed between two Pt electrodes. The electrical resistance was measured with an ohm-meter (Kontron, Model DMM 4021) for $1 \leq R \leq 2\times10^6$ ohm and with a teraohmmeter (Guildline Instruments Model 9520) for $10^6 \leq R \leq 10^{14}$ ohm. The electrical conductivity σ (ohm^{-1} cm^{-1}) was determined by the formula: $\sigma = (1/R)*(h/S)$, where R is the measured electrical resistance (ohm), S the cross sectional area of the electrodes (cm^2) and h the catalyst bed thickness (cm). The partial pressure of oxygen and temperature were varied. All the samples behaved as semiconductors, with σ varying exponentially with T according to $\sigma = \sigma_o \exp(-E_c / RT)$, where E_c is the activation energy of conduction (kJ/mol). The slope (1/n) of the various isotherms log σ = f(log PO_2) indicated the nature, n- or p-type, of the semiconductors (5). Reduction-reoxidation (redox) sequences were carried out at 500°C in the presence of reaction mixture (PC_3H_8 = 18 torr, PO_2 = 15 torr), then of oxygen (PO_2 = 15 torr) and finally of propane (PC_3H_8 = 18 torr). Steady-state was attained between each step.

Catalytic Test. The catalysts (30 mg) were tested at 500°C in a fixed bed reactor under atmospheric pressure. The reacting feed consisted of a C_3H_8/O_2/He mixture with a total flow rate varying between 20 and 100 ml/min and the C_3H_8/O_2 ratio between 1:0.66 and 1:8. The gas composition was determined by automated gas chromatography (TCD and FID).

In situ Diffuse Reflectance Fourier Transform Infra-Red Spectroscopy (DRIFT). The experiments were carried out with an *in situ* DRIFT cell connected to the catalytic testing set-up. The catalyst was loaded into the DRIFT cell after *ex situ* calcination. Catalyst loading varied from 20 to 60 mg depending on the sample. The sample was pretreated under an argon flow at increasing temperature (5°C/min) up to 550°C. Then the Ar stream was switched to the reaction mixture C_3H_8/O_2/He = 13.2/6.6/80.2 with a total flow rate of 50 ml/min. The spectra were recorded at each step of the standard testing procedure.

RESULTS

Catalytic performance optimization.

The effect of the vanadium loading on the catalytic performance under given conditions is illustrated in Fig. 1. A clear maximum of the specific activity (a) was observed at around 12 to 14 wt.% of V, whereas the selectivity to propene (c) remained almost constant (65-75%) over a wide range of V content (12-45 wt.%). For the intrinsic activity (b) a maximum was also found at the same low V content, but high values were also noted for high V content (35-45 wt.%). However, these latter values must be considered with care since very low surface areas were

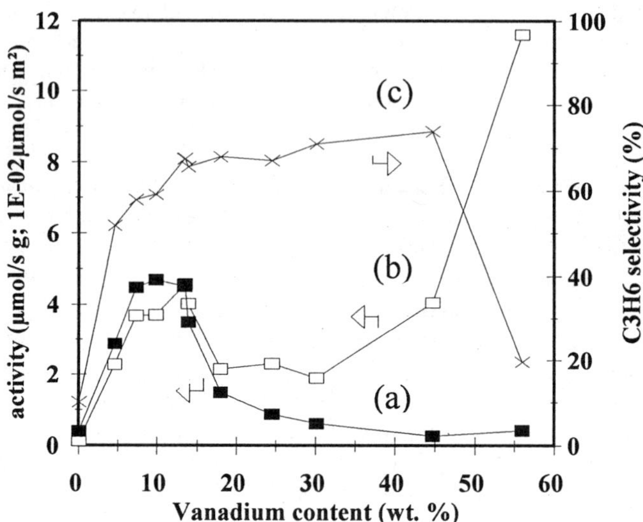

Fig. 1. Catalytic activity of the VMgO catalysts as a function of vanadium loading at 500°C; (a) specific activity, µmol/s g; (b) intrinsic activity, µmol/ s m^2; (c) propene selectivity, %.

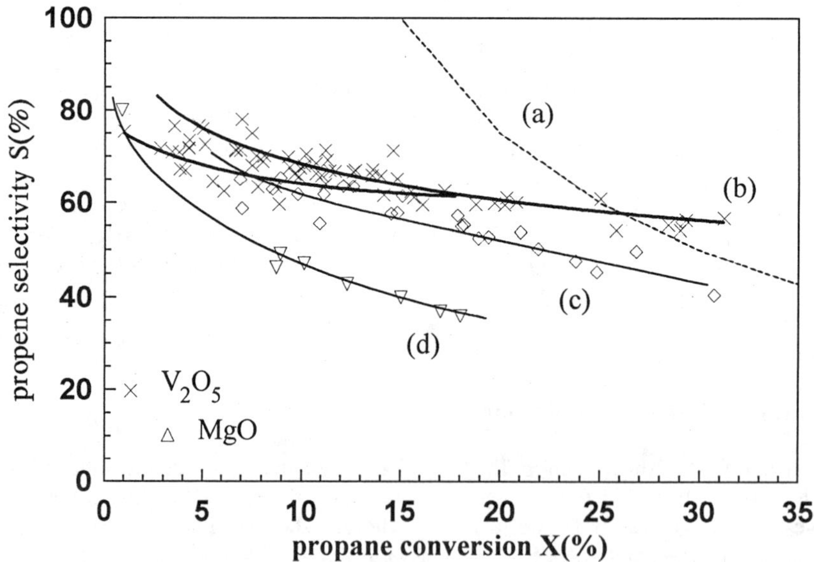

Fig. 2. Selectivity to propene versus conversion over the whole series of tested VMgO catalysts; (a) iso-yield at 15%; (b) 10-25 wt.% V; (c) 5-8 wt.% V; (d) 45V/VMgO.

measured for these samples (Table 2). Pure MgO exhibited poor activity and selectivity, while pure V_2O_5 showed high intrinsic activity but poor C_3H_6 selectivity with CO as the major product (S_{CO} = 65 %).

Fig. 2 summarized numerous experiments carried out with various reaction conditions (changes in contact time, reaction temperature, $P(C_3H_8)/P(O_2)$ ratio) aimed at optimizing the propene yield over the prepared series of catalysts. For the sake of clarity, the curves fitting the data are reported (curves b, c, and d). The data for pure magnesia and vanadia are also reported. These results clearly indicated that i) different catalytic behavior existed depending on domains of vanadium content, ii) the best performances corresponded to a rather large range of V content (10-25 wt.%) and iii) the selectivity was essentially controlled by the conversion level for each series of catalysts.

The surface area was found to decrease with increasing calcination temperature or time on stream. The influence of sintering on the catalytic performance of the 14V/VMgO catalyst is reported in Fig. 3. The intrinsic activity of the catalyst tended to increase with decreasing surface area (from 80 to 40 m^2/g). A slight increase in propene selectivity was observed. The latter can be ascribed to slight decreases in the conversion level. However, at the lowest surface area, both intrinsic activity and propene selectivity decreased significantly.

Catalyst characterization: vanadium content and calcination temperature effect.

The VMgO catalysts were characterized by XRD and ^{51}V NMR as summarized in Table 1. As vanadium content increased the MgO bands (2θ = 42.9 and 62.3) disappeared progressively while the VMgO bands increased, with a change from the $Mg_3V_2O_8$ orthovanadate (2θ = 35.2) to the α-$Mg_2V_2O_7$ pyrovanadate phase (2θ = 28.5) and finally to the MgV_2O_6 metavanadate phase (2θ = 29.1). At high vanadium content, only the VMgO bands were observed (Fig. 4, a-c). The coexistence of a $Mg_3V_2O_8$ orthovanadate phase with an excess of magnesium oxide was evidenced at low vanadium loading where the maximum activity was observed (Fig. 4, d). Note that the XRD peaks of the $Mg_3V_2O_8$ phase were very poorly resolved with the peaks very broad, indicating a high dispersion and a possible distorted structure of the orthovanadate. As calcination temperature increased these peaks narrowed, indicating an improvement in the solid crystallinity (Fig. 4, d-e). For the 14V/VMgO catalyst, the width of the 2θ = 35.2 peak ($Mg_3V_2O_8$) varied from 0.33 to 0.20 degrees and that of the 2θ = 42.9 peak (MgO) varied from 0.68 to 0.24 degrees, as calcination temperature increased from 550 to 900°C.

The ^{51}V NMR specta of the orthovanadate phase presented a narrow signal at δ_{iso} = -536.0 ppm; the pyrovanadate presented two signals at δ_1 = -534.4 ppm and δ_2 = -595.2 ppm; and the metavanadate phase showed four signals at δ_1 = -311.4 ppm, δ_2 = -472.7 ppm, δ_3 = -585.7 ppm, and δ_4 = -704.0 ppm, in accordance with literature data (6). The identification of phases for each VMgO catalyst was based on the spectra of these reference phases by means of deconvolution. By comparing peak widths of catalysts with those of the reference phases it was observed that

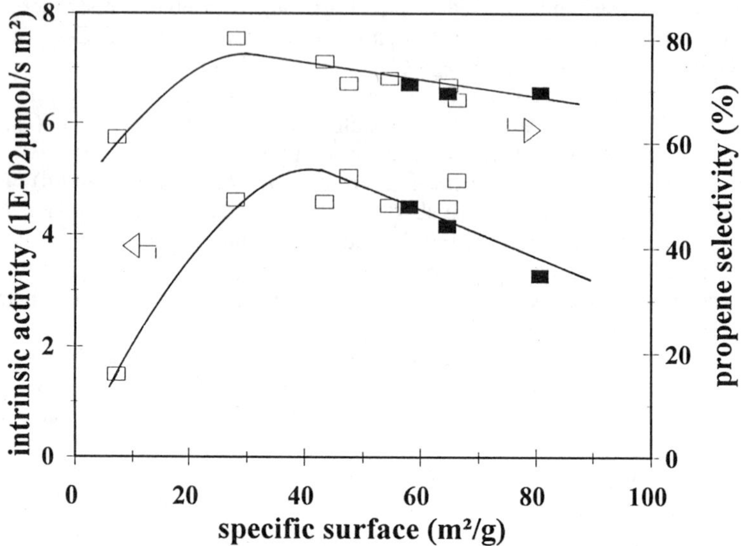

Fig. 3. Effect of sintering due to increasing calcination temperature (□; from 550 to 900°C) and time on stream (■; 5, 15 and 50 h) on the catalytic performances of the 14V/VMgO catalyst.

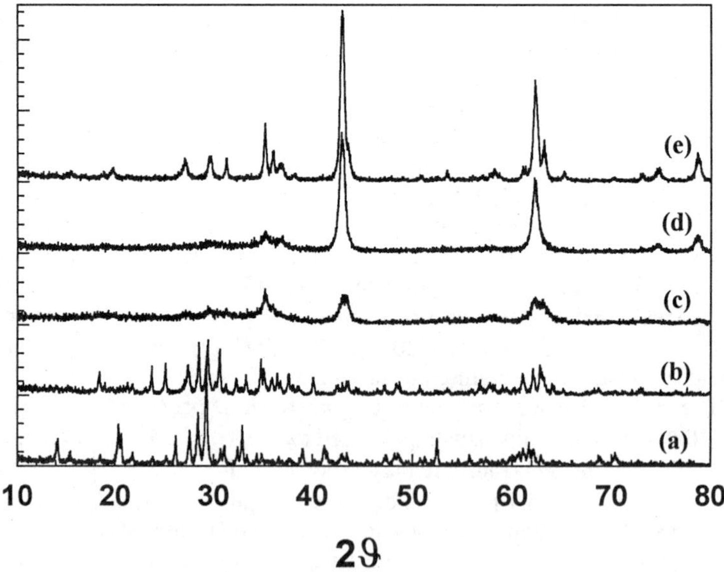

Fig. 4. XRD spectra for the (a) 45V/VMgO, (b) 30V/VMgO, (c) 25V/VMgO, and the (d-e) 14V/VMgO catalysts calcined at (d) 550°C, and (e) 800°C.

crystallinity improved as vanadium loading increased and/or calcination temperature was higher (Table 1).

Table 1. Identification of the VMgO catalysts by XRD and ^{51}V NMR.

Catalyst	$T_{calc.}$ (°C)	Degree of crystallization (^{51}V NMR)	Identified phases (XRD +^{51}V NMR)
14V/VMgO	800	93%	$MgO + Mg_3V_2O_8$
14V/VMgO	550	47%	$MgO + Mg_3V_2O_8$
25V/VMgO	550	46%	$Mg_3V_2O_8$
30V/VMgO	550	80%	α-$Mg_2V_2O_7$
45V/VMgO	550	-	MgV_2O_6+α-$Mg_2V_2O_7$+V_2O_5

As observed in Table 2, a major decrease in the BET surface area was observed after V addition, the lowest value corresponding to the pure vanadate (4 m^2/g for V_2O_5). The electrical conductivity (σ) was very high for the unpromoted magnesia, markedly decreased after V addition (almost two orders of magnitude) and remained more or less constant up to 30 wt.% V. For higher V loading σ increased again (one order of magnitude).

Table 2. Effect of VMgO catalyst composition on electrical conductivity σ.

Catalyst	$T_{calc.}$ (°C)	S_{BET} (m^2/g)	σ (ohm cm)$^{-1}$	-1/n
MgO	550	270	2.22 E-07	1/5.6
14V/VMgO	800	43	5.04 E-09	1/18.9
14V/VMgO	550	81	2.25 E-09	1/4.4
25V/VMgO	550	38	1.85 E-09	1/4.9
30V/VMgO	550	14	3.47 E-09	n. d
45V/VMgO	550	7	4.07 E-08	1/6.2

All the slopes (1/n) of the various isotherms log σ = f(log PO_2) were found negative, indicating that all samples behaved as n-type semiconductors. The physical meaning of the 1/n values is not obvious. Values of n close to 6 correspond to the model of doubly ionized anionic vacancies ($V_{O^{2-}}^{2+}$ or □) (5). This would be the case for most of the tested samples. Much higher values (n>10) as observed for the 14V/VMgO sample calcined at 800°C may imply the contribution of other types of surface defects besides the above anionic vacancies (5).

Fig. 5 presents the DRIFT spectra obtained at 550°C under an Ar flow. The unpromoted magnesia (a) revealed a strongly carbonated surface, mostly in the form of a unidentate species (1483-1438 and 1310 cm^{-1}) with some free carbonate at 1420 cm^{-1}. The surface was also partly hydroxylated (mainly basic OH groups at 3740 cm^{-1}) (7). After promotion with vanadium (b) the overall carbonate concentration decreased significantly, especially the unidentate species. However, other species such as bidentate carbonates or hydroxycarbonates (1660-1620 and 1320-1310 cm^{-1}) tended to appear or at least to remain on the surface, if already existing before V promotion. The basic OH groups decreased and two types of more acidic hydroxyl

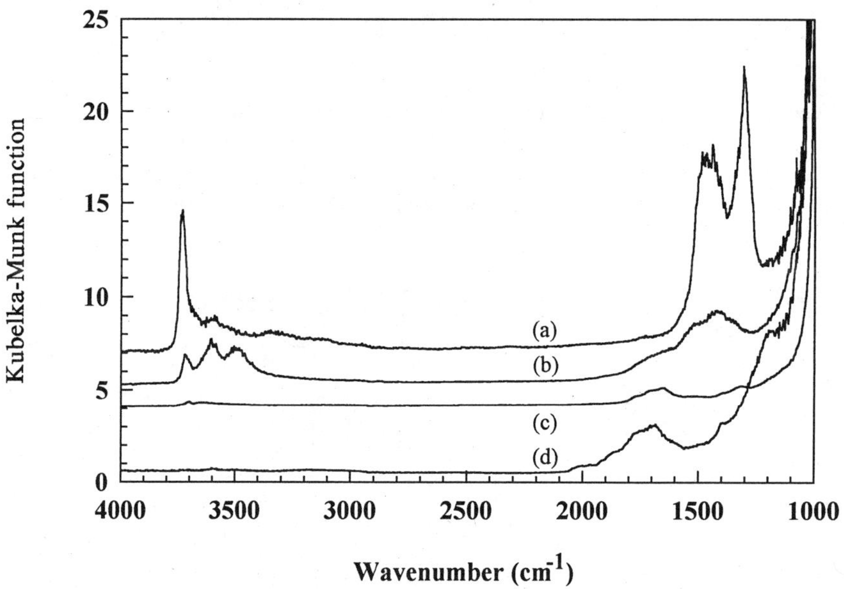

Fig. 5. Diffuse reflectance IR spectra recorded in situ at 550°C under Ar for (a) MgO, (b) 14V/VMgO, T_c = 550°C, (c) 14V/VMgO, T_c = 800°C, (d) 45V/VMgO.

groups appeared at 3600 and 3505 cm^{-1}. After calcination at 800°C (c) these OH groups almost vanished while most of the remaining traces of carbonates were bidentate (1660-1620 and 1320-1310 cm^{-1}), as expected from a highly decarbonated surface. At large V content (d), almost no hydroxyl groups were detected while bidentate carbonate or most likely acid carbonate groups (in line with the increase in acidity measured by calorimetry and SO_2 titration as reported below) were observed.

Table 3. Irreversible amounts of SO_2 and NH_3 adsorbed per surface unit as a function of V content.

Catalyst	NH_3 (μmol/m^2)	SO_2 (μmol/m^2)
MgO	0.25	4.17
5V/VMgO	0.93	1.71
7V/VMgO	1.11	n. d.
14V/VMgO	1.24	2.17
18V/VMgO	1.40	1.88
25V/VMgO	1.58	2.23
30V/VMgO	1.64	0.81
45V/VMgO	1.97	0

Changes in surface acidity and basicity as a function of vanadium content were investigated by SO_2 and NH_3 adsorption carried out at 80°C (8) (Table 3). Intrinsic surface basicity, measured by the amount of irreversibly adsorbed SO_2 per surface unit, was found to markedly decrease from unpromoted to promoted magnesia, then to be stable up to 25 wt.% V addition, then to decrease again and to vanish at high V content. In contrast, intrinsic acidity (irreversible NH_3 adsorption per surface unit) increased regularly over the range of V addition.

The changes in composition of the first layers of the 14V/VMgO samples were followed by XPS as a function of their surface area (depending on calcination temperature or time on stream, Table 4). In all cases, the Mg/V surface ratio given by XPS measurements was higher than the Mg/V bulk ratio (equal to 6.6) obtained by chemical analysis. This indicated either an Mg-enrichment of the surface or a V-depletion arising from vanadium incorporation into the catalyst subsurface. A significant increase in the Mg/V and O/V ratios was observed with decreasing surface area until a constant level was achieved, while the O/Mg ratio remained practically constant.

Table 4. Effect of surface area on the 14V/VMgO catalyst structure determined by XPS.

BET surface (m^2/g)	Atomic ratios			Concentration (%)			
	Mg/V	O/V	O/Mg	O^{2-} (MgO)	O^{2-} (VMgO)	OH^- ($Mg(OH)_2$)	$(CO_3)^{2-}$ ($MgCO_3$)
81	8.4	9.7	1.16	51.8	34.2	13.4	0.96
64	11.4	12.8	1.13	58.7	26.7	12.6	2.10
43	12.3	13.4	1.09	61.0	25.2	13.3	0.42
41	12.6	14.2	1.13	63.3	24.9	10.9	0.89

The O_{1s} peak deconvolution (O^{2-}, OH^- and $(CO_3)^{2-}$) in combination with the expected atomic ratios of the different phases (C_{1s} to O_{1s} for the carbonate species, Mg_{2p} to O_{1s} ratio for MgO and $V_{2p\ 3/2}$ to O_{1s} for $Mg_3V_2O_8$) led to a tentative evaluation of the concentration of phases on the catalyst surface (Table 4). The following features were outlined: (i) the surface was only weakly carbonated, but slightly more after reaction, in agreement with DRIFT data, (ii) hydroxyl groups were equally present before and after reaction, and (iii) the surface was enriched in MgO (or depleted in vanadate) after reaction and/or calcination at high temperatures.

In order to determine the V^{4+}/V^{5+} ratio, the $V_{2p\ 3/2}$ peak was deconvoluted and the corresponding surfaces integrated. Table 5 reports the corresponding binding energies (B.E.) and the calculated surface V^{4+}/V^{5+} ratios for the 14V/VMgO catalyst. The bulk V^{4+}/V^{5+} ratios determined by ESR (see experimental section) are also reported for comparison. The B.E. values are consistent with those reported in the literature (3-4). In all cases the total concentration of V^{4+} was smaller than the concentration of V^{5+}. Furthermore, the surface tended to be more reduced than the bulk vanadate. After reaction this difference between the surface and bulk state of reduction was more pronounced indicating a stronger reducing power of the reaction atmosphere in comparison with the calcination one.

Table 5. XPS and ESR results of the optimized 14V/VMgO catalyst.

Catalyst	$T_{calc.}$	Atomic ratio V^{4+}/V^{5+}		Binding energies $V_{2p\ 3/2}$ (eV)	
14V/VMgO	(°C)	Surface XPS	Bulk ESR	V^{5+}	V^{4+}
pretreatment	550	0.26	0.04	517.3	515.9
reaction	550	0.45	0.05	517.1	515.6
pretreatment	800	0.18	0.04	517.3	515.9
reaction	800	0.26	0.04	517.2	515.8

Catalyst characterization under reaction conditions.

The DRIFT experiments carried out under reaction conditions (Fig. 6) indicated that the surface of the optimized 14V/VMgO sample was considerably rehydroxylated (OH groups at 3300-3600 cm^{-1} and at 1640 cm^{-1}) without affecting the basic groups at 3700 cm^{-1}. No clear evidence of adsorbed hydrocarbons was found. The only intense bands were assigned to gaseous propane (2900-3020 cm^{-1}) and carbon dioxide (2360-2370 cm^{-1}). The small bands in the range 1560-1437 cm^{-1} indicated the presence of carbonate species as well as the expected C-H bending vibrations of gaseous propane and propene (7).

The above observations were confirmed by non steady-state and isotopic transient steady-state analysis, as detailed in (9), which gave a description of the surface occupancy under reaction conditions. The main conclusions were: i) no significant accumulation of hydrocarbon or coke deposits occurred under reaction conditions, ii) the formation of propene did not directly involve gaseous oxygen, iii) no measurable reversible and dissociative oxygen adsorption occurred on the surface under reaction conditions, iv) gaseous carbon dioxide rapidly equilibrated with surface carbonate species by fast adsorption/desorption, and v) carbon dioxide precursors accumulated on the catalyst surface.

Kinetic measurements of electric conductivity were carried out at 500°C in the presence of propane or under reaction conditions. Fig. 7 reports the data obtained for the 14V/VMgO sample calcined at 800 and 550°C (a and b) and for the 45V/VMgO sample calcined at 550°C (c). At time 0, the catalysts were under oxygen at 500°C. Note that the σ value under oxygen for the 800°C calcined sample was twice the one for the 550°C calcined sample (Table 2). As the reacting mixture oxygen/propane was introduced (point A), σ increased by several orders of magnitude, indicating a severe reduction of the catalysts but to levels depending on the samples. The highest increase corresponded to the 45V/VMgO sample (x 490) and to the 14V/VMgO sample calcined at 550°C (x 620). A minor increase was observed for the 14V/VMgO sample calcined at 800°C (x 22). Note that the slope of the increasing curves indicated that the most easily reduced catalyst was the 45V/VMgO sample. A significantly slower reduction was observed for the two 14V/VMgO samples. When reintroducing oxygen (point B), the electrical conductivity dropped to the initial values for the 14V/VMgO samples calcined at 800 and 550°C, indicating that the catalyst was reoxidized to its initial state. In contrast, the 45V/VMgO sample remained partially reduced. When pure propane was introduced (point C), the obtained values were slightly higher than those obtained in the presence of the reacting mixture with the same ranking of catalysts. The catalysts were thus less reduced under reaction conditions than in the presence of pure propane.

DISCUSSION

From the catalytic data illustrated in Fig. 1-3, it was concluded that no major effect of V concentration or temperature of calcination was noted for the selectivity to propene. In contrast, important effects of V concentration and treatment conditions were observed on intrinsic activities. These results indicated that within the range of V explored, the intrinsic activity for the ODHP reaction on VMgO catalysts is structure sensitive but the selectivity does not depend on a particular structure or phase. The latter conclusion disagrees somewhat with literature proposals assuming that either the pyrovanadate phase (2) or the orthovanadate phase (1) was the selective phase for the ODHP. Actually, the selectivity is proposed to be related to a competition between propane and oxygen for the same active sites (9), therefore little dependent on the site environment. Detailed examination of the state of VMgO catalysts under reaction conditions allows more precise conclusions about the actual nature of the active catalytic sites for ODHP.

State of the catalyst.

Pure MgO bulk material is known to be a poor electrical conductor (5). However, a high conductivity was measured on the present MgO sample (Table 2). This result suggests a preferential surface conductivity favored by mobile adspecies, such as hydroxyl groups and surface carbonate or hydroxycarbonates (5), which were detected by DRIFT spectroscopy (Fig. 5a). The addition of even low amounts of

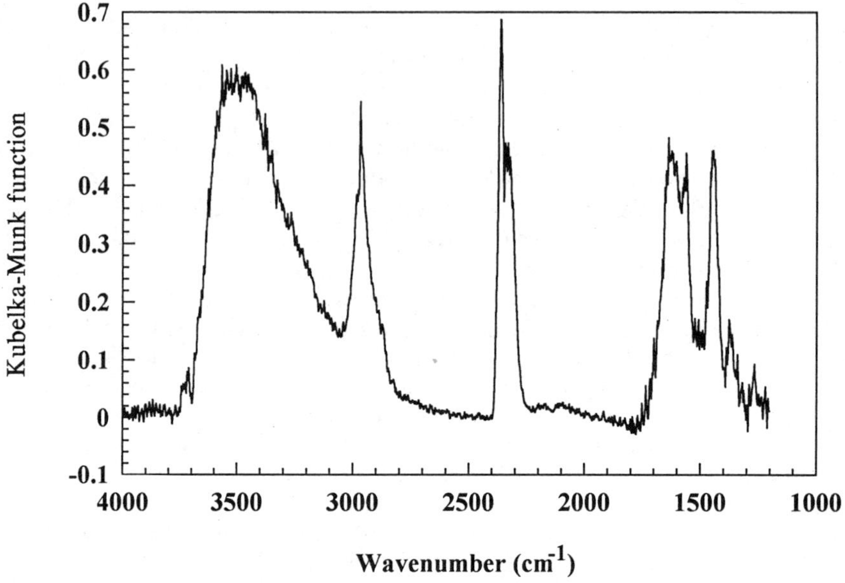

Fig. 6. DRIFT spectra of the 14V/VMgO catalyst (T_C = 800°C) recorded at 550°C under reaction conditions.

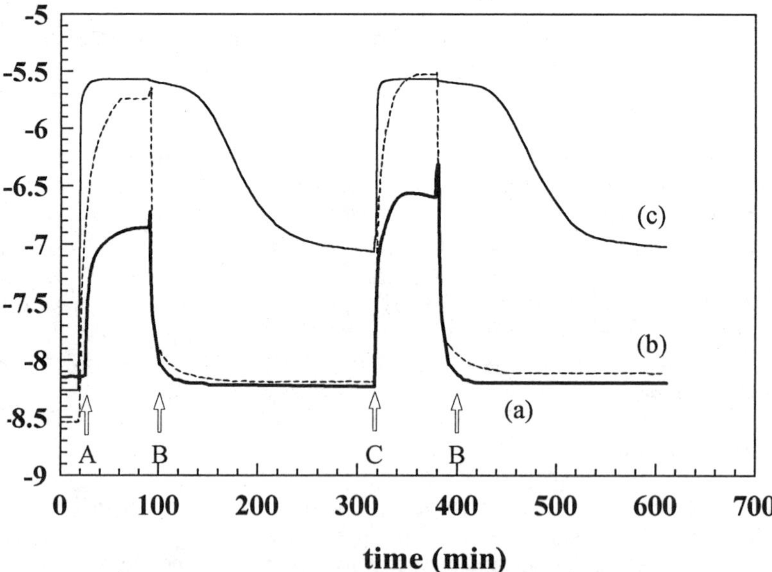

Fig. 7. Variation of electrical conductivity σ during redox sequences at 500°C on the 14V/VMgO catalyst calcined at 800°C (a) or at 550°C (b) and on the 45V/VMgO catalyst calcined at 550°C (c) initial atmosphere: oxygen, A: reaction mixture, B: vacuum, then oxygen, and C: propane.

vanadium oxide to magnesia induced a considerable loss in electrical conductivity (Table 2). Though significant, the loss of surface area observed after V addition (most likely due to a sintering effect induced by the easily melting vanadate) cannot explain by itself the almost two orders of magnitude decrease in electrical conductivity. Therefore, the observed decreased conductivity of the MgO material upon V addition is more likely to arise from changes in the surface composition of the MgO grains than in their morphology.

Various changes in surface composition were induced by V addition: (i) a strong decarbonation and dehydroxylation of the surface was observed by DRIFT spectroscopy (Fig. 5), (ii) a marked decrease in intrinsic basicity and a regular increase in intrinsic acidity (Table 3), and (iii) the development of the $Mg_3V_2O_8$ orthovanadate phase (Fig. 4, Table 1). A combination of the above effects probably explains the observed drastic changes in surface conductivity.

Another type of change in surface composition was observed by varying the surface area of the 14V/VMgO catalyst (Fig. 3-4, Tables 1, 4). XRD and ^{51}V NMR showed the progressive crystallization of the orthovanadate phase with increasing calcination temperature. This also corresponded to an increase in the O/V and Mg/V surface ratios determined by XPS, which indicated a surface enrichment in MgO and/or a depletion in vanadate possibly arising from V incorporation into the catalyst subsurface, in agreement with the model proposed by Corma et al. (10). Another effect of the calcination at 800°C was i) to increase the electrical conductivity (Table 2) and ii) to decrease significantly the reducibility of the vanadate phase under either the reacting mixture or a pure propane atmosphere (Fig. 7). Calcination under extreme conditions (900°C) led to an over-sintered material where uncovered MgO was now the dominant surface phase.

From the above results, some requirements for obtaining an active and selective catalyst under ODHP conditions can be proposed: i) a moderate V_2O_5 loading (from 15 to 44 wt%) and ii) a mild temperature calcination (800°C). This leads to a surface rather depleted in vanadium, which is only partially reduced under reaction conditions. Both acid and basic hydroxyl groups are present on the working system together with carbonates and hydroxycarbonate species.

Nature of the active sites.

The most active and selective catalysts were essentially formed of mixed MgO and $Mg_3V_2O_8$ orthovanadate phase. UV-vis DR studies (not reported here) showed that the surface concentration of isolated tetraedral species VO_4^{2-} (290 nm) seemed to be maximum at very low V content, i.e., out of the optimized V content range. The presence of distorted VO_4^{2-} units (320-340 nm) was observed in the active and selective mixed phases, indicating the formation of polymeric species within the optimum V content range. In addition, the positive effect of the calcination at 800°C on the intrinsic activity corresponded more to the insertion of vanadium into the MgO matrix rather than to a surface enrichment in the VO_4^{2-} species. Thus, the simple picture of isolated VO_4^{2-} surface species as active sites (1, 4, 10) remains therefore questionable. In more recent studies (11), it was suggested that the relatively small molecule of propane (as compared to butane) could interact only with one V unit for steric consideration. As a consequence, either VO_4 monomers, as observed in orthovanadate, or two corner-shared VO_4 polymers would act as active sites for ODHP. This enlarged concept fully agrees with our observations.

Several other physico-chemical properties also seem to contribute to the definition of active centers for ODHP. No direct relationship between total acidity and activity was found since intrinsic acidity was shown to increase regularly over the range of active and selective samples, while significant changes in intrinsic activity were noted. Acid sites could either be V-OH groups or V-O-COOH acid carbonates, as revealed by DRIFT spectroscopy. Though the latter can be discarded as active/selective centers (since they are essentially observed only at high V content), they could participate in non selective pathways, e.g. decomposition into gaseous CO. This would explain the higher selectivity to CO characteristic for V-rich samples. For instance :

$$\text{V-O-COOH} + \text{V-OH} \rightarrow \text{CO} + 2\ \text{V=O} + H_2O \qquad (1)$$

In contrast to acidity, basicity is obviously a prerequisite for catalysts with good performance. If one excepts the strong basicity observed on unpromoted MgO, a mild and constant basicity was measured over the whole range of active and selective catalysts. For catalysts with high V content, no basicity was measured, which corresponded to non selective samples. Two opposing roles of basic sites can be proposed from the *in situ* characterization studies:

i) A positive role in site regeneration. It was shown from non steady-state experiments that propene could be formed from propane in the absence of oxygen during a transient period, clearly demonstrating a Mars - Van Krevelen mechanism via lattice oxygen species (9). However, gaseous oxygen was shown to be necessary for regenerating the active surface, i.e. to reoxidize the sites reduced during the catalytic cycle. The required basicity could therefore be directly involved in the process of oxygen activation. As a matter of fact, the catalysts were shown to behave as n-type semiconductors. This means that gaseous oxygen could capture free electrons responsible for the n-type semiconductor behavior leading to the formation of O^- and O^{2-} adspecies according to the equations (5):

$$2e^- + 0.5\ O_2(g) \rightarrow O^{2-}_{(ads)} \qquad (2)$$

$$e^- + 0.5\ O_2(g) \rightarrow O^-_{(ads)} \qquad (3)$$

Equation (2) would correspond to site regeneration, since lattice oxygen was involved in the propane activation process. Equation (3) would rather be related to the generation of highly reactive oxygen species which might lead to CO_x production, as proposed in (9).

ii) A negative role in the non selective route to CO_2 formation. Carbonate species were shown to form under reaction conditions (Fig. 5) and to be in equilibrium with gaseous carbon dioxide according to (9):

$$CO_3^{2-} \rightleftharpoons CO_2 + O^{2-} \qquad (4)$$

As required in a Mars - Van Krevelen mechanism, the redox potential of the catalytic material appears as a key factor in the propane activation process. By investigating the changes in electrical conductivity under oxidative or reducing atmosphere, it was observed that catalysts which were easily reduced, such as 45V/VMgO, were not active under ODHP conditions. It was also noted that these catalysts could not be easily reoxidized under an O_2 atmosphere (Fig. 7c). In contrast, optimized catalyst performance for the ODHP, as observed using the 14V/VMgO sample calcined at 800°C, corresponded both to a totally reversible redox capacity and to a limited reduction under reaction conditions (Fig. 7a, Table 5). In order to understand these results, let us recall that the electrical conductivity was shown to involve doubly ionized lattice vacancies, i.e., empty vacancies (□). The electrical conductivity was shown to increase under reaction conditions. This effect can be accounted for by assuming that the electrons arising from the V^{5+} to V^{4+} reduction by propane become available for conductivity via the empty vacancies according to :

$$C_3H_8 + 2\{V^{5+} - \square + O^{2-}\} \rightarrow C_3H_6 + \{V^{4+} - \square\} + 2\ OH^- \quad (5)$$

Thus, for a very high V concentration, a fast and deep catalyst reduction occurred under reaction conditions (as observed for the 45V/VMgO sample, Fig. 7). This would hinder the redox cycles, and therefore decrease the site turn-over frequency. Other factors could also be considered, such as an excess of surface acidity and a lack of surface basicity.

For the case of a surface depleted of V^{5+} species (corresponding to the optimized 14V/VMgO sample calcined at 800°C), a fast and reversible reduction of the remaining surface units occurred under reaction conditions, which favored the catalytic redox process. Moreover, a relatively high concentration of oxygen vacancies is expected due to an enlarged VMgO/MgO interface, which could also account for the high intrinsic activity. By developing a larger interface between the MgO support and the vanadate phase, crystallographic defects such as "coherent interface" defects postulated by P. Courtine et al. (12) could be envisioned. When the calcination temperature is too high (900°C) or the surface concentration of V units becomes too low (even if a larger vacancy concentration is expected from the oversintering process), the site requirement is no longer fulfilled and the catalytic performance drops (Fig. 3). A modelling of the MgO/VMgO interface is in progress to test the validity of such a proposal.

CONCLUSIONS

At variance with literature proposals assuming that specific VMgO phases were responsible for ODHP, the present study revealed that the highest propene yields were achieved over MgO and VMgO mixtures (ortho and possibly pyro phases). Selectivity to propene was shown to be little dependent on the type of phase dispersed on magnesia over a wide range of V content. However severe decrease in selectivity was noted when the acid/base balance observed on selective materials was disrupted. In contrast with selectivity, intrinsic activity was shown to depend highly on the V content and on the catalyst surface area. Optimized catalyst performance was obtained for rather low concentrations of V^{5+} surface ions and for a high density of anionic vacancies. The latter are likely to arise from lattice defects created at MgO/VMgO interfaces, being favored after mild sintering of both phases. The combination of these two entities would constitute the active sites for ODHP. These sites would exist over a wide range of vanadium contents, which suggests that V^{5+} units may either be isolated or belong to polymeric units. This would account for the weak dependency of selectivity upon V content. This required balance between surface vanadium ions and anionic vacancies would control the redox potential of the active phase, a key parameter in the Mars-van Krevelen mechanism. These conclusions about the actual nature of the active sites for ODHP allow us to understand the efficiency of other types of supports which are able to generate the required mobile vacancies (e.g. niobia as demonstrated by Smits et al. (13)).

Acknowledgements : This work was supported by the EU network "Euroxycat". Thanks are due to A. Auroux, M. Brun, P. Delichère, J.M. Herrmann and A. Tuel for their valuable technical assistance and helpful discussions.

References

(1) M. Chaar, D. Patel and H. Kung, J. Catal. 109 (1988) 463.
(2) D. Siew Hew Sam, V. Soenen and J. C. Volta, J. Catal. 123 (1990) 417; A. Guerrero-Ruiz, I. Rodriguez-Ramos, J.L.G. Fierro, V. Soenen, J.M. Herrmann and J. C. Volta, in "Novel Development in Selective Oxidation by Heterogeneous Catalysis", P. Ruiz and B. Delmon, Eds., Elsevier, 72, (1992) 203.
(3) X. Gao, P. Ruiz, Q. Xin, X. Guo and B. Delmon, J. Catal. 148 (1994) 56.
(4) A. Corma, J. M. Lopez Nieto and N. Paredes, J. Catal. 144 (1993) 425.
(5) J.-M. Herrmann, *Applications of Electrical Conductivity Measurements in Heterogeneous Catalysis* in 'Catalyst Characterization' edited by B. Imelik and J. C. Vedrine, Plenum Press, New York (1994).
(6) O.B. Lapina, V.M. Mastikhin, A.A. Shubin, V.N. Krasilnikov, K.I. Zamaraev, Progress in NMR spectroscopy 24 (1992) 457.
(7) G. Socrates, *Infrared Characteristic Group Frequencies*, 2nd Ed., John Wiley & Sons, Chichester (1980).
(8) A. Pantazidis, A. Auroux, J.-M. Herrmann, C. Mirodatos, submitted to Catal. Today (1996).
(9) A. Pantazidis, C. Mirodatos, *Mechanistic Approach of the Oxidative Dehydrogenation of Propane over VMgO Catalysts by In Situ Spectroscopic and Kinetic Techniques*, 11th International Congress on Catalysis, Baltimore, 1996, in press.
(10) A. Corma, J.M. Lopez Nieto, N. Paredes, Appl. Catal.: General 104 (1993) 161.
(11) P.M. Michalakos, M.C. Kung, I. Jahan, H.H. Kung, J. Catal. 140 (1993) 226.
(12) P. Courtine, in J.F. Brazdil and R.K. Grasselli (Ed.), Solid State Chemistry in Catalysis, ACS Symposium Series 270 (1985) 370.
(13) R.H.H. Smits, K. Sheshan, J.R.H. Ross, in S.T. Oyama and J.W. Hightower (Ed.), Catalytic Selective Oxidation, ACS Symposium Series 523 (1992) 380.

Chapter 16

Promotion of Selectivity to Propene in $Mg_3V_2O_8$ Catalysts by Oxygen Spillover in the Oxidative Dehydrogenation of Propane

S. R. G. Carrazán[1], M. Ruwet[2], P. Ruiz[2], and B. Delmon[2]

[1]Departamento de Quimica Inorgánica, Facultad de Quimica, Plaza de la Merced s/n. Salamanca, Spain
[2]Unité de Catalyse et Chimie des Matériaux Divisés, Université Catholique de Louvain, Place Croix du Sud 2/17, B–1348 Louvain-la-Neuve, Belgium

The main objective of this work is to demonstrate that spill-over oxygen, Oso, promotes the selectivity to propene on $Mg_3V_2O_8$ during ODP via a remote control mechanism (RCM). Catalysts were prepared by a) mechanical mixtures of $Mg_3V_2O_8$+α-Sb_2O_4, b) calcining the mixtures at 873 K for 6 days and and c) by mechanical mixtures of $Mg_3V_2O_8$+α-$Mg_2V_2O_7$. A detailed physicochemical characterization (BET, XRD, XPS, TEM, AEM) of the solid state properties reveals that neither mutual contamination nor formation of a new phase occurs during catalytic reaction or calcination. α-Sb_2O_4 is inert in the reaction. Significant synergetic effects were observed in the mixtures: $Mg_3V_2O_8$+α-Sb_2O_4 exhibit an increase of selectivity by suppression of unselective reactions. $Mg_3V_2O_8$+α-$Mg_2V_2O_7$ mixtures exhibit an increase in the conversion, yield and selectivity and a decrease in the CO_2 formation. The role of Oso, which is produced by α-Sb_2O_4, would be that of inhibiting the non-selective sites. Oso is crucial in determining the catalytic properties of the magnesium vanadate catalyst. The existence of a cooperation via Oso between α-$Mg_2V_2O_7$ and $Mg_3V_2O_8$ could explain the discrepancies observed in the literature concerning the activity and selectivity of the pyro and ortho phases.

It is known that MgVO catalysts are relatively active and selective for the oxidative dehydrogenation of propane (ODP). However, no clear conclusion has yet been derived definig the role of the catalyst in this reaction.

Many studies attribute the catalytic performances to single MgVO phases. Kung's group (1-3) proposed magnesium orthovanadate, $Mg_3V_2O_8$, to be the active phase for the formation of alkenes. But Volta's group (4,5) attributed the highest selectivity to propene to pyrovanadate, α-$Mg_2V_2O_7$, considering the ability of this phase to stabilise V^{4+} ions associated with surface oxygen vacancies, and because of the easier reducibility of this phase compared to $Mg_3V_2O_8$.

0097–6156/96/0638–0223$15.00/0

This contradiction can arise from the phase purity which particularly depends on the preparation method (6-8) and which affects the catalytic behaviour of the MgVO phase, as has been observed by Delmon's group (9,10). Thus, the selectivity of $Mg_3V_2O_8$ can be improved by α-$Mg_2V_2O_7$ or excess of MgO in intimate contact. This suggests that some type of cooperation between separate phases must exist in these catalysts and that the presence of more than one phase is necessary to explain the catalytic properties of MgVO catalysts to promote selectivity to propene.

The main objective of this work is to demonstrate that spill-over oxygen, Oso, promotes the selectivity to propene on $Mg_3V_2O_8$ during ODP via a remote control mechanism (RCM).

This mechanism has been shown to operate on the oxidation of olefins to unsaturated aldehydes (11-12), the oxidative dehydrogenation of ethanol to acetaldehyde (13), the oxidative dehydrogenation of butene to butadiene (14), the oxidation of butane to maleic anhydride (15,16) and the dehydration of N-ethyl formamide to propionitrile (17). As α-Sb_2O_4 as an external phase has been shown to promote selectivity of the active phases in these reactions, we studied catalysts which contain mixtures of $Mg_3V_2O_8$+ α-Sb_2O_4.

As discussed above, MgVO catalysts in general do not consist of pure isolated phases. It is therefore interesting to study the synergetic effects existing between two different MgVO phases as, $Mg_3V_2O_8$+ α-$Mg_2V_2O_7$.

In order to demonstrate that the synergetic effects observed in these catalysts are due to a cooperation between two separate phases in contact, other possibilities which could explain those effects have been taken into account, such as contamination of one phase by small quantities of an element of the other or formation of a new phase. Thus, in order to minimize the contamination between the component oxides, the mechanical mixtures were obtained by dispersing separately prepared oxides in n-pentane, followed by evaporation of the solvent and drying without further calcination. The solid state evolution of $Mg_3V_2O_8$+α-Sb_2O_4 under harsh conditions was also studied.

A correlation of the catalytic activity and selectivity of the catalysts with a detailed physicochemical characterization of the solid state properties reveals that neither mutual contamination nor formation of a new phase occurs.

Experimental

Catalysts preparation.

Single phase catalysts: α-Sb_2O_4 was obtained by calcination of Sb_2O_3 in air at 773 K for 20 h. MgVO with Mg/V atomic ratios of 3/2, denoted as $Mg_3V_2O_8$, and 2/2, denoted as α-$Mg_2V_2O_7$, were prepared by the citrate method (18). A transparent solution of $Mg(NO_3)_2.6H_2O$ and NH_4VO_3 with the desired atomic Mg/V ratio was prepared. The citric acid was added in such a proportion to have 1.1 equivalent-gram of acid function per valence-gram of metal. After evaporation at 308 K, the viscous solution obtained was dried at 353 K for 15 hours. An amorphous solid organic compound was obtained. This compound was decomposed at 573 K for 16 hours. For the preparation of $Mg_3V_2O_8$ and α-$Mg_2V_2O_7$, different temperatures and times of calcination were used (Table I).

Two phase catalysts: a) mechanical mixtures of $Mg_3V_2O_8$/I+α-Sb_2O_4 were prepared by vigorously mixing the suspension of the powders in 200 ml of n-pentane for 3 minutes by means of a mixer (Ultra-Turrax from Janke & Kunkel) at room temperature. After evaporation of the n-pentane under reduced pressure with agitation, the mixtures obtained were dried in air at 323 K overnight, b) a mechanical mixture of $Mg_3V_2O_8$/I+α-Sb_2O_4 with 50% by weight of both oxides was calcined at 873 K for 6 days (this will be denoted calc. MM) and c) the same mixing method as in a) was used to prepare the mechanical mixtures of $Mg_3V_2O_8$/II+α-$Mg_2V_2O_7$/II except that the suspension was only agitated magnetically for 20 minutes.

Table I. $Mg_3V_2O_8$ and α-$Mg_2V_2O_7$ preparation conditions. 0, I and II indicate preparation number. Calcination temperature in K and calcination time in hours, and BET surface areas (m^2/g).

Sample	Temperature (K)	Time (h)	BET m^2/g
α-$Mg_2V_2O_7$/0	773	10	7.6
	823	6	
	873	6	
	923	6	
α-$Mg_2V_2O_7$/II	873	15	3.4
$Mg_3V_2O_8$/0	773	10	5.2
	823	6	
	873	6	
	923	6	
	973	6	
$Mg_3V_2O_8$/I	873	20	16.0
$Mg_3V_2O_8$/II	873	15	3.4
	973	7	
	1073	20	

The composition of the mechanical mixture is expressed as mass ratio by, Rm=weight A/(weightA+weightB) where A is $Mg_3V_2O_8$ and B is α-Sb_2O_4 or α-$Mg_2V_2O_7$. The Rm values are 0, 0.25, 0.50, 0.75 and 1 for the $Mg_3V_2O_8$/I+α-Sb_2O_4 system and 0.50 for $Mg_3V_2O_8$/II+α-$Mg_2V_2O_7$/II.

Catalysts characterization. All catalysts were characterized before and after the catalytic test using:

BET. the catalyst surface areas were measured by adsorption of krypton at 77 K on 200 mg of samples previously degassed at 423 K for 2h using the BET method in a Micromeritics Asap 2000 instrument.

XRD. XRD patterns were obtained with a high resolution X-ray diffractometer Siemens D5000 using Ni-filtered Cu-K α 1 radiation (λ=1.541 A).

TEM and AEM. The catalysts were studied with a JEOL-JEM 100C TEMSCAM electron microscope equipped with a Kevex 5100C energy dispersive spectrometer for electron probe microanalysis (EPMA). The samples were ground, dispersed in n-pentane with an ultrasonic vibrator and deposited on a thin carbon film supported on a standard copper grid.

XPS. XPS analyses were performed with a SSX-100 model 206 X-ray photoelectron spectrometer from FISONS. The analysis chamber was operated under ultrahigh vacuum with a pressure close to $5x10^{-9}$ Torr. The C1s, V2p, O1s, Sb3d, Mg2s and C1s bands were swept successively (in some catalysts, the most intense Mg(Auger) peak was swept instead of Mg2s) . The binding energy (BE) values were calculated with respect to C1s (BE of C-C,H fixed at 284.8 eV). V_2O_5 standard calcined in O_2 for 13 h at 773 K, as well as a V_2O_4, standard pretreated in H_2 at 393 K for 1h, were also analyzed.

Catalytic test. The catalytic tests were carried out in a continuous gas-flow system. The catalyst (particle size between 500-800 µm) was deposited between two inerts beds (each bed formed by small glass balls of 1 mm diameter and 10 cm^3 volume) and supported on a fixed quartz chip bed in a U-type quartz microreactor (internal diameter 12 mm) at atmospheric pressure. The reaction conditions were as follows: i) for catalysts α-$Mg_2V_2O_7$/0 and $Mg_3V_2O_8$/0, partial pressures for propane, oxygen and helium of 60.8, 182.4 and 516.8 Torr, respectively; total feed of 37 ml/min and reaction temperature of 823 K; amount of catalyst 70, 100 and 200 mg; volume of catalyst bed of 0.27, 0.39 and

0.77 cm^3 and contact times of 0.44, 0.63 and 1.26 s; ii) for all the other tests, the feed consisted of a mixture of propane, oxygen and helium with partial pressures of 42.4, 114.6 and 603 Torr, respectively; total feed of 33 ml/min and reaction temperature of 773 and 793 K. The amount of catalyst used was 500 mg in the case of $Mg_3V_2O_8$/I+α-Sb_2O_4 (non-calcined and calcined) and 1g was used for $Mg_3V_2O_8$/II+α-$Mg_2V_2O_7$/II. The volume of the catalyst bed was 0.9 and 3.4 cm^3 and the contact time 1.63 and 6.18 s, respectively. The range of rates of reaction per unit area was calculated from the lowest and the highest conversion values for each type of catalyst (see below, Table VI).

Analyses of reactants and products were carried out by gas chromatography. For the study of α-$Mg_2V_2O_7$/0 and $Mg_3V_2O_8$/0, analyses were made using two columns, a stainless steel VZ-07 (20 ft x 1/8", 60/80 mesh, He carrier flow of 30ml/min) for CO_2, propane and propene and a washed molecular sieve 5A (6ft x 1/8", 80/100 mesh, He carrier flow of 15 ml/min) for oxygen. For the tests with the other samples, a stainless-Steel Hayesep R column (1/8“ x 60/80 mesh, He carrier flow of 30 ml/min) was used to separate propane, propene and CO_2 at a temperature of 343 K. The main products were propene and CO_2. Oxygen and CO were not analyzed in these tests. Traces of another product situated after the peaks assigned to CO_2 were detected. This product could not be identified. No other oxygenate products was observed.

Expression of the synergetic effects. Our objective was the detection of synergy effects on the basis of measurements made under fixed conditions.

For simplification, we based the calculation of the synergetic effect on the simplified assumption of a zero order with respect to reactants. The synergetic effect observed for the mechanical mixtures is expressed as the increase of conversion or yield compared to the properly weighted average values obtained for the components of the mixture when alone.

The synergetic effects on the conversion are calculated by the following formula:

$$Syn^C = (C_{AB} - C_{(A+B)})/C_{(A+B)} \times 100, \qquad (1)$$

where C_{AB} is the conversion of the mixture and $C_{(A+B)}$ is the conversion in the absence of synergetic effect, namely $C_{(A+B)} = Rm.C_A+(1-Rm).C_B$, in which C_A and C_B are propane conversion of single phase catalysts A and B, respectively.

A similar equation can be obtained for the yield (Syn^Y). In this case Y_{AB} is the yield of the mixture and $Y_{(A+B)} = Rm.Y_A+(1-Rm).Y_B$ is the theoretical yield in the absence of synergetic effect, with Y_A and Y_B representing the yield in propene of single phase catalysts A and B, respectively. A similar equation can be obtained for the yield in CO_2.

For the selectivity, the synergetic effect is defined as:

$$Syn^S = \Delta S / S_{(A+B)} \times 100, \qquad (2)$$

in which $\Delta S = S_{AB} - S_{(A+B)}$, where S_{AB} is the selectivity of the mixture and $S_{(A+B)}$, the selectivity which would be observed in the absence of any synergetic effect, defined for a mixture with a given Rm such as $Y_{(A+B)} / C_{(A+B)}$.

Results.

Characterization. BET surface areas are presented in Table II.

The surface area is 16 m^2/g for $Mg_3V_2O_8$/I and is 3.4 m^2/g of $Mg_3V_2O_8$/II. All the $Mg_3V_2O_8$/I+α-Sb_2O_4 mechanical mixtures have surface areas corresponding to the properly averaged sum of those previously measured for pure oxides (see Table I and BETα-Sb_2O_4 = 1.0 m^2/g). No difference between fresh and used samples is detected.

However, a decrease in the BET value is observed for $Mg_3V_2O_8$/I+α-Sb_2O_4 after calcination at 823 K for 6 days.

Table II. BET surface areas in (m^2/g) for $Mg_3V_2O_8$/I+α-Sb_2O_4 (Rm=0.50) and $Mg_3V_2O_8$/II+α-$Mg_2V_2O_7$/II (Rm=0.50) mechanical mixtures.

Sample	$Mg_3V_2O_8$/I+α-Sb_2O_4	$Mg_3V_2O_8$/II+α-$Mg_2V_2O_7$/II
Fresh	9.2	3.4
Used	9.2	3.4
calc. MM	6.0	

The XRD pattern of MgVO with an Mg/V atomic ratio of 3/2 corresponds to the ortho phase, $Mg_3V_2O_8$, with a small amount of α-$Mg_2V_2O_7$ (5% for $Mg_3V_2O_8$/I and 3% for $Mg_3V_2O_8$/II). This percentage was estimated from the ratio of the integrated areas of the most relevant peaks of $Mg_3V_2O_8$ and α-$Mg_2V_2O_7$ (2θ = 35.22 for $Mg_3V_2O_8$ and 2θ = 28.03 for α-$Mg_2V_2O_7$). For MgVO with Mg/V atomic ratio of 2/2, the XRD pattern corresponds to α-$Mg_2V_2O_7$ with small quantities of impurities which could not be identified (see Figure 1). The XRD pattern of the antimony oxide corresponds to α-Sb_2O_4 (cervantite).

The XRD patterns of the fresh samples of mechanical mixtures corresponded to the superposition of those observed for the individual oxides. No change was observed in the position or intensities of the peaks and no new peak was detected after reaction or calcination.

TEM micrographs for the fresh and used $Mg_3V_2O_8$/I+α-Sb_2O_4 (Rm=0.50) are presented in Figures 2A and 2B, respectively.

The larger particles are α-Sb_2O_4 and the smaller ones, spheres of 0.05-0.1 μ, correspond to $Mg_3V_2O_8$. The difference in particle sizes corresponds to the BET results. The pictures show that α-Sb_2O_4 crystallites are surrounded by $Mg_3V_2O_8$. Some $Mg_3V_2O_8$ particles remain isolated. The morphology of $Mg_3V_2O_8$ particles in $Mg_3V_2O_8$/I+α-Sb_2O_4 is identical to that of fresh $Mg_3V_2O_8$/I. AEM analysis only detects isolated particles of α-Sb_2O_4 (point 1) and $Mg_3V_2O_8$ (points 4,5,6,7,8) (see Fig. 2B and Table III).

Table III. Analysis by electron microscopy of fresh and used $Mg_3V_2O_8$/I+α-Sb_2O_4.

Point	Fresh		Used	
	Mg/Sb	V/Sb	Mg/Sb	V/Sb
1	1.2×10^{-2}	9.6×10^{-3}	0.1	0.2
2	4.9	10.0	18.6	42.4
3	2.3	6.2	56.9	107.5
4	27.4	53.0	very high	very high
5			54.6	100.3
6			234.0	508.0
7			415.0	946.5
8			274.6	665.0

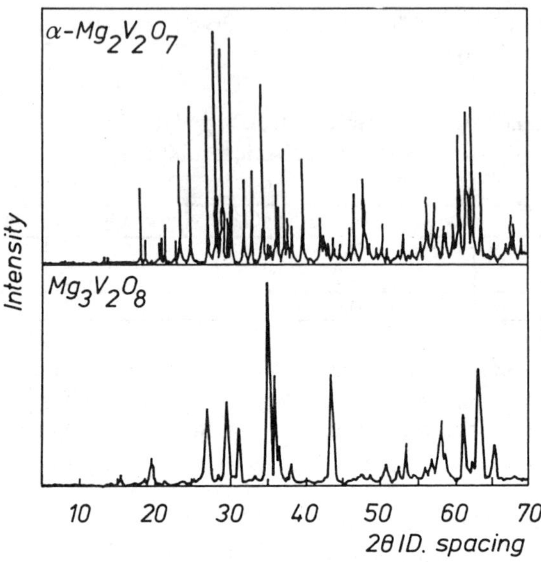

Figure 1. X-ray diffraction patterns of $Mg_3V_2O_8$, impurities: • α-$Mg_2V_2O_7$ and α-$Mg_2V_2O_7$, impurities: * not identified.

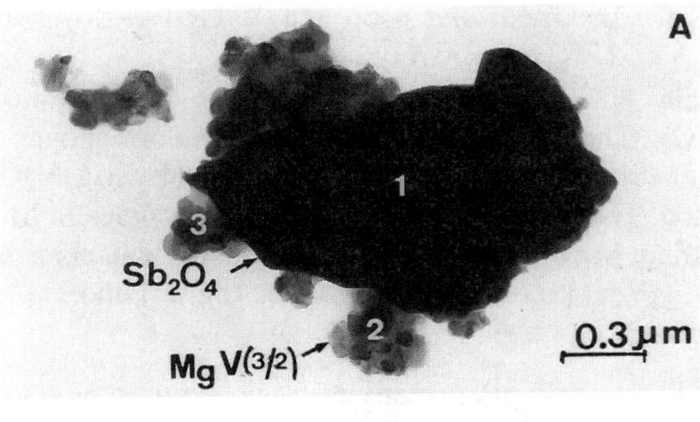

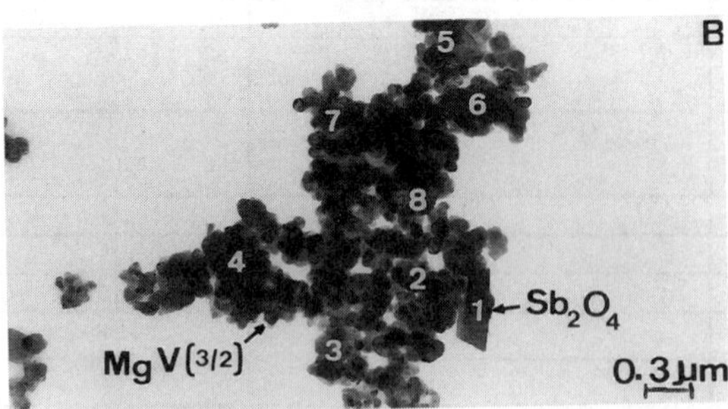

Figure 2. TEM micrographs of $Mg_3V_2O_8$/I+α-Sb_2O_4 (Rm=0.50). A) fresh and B) used.

XPS. Some remarks concerning the decomposition of the asymmetric V2p peaks must be made before examining the XPS results.

a) Difficulties were observed in the reconstruction of the V2p envelope using the value of 7.3 eV for the difference in BE between $V2p_{3/2}$ and $V2p_{1/2}$ peaks and the value of 2.0 for the $V2p_{3/2}/V2p_{1/2}$ ratio. The use of these constraints leads to an incorrect fitting which is evidenced by controlling the quality of the statistic parameter $\chi 2$ (see Fig. 3A). The fitting is especially bad for the $V2p_{1/2}$ peak. This is a result of the strong influence of the O1s (or O1s+$Sb3d_{5/2}$) peak in background subtraction.

b) A better fitting of the V2p envelope is achieved with no interval and area constraints compared to that obtained by using those constraints (see Fig. 3B). This leads a more accurate calculation of the quantity of total vanadium. In the following, this quantity will be used to calculate the XPS atomic ratios. Through use the contraints or not, calculations show a discrepancy of 6-9% for the total vanadium quantity. The position (517 ± 0.2 eV) and the FWHM (1.10 eV) for the $V2p_{3/2}$ peak of V^{5+} could also be accurately determined from the deconvolution of the spectrum of the pretreated V_2O_5 sample.

c) For the estimation of the XPS V^{5+}/V^{4+} atomic ratios in the mechanical mixtures, it is necessary to analyze the V_2O_5 and the V_2O_4 standards. In the case of V_2O_5, the FWHM of the $V2p_{3/2}$ peak of V^{5+} is 1.10. However, in spite of the precautions taken in preparing the V_2O_4 standard (see section 2.2), the most dominant peak is that of V^{5+}. This prevents the accurate determination of the position and FWHM of $V2p_{3/2}$ of V^{4+}.

Taking into account all these considerations, the estimation of the XPS V^{5+}/V^{4+} atomic ratios was made by analyzing the V2p envelope with the interval and area constraints and fixing the width of the $V2p_{3/2}$ peak of V^{5+} at 1.10 (see Fig. 3C). The use of both constraints (especially the area constraint) for the $V2p_{3/2}$ and $V2p_{1/2}$ peaks leads to a more realistic XPS V^{5+}/V^{4+} atomic ratio. Finally, the XPS V^{5+}/V^{4+} atomic ratios in $Mg_3V_2O_8$/I+α-Sb_2O_4 mechanical mixtures were calculated from the most intense $V2p_{3/2}$ peak (see Fig.3D).

d) The BE value of 517.3 ± 0.2 eV for the $V2p_{3/2}$ peak observed for all catalysts corresponds to that found in the V_2O_5 standard. This value is also in agreement with the literature data (19). Another $V2p_{3/2}$ peak, at lower BE value is also observed for all catalysts in the range 515.1-516.1 eV. This may be ascribed to V^{4+}. A similar BE range is reported in the literature (19, 20).

e) In spite of the fact that the BE separation between the Mg2s and $V2p_{3/2}$ peaks is greater than between Mg(Auger) and $V2p_{3/2}$ and that the Mg2s atomic sensitivity factor is smaller (0.575) than that of Mg(Auger) (4.1), similar V/Mg atomic ratios are obtained when using one or the other. Similar BE values of $V2p_{3/2}$, O1s, Mg (Auger) and $Sb3d_{3/2}$ are observed in all mechanical mixtures, fresh and used. These BE values correspond to those found in $Mg_3V_2O_8$ and α-Sb_2O_4 oxides (see Table IVa). The BE for O1s peak could not be accurately determined because it is superimposed on the $Sb3d_{5/2}$peak.

The XPS V/Mg atomic ratios for $Mg_3V_2O_8$ are lower than those of the bulk (see Table IVb). The XPS V/Mg, V/Sb and Sb/(V+Mg+Sb) atomic ratios for the $Mg_3V_2O_8$/I+α-Sb_2O_4 mechanical mixtures, fresh, used and calcined are presented in Table IVb. The XPS V/Mg atomic ratios are different from those of the bulk and remain unchanged after test. The XPS V/Sb and Sb/(V+Mg+Sb) atomic ratios are also different to those of the bulk. The XPS V/Sb atomic ratio decreases after test, whereas the XPS Sb/(V+Mg+Sb) atomic ratio remains nearly unchanged. When $Mg_3V_2O_8$/I+α-Sb_2O_4 is calcined a decrease in the XPS V/Sb atomic ratio is observed and all the XPS atomic ratios remain unchanged after reaction. The XPS C/V, C/Mg and C/Sb atomic ratios are not reported because no change in these ratios was observed after reaction or calcination.

V^{5+} and V^{4+} are found in $Mg_3V_2O_8$/I and in $Mg_3V_2O_8$/I+α-Sb_2O_4. Fresh and used $Mg_3V_2O_8$/I+α-Sb_2O_4 show similar XPS V^{5+}/V^{4+} atomic ratios (1.51 and 1.48, respectively), this value increasing for the calcined sample up to 1.95.

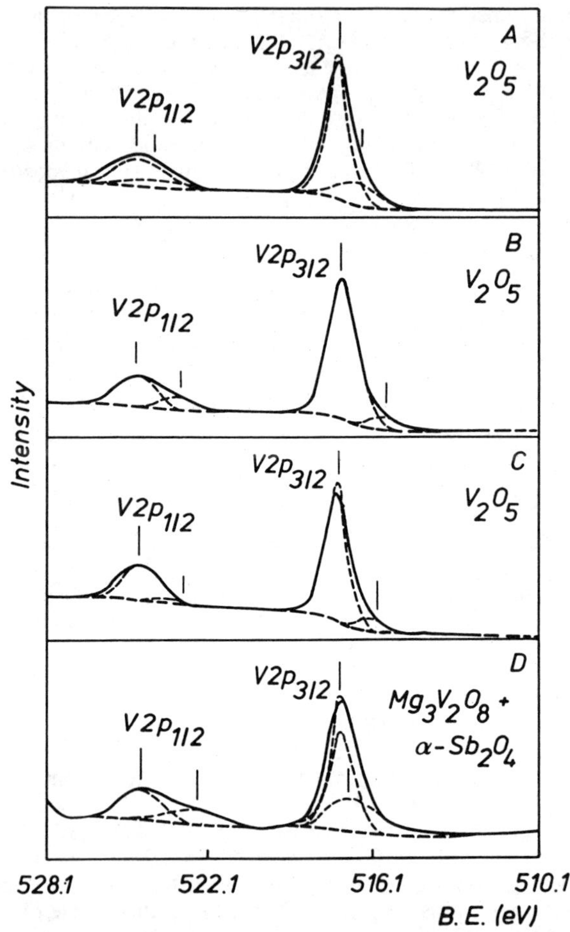

Figure 3. Analysis of XPS spectra of i) pretreated V_2O_5, A) with interval and area constraints, B) without constraints and C) with interval and area constraints and fixing the width at 1.10 for the V2$p_{3/2}$ peak of V^{5+} and ii) $Mg_3V_2O_8$/I+α-Sb_2O_4 (Rm=0.50) (D) using the constraints as in C).

Table IVa. XPS results. Binding energies for $V2p_{3/2}$, O1s, Mg2s, Mg (Auger), and $Sb3d_{3/2}$ peaks in $Mg_3V_2O_8$ and α-Sb_2O_4 oxides.

Sample	BE ($V2p_{3/2}$) V^{5+}	BE ($V2p_{3/2}$) V^{4+}	BE Mg2s	BE (Mg Auger)	BE ($Sb3d_{3/2}$)
$Mg_3V_2O_8$	517.0 ± 0.2	516.0 ± 0.2	89.5 ± 0.2	304.0 ± 0.2	
α-Sb_2O_4					539.9 ± 0.2

Table IVb. XPS results. V/Mg, V/Sb and Sb/(V+Mg+Sb) atomic ratios for fresh, used and calcined $Mg_3V_2O_8$/I+α-Sb_2O_4 mechanical mixtures. *Bulk atomic ratio.

Sample	V/Mg XPS			V/Sb XPS			Sb/(V+Mg+Sb) XPS		
	Fresh	Used	*Bulk	Fresh	Used	*Bulk	Fresh	Used	*Bulk
$Mg_3V_2O_8$/I	0.53	0.51	0.67	-	-	-	-	-	-
$Mg_3V_2O_8$/I +α-Sb_2O_4	0.48	0.49	0.67	3.57	2.48	1.02	0.08	0.12	0.29
calc. MM	0.55	0.59	0.67	2.40	2.40	1.02	0.13	0.13	0.29

Catalytic activity measurements. Some remarks concerning the development of the experimental work must be made before analysing the catalytic results. Different sample preparations and catalytic conditions (see sections 2.1, 2.3) were used in order to achieve a better understanding of the role played by $Mg_3V_2O_8$ and α-$Mg_2V_2O_7$ in the ODP of propane, according to the strategy proposed in the Introduction.

The activity of $Mg_3V_2O_8$/0 and α-$Mg_2V_2O_7$/0 as a function of the amount of catalyst (70, 100 and 200 mg) was also analyzed (Table V). The conversion is not a linear function of the mass of catalyst. The catalysts show a variation of 8% in the conversion while approximately constant values are observed for the selectivity.

Table V. Conversion and selectivity, for α-$Mg_2V_2O_7$/0 and $Mg_3V_2O_8$/0 as function of the amount of catalysts.

Mass of catalyst	75	100	200	Sample
C%	6.0	9.0	14.5	α-$Mg_2V_2O_7$/0
	12.0	14.6	19.0	$Mg_3V_2O_8$/0
S%	66.0	65.5	64.5	α-$Mg_2V_2O_7$/0
	46.7	45.8	45.3	$Mg_3V_2O_8$/0

The results for $Mg_3V_2O_8$/I+α-Sb_2O_4 and $Mg_3V_2O_8$/II+α-$Mg_2V_2O_7$/II are presented in Tables VIa and VIb.

α-Sb_2O_4 (Rm=0) is inert in this reaction. $Mg_3V_2O_8$/I+α-Sb_2O_4 mechanical mixtures show a significant decrease in CO_2 formation and a decrease in the conversion (see Table VIa). The intensity of this effect slightly decreases with increasing temperature. No synergy effect in the yield is observed, except for Rm=0.25. However, a synergy effect in the selectivity to propene is observed for all Rm. In conclusion, the $Mg_3V_2O_8$/I+α-Sb_2O_4 system exhibits an increase of selectivity by suppression of unselective reactions.

Significant synergy effects in the conversion, yield and selectivity and a decrease in CO_2 formation are observed in the $Mg_3V_2O_8$/II+α-$Mg_2V_2O_7$/II catalysts (Rm=0.5) (see Table VIb).

Table VI. Catalytic results. Table VIa) $Mg_3V_2O_8$/I+α-Sb_2O_4 mechanical mixtures. Range of rates of reaction of 7.78 x10^{-9} - 1.20 x 10^{-8} moles s^{-1} $m^{-2}$and Table VIb) $Mg_3V_2O_8$/II, α-$Mg_2V_2O_7$/II and $Mg_3V_2O_8$/II+α-$Mg_2V_2O_7$/II catalyst (Rm=0.5). Range of rates of reaction of 1.71 x 10^{-7} - 2.75 x 10^{-6} moles s^{-1} m^{-2}. Rm is the mass ratio (defined in section 2.1), C% is the conversion of propane, Y% is the yield of propene, S% the selectivity to propene and CO_2% the yield of CO_2, expressed in percentage. In parenthesis theoretical values in the absence of synergy the synergy effect (defined in section 2.4) is underlined.

Table VIa

Sample	Rm	773 K C%	773 K Y%	773 K S%	773 K CO_2%	793 K C%	793 K Y%	793 K S%	793 K CO_2%
$Mg_3V_2O_8$/I+ α-Sb_2O_4	0.25	1.4 (3.1) -55	1.3 (1.4) -7	91.8 (45.2) 103	1.0 (1.85) -46	2.8 (3.75) -24	2.5 (1.9) 32	90.2 (50.7) 76	1.8 (2.75) -34
'	0.50	3.5 (6.2) -43.5	2.1 (2.8) -25	60.6 (45.2) 34	1.2 (3.7) -67	4.3 (7.5) -43	2.9 (3.8) -24	68.2 (50.7) 34.5	1.7 (5.5) -69
'	0.75	4.4 (9.3) -53	3.5 (4.2) -17	80.0 (45.2) 77	3.0 (5.5) -45	5.9 (11.25) -47	5.0 (5.7) -12	85.5 (50.7) 68	5.1 (8.25) -38
$Mg_3V_2O_8$/I	1.00	12.4	5.6	45.2	7.4	15.0	7.6	50.7	11

Table VIb

Sample	773K C %	773K Y %	773K S %	773K CO_2%	793 K C %	793 K Y %	793 K S %	793 K CO_2%
α-$Mg_2V_2O_7$/II	41.3	6.2	15.1	48.3	46.5	6.2	13.3	53.4
$Mg_3V_2O_8$/II+ α-$Mg_2V_2O_7$/II	29.0 (25.0) 16	5.5 (3.5) 57	19.0 (14.0) 33	20.7 (27.8) -25	34.2 (28.0) 22	7.0 (3.8) 84	20.5 (13.7) 50	28.8 (31.3) -8
$Mg_3V_2O_8$/II	8.7	0.9	10.7	7.2	9.5	1.5	15.7	9.1

Discussion

The examination of two parameters such as deposition of coke and sintering are important to obtain a detailed characterization of these systems.

There is no deposition of coke on α-Sb_2O_4. For $Mg_3V_2O_8$/I+α-Sb_2O_4 the XPS C/V (or C/Mg) atomic ratios and BET surface areas remain nearly constant after test. No significant deposition of coke takes place on $Mg_3V_2O_8$ oxide during the test when it is mixed with α-Sb_2O_4. No sintering of $Mg_3V_2O_8$ in the $Mg_3V_2O_8$/I+α-Sb_2O_4 mechanical mixtures seems to occur as is revealed by the constancy of the BET surface areas after test. The large contribution of $Mg_3V_2O_8$ to the BET surface areas would make a sintering of this phase easily detectable.

The decrease of the XPS V/Sb atomic ratio for $Mg_3V_2O_8$/I+α-Sb_2O_4 calcined at 823 K during 6 days could be explained by sintering of $Mg_3V_2O_8$ at the surface. On the other hand, the XPS C/V and C/Mg atomic ratios remain unchanged after test. Then no sintering of bulk $Mg_3V_2O_8$ seems to occurs in the calcined mechanical mixture.

$Mg_3V_2O_8$+α-Sb_2O_4 system. There is no indication from XRD nor from BET of the formation of a new phase after testing or calcination.

AEM analyses show that $Mg_3V_2O_8$ has not been contaminated by Sb ions during the catalytic test. In fact, the AEM analysis is the same for the fresh and used sample. The fact that the AEM Mg/Sb and V/Sb peak intensity ratios are not the same for all the particles can be explained by the greater influence of the large α-Sb_2O_4 particles over the small $Mg_3V_2O_8$ particles in the near neighborhood (points 2,3; Fig. 2A) with respect to those separated from α-Sb_2O_4 (point 4, Fig. 2B). Thus, a contamination seems to be excluded. The principal argument to exclude the contamination is that the same AEM results are observed for the fresh and used samples. In both cases, $Mg_3V_2O_8$ and α-Sb_2O_4 particles are observed with the same variation in the AEM Mg/Sb and V/Sb peak intensity ratios.

The high XPS V/Sb and the low XPS Sb/(V+Mg+Sb) atomic ratios, compared to those of the bulk, is explained by the difference in particle sizes between the phases (S_{BET} $Mg_3V_2O_8$ =16 m^2/g and S_{BET} α-Sb_2O_4 = 1 m^2/g).

A possible contamination of α-Sb_2O_4 by $Mg_3V_2O_8$ is discarded because of the similarity of the XPS results before and after test. If a contamination at room temperature exists, it is highly improbable that it will remain stable after the sample has been tested at 793 K. In the calcined sample, the contamination is also excluded due to the identical XPS V/Sb atomic ratios before and after testing. A possible tendency of mutual contamination between $Mg_3V_2O_8$ and antimony was also examined by artificial contamination of $Mg_3V_2O_8$ with Sb ions. In this case highly dispersed Sb_xO_y species formed in $Mg_3V_2O_8$ and this new phase sintered and detached from the $Mg_3V_2O_8$ surface (21).

Thus, the physicochemical characterisation does not detect any change after test or calcination: these catalysts are composed of two separate phases in contact.

The higher proportion of V^{5+} in the used calcined samples is probably due to Oso species produced by α-Sb_2O_4 during the calcination.

$Mg_3V_2O_8$+α-$Mg_2V_2O_7$ system. The physicochemical characterisation does not reveal a formation of a new phase or contamination. This is logical for the first possibility. The second observation indicates that there is no preferential migration of one element from one phase to the other. This is also logical because both oxide phases contain the same metallic atoms in almost the same ratio.

Interpretation of the catalytic results. A comparison of the catalytic results in terms of selectivity to propene is possible taking into account that i) the selectivity is approximately constant, when varying the amount of the single phase catalysts, while conversion more than doubles (6 to 14.5%, Table V) and ii) for $Mg_3V_2O_8$/I+α-Sb_2O_4, conversion levels are approximately similar (range between 1.5 to 6%) for Rm=0.25, 0.50 and 0.75 (see Table Va) (in the range selected for the single phase catalyst, $Mg_3V_2O_8$, however, the selectivities are quite different) and iii) a similar behaviour is observed in the $Mg_3V_2O_8$/II+α-$Mg_2V_2O_7$/II system.

This fact cannot be explained by considering the different amounts of $Mg_3V_2O_8$ or α-$Mg_2V_2O_7$ in the mixtures. Thus, another phenomenon might be considered to explain the trends of the catalytic results showed by those systems.

The non-linearity of the conversion with the mass of catalyst is explained by the inhibitory role of reaction products. Propene, water, CO and CO_2 very likely compete with propane for the catalytic sites. As the amount of active phase increases, more reaction products are present in the gas phase, and the inhibition increases, thus diminishing the contribution of additional quantities of catalysts.

The above discussion shows that the catalytic activity of the mechanical mixture of $Mg_3V_2O_8$/I+α-Sb_2O_4 must be explained by a cooperative mechanism involving two separate, non-contaminated phases. We are thus led to the conclusion that some cooperation via a mobile surface species takes place.

As in most of the reactions involving oxygen, α-Sb_2O_4 is not active but it is known as a donor of mobile oxygen species (22, 23). The role of α-Sb_2O_4 in the present system would also be to produce mobile oxygen species which would migrate to the surface of $Mg_3V_2O_8$. This would be responsible for the modification of the catalytic performance. Mobile oxygen species can play two roles: a) react with propane adsorbed on $Mg_3V_2O_8$ (Oso used as a reactant) and b) migrate to the surface of $Mg_3V_2O_8$ and react with it to modify the properties of the catalytic sites and improve the selectivity (Oso used as a control species).

In the first case, the principal role of α-Sb_2O_4 would be that of increasing propane conversion. Our results show significantly lower conversions when $Mg_3V_2O_8$ is mixed with α-Sb_2O_4, compared to that observed in pure $Mg_3V_2O_8$. We therefore come to the conclusion that the role of spill-over oxygen is to modify the catalytic sites.

The active $Mg_3V_2O_8$ would not only contain the catalytic sites for the ODP of propane but also, in a large proportion, those for the total oxidation. Oso would inhibit the sites responsible for total oxidation. This explains why the addition of antimony while maintaining propene yield diminishes the overall formation of CO_2 (and, consequently, propene conversion). The remote control mechanism (RCM) then operates in these catalysts.

The cooperation between $Mg_3V_2O_8$ and α-$Mg_2V_2O_7$ has to be explained, like that for $Mg_3V_2O_8$+α-Sb_2O_4, by the contact between two uncontaminated phases via the RCM.

The existence of a cooperation in the $Mg_3V_2O_8$+α-$Mg_2V_2O_7$ system gives arguments to explain the differences observed in the literature concerning the role of the activity and selectivity of the pyro and ortho phases. Our results show that small impurities of one phase on the other can lead to a dramatic change in the catalytic performances of the pure phases. The selectivity to propene at high conversion is higher than the one obtained at lower propane conversion (compare $Mg_3V_2O_8$/I and $Mg_3V_2O_8$/II in Table VI). One pausible explanation of this apparent contradiction is that $Mg_3V_2O_8$(I) contains impurities (α-$Mg_2V_2O_7$), probably well dispersed (high BET surface area), facilitating the cooperation between phases, thus increasing conversion and selectivity. The respective roles of $Mg_3V_2O_8$ and α-$Mg_2V_2O_7$ in the donor-acceptor pair constituting the actors in the remote control mechanism must be identified.

α-$Mg_2V_2O_7$ shows both high propane conversion and CO_2 production, which increases as a function of temperature. On the contrary, $Mg_3V_2O_8$ shows both low propane conversion and CO_2 production, which remains nearly constant with temperature. As a whole, mixtures with Rm=0.5 exhibit a gain in conversion of propane, in propene yield and selectivity and by a diminution of CO_2 formation.

Conclusions

A remote control mechanism is the most plausible explanation for the synergy effects observed in $Mg_3V_2O_8$+α-Sb_2O_4 catalysts which contain two separate phases in contact. The role of Oso produced by α-Sb_2O_4 would be that of inhibiting the non-selective sites and is crucial in determining the catalytic properties of the catalyst.

The existence of a cooperation effect between α-$Mg_2V_2O_7$ and $Mg_3V_2O_8$ could explain the discrepancies observed in the literature concerning the activity and selectivity of the pyro and ortho phases. It seems very likely that both phases have a mixed role, acting both as donor and acceptors of spillover oxygen (especially $Mg_3V_2O_8$) but it is not clear which role is predominant for each species.

References

1. Chaar, M. A., Patel, D., Kung, M. C., and Kung, H. H. J., *J.Catal.*, **1987**, *105*, 483.
2. Chaar, M. A., Patel, D., Kung, M. C., and Kung, H. H. J., *J.Catal.*, **1988**, *109*, 463.

3. Michalakos , P. M. , Kung, M. C. , Jahan, I. and Kung, H. H. , *J. Catal.* , **1993**, *140*, 226.
4. Siew Hew Sam, D., Soenen, V., and Volta, J. C., *J. Catal.*, **1990**, *123*, 417.
5. Ruiz, A. G., Ramos, I, R., Fierro, J. L. G., Soenen, V. J., Herman, M., and Volta, J. C., in *Studies in Surface Science and Catalysis* (P. Ruiz and B. Delmon, Eds.), Elsevier, Amsterdam, 1992, 72, p. 203.
6. Corma, A., Lopez Nieto, J.M. and Paredes, N., *Appl. Catal. A*, **1993,** *104*, 161.
7. Corma, A., Lopez Nieto, J.M. and Paredes, N., *J. Catal.*, **1993,** *144*, 425.
8. Clark, G. M. and Morley, R. , *J.Solid State. Chem.*, **1976**, *16*, 429.
9. Gao, X., Ruiz, P., Xin, Q., Guo, X., and Delmon, B., *Catal. Lett.* **1994,** *23*, 32.
10. Gao, X., Ruiz, P., Qin, X., Guo., X, and Delmon, B., *J. Catal.*, **1994**, *148*, 56.
11. Weng, L. T., Spitaels, B., Yasse, B., Ladriere, J., Ruiz, P., and Delmon, B., *J. Catal.* **1991,** *132*, 319.
12. Cadus, L. E., Xiong, Y. L., Gotor, F. J., Acosta, D., Naud, J., Ruiz, P., and Delmon, B., in *Studies in Surface Science and Catalysis* (V. Cortés and S. Vic, Eds.), Elsevier, Amsterdam, 1994, 82, p. 41.
13. Castillo, R., Awasarkar, P. A., Papadopoulou, Ch., Acosta, D., and Ruiz, P., in *Studies in Surface Science and Catalysis* (V. Cortés and S. Vic, Eds.), Elsevier, Amsterdam, 1994, 82, p. 795.
14. Qiu, F. Y., Weng, L. T., Ruiz , P., and Delmon, B., *App. Catal.* **1989**, *47*, 115.
15. Bastians, Ph., Genet, M., Daza, L., Acosta, D., Ruiz, P., and Delmon, B., in *Studies in Surface Science and Catalysis* (P. Ruiz and B.Delmon, Eds.), Elsevier, Amsterdam, 1992, 72, p. 267.
16. Ruiz, P., Bastians, Ph., Caussin, L., Reuze, R., Daza, L., Acosta, D., and Delmon, B., *Catal. Today*, **1993**, *16*, 99.
17. Zhou, B., Sham, E., Bertrand, P., Machej, T., Ruiz, P., and Delmon, B., *J. Catal.*, **1991**, *132*(1), 157.
18. Courty, Ph., Ajot, H., Marcilly, Ch., and Delmon, B., *Powder Technol.*, **1973**, *7*, 21.
19. Nogier, J. P., and Delamar, M., *Catal. Today*, **1994,** *20*, 109.
20. Horvarth, B., Strutz, J., Geyeer-Lippmann, J., and Horvarth, E. G., Z. *Anorg. Allg. Chem.* , **1981**, *483*, 181.
21. Carrazán, S.R.G., Peres, C., Bernard, J.P., Ruwet, M., Ruiz, P., and Delmon, B., *J.Catal.*, in press.
22. Weng L. T., Duprez, D., Ruiz, P., and Delmon, B., *J. Mol. Catal.*, **1989**, *52*, 349.
23. Mestl, G., Ruiz, P., Delmon, B., and Knözinger, H., *J. Phys. Chem.*, **1994**, *98*, 11283.

Chapter 17

Vanadium Phosphate Catalyst: Ideal Structure, Real Structure, and Stability Region

P. T. Nguyen and A. W. Sleight[1]

Department of Chemistry, Oregon State University, 153 Gilbert Hall, Corvallis, OR 97331–4003

Highly perfect single crystals of $(VO)_2P_2O_7$ were grown for the first time. X-ray diffraction showed that the structure of these crystals had essentially no disorder of any kind. Both the space group and the environment around vanadium were found to be different than indicated in previous works. Peak broadening in X-ray diffraction patterns of actual VPO catalysts is shown be due to extended defects in addition to crystallite size and strain. The types of defects and their concentrations were quantitatively determined and correlated with the V^{5+} concentration in VPO catalysts. We conclude that most of the V^{5+} in VPO catalysts is present in certain of these extended defects. This conclusion is supported by magnetic susceptibility data, recent NMR experiments, and titrations of actual VPO catalysts. The stability field of $(VO)_2P_2O_7$ was determined as a function of temperature and oxygen partial pressure. This is of critical importance in defining the conditions under which the $V_2P_2O_9{\cdot}xH_2O$ precursor is optimally converted to a homogeneous VPO catalyst.

The vanadium phosphate catalyst (VPO) used for the oxidation of n-butane to maleic anhydride is remarkable. Selectivities can exceed 80%, with the only significant byproducts being CO, CO_2, and H_2O (*1*). No other catalyst is known which gives comparable performance. Despite the scientific interest and the commercial importance of this catalyst, it is poorly understood. We do know, however, that VPO catalytic properties are highly sensitive to the method by which VPO is prepared.

The VPO catalyst is essentially $(VO)_2P_2O_7$. However, both the ideal structure of $(VO)_2P_2O_7$ and the actual structure of VPO catalysts have been elusive. This paper addresses both of those structures. In addition, this paper addresses the issue of better

[1]Corresponding author

0097–6156/96/0638–0236$15.00/0

controlling VPO synthesis in order to achieve optimum catalysts in an efficient and reproducible fashion.

Ideal Structure of $(VO)_2P_2O_7$

There have been several previous structure determinations (*2-4*) of $(VO)_2P_2O_7$ by single crystal X-ray crystallography. All such studies have been hampered by poor quality crystals. The high defect concentration in previously available crystals has obscured certain details of the ideal structure of $(VO)_2P_2O_7$.

Crystals of $(VO)_2P_2O_7$ were prepared by slow cooling from the melt under controlled oxygen pressure. Crystals prepared when using the lower oxygen pressure (~0.001 atm) were green, but those found in preparations at higher oxygen pressures were red-brown. The titration method of Nakamura et al. (*5*) was used to determine the oxidation state of vanadium in the crystals. The result was 4.00 for the green crystals and 4.01 for the red-brown crystals. Diffraction patterns taken on a precession camera indicated that the green crystals were of higher quality from a diffraction point of view. Therefore, only those crystals were further studied. Ebner and Thompson (*4*) also reported both red-brown and green crystals. However, they did not report on the average oxidation state of vanadium in their crystals, and they indicated that all of their crystals had very high defect concentrations. We believe that it is control of oxygen partial pressure during crystal growth that gave us crystals that are essentially defect free.

The $(VO)_2P_2O_7$ crystal chosen for data collection was first examined on a precession camera. The diffraction pattern indicated a high quality crystal, but reflections were observed that are forbidden by the Pcam and $Pca2_1$ space groups assumed by others (*2-4*). The only space group consistent with our diffraction data is $P2_1$. A full sphere of data was collected on a Rigaku AFC6R diffractometer, using the unit cell parameters $a = 7.28$ Å, $b = 16.588$ Å, $c = 9.58$ Å, and $\beta = 89.875°$. After merging equivalent reflections, there were 4254 unique reflections for use in data analysis. A final agreement factor (R) of 3.1% was obtained, and the structure found is shown in Figure 1a. The refined positional and thermal parameters are given in Table I. All atomic positions were fully occupied, and no significant peaks appeared in a final difference Fourier map. More details on this structure refinement have been published elsewhere (*6*). The structure, in fact, is very similar to that reported by previous authors (*2-4*). The way in which distorted VO_6 octahedra are connected to pyrophosphate groups is in agreement with the earlier studies. There are, however, subtle symmetry differences which cannot be seen on the scale of this figure. Furthermore, we find significantly different V–O bond distances (Table II) than previously reported. Earlier work had suggested that the various vanadium atoms of the $(VO)_2P_2O_7$ were not all chemically similar. However, bond valence calculations (*7*) using our more accurate structure show that all vanadium atoms are essentially equivalent in defect-free crystals. The calculated bond valences for all eight vanadium atoms are in the range 4.02 to 4.18.

We thus now know that it is possible to prepare $(VO)_2P_2O_7$ which has no greater defect concentration than that typical of many other materials. This ideal structure should not, however, be considered the structure of a VPO catalyst. An optimum VPO catalyst has a very high defect concentration.

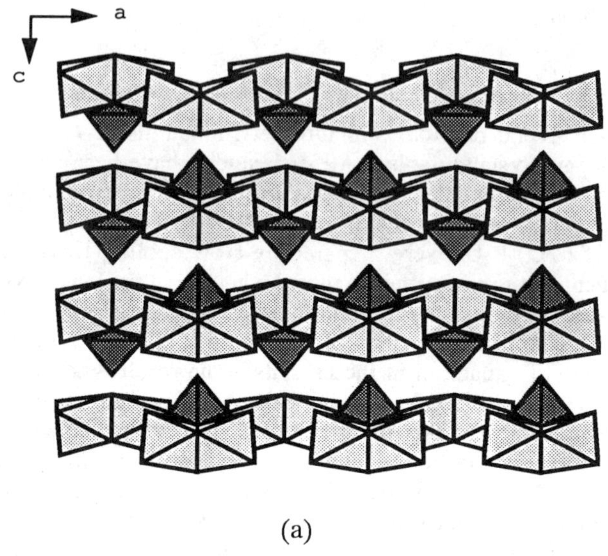

(a)

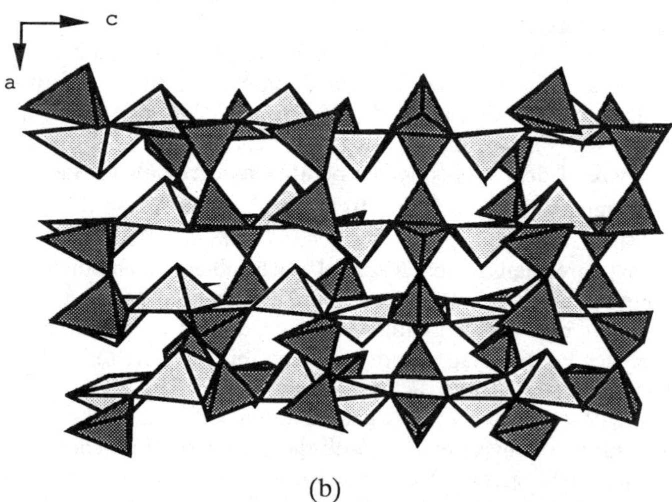

(b)

Figure 1. Dark polyhedra are PO_4 or P_2O_7 groups. Lighter polyhedra are VO_6 or VO_5 groups. (a) Layered precusor structure. (b) Ideal $(VO)_2P_2O_7$ structure. (c) $(VO)_2P_2O_7$ structure with **bc** fault. (d) $(VO)_2P_2O_7$ structure with **ab** fault.

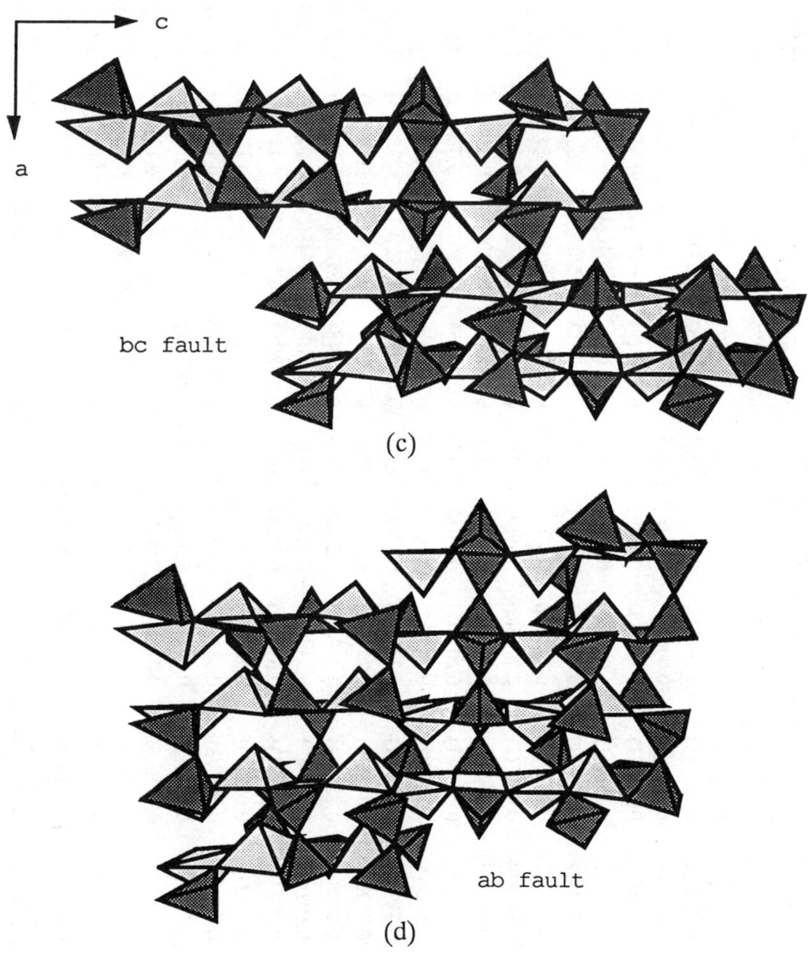

Figure 1. *Continued*

TABLE I. Atomic and Thermal Parameters for $(VO)_2P_2O_7$

Atom	x	y	z	Beq ($Å^2$)*	Atom	x	y	z	Beq($Å^2$)*
V1	0.2973(2)	0.0958	0.4912(1)	0.47(5)	O11	0.4960(7)	-0.2162(3)	0.2871(4)	0.59(7)
V2	0.2039(2)	0.5960(1)	0.4909(1)	0.36(5)	O12	0.0066(8)	0.2846(3)	0.2862(5)	0.79(8)
V3	0.2087(2)	-0.09442(9)	0.5023(1)	0.33(5)	O13	0.3007(9)	0.1734(3)	0.6421(5)	1.0(1)
V4	0.2923(2)	0.4057(1)	0.5017(1)	0.25(5)	O14	0.2017(8)	0.6734(3)	0.6410(5)	0.71(9)
V5	0.2055(2)	-0.1548(1)	-0.0005(2)	0.45(6)	O15	0.2133(8)	0.1761(3)	0.3613(5)	0.44(8)
V6	0.2935(2)	0.3453(1)	-0.0002(1)	0.43(6)	O16	0.2943(9)	0.6756(3)	0.3600(5)	0.9(1)
V7	0.2081(2)	0.1561(1)	-0.0014(1)	0.34(5)	O17	0.281(1)	0.0010(4)	0.6307(6)	0.9(1)
V8	0.2914(2)	0.65607(7)	-0.0021(1)	0.33(5)	O18	0.2200(8)	0.5007(3)	0.6300(5)	0.3(1)
P1	0.2952(3)	-0.0003(1)	0.7924(2)	0.3(1)	O19	0.2221(9)	0.0017(4)	0.3639(5)	0.4(1)
P2	0.2026(4)	0.4997(2)	0.7925(2)	0.39(8)	O30	0.276(1)	0.5012(4)	0.3619(6)	0.7(1)
P3	0.1948(3)	0.0005(2)	0.2012(2)	0.3(1)	O31	0.2860(9)	-0.1795(3)	0.6294(6)	0.7(1)
P4	0.3031(4)	0.5008(1)	0.2012(2)	0.24(8)	O32	0.2117(9)	0.3207(3)	0.6287(5)	0.6(1)
P5	0.1920(4)	-0.2447(2)	0.7064(2)	0.4(1)	O33	0.199(1)	-0.1772(3)	0.3578(5)	0.8(1)
P6	0.3106(4)	0.2556(1)	0.7061(2)	0.37(9)	O34	0.298(1)	0.3230(3)	0.3572(5)	0.7(1)
P7	0.3020(3)	-0.2436(1)	0.2907(2)	0.4(1)	O35	0.2659(9)	0.0775(3)	0.1395(5)	1.0(1)
P8	0.2008(4)	0.2562(1)	0.2906(2)	0.27(8)	O36	0.2355(8)	0.5768(3)	0.1390(5)	0.67(9)
O1	0.5014(8)	0.5914(3)	0.5399(4)	0.59(8)	O37	0.2274(8)	0.0749(3)	-0.1455(5)	0.79(9)
O2	-0.0024(8)	0.0933(3)	0.5397(4)	0.76(8)	O38	0.2809(9)	0.5743(3)	-0.1465(5)	1.05(9)
O3	0.005(1)	-0.0904(3)	0.5313(5)	0.8(1)	O39	0.2727(8)	-0.0754(3)	0.1391(6)	0.6(1)
O4	0.496(1)	0.4121(3)	0.5320(4)	0.41(9)	O50	0.2329(9)	0.4247(3)	0.1390(6)	0.7(1)
O5	-0.001(1)	-0.1597(3)	0.0131(6)	0.59(8)	O51	0.2234(9)	-0.0768(3)	-0.1513(5)	0.7(1)
O6	0.500(1)	0.3435(4)	0.0115(7)	0.9(1)	O52	0.2823(9)	0.4233(3)	-0.1514(6)	0.8(1)
O7	-0.0002(9)	0.1599(3)	0.0125(6)	0.67(8)	O53	0.2510(8)	-0.2504(3)	0.1328(6)	0.3(1)
O8	0.500(1)	0.6595(3)	0.0108(6)	0.88(8)	O54	0.255(1)	0.2497(4)	0.1352(6)	0.8(1)
O9	0.4975(8)	-0.0014(3)	0.8272(6)	1.55(9)	O55	0.2496(9)	-0.2482(4)	-0.1394(6)	0.6(1)
O10	0.0044(8)	0.4983(3)	0.8252(6)	1.41(9)	O56	0.2535(9)	0.2517(4)	-0.1382(6)	0.5(1)

*Beq = $(8\Pi 3/3)\Sigma i \Sigma j U_{ij} a^*_i a^*_j a_i a_j$

TABLE II. Bond Distances for $(VO)_2P_2O_7$

V1	O1	1.585(6)	V8	O7	2.253(7)
	O2	2.337(6)		O8	1.619(7)
	O13	1.936(5)		O36	1.934(5)
	O15	1.935(5)		O38	1.939(5)
	O17	2.067(6)		O53	2.043(6)
	O19	2.065(6)		O55	2.088(6)
V2	O1	2.307(6)	P1	O9	1.599(7)
	O2	1.585(6)		O17	1.554(6)
	O14	1.929(5)		O37	1.478(5)
	O16	1.951(6)		O51	1.486(6)
	O18	2.071(6)	P2	O10	1.564(7)
	O30	2.075(7)		O18	1.563(5)
V3	O3	1.597(8)		O38	1.497(6)
	O4	2.307(7)		O52	1.509(6)
	O17	2.081(7)	P3	O10	1.560(7)
	O19	2.076(6)		O19	1.573(5)
	O31	1.957(6)		O35	1.510(6)
	O33	1.951(5)		O39	1.518(6)
V4	O3	2.323(8)	P4	O9	1.565(7)
	O4	1.606(7)		O30	1.554(6)
	O18	2.075(6)		O36	1.489(6)
	O30	2.079(6)		O50	1.497(6)
	O32	1.964(6)	P5	O12	1.612(6)
	O34	1.949(5)		O14	1.498(5)
V5	O5	1.602(8)		O31	1.498(6)
	O6	2.277(9)		O55	1.544(6)
	O39	1.948(6)	P6	O11	1.568(6)
	O51	1.945(5)		O13	1.497(6)
	O53	2.066(6)		O32	1.516(6)
	O55	2.070(6)		O56	1.557(6)
V6	O5	2.266(8)	P7	O11	1.567(6)
	O6	1.601(9)		O16	1.496(5)
	O50	1.932(6)		O33	1.506(6)
	O52	1.943(5)		O53	1.568(6)
	O54	2.071(6)	P8	O12	1.574(6)
	O56	2.062(6)		O15	1.495(5)
V7	O7	1.617(7)		O34	1.484(6)
	O8	2.257(7)		O54	1.549(6)
	O35	1.930(5)			
	O37	1.935(5)			
	O54	2.062(6)			
	O56	2.087(6)			

Real VPO Structure

The oxidation state of vanadium in VPO catalysts is always slightly greater than four (e.g., +4.01). Also, the P-to-V ratio is somewhat greater than one (e.g., 1.05). These facts have led some to suggest that VPO catalysts are not single phase. Certainly many VPO catalysts are not single phase, but it does appear that an optimum VPO catalyst is usually single phase.

There is strong evidence that the extra phosphorous in a VPO catalyst is accommodated at the surface (*8*). Experimental evidence does not, however, support surface V^{5+} as the explanation of a vanadium oxidation state greater than four in a typical VPO catalyst. One argument against V^{5+} being predominantly at the surface is that the average vanadium oxidation state in a VPO catalyst is not perceptibly lowered when the catalyst is exhausted of its ability to selectively oxidize butane (*9,10*). Even more convincing arguments come from physical characterization of VPO.

X-ray diffraction patterns of optimum VPO catalysts that have come to steady state in a reactor generally show no evidence of a phase other than $(VO)_2P_2O_7$. This is, however, not conclusive evidence for VPO being single phase. Small amounts of a second phase might be undetected, especially if this phase were poorly crystalline or amorphous.

Magnetic susceptibility data (*11*) offer more convincing evidence that the V^{5+} present in VPO is, in fact, incorporated into VPO crystallites as a defect. In $(VO)_2P_2O_7$ with the ideal structure, all vanadium is present as V^{4+} cations. These cations, all with one *d* electron each, occur in pairs (Figure 1a). The spins of these electrons tend to align antiparallel, decreasing the magnetic susceptibility in a manner easily explained by conventional theory. However, susceptibility measurements show that some V^{4+} present in VPO is not part of a V^{4+}-V^{4+} dimer (*11*). Some defect or second phase must be present. A likely explanation is that a small percentage of the vanadium dimers are V^{4+}-V^{5+} dimers instead of V^{4+}-V^{4+} dimers. The magnetic susceptibility data could also be explained on the basis of a small impurity of a second phase containing isolated V^{4+} cations, but this does not help explain why the average vanadium oxidation state is greater than four. The conclusion from magnetic susceptibility is confirmed by NMR studies. Spin echo mapping studies (*12*) show that the V^{5+} present in VPO is closely associated with V^{4+} and could not arise from a second phase. The questions now become what are the defects in VPO, how can their concentration be controlled, and are these defects related to catalytic properties.

X-ray diffraction (XRD) patterns of VPO catalysts also show evidence of a defect, but there has been no previous attempt to extract information about the nature of this defect. Some of the peak broadening observed in VPO XRD patterns (Figure 2) can be attributed to strain and to the small size of the crystallites. However, reflections with odd *hkℓ* indices are typically much broader than those with even *hkℓ* indices. This effect is so dramatic that certain of these reflections with odd indices seem to disappear in samples prepared under conditions appropriate for obtaining a good catalyst (Figure 2). This observation can only be explained in terms of extended defects in VPO.

We have modeled (*12*) the extended defects in VPO using Diffax software (*13*) which has been previously used to model extended defects in zeolites. The fit to the observed patterns is shown in Figure 2. The two types of defects used in the model

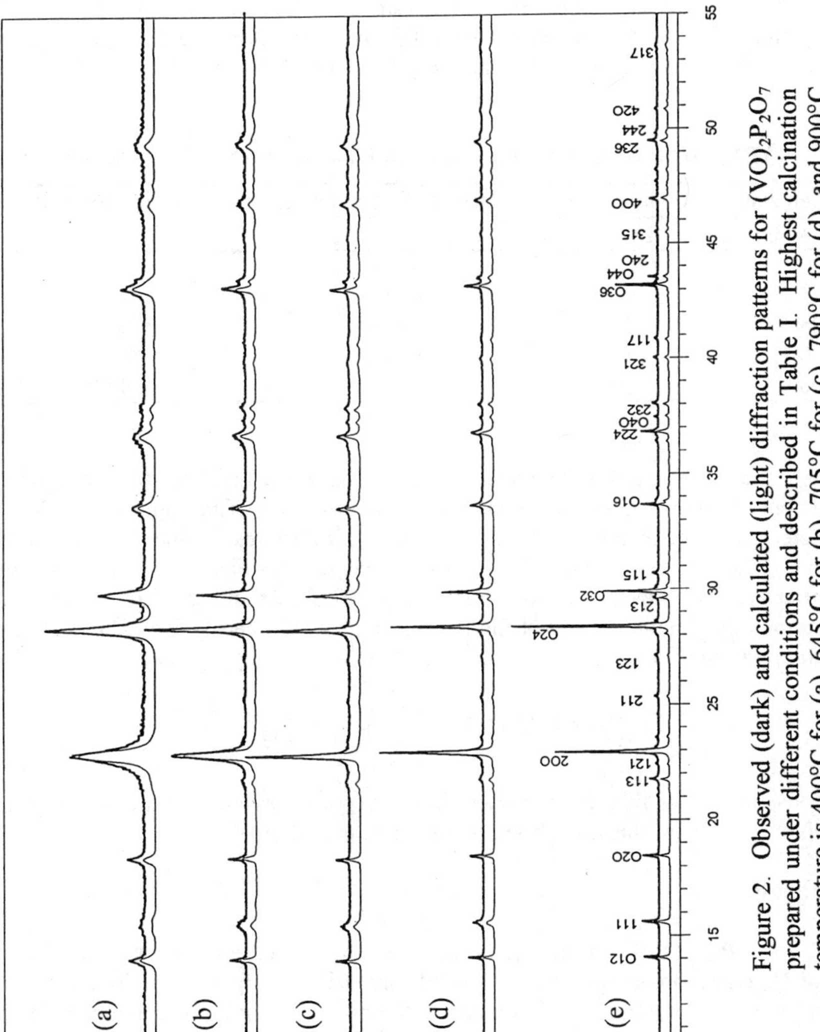

Figure 2. Observed (dark) and calculated (light) diffraction patterns for $(VO)_2P_2O_7$ prepared under different conditions and described in Table I. Highest calcination temperature is 400°C for (a), 645°C for (b), 705°C for (c), 790°C for (d), and 900°C for (e).

are shown in Figures 1b and 1c. Table III gives the conditions at which the $V_2P_2O_9{\cdot}xH_2O$ precursor was calcined, the overall defect concentrations used to calculate the patterns in Figure 2, and the vanadium oxidation states determined by titrations. More details on VPO synthesis are given in elsewhere (*12*). The oxygen partial pressure during calcination was controlled, as described in the next section, so that neither oxidation to VPO_5 nor reduction to VPO_4 could occur regardless of heating time. We also successfully used Diffax to simulate the diffuse scattering reported in electron diffraction and single crystal X-ray diffraction patterns (12).

TABLE III. Defect Concentration and Vanadium Oxidation State

Calcination Conditions		Defect Concentration	Oxidation State
Temp.	Time		
400°C	7 days	1.3 %	4.10
645°C	3 days	0.1%	4.06
705°C	24 hrs	0.006%	4.03
790°C	12 hrs	0.002%	4.01
900°C	12 hrs	0.000%	4.00

For the VPO samples shown in Table III, there is a good correlation between defect concentration and average vanadium oxidation state. This suggests that V^{5+} is associated with one or both of the defects shown in Figures 1b and 1c. The defect shown in Figure 1c would not give rise to any change in stoichiometry or vanadium oxidation state. However, the more dominant defect shown in Figure 1b would cause a stoichiometry change. Many of the P_2O_7 groups would necessarily convert to two PO_4 groups at this defect:

$$\left(V^{4+}_{1-2x}V^{5+}_{2x}O\right)_2\left(P_2O_7\right)_{1-x}\left(PO_4\right)_{2x}.$$

The consequence of this extra oxygen is an increased average oxidation state for vanadium. Details of this modeling are presented elsewhere (*12*).

Synthesis of VPO

The best VPO catalysts are prepared by the topotactic dehydration of a $(VO)_2P_2O_9{\cdot}xH_2O$ precursor which is synthesized in an organic medium using an organic reducing agent. The topotactic dehydration is normally conducted in air. This is a highly uncontrolled process leading to inhomogeneous products. The inhomogeneous catalysts obtained may, however, become homogeneous after coming to steady state in a reactor. There are two serious problems with VPO synthesis as normally practiced. One is that organic residue in the precursor acts as a reducing agent during the topotactic dehydration, and this may produce a V^{3+} containing phase such as VPO_4. The other problem is that $(VO)_2P_2O_7$ will oxidize to $VOPO_4$ when heated in air. Thus, the normal uncontrolled synthesis of VPO leads to simultaneous reduction of V^{4+} to V^{3+} by organic residue and oxidation of V^{4+} to V^{5+} by air. There is some fortuitous balancing of these two reactions, but this is not sufficient to give homogeneous products. The nature of the VPO obtained is very dependent on reaction time, gas flows, sample size, furnace configuration, etc. Conversion of the precursor under an inert gas does not solve the problem, because this is actually a reducing situation.

We have controlled VPO synthesis in two ways. First, we wash the precursor with a suitable solvent such as acetone to eliminate nearly all the organic residue. Second, the oxygen partial pressure is controlled during the topotactic dehydration. Regardless of the heating time, no oxidation of $(VO)_2P_2O_7$ to $VOPO_4$ can occur and no reduction of $(VO)_2P_2O_7$ to VPO_4 can occur.

The stability field for $(VO)_2P_2O_7$ was determined by heating samples at various temperatures under various oxygen pressures. The gas exiting the treatment chamber was continuously monitored for oxygen fugacity. When the oxygen fugacity achieved a constant value, the sample was assumed to have reached equilibrium at that particular temperature and oxygen fugacity. Apparent equilibrium was achieved in as quickly as 12 hours for samples prepared at relatively high temperature (790°C and 705°C) and as slowly as 6-7 days for the samples prepared at lower temperatures (580°C). For samples prepared at 645°C, apparent equilibrium was achieved only after 3-4 days. The average vanadium oxidation state is plotted against the oxygen partial pressure for various temperatures in Figure 3.

For samples prepared at 790°C, single phase $(VO)_2P_2O_7$ was observed from 6.48×10^{-4} atm to 2.09×10^{-2} atm. Above the latter value, β-VPO_5 phase was observed in X-ray powder patterns. Samples prepared at 790°C and with an oxygen fugacity greater than 3.1×10^{-2} atm melted. This is presumably due to β-VPO melting, which is known to occur at that temperature.

For samples prepared at 705°C, single phase $(VO)_2P_2O_7$ was observed when using 6.31×10^{-4} to 1.26×10^{-2} atm of oxygen. For samples prepared at 645°C, the stability region for single phase $(VO)_2P_2O_7$ was from 1.31×10^{-3} to 4.79×10^{-3} atm of oxygen. For samples prepared at 580°C, the stability region for single phase $(VO)_2P_2O_7$ was from 1.91×10^{-3} to 5.01×10^{-3} atm of oxygen. Similar to samples prepared at 790°C, if these samples were prepared at a higher oxygen fugacity, a mixture of $(VO)_2P_2O_7$ and β-VPO_5 was observed in the X-ray powder patterns. No melts were ever observed for samples prepared at these lower temperatures. For samples prepared at 705°C with an oxgyen fugacity value of 3.80×10^{-2} atm or more, single phase β-VPO_5 samples were observed. For samples prepared at 645°C, anything above 2.75×10^{-2} atm of oxygen resulted in single phase β-VPO_5 as the product. This boundary for single phase β-VPO_5 changed to 1.26×10^{-2} atm or more for samples prepared at 580°C.

With the information in Figure 3, we can be certain of the conditions of temperature and oxygen pressure where $(VO)_2P_2O_7$ will neither reduce nor oxidize to another phase region. According to the phase rule, two different oxides with a P-to-V ratio of one should not generally be observed. Our observations of two phases may indicate a lack of true equilibrium, or the two phase may be related to a P-to-V ratio that is not exactly one. Another study of the $(VO)_2P_2O_7$ stability field (*14*) found results similar to ours, including two phase regions in apparent violation of the phase rule. Two differences between the two studies are, however, significant. The

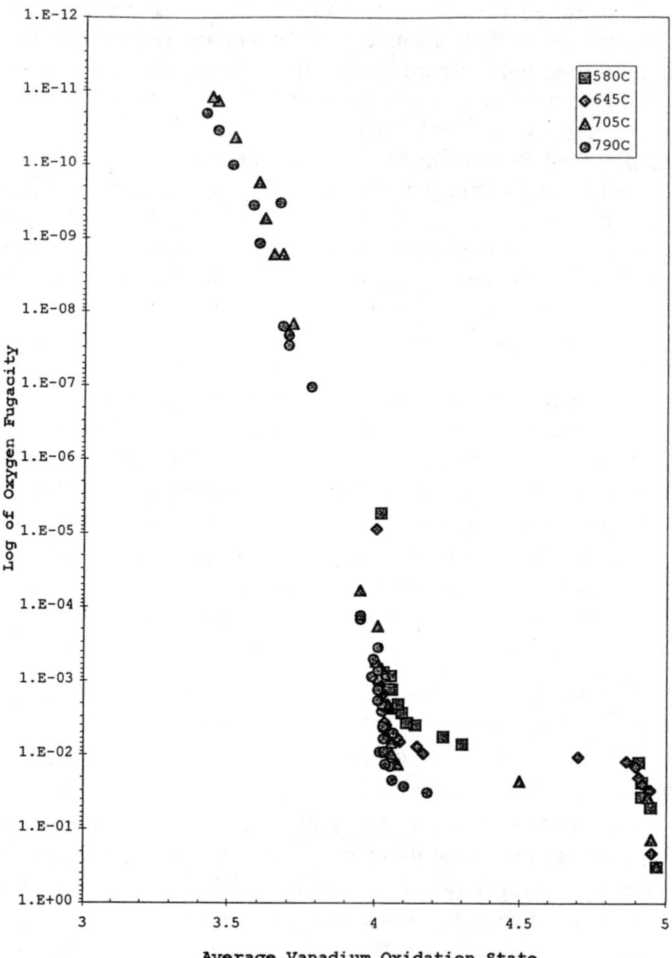

Figure 3. Average oxidation of vanadium from titration vs. oxygen fugacity after apparent equilibrium had been achieved.

previous study (*14*) reported evidence for a phase intermediate between $(VO)_2P_2O_7$ and VPO_5; we found no evidence for such a phase. Also, the previous study reported hysteresis effects on oxidation versus reduction which we did not observe.

Depending on the conditions of topotactic transformation of precursor $(VO)_2P_2O_7 \cdot xH_2O$ to VPO, the average oxidation state of vanadium in the VPO product varies in a systematic manner (Table I). Some of the samples in Table I were subsequently annealed at lower and higher oxygen partial pressures which were neither low enough to produce VPO_4 nor high enough to produce $VOPO_4$. Such annealing conditions might have been expected to alter the average oxidation of vanadium in these VPO samples. In fact, titrations showed no significant change in the vanadium oxidation state. It appears that the defect concentration is determined during the topotactic transformation. This defect concentration, in turn, fixes the vanadium oxidation state. Subsequent annealing at lower or higher oxygen pressures does not significantly change the vanadium oxidation state, because at the temperature of annealing, the defect concentration does not change significantly over several days.

Relationship to Catalysis

Our results have better defined the structure of $(VO)_2P_2O_7$, and new issues are raised. In particular, the possible impact of defects on catalytic properties needs further consideration. Our single crystal XRD study shows a low-symmetry acentric monoclinic structure, but there is much higher pseudosymmetry. This suggests that $(VO)_2P_2O_7$ will undergo phase transitions to higher symmetry phases at high temperatures. Catalytic properties may change in the vicinity of such phase transitions. Furthermore, it is likely that $(VO)_2P_2O_7$ crystallites have twin boundaries associated with the pseudosymmetry. When such boundaries intersect the surface, they could give sites with catalytic properties different than the usual VPO surface.

The extended defects that we have modeled could also be important for VPO catalytic properties. These defects will intersect the surface and create surface sites which otherwise would not be present. The concentration of the extended defects is very dependent on the way in which VPO is prepared. Possibly, the need for a special synthesis recipe to prepare optimum VPO catalyst is then related to these defects. In any event, it is clear that the concentration of these defects is very high in VPO catalysts before and after coming to steady state in a reactor.

Acknowledgments

This work was supported by E. I. du Pont de Nemours, Inc. and by NSF DMR-9305780.

Literature Cited

1. Contractor, R. M.; Bergna, H. E.; Horowitz, H. S.; Blackstone, C. M.; Chowdhry, U.; Sleight, A. W. In *Catalysis 1987*; Elsevier: Amsterdam, 1988, p. 645.
2. Middlemiss, N. E. *PhD. Thesis*, McMaster University, Hamilton, Ontario, Canada, 1978.
3. Gorbunova Yu. E.; Linde S. A. *Sov. Phys. Dokl.* **1979,** *24*, 138 .

4. Ebner, J. R.; Thompson, M. R. In *Studies in Surface Science and Catalysis*; Grasselli, R. K.; Sleight, A. W., Eds; Elsevier: Amsterdam, 1992, p. 18.
5. Nakamura, M.; Kawai, K.; Fujiwara, Y. *J. Catal.*, **1974**, *34*, 345.
6. Nguyen, P. T.; Hoffman, R. D.; Sleight, A. W. *Mater. Res. Bull.* **1995**, *30*, 1055.
7. Brown, I. D.; Wu, K. K. *Acta Cryst.* **1976**, *B32*, 1957.
8. Bergeret, G.; Broyer, J. P.; David, M.; Gallezot, P.; Volta, J. C.; Hecquet, G. J. *Chem. Soc. Chem. Commun.* **1986**, 825.
9. Contractor, R. M.; Sleight, A. W. *Catal. Today* **1988,** *3*, 175.
10. Contractor, R. M.; Bergna, H. E.; Horowitz, H. S.; Blackstone, C. M.; Malone, B.; Torardi, C. C.; Griffiths, B.; Chowdhry, U.; Sleight, A. W. *Catal. Today* **1987**, *1*, 49.
11. Johnson, J. W.; Johnston, D. C.; Jacobson, A. J.; Brody, J. F. *J. Am. Chem. Soc.* **1984**, *106*, 8123.
12. Nguyen, P. T.; Sleight, A. W.; Roberts, N.; Warren, W. W. *J. Solid State Chem.*, in press.
13. Treacy, J. M. M.;Newsam, J. M.; Deem, M. W. *Proc. R. Soc. Lond., A* **1991**, *433*, 499.
14. Bordes, E.; Courtine, P.; Johnson, J. W. *J. Solid State Chem.* **1984**, *55*, 270.

Chapter 18

Mechanism of Selective Oxidation of Butane to Maleic Anhydride on Vanadyl Pyrophosphate Catalysts

Quantum Chemical Description

J. Haber, R. Tokarz, and M. Witko

Institute of Catalysis and Surface Chemistry, Polish Academy of Sciences, ul.Niezapominajek, 30239 Cracow, Poland

Vanadyl pyrophosphate, the active and selective catalyst in the oxidation of butane to maleic anhydride, is characterized by the presence of pairs of edge-sharing vanadium-oxygen pyramids with V=O groups in trans positions, linked together through six pyrophosphate groups. Quantum-chemical INDO-type calculations have been carried out to obtain the optimum atomic positions at energy minima along the approach of butane to such an active site. Upon approach of butane with its plane perpendicular to the plane of the site, insertion of oxygen atoms from two opposite phosphate groups into C-H bonds of terminal carbon atoms and formation of a C-O-C bridge by vanadyl oxygen atom takes place. Thus, dihydroxy-dihydrofurane is formed as the first intermediate without activation energy. This intermediate may transform by losing two hydrogen atoms twice to form maleic anhydride.

Heterogeneous catalytic oxidation of butane to maleic anhydride is the only process of selective oxidation of a paraffin which has been commercialized (*1*). There are three striking features of this process: i). the only catalytic system found to be active and selective for this reaction is V-P-O, ii). no intermediates are observed among the reaction products under usual reaction conditions (*2,3,4*) although a complex 14-electron oxidation process is involved, in which 8 hydrogen atoms are abstracted from and 3 oxygen atoms are inserted into the butane molecule, iii). only the lattice oxygen ions of the catalyst, and not the adsorbed oxygen species, participate in the reaction (*5*). The generally accepted reaction pattern postulates butene, butadiene and furane to be formed as intermediates, which, however, have only been detected in the oxidation of n-butane over V-P-O catalysts under very unusual conditions in the TAP reactor (*6*) or in the oxidation of n-butane under aerobic conditions in a pulse reactor (*7*). Interesting conclusions can be drawn from the comparison of the oxidation of butane with that of butene. It is now generally accepted that selective

0097–6156/96/0638–0249$15.00/0

oxidation of butene starts with the interaction of the C-H bond of the methyl or the methylene group in the α-position (with respect to the double bond) with an active site on the surface of the oxide catalyst and results in the formation of an allyl species and a surface hydroxyl group (*8,9*). The allyl species then reacts along one of several possible reaction pathways: i). the abstraction of hydrogen from an α-position (with respect to the π-electron system) of the allyl may be repeated and butadiene formed, ii). if the catalyst surface contains active sites that can promote the nucleophilic addition of an oxide ion to the allyl species, croton aldehyde is formed by addition to the C1 position, or methyl-vinyl ketone is formed by addition to the C3 position, iii). if reactive electrophilic oxygen species are present at the catalyst surface, total oxidation to carbon oxides may occur. Usually on oxide catalysts all three parallel reaction pathways are followed and a mixture of products is formed, their selectivities depending on the properties of the catalyst. Indeed, in oxidation of butane on vanadia-titania monolayer catalysts, different oxygenated molecules are formed besides maleic anhydride. The fact that no intermediate products are observed during butane oxidation on vanadyl pyrophosphate catalysts seems to indicate that the reaction is not proceeding through consecutive steps of butene formation followed by further transformations to butadiene, furane, etc., but that a completely different mechanism operates on the surface of this catalyst.

Model and Method

A large number of papers attribute catalytic activity and selectivity to the presence of the bulk crystalline phase of vanadyl pyrophosphate $(VO)_2P_2O_7$ (*10-12*). It is now generally accepted that the crystal planes parallel to the (100) plane of vanadyl pyrophosphate contain the active sites (*13*). These sites are composed of two edge-linked vanadium-oxygen square pyramids with trans orientation of vanadyl groups, surrounded by phosphate groups corner-linking the dimeric vanadium square pyramids into sheets, and sheets into a tridimensional lattice by formation of pyrophoshpate bonds (*14*). A schematic representation of such a composite active site is shown in Figure 1. Optimal catalyst composition presents a slight stoichiometric excess of phosphorus with the surface P/V ratio varying in the range 1.5 to 3 (*15,16*) according to XPS analysis. This may indicate that the surface terminates with pendant groups of pyrophosphate, which surround the pairs of V-O square pyramids (*17*). In this type of truncation of the crystallites, vanadium ions are buried inside a cavity or cleft in the walls formed by pyrophosphate groups. In all other possible truncation, the termination is with orthophosphate groups (*18*). In this case vanadium ions are more easily accessible for the reactant molecules from the gas phase. Results of the determination of maximum yield in maleic anhydride, as a function of the P/V ratio, seem to indicate (*19*) that either of these two truncations or both may be present at the surface of the most active and selective catalyst. We have thus selected as the model of the active site the simpler case of the pair of edge-linked V-O square pyramids surrounded by phosphate groups in which the terminal oxygen atoms at the edges are saturated by hydrogen atoms, described by the formula $(VO_4)_2(PO_3)_6H_{12}$. This cluster is used for quantum chemical calculations of the interaction with the butane molecule.

The catalytic performance of vanadyl pyrophosphate is strongly related to the method of preparation employed (*2,12,20*). Crystallographic refinement of X-ray diffraction data indicates that structural differences can be described in terms of defects associated with linear disorder of the vanadium atoms, consisting in variability in the directional orientation of the vanadyl columns running perpendicular to the (100) surface. Two polytypes have been distinguished (*14*). Within these polytypes, the pyrophosphate groups are placed in various positions, giving "networked" or "layered" systems. Therefore, four different spatial arrangements of surface oxygens at the apices of phosphate and vanadyl groups were considered (Figure 2) in order to mimic the defects observed in the stacking of such elementary units into real three dimensional structures of vanadyl pyrophosphate.

Moreover, some data published recently seem to indicate that the active phase is amorphous vanadium phosphate of P/V ratio of about 2, supported on vanadyl pyrophosphate (*21,22*). This raises the question as to whether the long range order is really necessary for the catalyst to be active and selective, or whether short range order is sufficient for the transformation of the molecule to take place. These observations provide further arguments for the validity of using the elementary clusters shown in Figure 2 as models of the active sites.

The calculations were carried out by means of a semiempirical INDO type method using the ZINDO program (*23-25*). This program permits the optimization of coordinates for chosen atoms in the studied system, composed of the cluster and the incoming hydrocarbon molecule. The butane molecule was allowed to approach the cluster with its plane parallel or perpendicular to the basic plane of the cluster, and in the latter case the terminal carbon atoms were pointing to the cluster along the common edge of the two vanadyl-oxygen square pyramids. In this way local minima were identified along the pathway of the approach.

Results of the Calculations

Approach with Butane Molecular Plane Perpendicular to the Plane of the Cluster. When the butane molecule is allowed to approach the cluster represented by Model 1 in Figure 2 along the perpendicular reaction pathway pointing to the center of the edge common to the two vanadium-oxygen square pyramids, and the molecule is oriented perpendicularly with the terminal carbon atoms directed downwards, the two surface oxygen atoms of PO_4 tetrahedra become inserted into the C-H bonds of C1 and C4 carbon atoms, forming terminal C-OH groups. Simultaneously, the terminal carbon and hydrogen atoms interact with the oxygen atom of the vanadyl group, which results in splitting of the C-H bonds and formation of an oxygen bridge between the two carbon atoms, two hydrogen atoms being split off and forming the H_2 molecule (Figure 3). This is a spontaneous process experiencing no activation barrier. A possible transition state of such a concerted reaction is shown in Figure 4.

In Model 1, discussed above, the three oxygen atoms taking part in the reaction were sticking out above the surface of the cluster from adjacent vanadium and phosphorus polyhedra situated along the common edge of the two vanadium square pyramids. A different situation is encountered in the case of Model 2, where

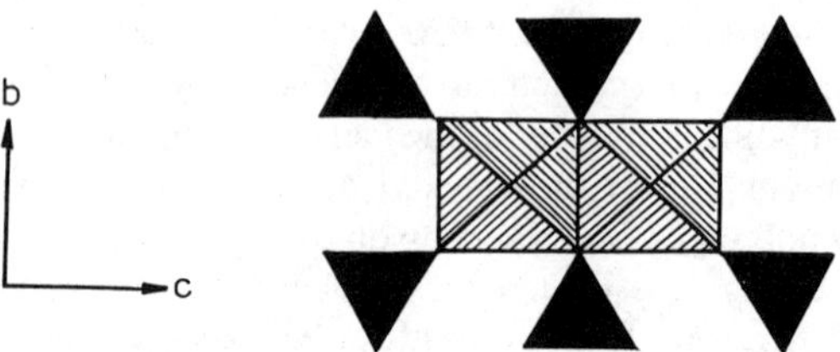

Figure 1. Schematic representation of the elementary unit of the $(VO)_2P_2O_7$.

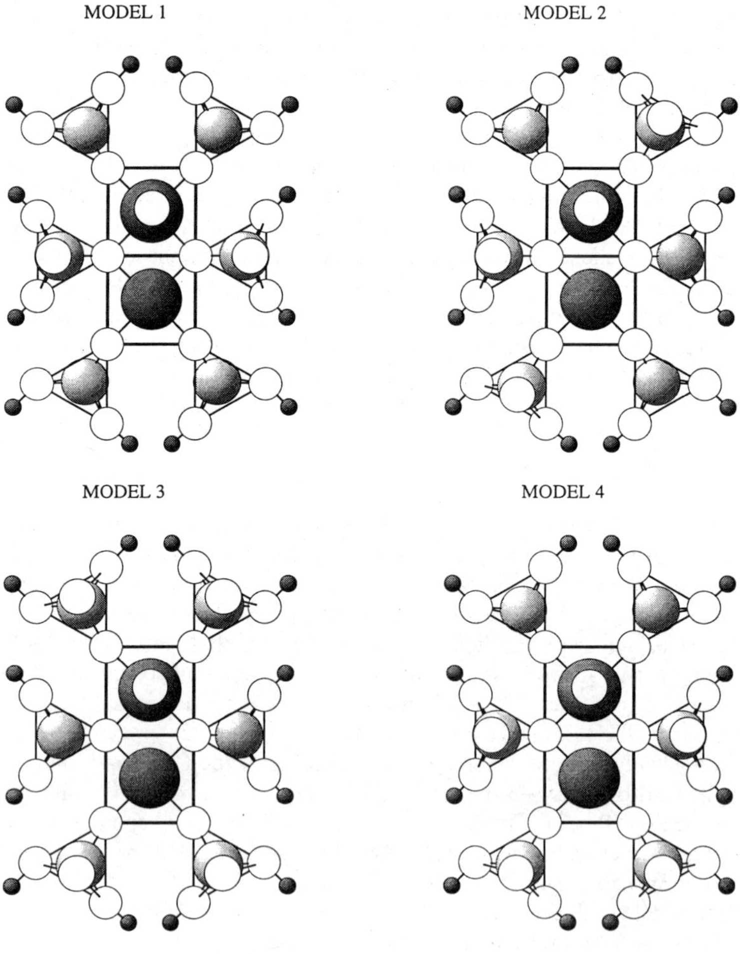

MODEL 1

MODEL 2

MODEL 3

MODEL 4

Figure 2. Cluster models used to mimic the active sites of the catalysts surface.

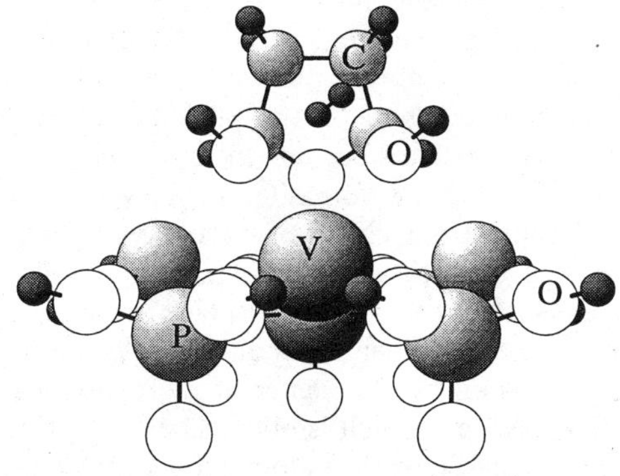

Figure 3. Precursor of maleic anhydride formed as the result of the perpendicular approach of the butane molecule to the cluster represented by Model 1.

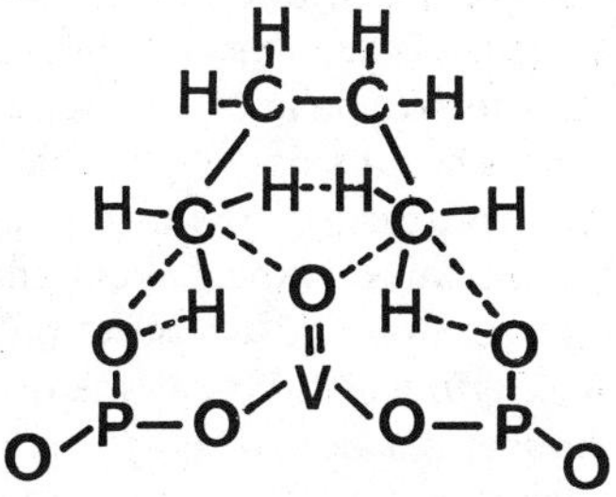

Figure 4. Transition complex of the reaction leading to the formation of the precursor of maleic anhydride represented in Figure 3.

the three oxygens above the plane of the cluster are arranged along the diagonal of the vanadium square pyramid, with the oxygen atoms being further apart than in Model 1. The butane molecule, approaching the same way as in Model 1, picks up only two adjacent oxygen atoms located at phosphorus and vanadium. A geminal dialcohol is formed because only one carbon atom is close enough to the phosphorus oxygen to become activated enough to form a bond with vanadyl oxygen. At the surface of Model 3 only terminal monoalcohol is formed, because the incoming molecule "sees" only one oxygen atom protruding from the surface, linked to vanadium. However, if the butane molecule is allowed to approach along the perpendicular pathway pointing to the center of the diagonal linking the three oxygen atoms in Model 2, or along the pathway pointing to the center of the external edge of the vanadium square pyramid, where the two phosphate groups have oxygen atoms sticking upwards, three oxygen atoms become incorporated into the butane molecule as was the case in Model 1, with the formation of the C-O-C bridge.

Model 4 has the same arrangement of three oxygen atoms pointing upwards along the common edge of the two vanadium square pyramids, as in Model 1, and two more oxygen atoms sticking out from phosphate groups at the external edge of the vanadium square pyramids. When the butane molecule approaches along the perpendicular pathway pointing to the center of the common edge of square V-O pyramids, as in Model 1, the same result is obtained as with Model 1, indicating that the presence of additional surface oxygen atoms in phosphate groups located further away does not influence the course of the reaction. Apparently, the given reaction pathway on the potential energy hypersurface seems to be dependent almost entirely on only the closest surroundings.

Approach with the Butane Molecular Plane Parallel to the Plane of the Cluster. An interesting reaction takes place when the the butane molecule is allowed to approach with its plane parallel to the plane of the cluster along the perpendicular pathway pointing to the center of the common edge of the two square V-O pyramids in Model 1. Namely, under the influence of the incoming the butane molecule, the three adjacent oxygen atoms (one linked to vanadium and two linked to phosphate groups) form the ozonide type of structure whose terminal oxygen atoms become inserted into C-H bonds of the terminal carbon atoms of the butane molecule. This is shown in Figure 5. A similar reaction takes place in Model 4. In this case, however, the two oxygens sticking out of phosphate groups at the terminal edge of the square pyramid become inserted into C-H bonds at the C2 and C3 positions of the butane, forming OH groups.

The tendency to form bonds between oxygen atoms sticking above the plane of the cluster when the butane molecule approaches parallel to the plane of the cluster is also seen in the case of Model 2, where only two oxygen atoms are sticking out, and become inserted into the C-H bond of one of the terminal atoms of the butane in the form of a peroxide bridge. In the case of Model 3, where only one oxygen atom is located above the plane, only this one oxygen is inserted into the C-H bond of the terminal carbon atom of the butane molecule.

Discussion

The results from modelling the approach of a butane molecule to the cluster which mimics the active sites of vanadyl pyrophosphate catalysts clearly show that the first intermediate, which is formed at the surface, depends on the orientation of approach of the butane molecule and the local arrangement of surface oxygen atoms. The different intermediates described for the models considered in this study represent local minima on the energy hypersurface. In the modelling procedure, we have allowed only the atoms of the the butane molecule and of oxygen atoms (sticking above the plane of the cluster) to move freely, and we have kept the geometry of the cluster frozen. This approach does not permit a determination of the absolute minima of the observed intermediates, and therefore it is not possible to estimate their relative probabilities of formation. Moreover, the reaction pathways for their further transformations, and the energy barriers encountered on these pathways, remain as yet unknown, and therefore we are unable to estimate the relative stabilities of these intermediates. One general conclusion can, however, be formulated that the intermediate formed strongly depends on the local arrangement of surface oxygen atoms.

A question may be raised at this point as to whether the presence of phosphate groups plays any role in the formation of the C-O-C bridge. In order to answer this question, the calculation has been repeated for an analogous system, in which the phosphorus atoms were substituted by vanadium atoms. It turned out that in such a case the terminal C-OH groups are formed similarly as on the vanadyl phosphate unit, but no formation of the C-O-C bridge takes place (Figure 6). It may thus be concluded that the arrangement of vanadium-oxygen square pyramids, surrounded by corner-linked phosphate groups, shows a unique property of forming the C-O-C bridge. The calculations carried out with a $V_{10}O_{31}$ cluster, cut from the real V_2O_5 structure gave the same result of forming only terminal C-OH groups, confirming that the ability to form a C-O-C bridge is uniquely connected with the simultaneous presence of both V-O and P-O polyhedra linked together through corners. It is interesting that recent "in situ" studies in HREM (*10*) revealed that, on exposure of $(VO)_2P_2O_7$ to the butane at 400^0C, shear planes are formed by removal of oxygen atoms from the phosphate groups bridging the vanadium-oxygen square pyramid dimers.

Experimental data reported recently indicate that the butane may also be oxidized to maleic anhydride with moderate selectivity on catalysts composed of submonolayer of vanadium oxides supported on titania. This observation raises the question as to whether the two edge linked square pyramids of vanadia pyrophosphate, considered as the active site, are indeed the indispensable requirement for the selective oxidation of the butane to take place. In order to answer this fundamental question we have carried out the modelling using a simpler model of the active site composed of one V-O square pyramid surrounded by four corner linked phosphate groups. The results shown in Figure 7 clearly demonstrate that the same intermediate is being formed as in the case of Model 1, in which two oxygen atoms are incorporated into the C-H bonds of the terminal carbon atoms, and the third oxygen forms the bridge between these atoms. This indicates that at least the first

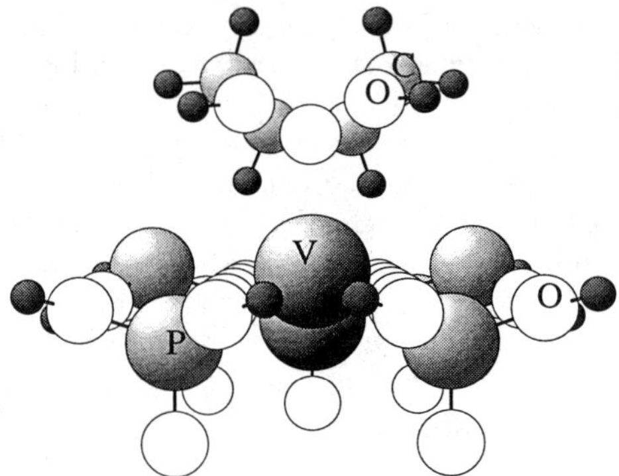

Figure 5. Complex of C4 fragment with O3 fragment formed as the result of the parallel approach of the butane molecule to the cluster represented by Model 1.

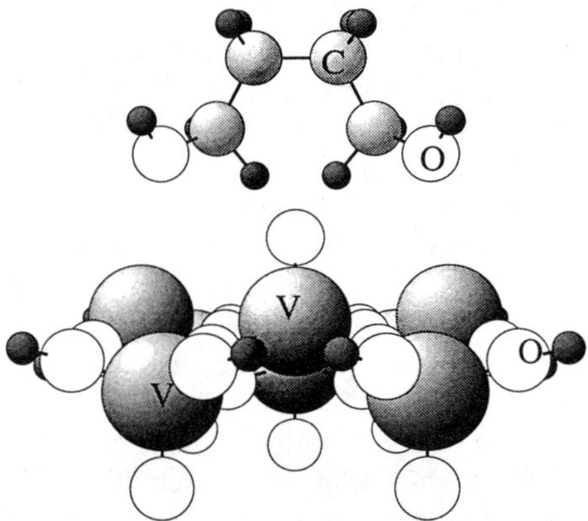

Figure 6. Butandiol resulting from the perpendicular approach of the butane molecule to the cluster represented by Model 1, in which phosphorus atoms have been substituted by vanadium atoms.

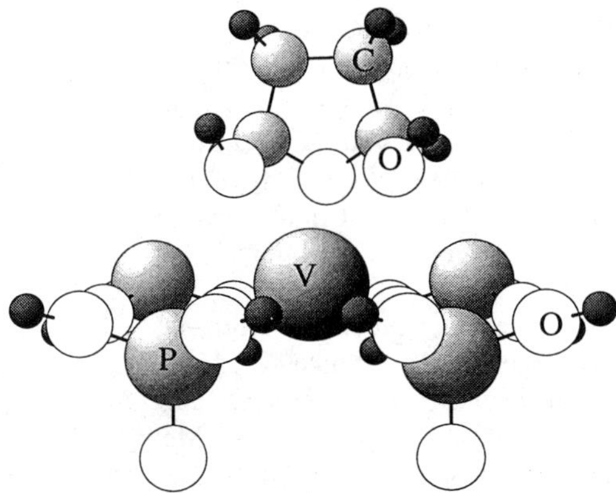

Figure 7. Precursor of maleic anhydride formed as the result of the perpendicular approach of the butane molecule to the active site composed of one V-O square pyramid surrounded by four corner linked phosphate groups.

step of the reaction of the butane oxidation can take place on an active site composed of single V-O polyhedron. A question remains open as to whether further transformations of this intermediate can also proceed on this site.

All examples discussed in this paper illustrate the critical importance of the presence of three surface oxygen atoms in the appropriate sequence, P-O, V=O, P-O, for the formation of an oxygen bridge between terminal carbon atoms of the butane molecule. It is noteworthy that all three oxygen atoms required to transform the butane into maleic anhydride may become incorporated in the first step of the reaction by a concerted transformation proceeding without any energy barrier. Only one V-O polyhedron seems to be involved in this transformation. The question remains as to how the four hydrogen atoms are removed from this intermediate in subsequent steps of the reaction to give the maleic anhydride molecule.

Literature Cited

(1) Cavani, F.; Trifiro, F. *Appl.Catal. A. General*, **1992**, *88*, 115.
(2) Centi, G.; Trifiro, F.; Ebner, J.R.; Franchetti, V.M. *Chem.Rev.*, **1988**, *88*, 55.
(3) Centi, G. *Catal.Today*, **1993**, *16*, 1.
(4) Cavani, F.; Trifiro, F. *Catalysis*, Specialist periodical Reports, Royal Society of Chemistry, London, UK, 1994, Vol.11; pp.224.
(5) Contractor, R.M.; Sleight, A.W. *Catal.Today*, **1988**, *3*, 175.
(6) Gleaves, J.T.; Ebner, J.R.; Kueckler, T.C. *Catal.Rev.-Sci.Eng.*, **1988**, *30*, 49.
(7) Szakacs, S.; Wolf, H.; Mink, G.; Bertoli, I.; Wüstnek , N.; Lücke, B.; Seeboth, H.; *Catal.Today*, **1987**, *1*, 27.
(8) Haber, J. *Stud.Surf.Sci.Catal.*, **1989**, *48*, 447.
(9) Bielański A.; Haber, J. *Oxygen in Catalysis*, Dekker, M. inc., Ed.: New York, 1991.

(10) Bordes, E. *Catal.Today*, **1988**, *3*, 163.
(11) Okuhara, T.; Misono, M. *Catal.Today*, **1993**, *16*, 61
(12) Ebner, J.R.; Thomson, M.R. *Stud.Surf.Sci.Catal.*, **1991**, *72*, 31.
(13) Thomson, M.R.; Ebner, J.R. *Stud.Surf.Sci.Catal.*, **1992**, *72*, 353.
(14) Thomson, M.R.; Hess, A.C.; White, J.C.; Anchell, J., Nicolas, J.B.; Ebner, J.R.; Lytle, F.W. *Stud.Surf.Sci.Catal.*, **1994**, *82*, 167.
(15) Stefani, G.; Budi, F.; Fumageli, C.; Sucin, D. *Stud.Surf.Sci.Catal.*, **1990**, *67*, 537.
(16) Zazhigalov, V.; Haber, J.; Stoch, J.; Komashko, G.A.; Pyatnitskaya, A.I.; Belousov, V.M. *Appl.Catal.A.,* **1993**, *96*, 135.
(17) Ebner, J.R.; Thomson, M.R. *Catal.Today*, **1993**, *16*, 51.
(18) Kubias, B.; Rodemerck, V.; Wolf, G.V.; Meisel, N.; Schaller, W. *Proc. DGMK Conference on Selective Oxidation in Petrochemistry*, Baerns, M.; Weitkamp, J. Eds., Tagungsbericht, 1992, Vol. 9204; pp.303.
(19) Cavani, F.; Centi, G.; Trifiro, F. *La Chimica & L'Industria*, **1992**, *74*, 182.
(20) Horowitz, H.S.; Blackstone, C.M.; Sleight, A.W.; Tenfer, G. *Appl.Catal.*, **1988**, *38*, 193.
(21) Bergeret, G.; David, M.; Broyer, J.P.; Volta, J.C.; Hecquet, G. *Catal.Today*, **1987**, *1*, 37.
(22) Yamazoe, N.; Morishige, H.; Tamaki J.; Miura, N. *Stud.Surf.Sci.Catal.*, **1993**, *75*, 1979.
(23) Ridley, J.; Zerner, M.C. *Theoret.Chim.Acta*, **1973**, *32*, 111.
(24) Bacon, A.D.; Zerner, M.C. *Theoret.Chim.Acta*, **1979**, *53*, 21.
(25) Head, D.; Zerner, M.C. *Chem.Phys.Lett.*, **1985**, *122*, 264.

Chapter 19

The Mechanism of Catalytic Ammoxidation of Propane and Propene over Vanadium Antimony Oxides

Sigurd A. Buchholz and Horst W. Zanthoff

Lehrstuhl für Technische Chemie, Ruhr-Universität Bochum, Universitätstrasse 150, D–44780 Bochum, Germany

The ammoxidation of propane and propene over V-Sb-O catalysts was studied using a vacuum transient technique. From the results gained it was concluded that the selective reaction to acrylonitrile occurs via the intermediate formation of propene, which is subsequently transformed into acrolein. Acrolein is converted into the nitrile via short-lived NH_x-species (NH_4^+ or $NH_{3,ads}$). These NH_x-species, as well as NO_x, are also intermediates in the non-selective N_2 formation pathway.

The ammoxidation of C_3 hydrocarbons produces acrylonitrile, an important raw material for the chemical industry. Acrylonitrile is presently produced from propene, mainly with the SOHIO/BP process which achieves acrylonitrile yields of about 80 % at a propene conversion of 98 % (*1*). However, in recent years much attention has been paid to the development of alternative processes which allow propane to be used as the feed gas. Such a new process could be economically viable due to the price difference between propane and propene (*2*). Conventional propene ammoxidation catalysts, Bi-Mo oxides or Fe-Sb oxides, show no or only little activity and/or selectivity to acrylonitrile in the conversion of propane (*1,3*). Therefore, new catalysts have had to be developed which offer active sites for both dehydrogenation and nitrogen insertion. Among the different catalytic systems investigated for the ammoxidation of propane into acrylonitrile, V-Sb oxides exhibited promising catalytic results (*1,4*), but maximum yields of around 40 % do not allow their use as industrial catalysts.

In order to develop more selective and active catalysts much attention has been paid in the past to elucidating the structure and mode of operation of V-Sb-O catalysts for the ammoxidation of propane. It is commonly suggested that propene is the primary intermediate in the reaction (*5*). However, the detailed reaction mechanism and the nature of the active sites of the nitrogen insertion step to acrylonitrile from propene is still a matter of discussion. Considering data from differential and integral kinetic investigations over V-Sb-Al-oxides (*4*) and V-Sb-W-Al-oxides (*6*), it has been concluded that direct formation of acrylonitrile from propene occurs, while acrolein is formed in a parallel reaction as an intermediate to CO_x and HCN, which are undesired

0097–6156/96/0638–0259$15.00/0

products. The direct formation of acrylonitrile without the participation of acrolein has also been proposed for the ammoxidation of propene over Fe-Sb-oxides (*7*). In contrast to this hypothesis, recent results published by Nilsson et al. (*8*) gained from kinetic investigations using V-Sb-oxides indicate that acrolein might be an intermediate in acrylonitrile formation during the ammoxidation of propene. Moreover, for the ammoxidation of methylaromatics over V-containing catalysts, i.e., V-P-O catalysts, the respective aldehydes are assumed to be intermediates in the nitrile formation (*9*). Furthermore, the nature of the active surface species for N-insertion is still not clear. For catalysts of the V-Sb-O, Ga-Sb-O and $(VO)_2P_2O_7$ type, different N-species are mentioned in literature as being active in ammoxidation reactions, i.e. NH_4^+, $NH_{3,ads}$, NH_2^- or M=NH (*5,10*). For Fe-Sb-oxides and V-Sb-Al the formation of Sb-NH-Sb groups has also been proposed (*4,6*).

Against this background it was the aim of the present work to elucidate the reaction mechanism of propane ammoxidation over VSb_yO_x (y = 2 and 5) catalysts. For this purpose, the interaction of propane, propene, acrolein and ammonia with the catalyst was investigated appyling a vacuum transient technique in the Temporal Analysis of Products reactor (TAP). Furthermore, the role of ammonia on the reaction pathways was investigated.

Experimental

The VSb_2O_x and VSb_5O_x catalysts were prepared according to Catani et al. (*6*) by redox reaction in aqueous medium. NH_4VO_3 and Sb_2O_3 were added in stoichiometric portions to a 1-N aqueous ammonia solution at 353 K. The suspension was stirred continously under reflux for 24 h. The water was evaporated and the remaining solid was dried at 413 K for 12 h and calcined in air at 623 K for 24 h, at 773 K for 3 h and finally at 900 K for another 3 h with an intermediate heating rate of 50 K/h. The resulting catalyst was pelletized and granulated. Particles with a diameter of 255 to 350 μm were used in the experiments.

VSb_1O_x and Sb_2O_4 were prepared as reference substances. VSb_1O_x was prepared by redox reaction of V_2O_5 and Sb_2O_3 as described by Berry et al. (*11*). Both metal oxides were finely ground, heated within 12 h to 933 K and kept at this temperature for 12 h. Subsequently the catalyst was heated at a rate of 15 K/h to 1023 K and kept at this temperature for an additional 24 h. Sb_2O_4 was prepared by heating Sb_2O_3 in air at a rate of 50 K/h to 773 K and then maintaining it at this temperature for 5 h.

These catalysts were characterized by the use of different methods: The BET surface of the catalysts was determined by N_2-adsorption at 77 K in accordance with Haul and Dümbgen. XRD spectra were recorded using an INEL CPS 120 powder diffractometer, x-ray reflection method and Cu-Kα radiation (1.5406 Å). IR-spectra were recorded using a Perkin-Elmer 1710 Fourier Transform Infrared Spectrometer with a DRIFT cell. The catalyst structure was also investigated by TEM employing a JEOL JEM 1200 EX microscope.

The vacuum transient experiments were performed in the Temporal-Analysis-of-Products (TAP) reactor (see Figure 1). Detailed descriptions of the equipment used and the operating conditions are given elsewhere (*12,13*). Therefore, only information specific to the present work are given below. During the transient experiments, small gas pulses (approx. size 10^{15}-10^{17} molecules/pulse) produced by two high speed pulse valves (opening time in the ms range) enter a catalytic microreactor containing 100 to

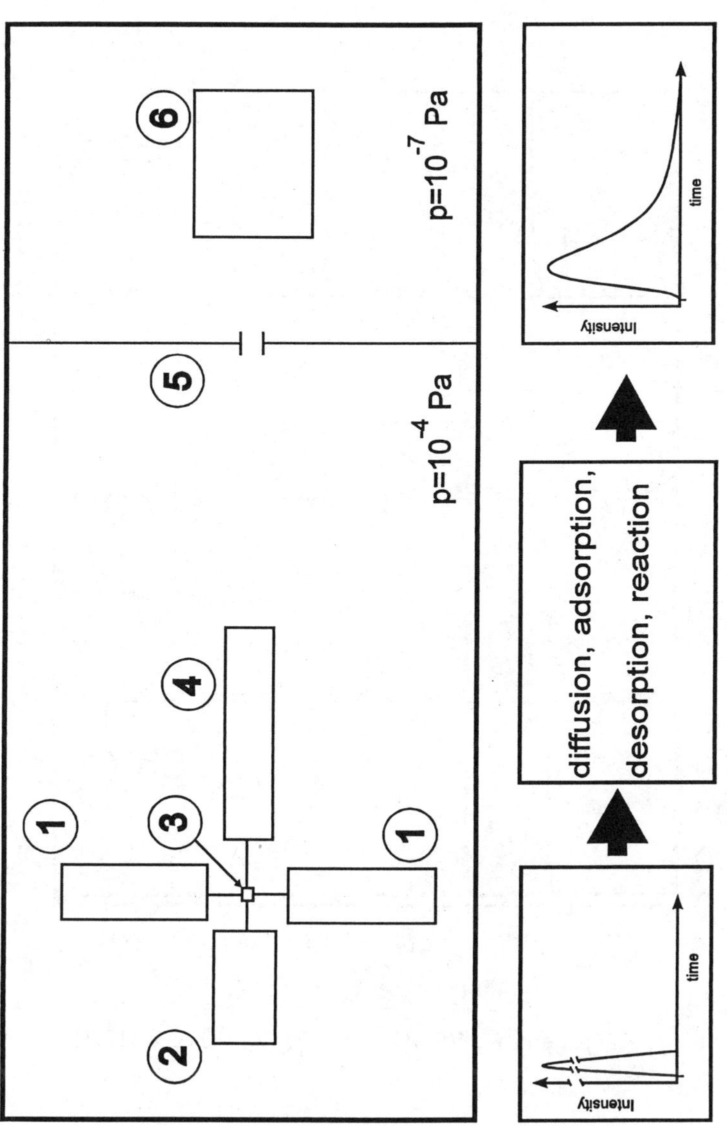

Figure 1. Scheme of the Temporal-Analysis-of-Products (TAP) reactor: (1) High-speed pulse valves, (2) continuous feed valve, (3) zero volume manifold, (4) fixed bed reactor, (5) focussing slits, (6) quadrupole mass spectrometer.

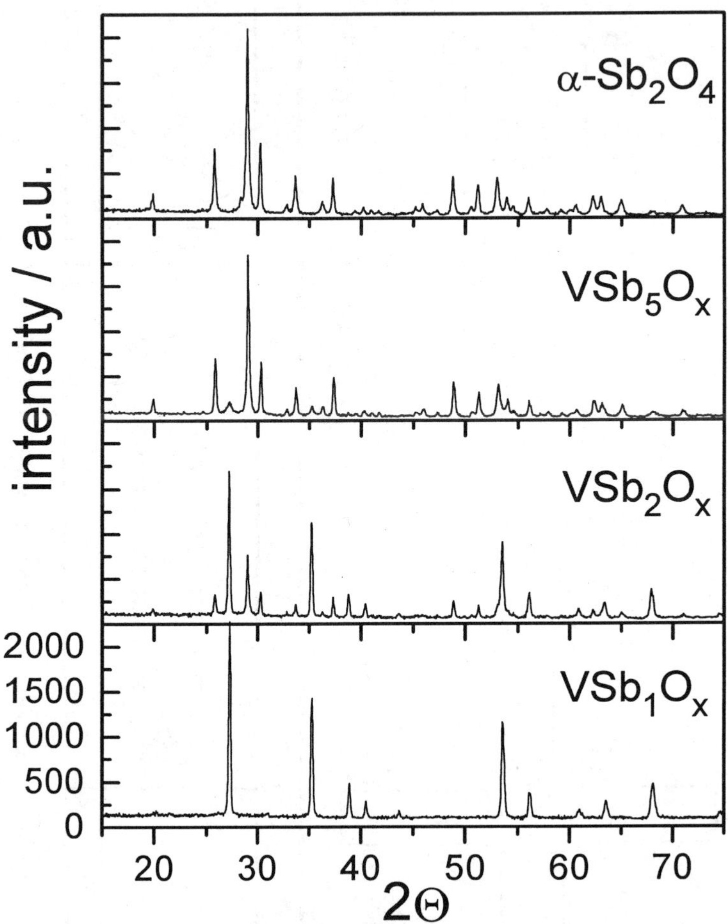

Figure 2. XRD patterns for VSb_1O_x, VSb_2O_x, VSb_5O_x and Sb_2O_4.

250 mg of catalyst. Transport of the molecules occurs by diffusion processes mainly because the reactor is kept in a vacuum ($\approx 10^{-4}$ Pa). The response to the inlet pulse is analyzed at the reactor outlet by using a quadrupole mass spectrometer with high time resolution (max. 10 μs). Using both pulse valves experiments can be performed in which the catalyst is first treated with a gas from one valve, followed by a gas pulse from the other valve after a defined interval of time Δt and analyzing the response signal at the reactor outlet. In this case the reactant pulsed first interacts with the catalyst, forming an adsorbed species or an adsorbed intermediate. The second reactant, pulsed with a time difference, interacts with this species on the surface forming a product. Further, one pulse from a single valve or one pulse from each valve during a sequential pulse-experiment is called "one cycle". The total cycle time was varied within a range of 0.9 to 20 seconds, the time difference between two sequential pulses was varied between 0.1 to 1.2 seconds. In order to improve the signal/noise ratio 6 to 20 cycles were averaged. All feed gas mixtures contained an inert reference gas (He or Ne) as an internal standard in order to assist calculation of conversions and yields. Substances were detected in the mass spectrometer at the following m/e ratios (amu): amu = 56 (acrolein), amu = 53 (acrolein, acrylonitrile), amu = 44 (CO_2, N_2O, C_3H_8), amu = 43 (C_3H_8), amu = 41 (C_3H_6, C_3H_8), amu = 32 (O_2), amu = 30 (NO, C_3H_6, C_3H_8) and amu = 28 (C_3H_8, C_3H_6, N_2, CO_2, CO). The contribution of different substances to a measured response signal at one single amu was calculated by the experimentally determined fragmentation patterns of the pure substances. A detailed description of the different experimental conditions during the TAP-experiments, such as reaction temperature, number of molecules per pulse and amount of catalyst is given in the Result and Discussion section of this paper.

Results and Discussion

Catalyst Characterisation. Table I summarizes the BET surfaces determined for the different catalysts. The catalytic surface of all substances were within the range of 2 to 3.6 m^2/g.

Table I. BET surface of the different V-Sb-O catalysts

catalyst	VSb_1O_x	VSb_2O_x	VSb_5O_x	Sb_2O_4
S_{BET} / m^2/g	2.0	2.2	3.6	2.2

XRD and electron diffraction analysis revealed the existence of only two crystalline phases in the catalysts: $VSb_{1-x}O_{4-1.5x}$ and Sb_2O_4. Figure 2 shows the XRD patterns for the catalysts investigated. While the VSb_1O_x exhibited only the diffraction patterns of $VSb_{1-x}O_{4-1.5x}$ (JCPDS No. 35-1485) the other V-Sb-O catalysts also showed additional patterns for Sb_2O_4 (JCPDS No. 11-0694, syn. Cervantite) increasing with increasing Sb content. TEM pictures for the three V-Sb oxides are shown in Figure 3. The granular structures originated from Sb_2O_4, while the cubic crystals were identified as $VSb_{1-x}O_{4-1.5x}$. The XRD patterns of Sb_2O_4 revealed small impurities of Sb_2O_3 (JCPDS No. 43-1071). From DRIFT spectroscopic investigations on the adsorption of pyridine it can be concluded that the three V-Sb-O catalysts exhibit only a low level of Lewis and Bronstedt acidity.

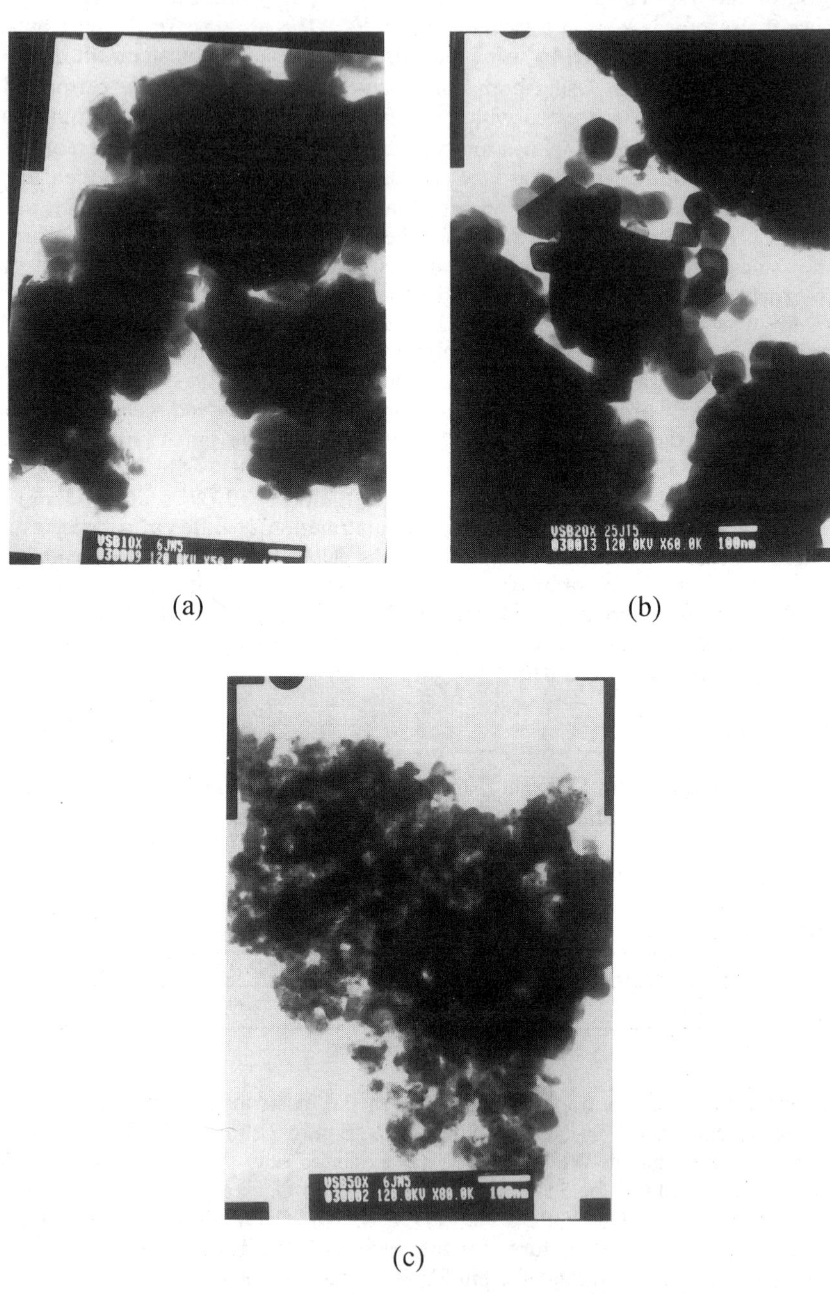

Figure 3. TEM of (a) VSb_1O_x, (b) VSb_2O_x and (c) VSb_5O_x.

Interaction of Pure Ammonia with the Catalyst. When NH_3 was pulsed over the three different V-Sb-O catalysts at a temperature of 673 K, on all solids N_2, NO and N_2O were observed as products. NO_2 was not detected. The normalized response signals obtained at the reactor outlet employing a VSb_5O_x catalyst are shown in Figure 4. The strong tailing observed for mass 17 is due to the OH-fragmentation of water which is formed in the reaction of NH_3 to NO, N_2O and N_2. Conversion of NH_3 and yields to NO, N_2O and N_2 for all catalysts are summarized in Table II. On all V-Sb-O catalysts studied, N_2 was observed as the main product. Sb_2O_4 was found to be inactive towards the conversion of NH_3 under the conditions applied, but ammonia was irreversibly adsorbed, but could be desorbed as NH_3 at higher temperatures under vacuum.

Table II. Conversion of NH_3 and yields to NO, N_2O and N_2 obtained when pulsing a mixture of NH_3/Ne over a Sb_2O_4 and the V-Sb-O catalysts (m_{Cat}= 150 mg, T = 673 K, pulse size = $6.2 \cdot 10^{16}$ molecules/pulse)

Catalyst	X (NH_3) %	Y (NO) %	Y (N_2O) %	Y (N_2) %
VSb_1O_x	48	9	4	35
VSb_2O_x	94	6	23	65
VSb_5O_x	93	8	23	62
Sb_2O_4	51*	-	1	3

* - NH_3 irreversibly adsorbed on the solid

The formation of NO, N_2O and N_2 shows that ammonia is able to reduce the catalyst at temperatures usually applied for the ammoxidation reaction. For a continuous flow of an ammoxidation feed gas ($C_3H_8/NH_3/O_2$=1:2.5:2.5) the formation of NO_x in addition to N_2 was also observed under vacuum conditions (*15*). However, under atmospheric pressure only N_2 can be detected (*6*). It has to be assumed that under atmospheric conditions NO and N_2O are reduced to N_2 by excess NH_3 in the selective catalytic reduction (SCR) of NO as is known to occur on vanadium oxides (*16*). In order to show that this reaction can also take place on the V-Sb-O surface, ammonia was pulsed over the catalyst, followed by NO. N_2 was the major product in the response signal to the NO pulses, which can be attributed to the SCR reaction. A direct conversion of NO and N_2O to N_2 on the catalyst can be excluded, because no reaction was observed when NO and N_2O were pulsed separately over the catalyst in the vacuum.

Interaction of Propane and Ammonia with the Catalysts. When a mixture of C_3H_8/NH_3/Ne=10:25:40 (pulse size: $2.7 \cdot 10^{16}$ molecules/pulse; T = 693 K) was pulsed over the V-Sb-O catalyst in vacuum, propene, acrolein, acrylonitrile, CO and CO_2 were detected as products. Since gas phase reactions can be neglected under these conditions, it can be concluded that the ammoxidation of propane on V-Sb oxides under these conditions is a heterogeneous reaction.

In a previous paper (*14*) we already showed by means of a DRIFT spectroscopic investigation that ammonia interacts with the V-Sb-O catalysts forming

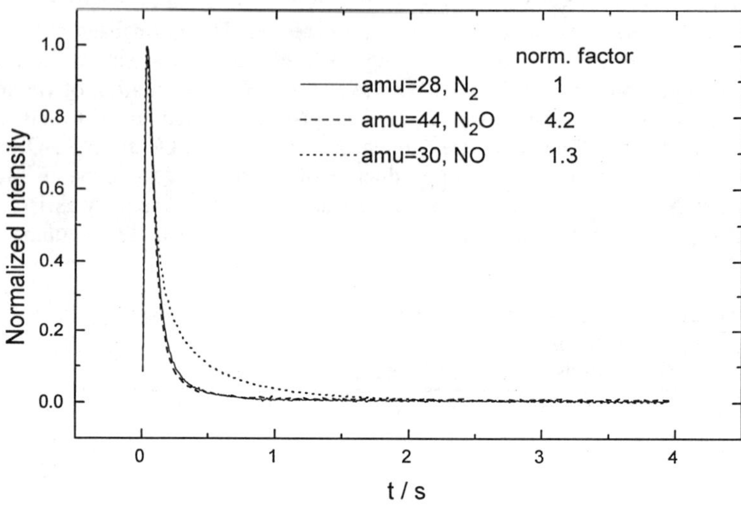

Figure 4. Response signals for N_2, NO and N_2O to pulses of NH_3 over a VSb_5O_x (m_{cat}= 250 mg, T_b= 698 K, $1 \cdot 10^{17}$ molecules/pulse).

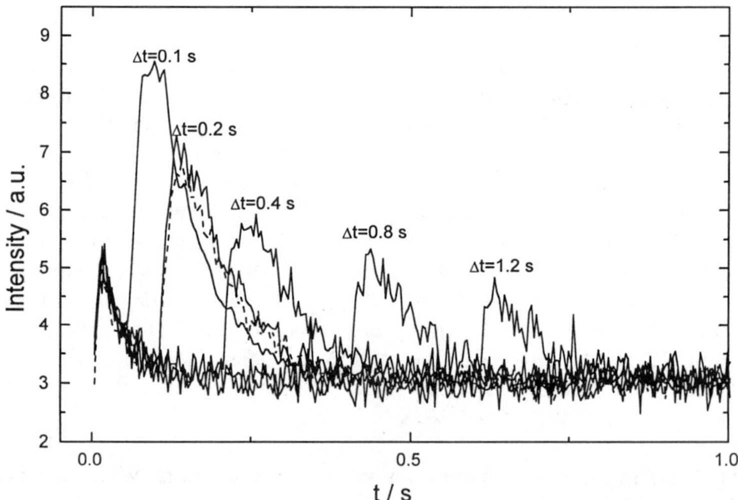

Figure 5. Sequential pulsing of 1. NH_3 and 2. C_3H_8 over a VSb_5O_x catalyst - Dependence of the amount of acrylonitrile observed on the time difference between the pulses (m_{cat}=250 mg, T_b=723 K, NH_3/He=1:3; $4 \cdot 10^{16}$ molecules/pulse; C_3H_8/Ne=1:9; $2 \cdot 10^{16}$ molecules/pulse).

NH_4^+ (bands at 1405, 1430 and 1465 cm^{-1}) and $NH_{3,coord.}$ (band at 1622 cm^{-1}) on the surface. In order to obtain information about the reaction behaviour, of these species sequential pulse experiments were performed in which the catalyst was treated with NH_3 first, in order to form such kinds of NH_x species, followed by a pulse of propane after a time interval Δt. Figure 5 shows the response signals of acrylonitrile (amu=53) at the reactor outlet for different time intervals $\Delta t = 0.1$ to 1.2 seconds. With increasing time difference Δt between the pulses of NH_3 and propane the amount of acrylonitrile (which is proportional to the area under the response curves) decreases.

Since the amount of pulsed propane was constant, this decrease has to be attributed to a decrease of the amount of active N-species on the surface. This depletion of active N-species can be explained by two processes. Either the preadsorbed ammonia desorbs slowly from the catalyst or it is oxidized to NO_x and/or nitrogen by surface lattice oxygen. The present results do not allow us to distinguish between these possibilities. However, the experiment proves that N-species with a short lifetime are active in the formation of the nitrile. The half time of these active N-species can be estimated to about 0.7 sec under these conditions (m_{cat}= 200 mg, T=723 K).

Interaction of propene and ammonia with the catalyst. During the ammoxidation of propane acrolein is observed as a product. However, under the applied vacuum conditions of the TAP reactor its amount is small and quantification is difficult. Therefore, the interaction of the primary intermediate propene with the catalyst in the presence of ammonia was studied in order to investigate the role of acrolein in the ammoxidation of propene. If C_3H_6 and NH_3 were pulsed simultaneously over VSb_5O_x at a temperature of T_{cat}= 700 K, the desorbable products occured at the reactor outlet in the following sequence: acrolein $\rightarrow$ acrylonitrile $\rightarrow$ CO_2. This indicates a sequential formation of acrylonitrile from acrolein as an intermediate product. However, from these experiments alone a parallel formation cannot be excluded.

In order to investigate the role of acrolein in the formation of acrylonitrile, sequential pulse experiments were applied. In this series of experiments the VSb_5O_x catalyst was first treated with C_3H_6 followed by pulses of NH_3 with a time difference of Δt = 0.1 s (T = 700 K). Response signals for acrolein (amu=56, solid line) and acrylonitrile (amu = 53, dotted line) were obtained at the mass spectrometer as shown in Figure 6. In addition the amu = 56 signal for the situation when C_3H_6 only is pulsed over the catalyst (broken line) is shown, too. When propene is pulsed over the catalyst at t = 0 s acrolein is formed, but when NH_3 is added at t = 0.1s the response signal for acrolein decreases rapidly compared to the amu = 56 signal for pulsing C_3H_6 only and the formation of acrylonitrile starts. It can therefore be concluded that adsorbed acrolein species react with short lived NH_x species on the catalyst surface to form acrylonitrile.

If acrolein is the intermediate in the formation of acrylonitrile, the product distribution of acrolein and acrylonitrile should vary in accordance with the amount of active NH_x species on the catalyst´s surface. Therefore, sequential pulse experiments were performed in which the VSb_5O_x catalyst was first treated with NH_3 followed by pulses of C_3H_6 at different time intervals Δt between 0.1 s and 1.2 s. Similar observations were made as was described above for propane. Figure 7 shows the yields of acrolein and acrylonitrile obtained in accordance with the time difference Δt. While the acrylonitrile yield decreases with increasing time interval from Y_{ACN}= 6.5 % to 1.7 %, the acrolein yield increases from $Y_{Acrolein}$= 6.6 % to 16.5 %. The catalyst´s activity de

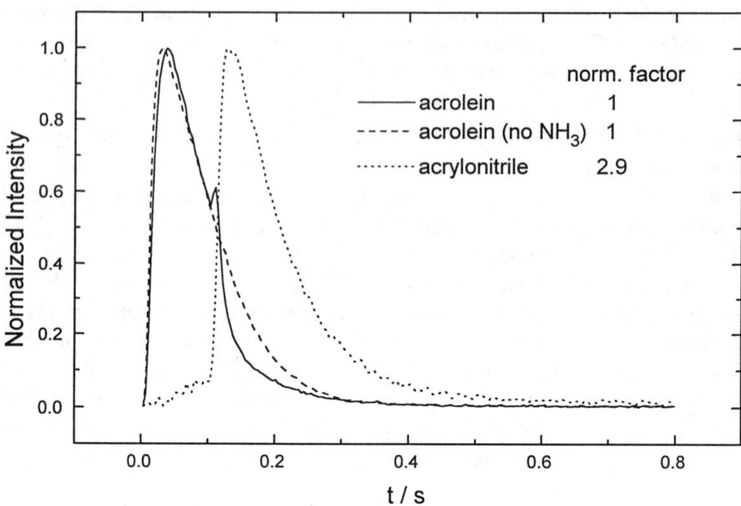

Figure 6. Sequential pulsing of 1. C_3H_6 and 2. NH_3 over a VSb_5O_x catalyst - Response signals for acrolein (solid line, with NH_3), acrolein (broken line, without NH_3) and acrylonitrile (dotted line) (m_{cat}= 250 mg, T = 703 K; Δt = 0.1 s; C_3H_6/Ne = 1:3; NH_3/He = 1:3).

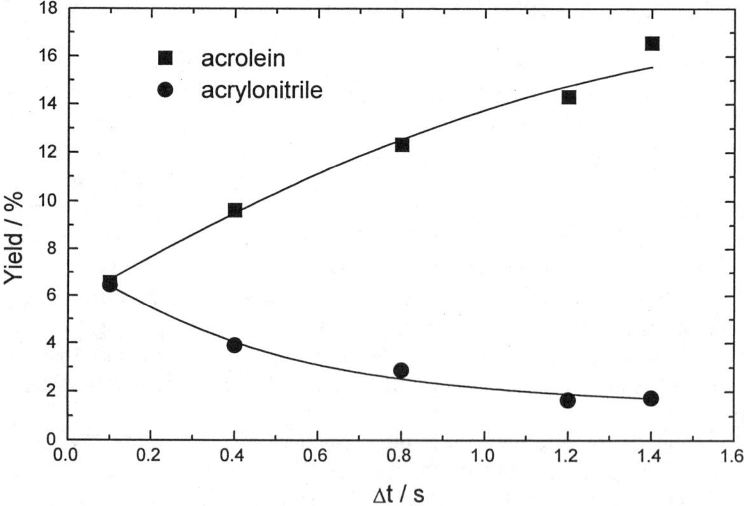

Figure 7. Dependence of the yields of acrolein and acrylonitrile on the time difference Δt between the pulses when sequentially pulsing 1. NH_3 and 2. C_3H_6 over a VSb_5O_x catalyst (m_{cat}= 150 mg, T_{cat}= 697 K, C_3H_6/Ne = 1:2; $5 \cdot 10^{15}$ molecules/pulse; NH_3/He = 1:2; $9 \cdot 10^{16}$ molecules/pulse).

creases only slightly during the measurement from X_{C3H6}= 88 % to 85 %, respectively. These observations confirm the assumption that acrolein is the intermediate in the formation of acrylonitrile; with a decreasing amount of NH_x on the surface, less acrolein is transformed into acrylonitrile and the yield of acrolein observed at the reactor outlet increases.

Furthermore, the assumption of acrolein being an intermediate in acrylonitrile formation is supported by steady state kinetic investigations of the ammoxidation of propene on a VSb_2O_x catalyst obtained by Nilsson et al. (*8*). With decreasing propene conversion they observed a decrease in selectivity to acylonitrile, while the acrolein selectivity increases. This indicates that acrolein is a primary product from propene, while acrylonitrile is formed consecutively. Centi et al. (*19*) observed acrylate and acrolein structures in IR spectroscopic investigations during ammoxidation of propene on V-Sb oxides. From these results they also concluded acrolein to being the intermediate for acrylonitrile formation.

However, although the reported results most probably indicate consecutive formation of acrolein and acrylonitrile over V-Sb-O catalysts, the existence of a parallel pathway via the formation of an allyl radical intermediate cannot be totally ruled out. Sokolovskii et al. (*5*) concluded in their recent review that both routes should be regarded and that the nucleophilicity of surface oxygen determine the ratio between both routes. The direct allylic route should be preferable on more acidic catalysts, while the aldehyde one should be typical for base catalysts (*5*). In fact, the acidity of V-Sb-Me-O catalysts (as determined by NH_3-TPD) (*17*) is more than 10 times lower than for molybdenum based catalysts (*20*). Therefore, the assumption that the consecutive mechanism occurs predominatly on V-Sb oxides seems to be reliable.

Interaction of Acrolein and Ammonia with the catalyst. Sequential pulse experiments were also used to investigate whether acrolein can be converted to acrylonitrile under the reaction conditions used for the VSb_5O_x catalyst. In these experiments the catalyst was treated first with NH_3 followed by acrolein with a time interval of $\Delta t = 0.1$ s. Figure 8 shows the response signals obtained for the unconverted acrolein (amu = 56) and the acrylonitrile formed (amu = 53, corrected for the acrolein fragment). It becomes obvious that acrylonitrile can be formed from acrolein and NH_x species adsorbed on the surface. However, the formation of acrylonitrile was also observed when the catalyst was first treated with acrolein followed by NH_3. Therefore, whether acrolein reacts from the gas phase or if it has to be preadsorbed on the catalyst surface cannot be distinguished.

Influence of the Redox State of the Catalyst on the Conversion of Propene and the Yields of Acrolein and Acrylonitrile. Since it was known from the literature (*17,18*) that V-Sb-O catalysts operate in a reduced state under steady state ammoxidation reaction conditions, the influence of the degree of reduction on the conversion of C_3H_6 and the yield of acrolein and acrylonitrile was studied. In a series of sequential pulse experiments the VSb_5O_x catalyst was treated with NH_3 followed by C_3H_6 at a constant time difference of $\Delta t = 0.1$ s. Responses at amu = 56, 53, 41 and amu = 20 signals were used for the calculation of propene conversion and yields of acrolein and acrylonitrile. After each experiment the catalyst was reduced by treatment with a definite amount of ammonia as a reducing agent. Immediately after this reduction, a sequential pulse experiment as described above was performed. Figure 9

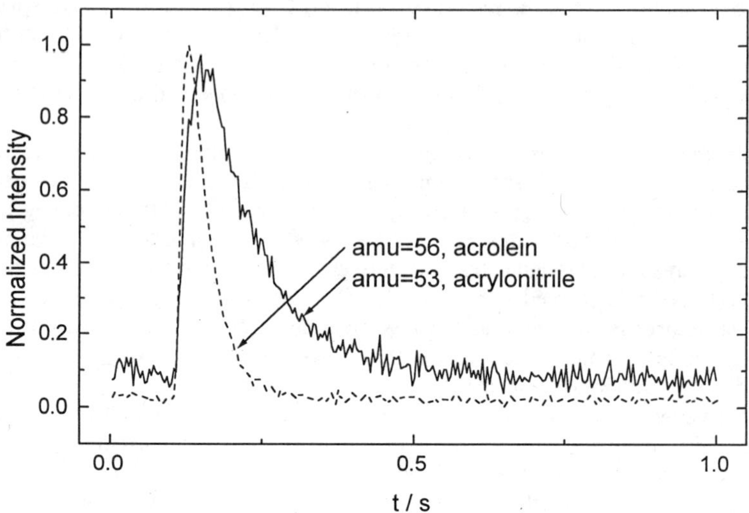

Figure 8. Response signals for unconverted acrolein (amu 56) and product acrylonitrile (amu 53, corrected for the acrolein fragment) for a sequential pulse experiment in which 1. NH3 and 2. acrolein was pulsed over the VSb_5O_x catalyst (m_{cat}= 250 mg, T_{cat}= 703 K).

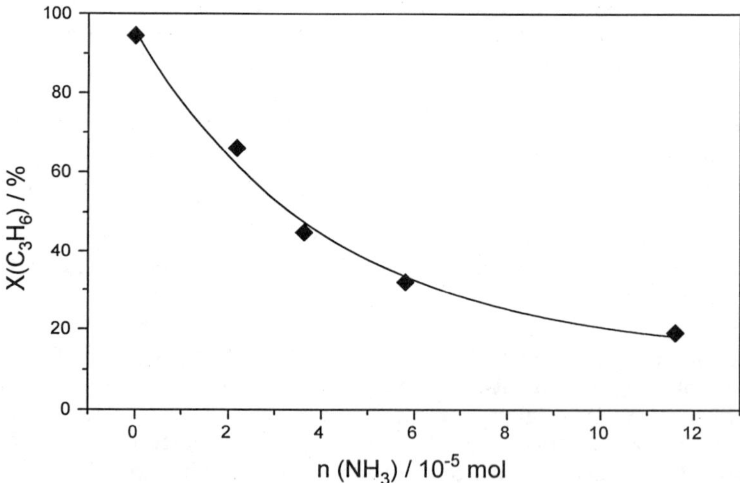

Figure 9. Dependence of the conversion of propene on the amount of NH_3 pulsed over the VSb_5O_x catalyst ($NH_3 \rightarrow C_3H_6$; $\Delta t = 0.1$ s ; m_{cat}= 150 mg, T_{cat}= 697 K, C_3H_6/Ne=1:2; $5 \cdot 10^{15}$ molecules/pulse, NH_3/He=1:2; $9 \cdot 10^{16}$ molecules/pulse).

shows the propene conversion obtained as a function of the amount of ammonia pulsed over the catalyst. With an increasing amount of ammonia pulsed, i.e., increasing the degree of reduction, the conversion of propene decreased. The yields of the products, acrolein and acrylonitrile decrease as shown in Figure 10. While propene conversion decreases from X_{C3H6}= 94.6 % to 19.1 % the yields of acrolein and acrylonitrile decrease from Y_{ACR}= 10 % and Y_{ACN}= 9 % to Y_{ACR}= 4 % and Y_{ACN}= 5.5 %, respectively. Also shown in Figure 10 is the initial yield to acrolein, when only propene is pulsed over the catalyst (open square). As can be deduced, the initial acrolein yield is lower compared to the yields obtained after pulsing about $2 \cdot 10^{-5}$ mol NH_3. The acrolein yield, therefore, runs through a maximum, and if the catalyst is reduced further, the yields of acrylonitrile and acrolein also decrease. From these results it can be concluded that NH_3 at first removes oxygen from the fully oxidized surface, which would otherwise be responsible for the formation of total oxidation products, thus increasing the yields of selective products. With further NH_3 treatment, the active oxygen necessary for propene activation is also removed and the yield of all the products therefore decreases.

Conlusions

The interaction of NH_3, propane and propene with VSb_yO_x (y = 2, 5) catalysts was studied in order to identify intermediates in the ammoxidation of propane and propene into acrylonitrile and to elucidate details of the mechanism involved, in particular the N-insertion step into the primary intermediate propene. The catalysts were found to be very active for undesired NH_3 oxidation reactions to NO, N_2O and N_2 under vacuum conditions. Since no oxidation products were observed over pure Sb_2O_4 it must be assumed that NH_3 oxidation occurs on the $VSb_{1-x}O_{4-1.5x}$ phase detected in the catalysts, probably on vanadium sites. NO and N_2O were found to be intermediates in N_2 formation. The amount of total oxidation product can be decreased if the catalyst surface is partly reduced by the ammonia. Therefore, it can be concluded that the oxygen sites that are responsible for the undesired ammonia depletion also cause the total oxidation of hydrocarbons and/or the selective products acrolein and acrylonitrile.

The nitrogen insertion into propene occurs most probably via the intermediate formation of acrolein. Short lived N-species are involved in the transformation of acrolein into the desired product acrylonitrile. Since DRIFT spectroscopic investigations of ammonia adsorption on V-Sb-O catalysts (*14*) revealed only the presence of NH_4^+ or $NH_{3,coord.}$ on the catalyst surface, we assume these species to be the active ones in N-insertion into acrolein.

From the results obtained in this work, a mechanism can be proposed for the ammoxidation of propane to acrylonitrile on V-Sb oxides via acrolein and NH_x, shown in Figure 11.

Acknowledgements

This work was financially supported by the German Federal Ministry of Education, Science, Research and Technology (Contract No. 03D 0001A7). The authors also want to thank Prof. M. Baerns for supporting the work.

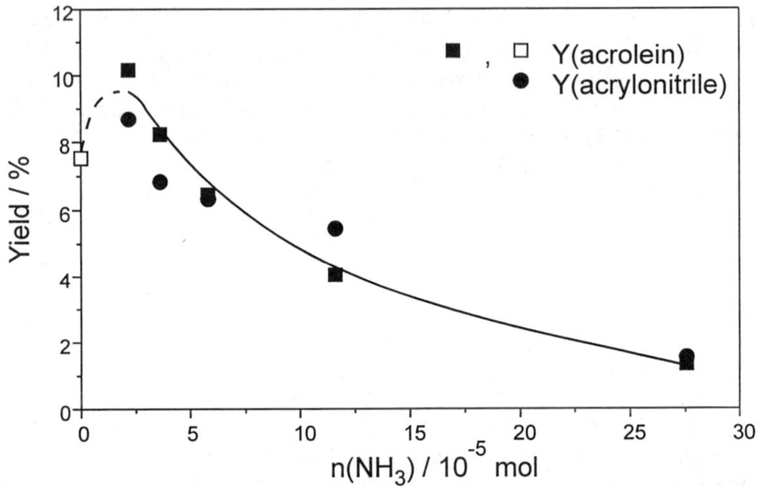

Figure 10. Dependence of the yields to acrolein and acrylonitrile on the amount of NH_3 pulsed over the VSb_5O_x catalyst ($NH_3 \rightarrow C_3H_6$; $\Delta t = 0.1$ s ; m_{cat}= 150 mg, T_{cat}= 697 K, C_3H_6/Ne=1:2; $5{\cdot}10^{15}$ molecules/pulse, NH_3/He=1:2; $9{\cdot}10^{16}$ molecules/pulse).

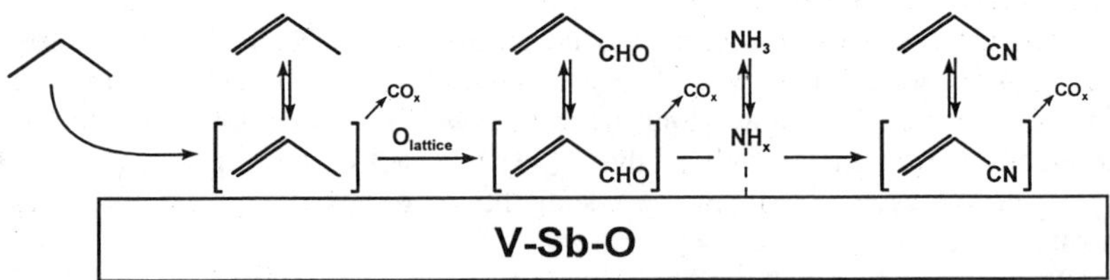

Figure 11. Proposed mechanism for the ammoxidation of propane over V-Sb oxides.

References

1. Moro-oka, Y.; Ueda, W. *Partial Oxidation and Ammoxidation of Propane: Catalysts and Processes,* in Catalysis, vol. 11, Royal Society of Chemistry, Athenaeum Press Ltd, Newcastle, GB **1994**.
2. Centi, G.; Grasselli, R. K.; Trifiro, F. *Catal. Today* **1994,** *13(4),* 661.
3. Centi, G.; Pesheva, D.; Trifiro, F.; *Appl. Catal.*, **1987,** 33, 343.
4. Andersson, A.; Andersson, S. L. T.; Centi, G.; Grasselli, R.K.; Sanati, M.; Trifiro, F. *Stud. Surf. Sci. Catal.* **1993,** *75,* 691.
5. Sokolovskii, V. D.; Davydov, A. A.; Ovsitser, O.Y. *Catal.Rev. - Sci. Eng.,* **1995,** *37(3),* 425.
6. Catani, R.; Centi, G.; Trifiro, F.; Grasselli, R.K. *Ind. Eng. Chem. Res.*, **1992,** *31,* 107.
7. Burrington, J. D.; Kartisek, V.; Grasselli, R. K. *J. Catal.*, **1984,** *84,* 363.
8. Nilsson, R.; Lindblad, J.T.; Andersson, A. *Catal. Lett.*, **1994,** *29,* 209.
9. Martin, A.; Zhang, Y.; Lücke, B.; Meisel, M. *Dechema Annual Meeting, Abstract Book,* 53 Wiesbaden, **1995**.
10. Centi, G.; Perathoner, S. *J. Catal,* **1993,** *142,* 84.
11. Berry, F. J., Brett, M.E.; Patterson, W.R. *J. Chem. Soc. Dalton Trans.* **1983,** 9.
12. Gleaves, J.T.; Ebner, J.B.;Küchler, T.C. *Catal. Rev. - Sci. Eng.*, **1988,** *30,* 49.
13. Buyevskaya, O.V.; Rothaemel, M.; Zanthoff, H.W.; Baerns, M. *J. Catal.,* **1994,** *146,* 366.
14. Zanthoff, H.W.; Buchholz, S.A.; Ovsitser, O.Y. *5th Europ. Workshop on Selective Oxidation by Heterogeneous Catalysis,* Berlin, 6.-7.11.1995 (accepted for publication in Catal. Today).
15. Ovsitser, O.Y.; Buchholz, S.A.; Zanthoff, H.W.; Baerns, M. to be published.
16. Topsoe, N.Y.; Topsoe, H.; Dumesic, J.A. *J. Catal.,* **1995**, *151,* 226; *ibid.* **1995**, *151,* 241.
17. Zanthoff, H.W.; Schäfer, S. *Dechema Annual Meeting, Abstract Book,* Wiesbaden, **1995,** 55.
18. Centi, G.; Foresti, E.; Guarnieri, F. *Stud. Surf. Sci. Catal.* **1994**, 82, 281.
19. Centi, G.; Marchi, F.; Mazzoli, P.; Perathoner, S.; *EUROPACAT-II-Symposium*, Maastricht, 3.-8.9.1995, Poster S3P16.
20. Zhou, B.; Chuang, K.T.; Guo, X. J. Chem. Soc., Faraday Trans., **1991**, *87(22),* 3695 and references therein.

Synthesis and Characterization of Advanced Materials for Oxidation

Chapter 20

Selective Partial Oxidation of α-Olefins over Iron Antimony Oxide

E. van Steen, M. Schnobel, and C. T. O'Connor

Catalysis Research Unit, Department of Chemical Engineering, University of Cape Town, Private Bag, Rondebosch 7700, South Africa

α-olefins in the range of ethene to 1-nonene were studied over an iron antimony oxide catalyst at temperatures between 350°C and 400°C in a tubular flow reactor. The primary reactions can be classified into five distinct classes, viz. double bond isomerization, partial oxidation, oxidative dehydrogenation, cracking and total oxidation. Ethene was unreactive and only total combustion products were formed. Propene and 1-butene formed conjugative aldehydes while 2-butenal produced 1,3 butadiene. On the basis of these results a mechanism for the partial oxidation and oxidative dehydrogenation of 1-butene is proposed. Increasing carbon number was found to increase the primary rate of total oxidation and the rate of formation of cracking products. Increasing carbon number decreased the rate of oxidative dehydrogenation and partial oxidation, except when going from C_3 to C_4. The observed carbon number dependencies might be explained in terms of ease of formation and of stability of the π-allyl intermediate.

Iron antimony oxide is a well known catalyst for the partial oxidation/ammoxidation of propene *(1)* and the oxidative dehydrogenation of n-butene *(2)*. Aso et al. *(3)* studied iron antimony oxide containing various Sb:Fe ratios for the partial oxidation of propene. The activity and selectivity to acrolein increased strongly with increasing Sb contents beyond a ratio of 1:1. In a recent study of the surface properties of $FeSbO_4$ *(4)*, it was concluded that the surface has a Sb-rich "skin" which is essential to the selective properties of the catalyst. Although iron antimony oxide is a well known catalyst for the selective oxidation of olefins, this catalyst is not selective for the partial oxidation of paraffins (van Steen, E.; Schnobel, M.; O'Connor, C. T., University of Cape Town, unpublished work).

The partial oxidation of higher olefins has not been studied to the same extend as the oxidation of ethene, propene and butene. The dependency of the

0097–6156/96/0638–0276$15.00/0

activity and selectivity of partial oxidation catalysts on carbon number would give valuable insight into the mechanisms involved in α-olefin oxidation and might yield valuable information regarding the partial oxidation of paraffins. Adams *(5)* studied the oxidation of a number of olefins over a bismuth molybdenum catalyst at 460° C at different conversions. The reactivity of the α-olefins (C_3-C_5) increased with increasing carbon number. Further it was observed that α-olefins were much more reactive than β-olefins. The products observed at low conversions were conjugated dienes and oxides of carbon. Only propene and isobutene yielded unsaturated aldehydes as primary products. The selectivity to the total oxidation products increased with increasing carbon number. The selectivity to the unsaturated aldehydes was higher in the oxidative conversion of branched olefins than for linear olefins of the same carbon number.

This paper reports on the effect of carbon number (C_2-C_9) on the activity and primary selectivity in the oxidative conversion of α-olefins over iron antimony oxide. Furthermore the influence of temperature on the selectivity of the partial oxidation of α-olefins was studied.

Experimental

The catalyst (Fe:Sb=1:1) was synthesized by mixing iron nitrate (Saarchem, 98%) with antimony trioxide (Saarchem, 98%) and aqueous ammonia (25% NH_3) according to the method described by Allen et al. *(1)*. The solid was recovered by filtration, dried at 120° C for 16 hours and then calcined in air at 900° C for 7 hours.

Catalytic oxidation was carried out in a Pyrex glass U-reactor mounted in a convection oven. The catalyst (0.5 g; d_P < 0.1 mm) was mixed with sand (4 g; d_P = 0.2 - 0.3 mm) to minimize temperature gradients. The space time was calculated from the ratio of the total volumetric feed rate at reaction conditions to the volume of the catalyst bed which was kept constant at 2.2 ml. The pre-mixed gases were fed via a preheated zone consisting of washed sand heated to reaction temperature. All experiments were performed using the same batch of catalyst. In the case of coke formation the activity of the catalyst was restored by flushing the catalyst bed overnight with an oxygen/nitrogen mixture at 400° C.

Flows of gaseous hydrocarbons (Fedgas,99+%), oxygen, and nitrogen were controlled by mass flow controllers. Liquid hydrocarbons (Aldrich, 98+%) were fed by saturating a nitrogen flow in a temperature-controlled fixed bed saturator filled with macroporous Chromosorb (Sigma). The partial pressures of hydrocarbon, oxygen and nitrogen at the inlet of the reactor were kept at 16, 34 and 130 kPa, respectively. The total volumetric feed rate varied between 75 and 300 ml (NTP)/min and the reaction temperature varied between 350 and 400° C or 400 and 470°C for the conversion of ethene. A constant flow of an internal standard, viz. iso-octane (Merck, 99.5%) was used in the studies of the partial oxidation of C_2-C_6 and cyclohexane (Saarchem, 98.5%) was used as an internal standard in the C_7-C_9 experiments. These were added to the effluent line of the reactor in order to obtain a qualitative momentary evaluation of the mass balance during the experiment. The organic compounds in the effluent were analysed by on-line capillary gas chromatography using a flame ionization detector. Liquid products were collected

Table I.: Partial Oxidation of linear α-olefins at different temperatures and space times over iron antimony oxide catalyst

α-olefin	T [°C]	τ [s]	X [mol-]	S_{PO} [1] [C-%]	S_{OD} [1] [C-%]	S_{DI} [1] [C-%]	S_{TO} [1] [C-%]	S_{Cr} [1] [C-%]
Ethene	450	0.81	4.5	0	0	0	100.0	0
	460	0.80	6.2	0	0	0	100.0	0
	470	0.79	6.3	0	0	0	100.0	0
Propene	350	0.31	3.3	87.9	0	0	12.1	0
		0.54	4.5	85.0	0	0	15.0	0
		1.25	8.1	84.0	0	0	16.0	0
	375	0.30	5.4	86.8	0	0	13.2	0
		0.52	8.0	84.8	0	0	15.2	0
		1.21	11.8	81.9	0	0	18.1	0
	400	0.29	8.3	81.7	0	0	18.3	0
		0.50	11.3	81.8	0	0	18.2	0
		1.16	22.0	79.7	0	0	20.3	0
1-Butene	350	0.54	7.3	33.3	36.1	10.4	20.2	0
		0.75	8.6	34.1	36.9	9.2	19.8	0
		0.94	13.3	32.2	38.3	9.6	19.9	0
	375	0.52	14.1	35.3	36.0	6.3	20.8	1.6
		0.72	17.6	34.6	36.0	6.2	22.6	0.6
		0.90	18.5	32.9	36.5	6.1	23.5	1.0
	400	0.50	22.9	33.7	35.3	4.7	23.8	2.5
		0.70	27.8	31.4	37.1	5.0	24.0	2.5
		0.87	35.1	29.9	40.2	5.3	23.3	1.3
1-Pentene	350	0.31	1.0	0	62.6	16.0	21.4	0
		0.47	3.9	0	62.9	18.3	18.8	0
		1.25	5.5	0	58.4	17.7	23.9	0
	375	0.30	5.6	1.0	54.9	21.8	21.3	1.0
		0.52	7.5	1.0	53.4	20.4	24.1	1.0
		1.21	10.8	0.6	42.2	24.4	31.2	1.7
	400	0.29	10.2	1.0	64.9	8.4	24.8	0.7
		0.44	12.0	0.9	60.4	8.7	28.9	1.1
		1.16	21.7	0.7	48.7	8.2	40.9	1.5
1-Hexene	350	0.42	2.6	0	54.7	18.6	26.7	0
		0.54	3.9	0	57.4	17.4	25.2	0
		0.75	6.3	0	54.5	15.8	29.7	0
	375	0.40	6.7	0	58.3	9.2	28.6	3.9
		0.52	6.6	0	51.8	8.8	34.5	5.0
		0.72	8.7	0	53.4	9.2	32.2	5.2
	400	0.39	18.8	0	21.2	3.2	56.4	19.2
		0.50	19.5	0	21.3	3.1	57.8	17.8
		0.70	19.0	0	30.5	4.1	50.9	14.5
1-Heptene	350	0.38	9.3	0	18.2	49.4	32.4	0
		0.54	4.0	0	18.8	45.0	36.2	0
		0.94	14.0	0	37.9	28.5	28.4	5.2
	375	0.36	19.7	0	3.9	9.3	53.2	34.0
		0.52	14.8	0	6.6	6.3	54.7	32.4
		0.98	23.7	0	4.1	5.3	55.2	35.4
	400	0.35	31.0	0	3.3	5.2	53.5	37.2
		0.50	36.9	0	3.5	3.8	50.9	41.8
		0.87	37.4	0	3.2	3.0	55.2	38.6
1-Octene	350	0.38	5.6	0	7.8	85.1	6.1	1.0
		0.54	4.2	0	8.9	71.0	20.1	0
	375	0.36	12.3	0	3.0	15.3	34.5	47.2
		0.52	11.5	0	2.3	15.9	37.2	44.6
1-Nonene	350	0.38	29.6	0	0	59.8	40.2	0
		0.54	14.5	0	0	60.5	39.5	0
	375	0.36	24.0	0	0	9.6	49.1	41.4
		0.52	16.9	0	0	5.2	44.4	50.4

[1] S_{PO}=Selectivity to partial oxidation; S_{OD}=Selectivity to oxidative dehydrogenation; S_{DI}=Selectivity to double bond isomerization; S_{TO}=Selectivity to total oxidation; S_{Cr}=Selectivity to cracking

in a cold trap and analysed using GC-MS. The total oxidation products CO and CO_2 were monitored continuously using an on line IR-analyzer. The carbon balance of the reported experiments was 100±2%. In this study the reactivity and primary selectivity of C_2-C_9 α-olefins for their oxidative conversion were investigated. All selectivities, conversions and rates of reaction were calculated on a carbon basis.

In order to observe the primary selectivity for these reactions, the influence of space time was studied for all hydrocarbons by changing the total flow while keeping the inlet partial pressures of each compound constant. Decreasing space time will result in a reduction of secondary reactions by decreasing the probability of readsorption. At short space times changes in selectivity become negligible indicating the primary formation of the observed product compounds.

Results

A selection of the experimental data is given in Table I. The selectivities have been grouped to different product classes, i.e. partial oxidation for oxygenates of the same carbon number, oxidative dehydrogenation for dehydrogenated compounds of the same carbon number, double bond isomers, cracking products for organic compounds with a lower carbon number and total oxidation for CO and CO_2. No products with higher carbon number than the feed were detected.
The only observed reaction in the ethene oxidation down to a conversion level of 2 C-% was the total oxidation yielding CO and CO_2. Propene yielded the partial oxidation product acrolein (2-propenal) and the total oxidation products CO and CO_2. In the 1-butene oxidation the formation of an aldehyde (2-butenal) and of total oxidation products, CO and CO_2, can be observed. In addition trans- and cis-2-butene and 1,3-butadiene, were formed. The main products observed in the 1-pentene oxidation were trans- and cis-1,3-pentadiene, trans-and cis-2-pentene, CO and CO_2. Only a small amount of C_5-oxygenates, viz. 2-methyl furan, was formed during the partial oxidation of 1-pentene at the higher reaction temperatures. The conjugated aldehyde 2-pentenal was not observed. 2-methyl-furan might be formed by ring closure of the conjugated aldehyde and subsequent oxidative dehydrogenation. The oxidative conversion of C_6 to C_9 α-olefins revealed new classes of products, viz. the dehydrogenated cyclic compounds and cracking products. The cracked products might originate from the oxidative decomposition of the olefin yielding stable intermediates on the catalyst surface which then desorb or are further oxidised to partial oxidation products, e.g. acrolein and furan or to the total oxidation products.

Ethene Oxidation. Ethene was rather unreactive and high temperatures (400 - 475° C) and high space times had to be applied to get detectable conversions of this compound. This low reactivity and high selectivity to total oxidation might be explained by the lack of stable C_2-intermediates on the catalyst surface (e.g. allylic species) and of stable C_2-oxygenates under the applied conditions.

Propene Oxidation. At temperatures between 350 and 400°C, acrolein was formed with a primary selectivity between 80 and 90 C-% at a propene conversion of 3 to 10 C-%. The primary acrolein selectivity decreased with increasing temperature

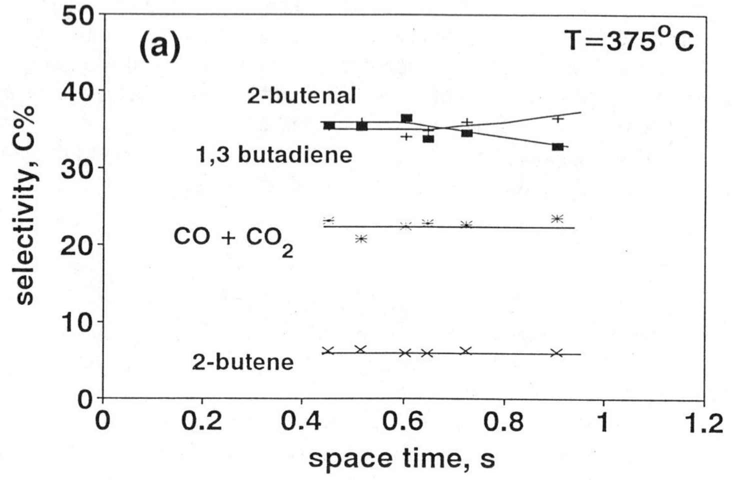

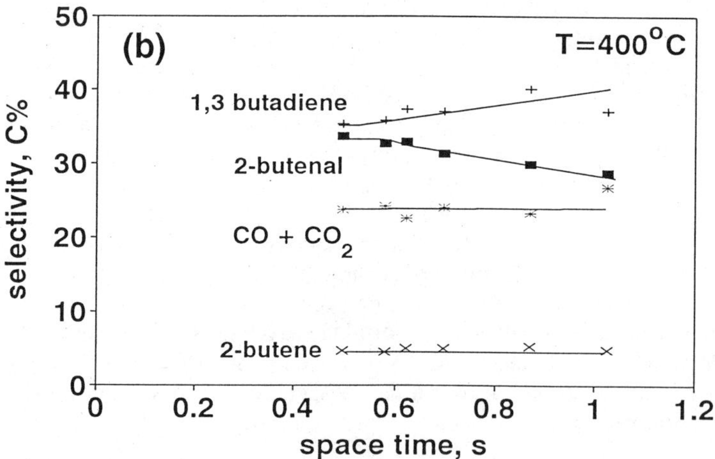

Figure 1. Influence of space time on the selectivity of the partial oxidation of 1-butene over iron antimony catalyst at 375°C (a) and 400°C (b) ($p_{1\text{-butene}}$= 16 kPa; p_{O2}=34 kPa; p_{N2}=130 kPa; m_{cat}=0.5 g diluted with 4g sand)

from 88 C-% at 350°C to 82 C-% at 400°C and thus the selectivity to total oxidation increased. This indicates that the activation energy for total oxidation is higher than for partial oxidation. No change in the CO content of the total oxidation products was observed with changing space time, thus indicating the parallel formation of CO and CO_2. However, the CO content in the fraction of CO plus CO_2 decreased with increasing temperature, viz. from 27 C-% at 350°C to 19 C-% at 400°C indicating that the activation energy for CO_2 formation is higher than for CO formation.

1-Butene Oxidation. Figures 1a and 1b show the influence of space time on the selectivity in the oxidative conversion of 1-butene at 375 and 400° C, respectively. With decreasing space time a slight increase in the selectivity to 2-butenal was observed in conjunction with a decrease in the selectivity of 1,3 butadiene, whereas the selectivity of both the double bond isomers cis- and trans-2-butene and the total oxidation products CO and CO_2 remained constant. This shows that 2-butenal is a primary product. Because the selectivity to 1,3 butadiene increased, 2-butenal selectivity decreased and the other product selectivities remained constant with increasing space time, it can be concluded that 2-butenal can be converted in a secondary reaction step into 1,3-butadiene. The selectivity to 2-butene remains totally unchanged during the space time study which is to be expected on the basis of the findings of Adams [6] who showed that β-olefins have a lower reactivity than the α-olefins. The amount of trans-2-butene in the fraction of 2-butene remained unchanged at 45% with both an increase of space time and of temperature. This indicates that the more reactive and thermodynamically less stable cis-2-butene is preferentially formed from a common activated intermediate complex. Selectivities to cis- and trans-2-butene and to CO and CO_2 remained unchanged with increasing space times. Scheme 1 shows the proposed reaction pathway for the oxidative conversion of 1-butene.

Temperature did not affect significantly the primary selectivities of 2-butenal and 1,3 butadiene. However, the primary selectivity to the double bond isomers decreased with increasing temperature from 10.4 C-% at 350°C to 4.7 C-% at 400°C while, at the same time, primary selectivity of the total oxidation products increased indicating a relatively low activation energy for the double bond isomerization and a relatively high activation energy for total oxidation. The CO content in the fraction of total oxidation products decreased slightly with increasing temperature from 23.8 C-% at 350°C to 21.8 C-% at 400°C showing a slightly higher activation energy for CO_2 formation than for CO formation, which has already been shown for the total oxidation of propene.

1-Pentene Oxidation. Figure 2 shows the influence of space time on selectivity in the oxidative conversion of 1-pentene at 375° C. Decreasing space time showed an increase in the selectivity to the dienes with a corresponding decrease in the selectivity to the total oxidation products and a slight decrease in the selectivity to 2-pentene showing that 1,3-pentadiene is a primary product, but it can be converted more easily into total oxidation products than 1,3 butadiene. In this case an excess of the trans-isomer of 2-pentene was observed indicating that this conformation is the preferred one upon desorption of the C_5-intermediate. Trans- and cis- pentadiene

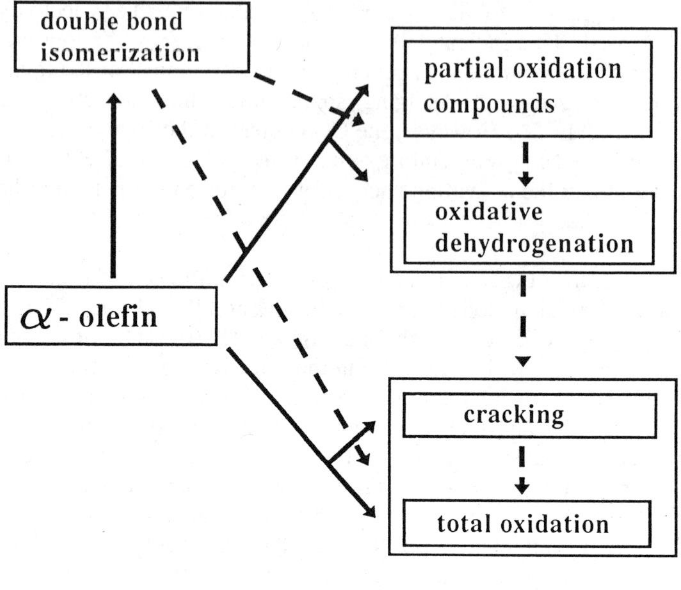

Scheme 1. Proposed general scheme for the oxidative conversion of α-olefins over iron antimony catalyst

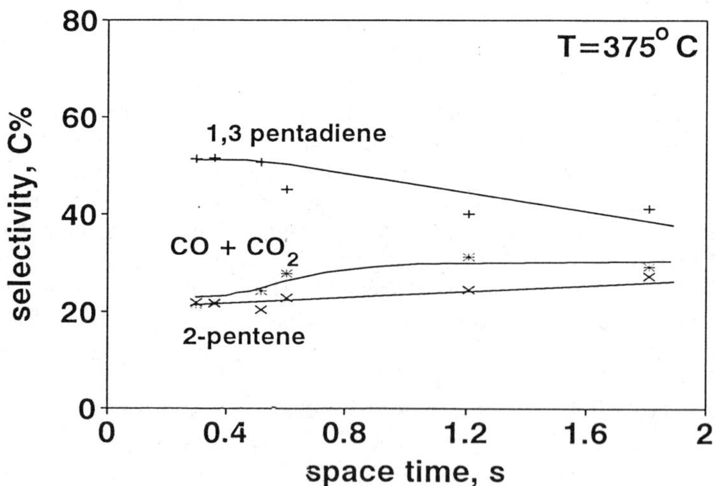

Figure 2. Influence of space time on the selectivity of the partial oxidation of 1-pentene over iron antimony catalyst at 375°C
($p_{1\text{-pentene}}$= 16 kPa; p_{O2}=34 kPa; p_{N2}=130 kPa; m_{cat}=0.5 g diluted with 4g sand)

were formed in almost identical amounts. Increasing temperature showed an increase in the primary formation of the trans-isomer in the fraction of both 2-pentene and 1,3 pentadiene showing a higher activation energy for the desorption of the trans-isomer than for the cis-isomer. The primary selectivity of 1,3 pentadiene increased from 63 C-% at 350°C to 65 C-% at 400°C showing that the activation energy is higher than for the formation of 2-pentene, for the primary selectivity to 2-pentene decreased (from 16 C-% at 350°C to 8 C-% at 400°C). The primary selectivity to total oxidation increased with increasing temperature. The CO content in the total oxidation products decreased with increasing temperature from 23% at 350°C to 20% at 400°C.

1-Hexene Oxidation. Figure 3a shows the influence of space time on the selectivities of partial oxidation of 1-hexene at 375 °C. The main oxidation products were hexadienes, 2-hexene, cyclohexene, cyclohexadiene, benzene and CO/CO_2. The selectivity to the hexadienes increased with decreasing space time, whereas the selectivities to the cracking products, CO/CO_2 and cyclic products decreased. The selectivity to the double bond isomers remained constant. The increase in selectivity to the hexadienes indicates their primary formation in the oxidative conversion of 1-hexene. The fraction of hexadienes consist of cis-and trans-1,3 hexadiene and different isomers of 2,4 hexadiene. Figure 3b shows the content of cis- and trans-1,3 hexadiene and the sum of 2,4 hexadienes in the fraction of hexadienes shown in Figure 3a. Trans 1,3 hexadiene content in this fraction remained constant indicating that it is relatively unreactive. Apparently, cis-1,3 hexadiene can be isomerized into 2,4 hexadienes.

Increasing temperature yielded an decrease in the primary selectivity to the hexadienes from 39 C-% at 350°C to 15 C-% at 400°C. The primary selectivity to total oxidation products CO and CO_2 increased from 27 C-% at 350°C to 56 C-% at 400°C as shown in Table I. This strong increase indicates a higher activation energy for the formation of total oxidation products. Products of cracking also increased from 0 C-% at 350°C to 19 C-% at 400°C. The primary selectivity to 2-hexene decreased from 19 C-% at 350°C to 3 C-% at 400°C, showing that the activation energy to form 2-hexene is lower than for the formation of any of the other products.

The selectivities of the cyclic products benzene and cyclohexadiene, and at a higher reaction temperature of 400°C cyclohexene, increased with increasing space time and increasing temperature. This shows that the ring closure is enhanced by re-adsorption of probably dienes and/or trienes. The activation energy for the desorption of an intermediate cyclic compound is higher than for the consecutive reaction which yields benzene, for the benzene content in the fraction of cyclic compounds decreased with increasing temperature.

1-Heptene Oxidation. The influence of space time and temperature on the oxidative conversion of 1-heptene are summarized in Table I. Increasing temperature yielded a decrease in the selectivities to products of the groups of oxidative dehydrogenation and double bond isomerization. At the same time the selectivities to products of the group of total oxidation and cracking increased with increasing temperature. The primary selectivity to cracking products increased most strongly with increasing

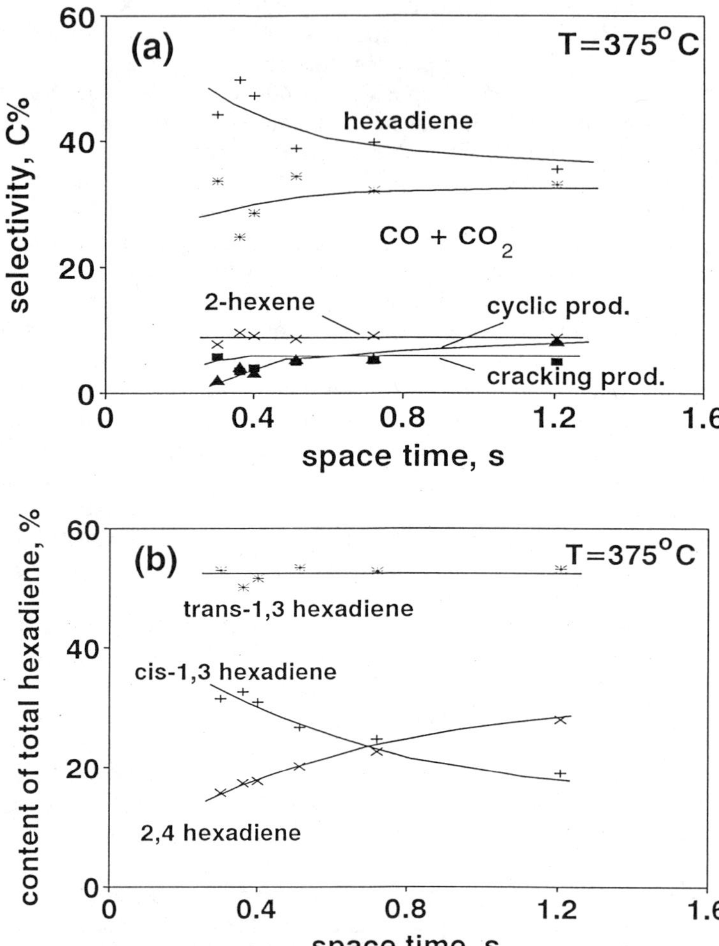

Figure 3. Influence of space time on the selectivity of the partial oxidation of 1-hexene over iron antimony catalyst at 375°C
($p_{1\text{-hexene}}$= 16 kPa; p_{O2}=34 kPa; p_{N2}=130 kPa; m_{cat}=0.5 g diluted with 4g sand)
a: Selectivity of major product groups as a function of space time
b: Content of some hexadienes in the fraction of hexadienes as a function of space time

temperature, viz. from 0 C-% at 350°C to 37 C-% at 400°C. The primary selectivity to total oxidation products increased to a lesser extend from 32 C-% at 350°C to 54 C-% at 400°C. The primary selectivities to oxidative dehydrogenation products and double bond isomerization products both decreased, however it decreased more strongly with increasing temperature to double bond isomerization products than to oxidative dehydrogenation products (from 49 C-% at 350°C to 5 C-% at 400°C vs. 18 C-% at 350°C to 3 C-% at 400°C).

1-Octene Oxidation. The influence of space time and temperature on the oxidative conversion of 1-octene are summarized in Table I. The product groups showed the same trends as for the oxidative conversion of 1-heptene. The primary selectivity to cracking products increased most strongly with increasing temperature, viz. from 1 C-% at 350°C to 47 C-% at 375°C, whereas the primary selectivity to total oxidation products increased from 6 C-% at 350°C to 35 C-% at 375°C. The primary selectivities to oxidative dehydrogenation products and double bond isomerization products both decreased, however it decreased more strongly with increasing temperature to double bond isomerization products than to oxidative dehydrogenation products (from 85 C-% at 350°C to 15 C-% at 375°C vs. 8 C-% at 350°C to 3 C-% at 375°C).

1-Nonene Oxidation. The influence of space time and temperature on the oxidative conversion of 1-nonene are summarized in Table I. No oxidative dehydrogenation products could be detected for the oxidative conversion of 1-nonene. The primary selectivity to cracking products increased most strongly with increasing temperature, viz. from 0 C-% at 350°C to 41 C-% at 375°C, whereas the primary selectivity to total oxidation products increased from 40 C-% at 350°C to 49 C-% at 375°C. The primary selectivity to double bond isomerization products decreased with increasing temperature from 60 C-% at 350°C to 10 C-% at 375°C.

Discussion

Based on the results obtained in the space time experiments of the oxidative conversion of C_2 to C_9, a general scheme of all possible reaction pathways for the oxidative conversion of α-olefins is presented in Scheme 2. Any product formed will fall into one of the following product categories, viz. double bond isomerization, partial oxidation, oxidative dehydrogenation, cracking and total oxidation products. As shown for the partial oxidation of 1-butene, the partial oxidation products can be interconverted into dehydrogenated compounds. Both of them can be either cracked or totally oxidized. The double bond isomers can also be converted into either product groups, but as been shown by Adams *(3)* have a lower reactivity than the α-olefins.

Figures 4a and b show the primary selectivities to the distinct product groups as a function of the chain length of the α-olefin for 350°C and 375°C, respectively. Higher temperatures could not be applied for the whole series of feedstocks (C_3-C_9 α-olefins), because thermal cracking, as determined by carbon balance, became quite severe for higher molar weight feedstocks especially in the case of 1-octene and 1-nonene. Ethene, as already mentioned, showed quite a lower reactivity and yielded

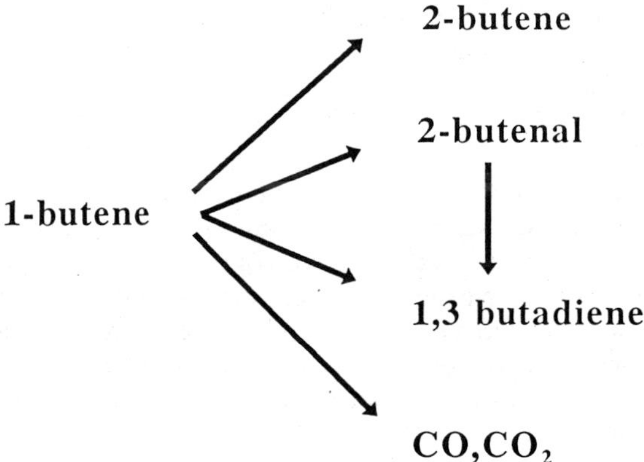

Scheme 2. Reaction scheme for the oxidative conversion of 1-butene over iron antimony oxide catalyst

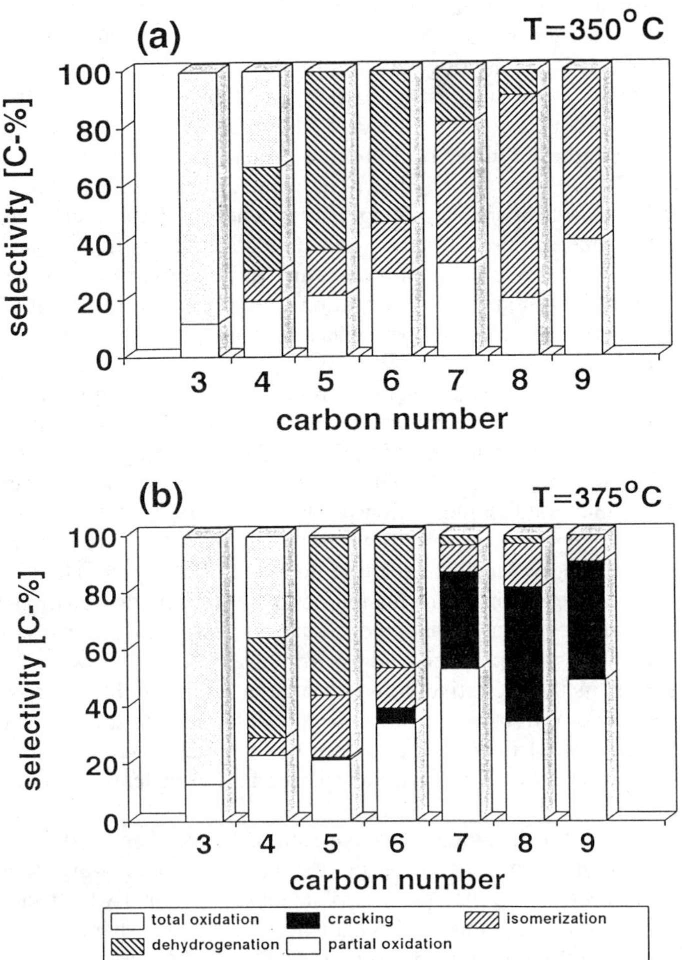

Figure 4. Influence of carbon number on the primary selectivity of α-olefin partial oxidation over iron antimony catalyst at 350°C (a) and 375°C (b) (p_{HC}= 16 kPa; p_{O2}=34 kPa; p_{N2}=130 kPa; τ<0.5s; m_{cat}=0.5 g diluted with 4g sand)

only total oxidation products. The formation of partial oxidation products appeared only to a significant extent for propene and 1-butene as feedstocks and the primary selectivity to this product class dropped dramatically with carbon number. At 350°C, the primary selectivity to double bond isomerization products increased steadily for carbon numbers of 4 and higher. The same applies to the primary selectivity to total oxidation products, which increased steadily with carbon number. The primary selectivity to oxidative dehydrogenation products increased from C_4 to C_5 and decreased steadily with increasing carbon number and at C_9 no oxidative dehydrogenation products could be detected anymore. At 375°C, cracking products appeared from a carbon number of 5 and the primary selectivity to those products increased steadily with carbon number. The primary selectivity to total oxidation products increased with increasing temperature, whereas the primary selectivity to double bond isomerization products and oxidative dehydrogenation products decreased, especially when cracking products are formed, viz. from hexene onwards. This leads to the conclusion that the activation energy for the formation of cracking and total oxidation products is higher than for the formation of double bond isomerization products and oxidative dehydrogenation products. The activation energy for the formation of total oxidation products increased with increasing carbon number because the primary selectivity to total oxidation products increased more strongly for higher carbon numbers of the feed. The same applies for the activation energy for the formation of cracking products. The activation energy for the formation of double bond isomerization products decreased with increasing carbon number because the primary selectivity to double bond isomerization products decreased more strongly for higher carbon numbers of the feed. The same applies for the activation energy for the formation of oxidative dehydrogenation products.
The mechanism of the partial oxidation of propene to acrolein has been well studied. On the basis of deuterated propene studies, Keulks and Lo *(6)* showed, that the rate determining step in the oxidative conversion of propene is the abstraction of an allylic hydrogen from propene to form a π-allyl intermediate. Subsequent addition of oxygen to the π-allyl intermediate before the abstraction of a second hydrogen yields a σ-allylic species. This σ-allylic species is subsequently converted to the product acrolein.

The space time runs for the oxidative conversion of 1-butene showed that a decrease in space time resulted in a decrease in the selectivity to 1,3-butadiene and an increase in the selectivity to the conjugated aldehyde 2-butenal which shows that 2-butenal can be converted into 1,3-butadiene. In a series-parallel "rake" type mechanism, where each intermediate compound adsorbed on the catalyst surface can desorb, this would imply that the precursor of the conjugated aldehyde forms prior to the precursor of the diene in the chain of reactions on the catalyst surface. This can be visualized as follows. In the first step hydrogen is abstracted from 1-butene to form an allylic intermediate. In the second step a C-O bond is formed to give a σ-O-alkene intermediate. The reversible α-hydrogen abstraction can then yield 2-butenal. 1,3 butadiene might then be formed via the desorption of the σ-O bonded π-allyl intermediate, which might be formed upon abstraction of the allylic hydrogen of the σ-alkene intermediate. The difference to the oxidative conversion of propene might be due to the fact that their is no stable C_3-diene. This proposed mechanism is similar to the proposed mechanism for the selective oxidation of propene *(6)* and for the selective oxidation of ethane, where the formation of a metal-allyl ether yields the selective partial oxidation products, whereas the formation of metal-alkyl bond yields total oxidation *(7)*. However, further investigation is necessary to proof this proposed reaction mechanism.

The rate of consumption of olefins in the oxidative conversion reaction equals the sum of the rate of formation of the individual product compounds. The rates for the formation of partial oxidation products and oxidative dehydrogenation products showed the same trend when increasing the carbon number of the olefin. Furthermore as outlined in the mechanism proposed for the partial oxidation of 1-butene, the formation of the partial oxidation product and of the dienes are thought to proceed along the same reaction pathway. Therefore they were investigated in one group. The same was observed for the rate of formation of cracking products and total oxidation products and both were therefore lumped together. Figure 5a shows the carbon number dependency of the sum of the rates of formation of partial oxidation and oxidative dehydrogenation for 350, 375 and 400° C. Both the dienes and the conjugated aldehyde might be formed from a π-allylic intermediate. A maximum rate for C_4 at the temperatures 375 and 400°C and for C_3 at 350°C can be seen, which steadily decreases towards almost zero for C_8 and C_9. The existence of a maximum might be explained by a dual effect, viz. reduction of the electron density in the π-allylic species by hyperconjugation, thereby stabilizing this species on the catalyst surface, and shielding of both the allylic hydrogen atom thus inhibiting the formation of the π-allylic species and the double bond thus inhibiting the initial olefin adsorption.

The rate of cracking and of total oxidation both increased with carbon number. The combined rate of cracking and of total oxidation increased dramatically with increasing carbon number (figure 5b). The reaction pathway of oxidative decomposition of α-olefins must involve the double bond via an adsorbed π-olefinic complex, because the oxidation of paraffins with the same catalyst and under identical reaction conditions was not feasible (van Steen, E.; Schnobel, M; O'Connor, C.T., University of Cape Town, unpublished work) proving that the activation of the paraffinic C-H bond was not possible under the conditions applied. The strong increase in the cracking and total oxidation rate with carbon number might be explained by the enhanced strength of adsorption of the olefin on the oxide surface. However, an electronic effect caused by the change in alkyl chain length is a short range effect which might well explain the observed increase from C_2 to C_3, but cannot explain the dramatic increase of these rates with carbon number. Increasing chain length causes a move effective shielding of the allylic hydrogen atom, which might therefore increase the probability of a direct attack on the double bond by a surface hydroxyl group thus yielding a surface alkyl species bonded to lattice oxygen resulting in C-C bond cleavage and the formation of both cracked products and total oxidation products *(8)*. The rate of isomerization on a carbon atom basis was hardly affected by both carbon number and temperature.

Conclusion

The investigation of a set of linear α-olefins showed a strong influence of the chain length of the olefin on the reactivity and product distribution. The formation of

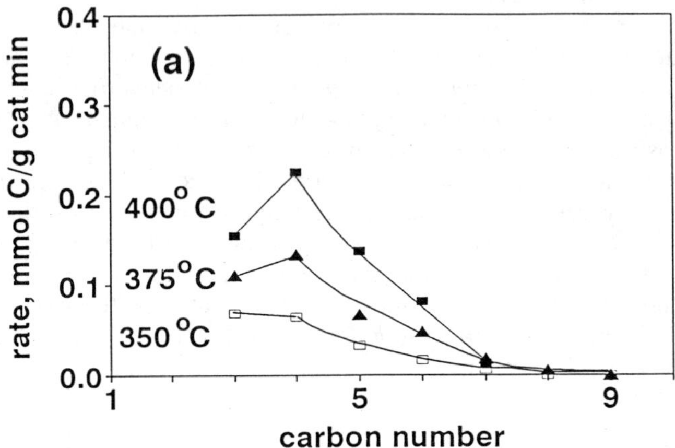

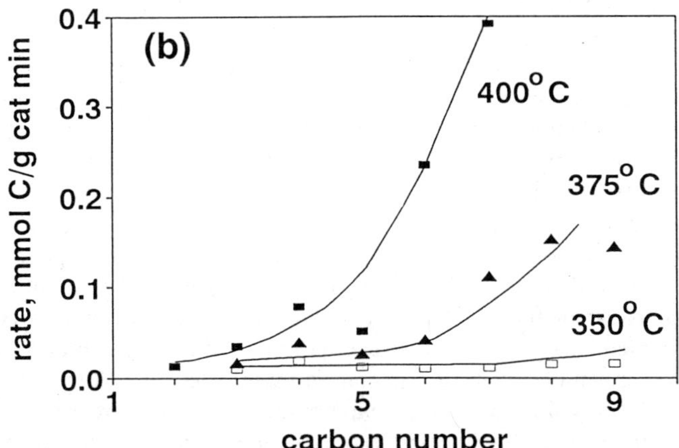

Figure 5. The rate of formation of some product groups as a function of carbon number at different temperatures over iron antimony catalyst
a: rate of formation of partial oxidation and oxidative dehydrogenation products
b: rate of formation of cracking and total oxidation products
(p_{HC}= 16 kPa; p_{O2}=34 kPa; p_{N2}=130 kPa; τ<0.5s; m_{cat}=0.5 g diluted with 4g sand)

partial oxidation products and oxidative dehydrogenation products are thought to proceed via a common reaction pathway. The rate of formation of partial oxidation products plus oxidative dehydrogenation products shows a maximum, which might be explained by a dual effect, i.e. the reduction of the electron density in the π-allylic radical and the shielding of both the double bond and the allylic hydrogen. An increase in the chain length of the α-olefin enhances the rate of formation of total oxidation and cracking products, probably due to inhibition of the formation of the π-allylic intermediate, which might be caused by shielding of the allylic hydrogen. The activation energy for the formation of cracking and total oxidation products is higher than for the formation of double bond isomerization products and oxidative dehydrogenation products. The activation energy for the formation of total oxidation products and cracking products increased with carbon number, whereas the activation energy for the formation of isomerization products and oxidative dehydrogenation products decreased with carbon number.

Acknowledgement

The authors gratefully acknowledge financial support from the University of Cape Town, FRD,SASOL and AECI.

References

1. Allen, M.; Betteley, R.; Bowker, M.; Hutchings, G.J. *Catal. Today* **1991,** *9,* 97.
2. Sala, F.; Trifirò, F. *J. Catal.* **1976,** *14,* 1.
3. Aso, I.; Furukawa, S.; Yamazoe, N.; Seiyama, T. *J. Catal.* **1980,** *64,* 29.
4. Allen, M. D.; Bowker, M.; *Catal. Letters* **1995**, *33,* 269.
5. Adams C.R.; In *Proc. 3rd Int. Congr. on Catalysis;* Amsterdam 1964, Sachtler, W. H. M.; Schuit, G. C. A.; Zwietering, P., Eds.; North-Holland Publishing Company: Amsterdam, 1965; pp. 240-251.
6. Keulks, G. W.; Lo, M.-Y. *J. Phys. Chem.* **1986,** *90,* 4768.
7. Oyama, S. T.; Desikan, A.N.; Zhang, W. In *Catalytic Selective Oxidation,* Oyama, S. T.; Hightower J. W., Eds.; ACS Symp. Ser. 523; American Chemical Society: Washington, D. C., 1993; pp. 16-30.
8. Finocchio, E.; Bucsa, G.; Lorenzelli, V. *J. Chem Soc. Faraday Trans.* **1994,** *90,* 3347.

Chapter 21

The Activity and Selectivity Properties of Supported Metal Oxide Catalysts During Oxidation Reactions

I. E. Wachs, G. Deo, J-M. Jehng[1], D. S. Kim[2], and H. Hu

Zettlemoyer Center for Surface Studies, Sinclair Laboratory, 7 Asa Drive, Lehigh University, Bethlehem, PA 18015

Structure-reactivity studies involving model supported metal oxide catalysts reveal that the efficiency of a particular selective oxidation reaction depends on the specific supported metal oxide species, the coverage of the surface metal oxide species, the oxide support, and the presence of secondary metal oxide additives. *In situ* Raman spectroscopy, activity, and selectivity of the partial oxidation of methanol, alkanes, and the selective catalytic reduction of NO with ammonia are discussed with respect to the above factors for supported metal oxide catalysts.

Supported metal oxide catalysts are extensively employed in the petrochemical and pollution control industries as oxidation catalysts. For example, titania supported vanadia catalysts are used for the selective oxidation of o-xylene to phthalic anhydride, ammoxidation of aromatic methyl groups and the selective catalytic reduction (SCR) of NO_x with ammonia. The active redox components (V, Mo, Cr or Re) of supported metal oxide catalysts are present as two-dimensional metal oxide overlayers on the high surface area oxide supports (e.g., titania, alumina, silica, niobia, zirconia, etc.). Such supported metal oxide catalysts are also ideal model systems to study the fundamental aspects of selective oxidation reactions because the active surface redox sites can be molecularly characterized with *in situ* Raman, IR, solid state NMR, DRS and EXAFS/XANES. Furthermore, the environment around the active redox sites can also be varied by changing the surface metal oxide coverage, changing the specific oxide support or by the addition of secondary metal oxide additives (e.g., oxides of W, Nb, P, etc.) to the two-dimensional overlayer. In the present investigation, Raman

[1]Current address: Department of Chemical Engineering, National Chung Hsing University, Taichung, Taiwan, Republic of China

[2]Current address: Research and Development Division, Daelim Engineering Company, Limited, Seoul, Korea

0097–6156/96/0638–0292$15.00/0

spectroscopy was primarily employed to obtain information about the molecular structures present in the two-dimensional overlayers and methanol oxidation was primarily used to chemically probe the surface redox as well as acid sites. Combination of this information with catalytic data for the selective oxidation of alkanes and the selective reduction of NO with ammonia has provided new insights into the origin of the activity and selectivity properties of supported metal oxide catalysts during oxidation reactions.

Molecular structures of the surface metal oxide species

The molecular structures of the oxidized and dehydrated surface metal oxide species have been extensively investigated with *in situ* Raman spectroscopy as well as other spectroscopies and have been reported elsewhere *(1-8)*. Surface rhenium oxide species are present as isolated, tetrahedrally coordinated ReO_4 species *(2)*. In contrast, surface chromium oxide species are primarily present as polymerized surface CrO_4 species *(3)*. Surface vanadium oxide species are also tetrahedrally coordinated and favor isolated surface species at low coverages and polymerized surface species at high coverages *(4-7)*. Surface molybdenum oxide species favor isolated, tetrahedral coordination at low coverages and polymerized, octahedral species at high coverages *(8)*. An exception to the above trends is when silica is employed as a support since the surface metal oxide species are usually not polymerized on the silica surface due to the low surface metal oxide densities achievable on this somewhat unreactive substrate. All the surface metal oxide species form highly distorted structures containing terminal M=O bonds and bridging M-O-M' bonds where M' represents the same or another surface metal oxide species or the oxide support *(1-8)*.

The influence of different reaction environments upon the state of the surface metal oxide species during oxidation reactions have recently been investigated with *in situ* Raman spectroscopy. During methanol oxidation, a significant portion of the surface metal oxide species are reduced and the fraction reduced is not a function of the specific oxide support *(9,10)*. In contrast, the fraction of the surface vanadia species reduced during alkane (methane and butane) oxidation is a function of the specific oxide support *(11,12)*. During alkane oxidation the surface vanadia species are not reduced for V_2O_5/SiO_2, but are partially reduced for V_2O_5/TiO_2, V_2O_5/ZrO_2, V_2O_5/Nb_2O_5, V_2O_5/CeO_2 and V_2O_5/SnO_2. The selective catalytic reduction of NO with ammonia at elevated temperatures appears to cause only slight or no reduction of the surface vanadia species on titania because of the very low concentrations of ammonia and the high concentrations of oxygen typically employed for this commercial reaction *(13)*. Thus, under oxidation reaction conditions, both fully oxidized and partially reduced surface metal oxide species are present and the extent of reduction appears to depend on the reducing power (defined as the ability to reduce the surface metal oxide active site) of the specific reactant molecule (methanol > butane ~ methane), the concentration of the reactant molecules (NO reduction with ammonia) and the specific oxide support.

The molecular structures of the secondary surface metal oxide additives (e.g., oxides of W, Nb and P) have also been determined *(14)*. The surface tungsten oxide and niobium oxide species are primarily present as highly distorted, octahedrally coordinated species and the surface phosphorous oxide species possesses tetrahedral coordination. These surface metal oxide additives are generally non-reducible under most reaction conditions, but may become partially reduced at elevated temperatures on certain oxide supports.

Catalytic activity

The reactivity properties of supported metal oxide catalysts during methanol oxidation have been recently reviewed and will only be briefly discussed here *(1)*. The two-dimensional nature of the active surface metal oxide species allows the reaction rates to be expressed as turnover frequencies (TOF), which is the number of methanol molecules converted per active redox atom per second. The methanol oxidation TOF was found to essentially be independent of the surface coverage of the surface redox sites which suggests that this oxidation reaction requires only one surface redox site to proceed. A similar conclusion was also reached for methanol oxidation over model Keggin metal oxide catalysts *(15,16)*. The methanol oxidation TOF was also not influenced by the presence of other non-reducible surface metal oxide species in the two-dimensional overlayer (e.g., oxides of W, Nb, Si and S) and is consistent with the conclusion that only one redox site is required for this reaction *(14)*. However, the specific oxide support had a pronounced effect on this reaction and the TOF varied by several orders of magnitude (Ce > Ti ~ Zr > Nb > Al > Si) *(1)* as shown in Table I. It has been proposed that the origin of this dramatic support effect is due to the involvement of the bridging M-O-support bonds in the rate determining step and the observed trend correlates with the basicity or electronegativity of the oxygens associated with the oxide support. Recent theoretical calculations are in agreement with this model *(17-19)*. The specific surface redox site also had an effect on the TOF and the surface V and Re oxides were approximately an order of magnitude more active than the surface Mo and Cr oxides. Thus, for oxidation reactions requiring only one surface redox site the TOF is only influenced by the specific oxide support and the specific surface redox site.

Table I. The TOF of methanol oxidation over supported vanadia catalysts as a function of the oxide support

1% V_2O_5 on oxide support	*TOF (s^{-1})*
CeO_2	~$1x10^1$
TiO_2	$2.0x10^0$
ZrO_2	$1.6x10^0$
Nb_2O_5	$8.0x10^{-1}$
Al_2O_3	$3.6x10^{-2}$
SiO_2	$3.9x10^{-3}$

The TOF for the oxidation of alkanes over supported vanadium oxide catalysts behaved very similarly to the TOF for methanol oxidation *(11,12,20)*: the TOF was independent of surface vanadia coverage and secondary non-reducible surface metal oxide additives, but significantly varied with the specific oxide support. This suggests that activation of alkanes requires only one surface redox site.

In contrast to the above oxidation reactions, the SCR of NO with ammonia exhibits a TOF that is a strong function of surface coverage and the presence of non-reducible surface metal oxide additives *(21)*. This TOF pattern suggests that the SCR reaction requires two surface sites to proceed: a surface redox site and an adjacent surface non-reducible metal oxide site *(21-23)*. The SCR TOF also varied with the specific oxide support and the specific surface metal oxide redox site since these parameters control the redox properties of the surface metal oxide species *(21,24,25)*.

Catalytic selectivity

The selectivity properties of supported metal oxide catalysts during methanol oxidation to formaldehyde depend on the specific oxide support, the specific surface metal oxide redox site and the surface coverage. Comparision of the Lewis acidity data from pyridine adsorption experiments (strength of Lewis sites Al > Nb > Ti ~ Zr and no acid sites on Si *(26)*) and the methanol oxidation activity and selectivity data of the oxide supports *(27)* suggests that the surface Lewis acid sites present on the oxide supports are responsible for the non-selective formation of dimethyl ether and the surface redox sites present on zirconia are responsible for the formation of methylformate as shown in Table II *(27)*. In all cases, the contribution of the support towards these side reactions is minimized as the surface coverage of the surface metal oxide redox sites is increased towards monolayer coverage as shown in Table III for supported vanadium oxide catalysts *(9,27)*. The highest formaldehyde selectivities were found for the titania supported metal oxide catalysts because titania possessed the weakest Lewis acid sites and did not possess significant surface redox sites. The selectivity towards formaldehyde was strongly dependent on the specific surface metal oxide redox site for titania supported catalysts (V (95-99%) > Mo (75-82%) > Cr (70%) ~Re (70%)) where the primary side reaction was the formation of methylformate *(28,29)*. The formation of methylformate correlated with the intrinsic acidic properties of these metal oxides *(30,31)*. The addition of non-reducible surface metal oxide species (W, Nb, Si and S) to the two-dimensional surface metal oxide overlayer did not influence the selectivity towards formaldehyde as shown in Table IV *(14,21)*. Thus, the most selective supported metal oxide catalyst for methanol oxidation to formaldehyde consisted of the surface metal oxide redox site that possessed the weakest acidic character, vanadium oxide, and the oxide support that possessed both weak surface acid and redox sites, titania.

The selectivity pattern for alkane oxidation over the supported vanadia catalysts depended on the specific alkane oxidation reaction. For methane oxidation over silica supported vanadia catalysts, the selectivity was a function of methane conversion since formaldehyde was further oxidized to carbon monoxide. At the high temperatures required to conduct this oxidation reaction, the fate of gas-phase methyl radicals may also be an important parameter on the reaction selectivity *(11)*. The selective oxidation of methane over supported vanadia catalysts was also very sensitive to the specific oxide support since formaldehyde could be further oxidized to carbon dioxide depending on the reducibility of the oxide support (Sn > Ti > Si) *(11)*. The selectivity of propane oxidation to propylene was not a function of the surface vanadia coverage or propane conversion *(20)*. The selectivity of butane oxidation to maleic anhydride/acid was not a strong function of the specific oxide support, but was influenced by the surface coverage *(12)*. The selectivity to maleic increased with surface coverage which reflects the higher efficiency of this reaction path when two adjacent surface vanadia species are present. The introduction of surface Lewis acid sites, surface niobia and tungsten oxide species, slightly increased the maleic selectivity. Thus, the selectivity of alkane oxidation reactions involving only the abstraction of hydrogen is less sensitive to their immediate environments than alkane oxidation reactions involving the insertion of oxygen.

Table II. The activities and selectivities of the oxide supports during methanol oxidation

Oxide Support	*Activity (mol CH_3OH conv./ g.hr)*	*Selectivity (%)* FA	MF	DMM	DME	CO_x
Al_2O_3	$3.7x10^0$	--	--	--	100	--
Nb_2O_5	$1.0x10^{-2}$	--	--	--	100	--
TiO_2	$2.0x10^{-3}$	--	--	--	91	9
ZrO_2	$2.0x10^{-2}$	--	86	--	Tr	14
SiO_2	$2.6x10^{-4}$	--	--	--	15	85

Table III. The methanol oxidation selectivities as a function of surface coverage for V_2O_5/Al_2O_3 catalysts

Catalyst	*Selectivity(%)* FA	MF	DMM	DME	CO_x
Al_2O_3	--	--	--	100	--
1% V_2O_5/Al_2O_3	<1	--	--	99	--
3% V_2O_5/Al_2O_3	2	--	--	97	Tr
10% V_2O_5/Al_2O_3	16	--	Tr	84	--
15% V_2O_5/Al_2O_3	22	--	1	77	--
20% V_2O_5/Al_2O_3	46	--	3	50	Tr

The selective formation of nitrogen during the SCR of NO with ammonia is dependent on the specific oxide support, the specific surface metal oxide redox site, surface coverage and the presence of non-reducible surface metal oxide species *(21,25)*. The selectivity pattern follows the same trend as observed for methanol oxidation to formaldehyde (V > Cr > Re) and suggests that it is related to the intrinsic properties of the specific surface metal oxide species. However, the selectivity decreases with increasing coverage due to oxidation of ammonia by too many adjacent surface redox sites at elevated temperatures. The selectivity may also be decreased by the addition of non-reducible secondary surface metal oxide additives that may become active for the oxidation of ammonia at elevated temperatures on a reducible oxide support such as titania. The selectivity is also dependent on the specific oxide support (Ti > Al > Si). Thus, the selectivity of the SCR reaction is sensitive to all variations since it is a reaction that requires two adjacent sites.

Table IV. Effect of additives on the selectivity of methanol oxidation

Catalyst	*Selectivity(%)*				
	FA	*MF*	*DMM*	*DME*	CO_x
1% V_2O_5/TiO_2	99+	--	--	Tr	--
6% Nb_2O_5/ 1% V_2O_5/TiO_2	97	--	2	1	--
7% WO_3/ 1% V_2O_5/TiO_2	93	--	5	2	--
3% SiO_2/ 1% V_2O_5/TiO_2	96	--	3	Tr	Tr

Conclusions

Under oxidation reaction conditions, the surface metal oxide species may exhibit both fully oxidized and partially reduced surface redox sites. The extent of reduction depends on the reducing power of the reactant (methanol >> butane ~ methane) as well as the concentration of the reducing agent (e.g., SCR of NO with ammonia). The specific oxide support may also influence the extent of reduction of the surface metal oxide species for certain reactions (e.g., during alkane oxidation Sn > Ti ~ Zr > Al > Si).

The TOF of a specific oxidation reaction depends on the specific oxide support and the specific surface redox site. In general, the more reducible oxide supports result in more active ligands for the surface metal oxide species (Sn > Ce > Zr ~ Ti > Nb > Al > Si). For oxidation reactions that require one surface redox site (e.g., methanol, methane, propane and butane), the TOF is generally not influenced by surface metal oxide coverage and the presence of non-reducible secondary surface metal oxide additives. For oxidation reactions that require more than one surface site, two surface redox sites or a surface redox site and an adjacent surface acid site (e.g., SCR of NO with ammonia), the TOF is influenced by surface metal oxide coverage and the presence of secondary metal oxide species.

The selectivity properties of a specific oxidation reaction may depend on the nature of the oxide support, the surface metal oxide species, the surface coverage of the surface metal oxide species and the presence of secondary surface metal oxide additives. Surface vanadium oxide species appear to be the most selective surface redox sites during oxidation reactions. For methanol oxidation to formaldehyde, the most selective supported metal oxide catalyst consists of the surface metal oxide redox site that possesses the weakest acidic character and the oxide support that possesses both weak surface acid and redox sites: V_2O_5/TiO_2. These same properties are also responsible for making the V_2O_5/TiO_2 catalyst system one of the most selective catalysts for the SCR of NO with ammonia and oxidation of butane to maleic anhydride. For methane oxidation to formaldehyde, which generally occurs at much higher temperatures than the other oxidation reactions, a relatively inert oxide support, essentially possessing no surface acid or redox sites, that will minimize the decomposition of gas-phase methyl radicals and the further oxidation of formaldehyde is required and the most effective catalyst is V_2O_5/SiO_2. Surface coverage of the surface metal oxide species is generally an important parameter in modulating the selectivity since it can suppress side reactions that are associated with the oxide support (e.g., methanol oxidation and methane oxidation) and affect reactions that require two adjacent sites (SCR of NO with ammonia and the selective oxidation of butane to maleic anhydride). Similarly, secondary surface metal oxide additives can also modulate the reaction selectivity by influencing side reactions or enhancing the desired reaction (especially if the reaction requires two adjacent surface sites). Thus, each type of oxidation reaction has specific selectivity requirements which depend on the functional groups to be oxidized.

Acknowledgments

The financial support of the Division of Basic Energy Sciences, Department of Energy (grant #DEFG02-93ER14350) is gratefully acknowledged.

Literature Cited

1. Wachs, I. E., Deo, G., Vuurman, M. A., Hu, H., Kim, D. S. and Jehng, J.-M., *J. Mol. Catal.* **1993**, *82*, 443.
2. Vuurman, M. A., Oskam, A., Stufkens, D. J. and Wachs, I. E., *J. Mol. Catal.* **1992,** *77*, 263.
3. Vuurman, M. A., Oskam, A., Stufkens, D. J. and Wachs, I. E., *J. Mol. Catal.* **1993,** *80*, 209.
4. Vuurman, M. A., Wachs, I. E. and A. M. Hirt, *J. Phys. Chem.* **1991,** *95*, 9928.
5. Vuurman, M. A. and Wachs, I. E. *J. Phys. Chem.* **1992,** *96*, 5008.
6. Das, N., Eckert, H., Hu, H., Wachs, I. E., Walzer, J. F. and Feher, F. J., *J. Phys. Chem.* **1993,** *97*, 8240.
7. Deo, G., Wachs, I. E. and Haber, J., *Critical Rev. Surf. Chem.* **1994,** *4*, 141.
8. Hu, H., Wachs, I. E. and Bare, S., *J. Phys. Chem.* **1995,** *99*, 10897.
9. Hu, H. and Wachs, I. E., *J. Phys. Chem.* **1995,** *99*, 10911.
10. Deo, G. and Wachs, I. E., unpublished data.
11. Sun, Q., Jehng, J.-M., Herman, R., Wachs, I. E. and Klier, K. and Bhasin, M. M., to be published (Technical Program Abstracts, 14th North American Meeting of the Catalysis Society, Snowbird, Utah, June 11-16, 1995)
12. Wachs, I.E., Jehng, J.-M., Deo, G., Weckhuysen, B.M., Guliants, V. V., Benziger, *Catal. Today*, submitted.
13. Went, G. T., Leu, L.-J., Rosin, R. R. and Bell, A. T., *J. Catal.* **1992,** *134*, 492.

14. Deo, G. and Wachs, I. E., *J. Catal.* **1994,** *146*, 335.
15. Sorensen, C. M. and Weber, R. S., *J. Catal.* **1993,** *142*, 1.
16. Bruckman, K., Tatibouet, J. M., Che, M., Serwicka, E. and Haber, J., *J. Catal.* **1993,** *139*, 455.
17. Weber, R., *J. Phys. Chem.* **1994,** *98*, 2999.
18. Bruckman, K., Haber, J. and Serwicka, *Faraday Discuss. Chem. Soc.* **1989,** *87*, 173.
19. Stiegman, A. E., *J. Am. Chem. Soc.* **1995,** *117*, 2618.
20. Watling, T.C., Deo, G., Seshan, K., Wachs, I.E., and Lercher, J.A., *Catal. Today,* submitted.
21. Wachs, I. E., Deo, G., Weckhysen, B., Andreini, A., Vuurman, M. A., de Boer, M. and Amiridis, M., *J. Catal.*, submitted.
22. Szakacs, S., Altena, G. J., Fransen, T., van Ommen, J. G. and Ross, J. R. H., *Catal. Today* **1993,** *16*, 237.
23. Topsoe, N.-Y., Topsoe, H. and Dumesic, J. A., *J. Catal.* **1995,** *151*, 226.
24. Amiridis, M., Wachs, I. E., Deo, G., Jehng, J.-M. and Kim, D. S., *J. Catal.*, submitted.
25. Andreini, A. A., de Boer, Vuurman, M. A., Deo, G. and Wachs, I. E., *Appl. Catal*, submitted.
26. Datka, J., Turek, A. M., Jehng, J.-M. and Wachs, I. E., *J. Catal.* **1992,** *135*, 186.
27. Deo, G. and Wachs, I. E., *J. Catal.* **1994,** *146*, 323.
28. Kim, D. S. and Wachs, I. E., *J. Catal.* **1993,** *141*, 419.
29. Kim, D. S. and Wachs, I. E., *J. Catal.* **1993,** *142*, 166.
30. Busca, G., Ramis, G. and Lorenzelli, V., *J. Mol. Catal.* **1989,** *50*, 231.
31. Kung, H., in "Transition metal oxide: Catalysis and Surface Chemistry", Elsevier, Amsterdam, 1989.

Chapter 22

Development of Hydrotalcite Catalysts in Heterogeneous Baeyer–Villiger Oxidation

Kiyotomi Kaneda and Shinji Ueno

Department of Chemical Engineering, Faculty of Engineering Science, Osaka University, Toyonaka, Osaka 560, Japan

Hydrotalcites catalyze the Baeyer-Villiger oxidation of various ketones using a combination oxidant system composed of molecular oxygen and benzaldehyde to give high yields of lactones and esters. The catalytic reaction depends on the basic character of the hydrotalcites. Multi-metallic hydrotalcites having the elements Fe, Ni, or Cu in the Brucite-like layers of the hydrotalcites give the highest yields of oxidation products.

In 1990, we reported that Ru compounds in the presence of molecular oxygen and aldehyde catalytically cleaved carbon-carbon double bonds of terminal olefins to give the corresponding carboxylic acids and ketones. We also showed that peracids and/or peroxyaldehydes generated *in situ* from the reaction of molecular oxygen with aldehyde oxidize RuO_2 to higher oxidation states of ruthenium, *e.g.*, RuO_4. (*1*) Using the above combined oxidant of molecular oxygen and aldehyde, many selective oxidations catalyzed metal compounds have been explored.

- Epoxidation (*2-18*)
- Baeyer-Villiger oxidation (*11, 16, 19-22*)
- Oxidation of alkanes and aromatic compounds (*11, 16, 18, 23-28*)
- Oxidation of aldehydes, alcohols and sulfides (*16, 29-31*)

We also found that the combined oxidant has potential for the epoxidation of olefins and the Baeyer-Villiger oxidation of ketones (Eq.1) even in the absence of metal catalysts. (*32, 33*)

$$R_1-\overset{O}{\overset{\|}{C}}-R_2 \;+\; C_6H_5CHO \;+\; O_2 \xrightarrow[20\sim40\ ^\circ C]{CCl_4} R_1-\overset{O}{\overset{\|}{C}}-O-R_2 \;+\; C_6H_5\overset{O}{\overset{\|}{C}}-OH \qquad (1)$$

0097–6156/96/0638–0300$15.00/0

Hydrotalcites consist of Brucite-like layers (*34*) with positive charge and anionic compounds in the interlayer to form neutral materials. $Mg_6Al_2(OH)_{16}CO_3$, a prototype of the hydrotalcites was originally found as a mineral clay. Recently, many kinds of hydrotalcites have been synthesized using various elements in the Brucite-like layer, *e.g.*, Al, Mg, Fe, and Ni, and anionic compounds in the interlayer, *e.g.* halogenated, metal complex, organic acid, and heteropolyacid anion. (*34*) Combination of the elements, change of element ratios in the Brucite-like layer, and selection of anionic compounds can tune the basicity of the hydrotalcites and the interlayer distance. (*35*) It has been known that base and acid compounds acted as promotors of the Baeyer-Villiger oxidation of ketones with organic peracids. (*36, 37*) The tunable basic character of the hydrotalcites has attracted considerable interest and the effect on the Baeyer-Villiger oxidation is the subject of this paper. The oxidation of various ketones in the presence of the combined oxidant is reported with emphasis on the following three cases.

- Baeyer-Villiger oxidation in the absence of metal catalysts. (*33*)
- Baeyer-Villiger oxidation using various kinds of hydrotalcites with different ratios of the Mg/Al and interlayer anions. (*38*)
- Catalysis of functionalyzed hydrotalcites containing various transition metal elements, *e.g*., Ni, Fe, and Cu. (*39*)

Experimental

1) General

NMR spectra were obtained at JNM GSX270 with tetramethylsilane as an internal standard. IR spectra were recorded on a Hitachi EPI-G. Analytical GLC was performed with OV-17, Silicone SE-30 and SPB^{TM}-1(Fused Silica Capillary, 1.0μmdf) columns on an instrument equipped with a flame ionization detector. Powder X-ray diffraction patterns were obtained using a Shimazu VD-1 diffractmeter using Cu Kα radiation.

Aldehydes were purified according to the literature (*40*) and stored under an argon atmosphere. Solvents, such as carbon tetrachloride and 1,2-dichloroethane *etc*., were purchased from Wako Chemicals as special grade and dried with $MgSO_4$, followed by distillation under a nitrogen atmosphere. Ketones were also purchased from Wako Chemicals and dried with $MgSO_4$, followed by distillation. Ketones of 4-*tert*-butylcyclohexanone, norcamphor, 2-adamantanone, p-methoxyacetophenone, and benzylphenylketone were recrystallized before use. m-CPBA was purchased from Nacalai tesque and used without further purification. $Al(NO_3)_3{\cdot}9H_2O$, $Al(OH)_3$, $Mg(NO_3)_2{\cdot}6H_2O$, $Mg(OH)_2$, $Ni(NO_3)_2{\cdot}6H_2O$, $Cu(NO_3)_2{\cdot}3H_2O$, and $Fe(NO_3)_3{\cdot}9H_2O$ were purchased from Wako Chemicals as special grade.

2) Preparation of various hydrotalcites

$Mg_6Al_2(OH)_{16}CO_3$, $Mg_3Al(OH)_8Cl$, $Mg_4Al_2(OH)_{12}(C_6H_4(CO_2)_2)$, and $Mg_2Al(OH)_6(p\text{-}SO_3CH_3C_6H_4)$ were prepared by literature procedures. (*34,41-43*) Various multi-metallic hydrotalcites were prepared by a modification of the above method. (*34*) A typical example is for $Mg_6Al_2Ni_{0.6}(OH)_{17.2}CO_3$. $Ni(NO_3)_2{\cdot}6H_2O$ (0.003mol), $Mg(NO_3)_2{\cdot}6H_2O$ (0.03mol), and $Al(NO_3)_3{\cdot}9H_2O$ (0.01mol) were dissolved in water (100ml). A second water solution (60 ml) of Na_2CO_3 (0.03mol) and NaOH (0.07mol) was prepared. The first solution was slowly added to the second. The resulting mixture was heated at 65°C for about 18 h with good mixing. The greenish slurry was then cooled to room temperature, filtered, washed with a large amount of water and dried overnight at 110°C. From the XRD spectrum, the basal spacing was 7.82Å.

3) General procedure for the Baeyer-Villiger oxidation

Into a three necked flask with a reflux condenser cooled at -15°C were placed hydrotalcite (25mg), benzaldehyde (12mmol), and 1,2-dichloroethane (15ml). Oxygen was bubbled into the stirred heterogeneous mixture at 40°C for 30 min. A 1,2-dichloroethane solution (5ml) of ketone (4mmol) was added and the resulting mixture was stirred with bubbling of oxygen at 40°C for 4.5h. Hydrotalcite was separated by filtration and the filtrate was treated with Na_2SO_3 and $NaHCO_3$. Removal of the solvent under reduced pressure afforded a clear liquid, which was subjected to column chromatography on silica gel (hexane / ethyl acetate, 3:1) giving the corresponding pure ester.

4) Baeyer-Villiger oxidation using m-CPBA

A typical example of the title reaction is for 2-methylcyclopentanone. Into a three necked flask with a reflux condenser cooled at -15°C were placed $Mg_{10}Al_2(OH)_{24}CO_3$ (25mg), 2-methylcyclopentanone (2mmol), m-chloroperbenzoic acid (3mmol), and 1,2-dichloroethane (15ml). The resulting mixture was stirred at room temperature for 5 h. GLC analysis of the reaction mixture showed 94% yield of 5-methylvalerolactone. A similar reaction procedure was operated in the absence of the hydrotalcite and the lactone was formed in 12% yield after 5 h.

5) Reuse experiments of various hydrotalcites

Into a three necked flask with a reflux condenser cooled at -15°C were placed $Mg_{10}Al_2(OH)_{24}CO_3$ (25mg), benzaldehyde (12mmol), and 1,2-dichloroethane (15ml). Oxygen was bubbled into the stirred mixture at 40°C for 30 min. A 1,2-dichloroethane solution (5ml) of cyclopentanone (4mmol) was added and the resulting mixture was stirred with bubbling of oxygen at 40°C for 4.5h. GLC analysis of the reaction mixture showed 90% yield of δ–valerolactone. The hydrotalcite was washed with 3x10ml saturated Na_2CO_3 solution and 5x10ml of water, and dried at 50°C in vacuum overnight. IR and XRD spectra of the washed hydrotalcite did not change appreciably. For a reuse experiment of the hydrotalcite catalyst, a half scale mole reaction of cyclopentanone (2mmol) with benzaldehyde (6mmol) in 1,2-dichloroethane (10ml) was carried out in the presence of the spent hydrotalcite (12.5mg). After 5 h, the lactone was formed in 80% yield. On the other hand, the spent Mg-Al-Fe-CO_3 hydrotalcite was used without the above aqueous Na_2CO_3 treatment for the Baeyer-Villiger oxidation of cyclopentanone and gave 81% yield of δ–valerolactone without appreciable loss of catalytic activity, *viz.* 83% yield for fresh catalyst.

6) Measurement of base strength distribution on hydrotalcites (*44*)

Measurement of base strength on hydrotalcites was carried out by indicator method. Distilled cyclohexane was poured into the flask containing hydrotalcite evacuated at 150°C under an Ar atmosphere. Then an indicator benzene solution of 2,4-dinitroaniline (pK_{BH}=15.0) or 4-chloro-2-nitroaniline (pK_{BH}=17.2) was added into the above solution. The colour of indicator changed to violet or orange from yellow in both indicators, respectively. The quantity of bases was measured by titration method using benzoic acid titres. The results are shown in Table I.

Results and Discussion

1) Baeyer-Villiger oxidation without metal catalysts

We have already reported that both epoxidation and oxidative cleavage of olefins using the system consisting of aldehydes and molecular oxygen were strongly dependent on the kinds of aldehydes used . (*1,32*) The effect of the aldehyde components in the oxidation system was examined on Baeyer-Villiger oxidation of cyclohexanone as a model substrate. Baeyer-Villiger oxidation of cyclohexanone with benzaldehyde smoothly occurred to give ε-caprolactone at 40 °C and benzoic acid was also formed as a co-product. Aliphatic aldehydes, *e.g.*, isobutyraldehyde and isovaleraldehyde having high reactivities for the epoxidation (*32*) showed lower yields of ε-caprolactone than benzaldehyde. It became clear that the effectiveness of aldehydes is different between the Baeyer-Villiger oxidation and the epoxidation of olefins. In the screening of solvents in the cyclohexanone oxidation with benzaldehyde, carbon tetrachloride was the best solvent. 1,2-Dichloroethane, ethyl acetate, and benzene were good, while alcohols and acetonitrile were poor for this Baeyer-Villiger oxidation.

The representative results of Baeyer-Villiger oxidation using the above best system of benzaldehyde and carbon tetrachloride under an oxygen atmosphere are shown in Table II. Monoalkyl substituted cyclohexanones were converted into the corresponding ε-caprolactones in high yields; 2-methylcyclohexanone was oxidized into 84% of 6-methylcaprolactone (Run 5). 3-Methylcyclohexanone gave two regioisomers of 5-methylcaprolactone and 3-methylcaprolactone (*ca.* 1:1, Run 9). In the case of menthone, dialkyl substituted cyclohexanone, the Baeyer-Villiger oxidation occurred regioselectively *via* migration of the isopropyl substituted carbon to give 6-isopropyl-3-methylcaprolactone (Run 10). Oxidation of cyclopentanone and norcamphor proceeded selectively in spite of slow rates, respectively (Runs 11 and 12).

As mentioned later, ^{13}C NMR analysis of the reaction mixture pretreated with oxygen gas showed formation of perbenzoic acid. Therefore, it is conceivable that this Baeyer-Villiger oxidation might occur mainly by an organic peracid generated from the reaction of an aldehyde with molecular oxygen. The three component system of aldehyde, molecular oxygen, and chlorohydrocarbon is a useful monooxygenation-type reagent for both Baeyer-Villiger oxidation of ketones and epoxidation of olefins even *in the absence of metal catalysts.* (*21, 32, 33*) A suitable

Table I. Base Strength Distribution of Hydrotalcites

Hydrotalcite[a]	benzoic acid titer(mmol/g)	
	$7.1 \leq pK_{BH} \leq 15.0$	$15.0 \leq pK_{BH} \leq 17.2$
Mg-Al-CO_3(Mg:Al=1:1)	0.282	0.301
Mg-Al-CO_3(Mg:Al=3:1)	0.521	0.416
Mg-Al-CO_3(Mg:Al=5:1)	1.098	0.23
Mg-Al-CO_3(Mg:Al=8:1)	0.359	0.32
Mg-Al-Ni-CO_3(Mg:Al:Ni=3:1:0.3)	0.485	0.318
Mg-Al-Fe-CO_3(Mg:Al:Fe=3:1:0.3)	0.473	? [b]
Mg-Al-Cu-CO_3(Mg:Al:Cu=3:1:0.3)	0.459	0.391

[a] Evacuated for 4 h. at 150 °C.

[b] Yellow colour of the Mg-Al-Fe-CO_3 hydrotalcite led to hard distinction in coloration with indicator.

Table II. Baeyer-Villiger Oxidation of Various Ketones Using Benzaldehyde under Oxygen Atmosphere[a)]

Run	Substrate	Products	Reaction Time(h)	Conv. (%)	Yield (%)[b)]
1	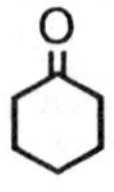	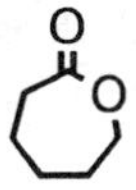	5	100	90 (82)
2[c)]			5	83	80
3[d)]			5	77	77
4[e)]			5	0	0
5	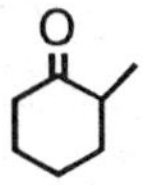	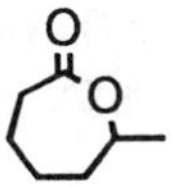	5	90	84 (80)
6		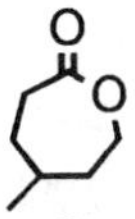	5	88	85
7[c)]			5	66	63
8		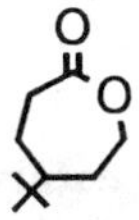	24	100	quantitative
9	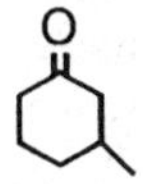	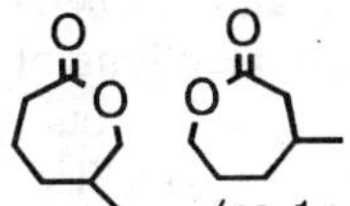 (*ca.* 1 : 1)	5	88	85 (80)
10	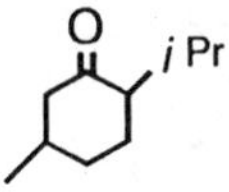 *i* Pr	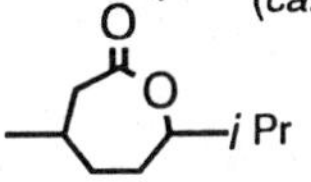 *i* Pr	24	70	70
11[f)]		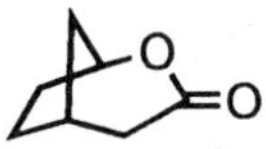	17	91	90
12	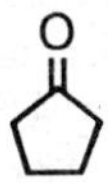	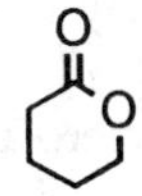	5	63	63
13	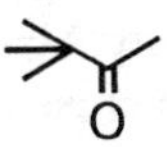	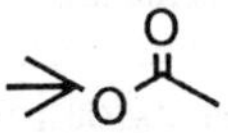	24	60	58

[a)]Reaction conditions: ketone 4 mmol, benzaldehyde 12 mmol, carbon tetrachloride 20 ml, O_2 bubbling, 40 °C. [b)]Yields were determined by GC and based on ketones used. Isolation yields are in parenthesis. [c)]1,2-Dichloroethane was employed instead of carbon tetrachloride. [d)]Ethyl acetate was used. [e)]Methanol was used.
[f)]3-Oxabicyclo[3.2.1]octa-2-one as a regioisomer (6%) was detected in NMR.

choice of aldehyde can lead to selective Baeyer-Villiger oxidation and epoxidation, respectively.

Addition of benzoyl chloride to the above oxidation system could accelerate the Baeyer-Villiger oxidation to give high yields of lactones at a lower temperature of 20 °C. Typical results are shown in Eqs. 2 and 3. We think that the role of benzoyl chloride might be to promote the yield of perbenzoic acid derived from oxidation of benzaldehyde. Notably, addition of benzoyl chloride does not lead troublesome workup because of its complete conversion to benzoic acid after the oxidation.

benzaldehyde, O_2 bubbling
CH_2ClCH_2Cl, 20 °C, 5h.
90% yield C_6H_5COCl(10mol%)
55% yield no addition (2)

benzaldehyde, O_2 bubbling
CH_2ClCH_2Cl, 20 °C, 24h.
94% yield C_6H_5COCl(10mol%)
70% yield no addition (3)

2) Baeyer-Villiger oxidation using hydrotalcite catalyst systems

2-1) Catalyst effect: First, in order to examine whether hydrotalcites have catalytic ability in the Baeyer-Villiger oxidation, reactions of cyclopentanone as a model substrate using various hydrotalcites were carried out in the presence of a combined oxidant of benzaldehyde and molecular oxygen (Eq. 4). Results of the oxidation

hydrotalcite, benzaldehyde, O_2
1, 2-dichloroethane, 40 °C (4)

with 1,2-dichloroethane as a solvent at 40 °C are shown in Table III. Baeyer-Villiger oxidation of cyclopentanone occurred selectively to give δ-valerolactone with a co-product of benzoic acid derived from benzaldehyde. Use of a mineral hydrotalcite, $Mg_6Al_2(OH)_{16}CO_3$ (Mg/Al=3) gave 46% of δ-valerolactone for 5 h. In a series of hydrotalcites containing CO_3^{2-} anion (Runs 2-5), $Mg_{10}Al_2(OH)_{24}CO_3$(Mg/Al= 5) showed the highest catalytic activity; cyclopentanone could be converted into δ-valerolactone in 90 % yield within 5 h. Other hydrotalcites of Mg/Al=1 and 8 were not effective catalysts for the Baeyer-Villiger oxidation at the present conditions. Next, oxidations of cyclopentanone were carried out in the presence of hydrotalcites containing interlayer anions of Cl^- and p-toluenesulfonate in place of CO_3^{2-} with keeping a ratio of Mg/Al=5, respectively. $Mg_5Al(OH)_{12}Cl$ gave 83% yield of the lactone, while the corresponding p-toluenesulfonate one was a poor catalyst for the oxidation. Further, other hydrotalcites of $Mg_4Al_2(OH)_{12}(C_6H_4(CO_2)_2)$ and $Mg_2Al(OH)_6(p\text{-}SO_3CH_3C_6H_4)$ were tested as the catalyst, but gave low yields of the lactone, respectively.

Table III. Baeyer-Villiger Oxidation of Cyclopentanone Using Various Hydrotalcites [a)]

Run	Catalyst	Conv. (%)	Yield (%)
1	No Catalyst	45	40
2	$Mg_2Al_2(OH)_8CO_3$	57	45
3	$Mg_6Al_2(OH)_{16}CO_3$	56	46
4	$Mg_{10}Al_2(OH)_{24}CO_3$	94	90
5	$Mg_{16}Al_2(OH)_{36}CO_3$	77	60
6	$Mg_4Al_2(OH)_{12}(TA)$ [b)]	56	37
7	$Mg_2Al(OH)_6(TS)$ [c)]	49	38
8	$Mg_5Al(OH)_{12}(TS)$	44	44
9	$Mg_5Al(OH)_{12}Cl$	93	83
10	$Mg(OH)_2$	68	50
11	$Al(OH)_3$	52	30

[a)] Reaction Conditions: cyclopentanone 4 mmol, benzaldehyde 12 mmol, 1, 2-dichloroethane 20 ml, Catalyst 0.025 g, O_2 bubbling, 40 °C, 5 h.
[b)] TA=C_6H_4-1,4-$(CO_2)_2$ [c)] TS=$CH_3C_6H_4SO_3$

In a series of hydrotalcites containing CO_3^{2-} anion (Runs 2-5), the base strength distributions were measured by titration with different indicators. (*44*) For the $Mg_{10}Al_2(OH)_{24}CO_3$(Mg/Al= 5), 1.098 mmol of benzoic acid/g for 7.1-15.0 pK_{BH} and 0.23 mmol of benzoic acid/g for 15.0-17.2 pK_{BH} were consumed, respectively. The hydrotalcite with Mg/Al=5 had the largest quantity of basic sites in the range of 7.1-15.0 pK_{BH} among the hydrotalcites.(see Table I.) It suggests that the quantity of basic sites at the low range of pK_{BH} might be related to the catalytic activity for the Baeyer-Villiger oxidation. It is concluded that selection of the Mg/Al ratio and of interlayer anions must be important factors in designing more active hydrotalcite catalysts for this oxidation.

2-2) Selection of reaction conditions: Oxidations of cyclopentanone using various aldehydes in the presence of $Mg_{10}Al_2(OH)_{24}CO_3$ were carried out under an oxygen atmosphere. Aromatic aldehydes had higher reactivity for the Baeyer-Villiger oxidation than aliphatic ones. p-Methylbenzaldehyde as well as benzaldehyde gave a high yield of δ-valerolactone.

In examination of the reaction temperature, it was found that 40 °C was suitable for this oxidation and while increasing the temperature to 70 ° C lowered the yield of the lactone in spite of increasing the conversion of cyclopentanone. Generally, it is recommended that Baeyer-Villiger oxidations using organic peracids should be carried out below room temperature. (*36*) Under our oxidation conditions, *in situ* generation of perbenzoic acid from benzaldehyde might need higher temperature operation.

Using benzaldehyde, the effect of various solvents on cyclopentanone oxidation was examined in the presence of $Mg_{10}Al_2(OH)_{24}CO_3$ at 40 °C. 1,2-Dichloroethane was the best solvent tried. Acetonitrile and ethylacetate gave moderate yields of δ-valerolactone, 62 and 57%, respectively, while acetone and methanol were poor solvents.

2-3) Reactivities of various ketones

Using the above best reaction conditions, $Mg_{10}Al_2(OH)_{24}CO_3$ / benzaldehyde / 1,2-dichloroethane / 40 ° C, oxidations of various ketones were carried out under an oxygen atmosphere.

a) cyclic ketones. Oxidation of cyclic ketones is summarized in Table IV. Cyclopentanone gave a high yield of δ-valerolactone.(*vide supra*) Similarly, Baeyer-Villiger oxidation of alkyl substituted cyclopentanones, 2-methylcyclopentanone and 2,2-dimethylcyclopentanone, occurred regioselectively to give 5-methylvalerolactone and 5, 5-dimethylvalerolactone as sole products, respectively. 94% yield of 5-methylvalerolactone could be obtained from reaction of 2-methylcyclopentanone for 5 h. The fact shows that hydrotalcites are excellent oxidation catalysts, when compared with other reagents. For example, oxidation of 2-methylcyclopentanone using m-CPBA gave 74% of 5-methylvalerolactone for 12 h (*45*) and use of H_2O_2 resulted in a low selectivity for the Baeyer-Villiger oxidation. (*46-48*) Norcamphor as a bicyclic five-membered ring ketone showed a high reactivity; norcamphor was completely consumed after 3 h and afforded 86% 2-oxabicyclo[3.2.1]octa-3-one, accompanied by 5% of a regioisomer of 3-oxabicyclo[3.2.1]octa-2-one (Eq. 5).

Table IV. Baeyer-Villiger Oxidation of Cyclic Ketones in the Presence and the Absence of $Mg_{10}Al_2(OH)_{24}CO_3$ [a)]

Run	Substrate	Product	Reaction Time (h)	Conv., Yield(%) $Mg_{10}Al_2(OH)_{24}CO_3$	Conv., Yield(%) No Catalyst
1			5	94, 90(82)	45, 40
2			5	96, 95	10, 8
3			24	65, 61	trace
4			3	100, 91(83)[b)]	26, 24
5			5	100, quantitative (85)	43, 43
6			5	82, 80	83, 80
7			3	100, 93	77, 74
8			5	56, 48	51, 47
9			3	63, 53	50, 49

[a)] Reaction Conditions: substrate 4 mmol, benzaldehyde 12mmol, 1, 2-dichloroethane20 ml, Catalyst 0.025g, 40 °C [b)] 3-Oxabicyclo[3.2.1]octa-2-one as a regioisomer (6%) was detected in ^{1}HNMR.

hydrotalcite, benzaldehyde, O_2 / 1, 2-dichloroethane, 40 °C, 3 h. (5)

86% 5%

Six-membered ring ketones were also oxidized to give the corresponding lactones in high yields, respectively. In a series of methyl substituted cyclohexanones (Runs 7-9), 2-methylcyclohexanone showed the highest reactivity for the Baeyer-Villiger oxidation to give one regioisomer product, while two regioisomers of 3-methylcaprolactone and 5-methylcaprolactone were obtained in the case of 3-methylcyclohexanone. Oxidation of 2-adamantanone, a cyclic six-membered ring ketone, was fast and afforded 4-oxahomoadamantan-5-one in an almost quantitative yield for 5 h (Eq. 6).

hydrotalcite, benzaldehyde, O_2 / 1, 2-dichloroethane, 40 °C, 5 h. (6)

quantitative

b) acyclic ketones. Oxidation of acyclic ketones are shown in Table V. Among p-substituted acetophenone derivatives (Runs 1-3), p-methoxyacetophenone had the highest reactivity and gave p-methoxyphenylacetate in an almost quantitative yield for 24 h, while oxidation of acetophenone was very slow. Similar phenomena on p-substituent effect of acetophenone derivatives are usually observed in Baeyer-Villiger oxidations with other reagents. (*49*) In reaction of pinacolone, 90 % yield of t-butylacetate was obtained as a sole Baeyer-Villiger product after 20 h (Eq. 7).

hydrotalcite, benzaldehyde, O_2 / 1, 2-dichloroethane, 40 °C, 20 h. (7)

90%

Such activity of the hydrotalcite can compete with other reported reagents for the oxidation of pinacolone, *e.g.*, Ni / O_2 / *iso*valeraldehyde (*19*), $PhAsO_3H_2$ / H_2O_2 (*50*), polymer-bound selenium complex / H_2O_2 (*46*), and magnesium monoperphthalate. (*51*) Further, in the case of 3-methyl-2-butanone, 89 % of isopropylacetate was obtained after 24 h, accompanied by a small amount of methyl *iso*butylate. Similar regioisomer distributions of the esters are observed also in Baeyer-Villiger oxidation of pinacolone and 3-methyl-2-butanone using perorganic acids, which can be explained by aptitude of alkyl migration in a Criegee intermediate as follows: tertiary > secondary > primary > methyl. (*52*) 1-Phenyl-2-butanone and benzyl phenyl ketone also could be oxidized to give the corresponding benzyl esters in high yields, respectively. On the other hand, reactions of 3-pentanone and 2,4-dimethyl-3-pentanone resulted in

Table V. Baeyer-Villiger Oxidation of Acyclic Ketones in the Presence and the Absence of $Mg_{10}Al_2(OH)_{24}CO_3$ [a)]

Run	Substrate	Reaction Time (h)	Yield(%)	
			$Mg_{10}Al_2(OH)_{24}CO_3$	No Catalyst
1	$CH_3O-C_6H_4-C(=O)-CH_3$	24	quantitative	15
2	$CH_3-C_6H_4-C(=O)-CH_3$	24	35	23
3	$C_6H_5-C(=O)-CH_3$	24	11	trace
4[b)]	O	24	26	14
5[b)]	O	24	89	63
6[b)]	O	20	90	38
7[b)]	O	24	21	8
8	$C_6H_5-CH_2-C(=O)-C_6H_5$	24	62	6
9	$C_6H_5-CH_2-C(=O)-C_2H_5$	24	89	59

[a)] Reaction conditions: Mg-Al-CO_3(Mg:Al=5:1) 0.05g, ketone 2 mmol, benzaldehyde 6 mmol, 1,2-dichloroethane 20 ml, 40 °C. [b)] Catalyst 0.025g.

low yields of the Baeyer-Villiger products, respectively, under the present conditions. In comparison with oxidation of cyclic ketones discussed in a former section, acyclic ketones necessitates long reaction times to achieve high yields of esters. In many oxidizing reagents, reactivities of acyclic ketones for the Baeyer-Villiger oxidation are lower than those of cyclic ketone. (*36*)

3) Baeyer-Villiger oxidation using m-CPBA

We think that this Baeyer-Villiger oxidation using a combination system of molecular oxygen and benzaldehyde consists of the two steps of forming perbenzoic acid and of the oxygen transfer from perbenzoic acid to ketone.(Eqs. 8 and 9) In

1) The step of generating peracid

$$R_3C(=O)H + O_2 \longrightarrow R_3C(=O)O{-}O{-}H \qquad (8)$$

2) The oxygen transfer step of peracid

$$R_1C(=O)R_2 + R_3C(=O)O{-}O{-}H \longrightarrow R_1C(=O)O{-}R_2 + R_3C(=O)O{-}H \qquad (9)$$

order to examine whether hydrotalcites accelerate the oxygen transfer step, the Baeyer-Villiger oxidations using m-CPBA instead of a combination oxidant system of molecular oxygen and aldehyde was carried out with and without the hydrotalcite, respectively. The reactions of five-membered ring ketones in the presence of $Mg_{10}Al_2(OH)_{24}CO_3$ gave higher yields of the corresponding lactones than those without the hydrotalcite. Particularly, in the presence of hydrotalcite, 2-methylcyclopentanone gave an almost quantitative yiled of 5-methylvalerolactone (Eq. 10), while the reaction

$$\text{2-methylcyclopentanone} + \text{m-}ClC_6H_4C(=O){-}OOH \xrightarrow[\text{1, 2-dichloroethane, r.t. 5 h.}]{Mg_{10}Al_2(OH)_{24}CO_3} \text{5-methylvalerolactone (98\%)} + \text{m-}ClC_6H_4C(=O){-}OH \qquad (10)$$

without the hydrotalcite led to 12% of the lactone under the same reaction conditions. Notably, the strong catalytic effect of hydrotalcite could not be observed in the case of six-membered ring ketones.

4) Catalysis of functionalyzed hydrotalcites

It has been reported that transition metals, *e.g.*, Ni, Fe, and Cu were effective catalysts for Baeyer-Villiger oxidation using a combination oxidant of molecular oxygen and aldehydes. (*19-21*) Various transition metal elements can be introduced into the Brucite-like layer in hydrotalcites to form multi-metallic hydrotalcites. (*34*)

If both basic and the transition metal active sites act cooperatively as a catalyst within hydrotalcites, it is expected that such designed hydrotalcites might have a highly catalytic activity for Baeyer-Villiger oxidation. Therefore, we prepared multi-metallic hydrotalcites having transition metals of Fe(III), Ni(II), and Cu(II) in Brucite-like layer. Baeyer-Villiger oxidations of various ketones using a combination oxidant of molecular oxygen and aldehydes were carried out in the presence of $Mg_6Al_2Fe_{0.6}(OH)_{17.2}(CO_3)_{1.3}$, $Mg_6Al_2Ni_{0.6}(OH)_{17.2}CO_3$, and $Mg_6Al_2Cu_{0.6}(OH)_{17.2}CO_3$. Typical results are shown in Table VI together with those using $Mg_6Al_2(OH)_{16}CO_3$. Three multi-metallic hydrotalcites had higher catalytic activities for the Baeyer-Villiger oxidation of many ketones than the $Mg_6Al_2(OH)_{16}CO_3$. In particular, the Mg-Al-Fe-CO_3 hydrotalcite could efficiently oxidize various cyclic ketones to give high yields of the corresponding lactones, while the Mg-Al-Cu-CO_3 hydrotalcite showed high activity for bicyclic ketones. In changing the ratio of Mg to Cu, the hydrotalcite having Cu / (Mg+Cu)=0.5 gave the highest catalytic activity.

5) Mechanism of Baeyer-Villiger oxidation

During pretreatment of benzaldehyde solution with molecular oxygen, autoxidation of benzaldehyde occurred to give perbenzoic acid. Analysis of the resulting solution by ^{13}CNMR showed formation of perbenzoic acid at 168.0 ppm due to O=COOH in $CDCl_3$, whose value was in fair agreement with that of an authentic sample. It can be thought that the oxidation using a combination system of molecular oxygen and benzaldehyde consists of the two steps of generating the peracid and of the oxygen transfer of the peracid to the ketone in Eqs. 8 and 9. Using m-CPBA as an oxidant occurred Baeyer-Villiger oxidation smoothly in the presence of hydrotalcite (*vide supra*). The above result shows that hydrotalcites contribute strongly to the oxygen transfer step to afford ester and benzoic acid. It has been reported that Baeyer-Villiger oxidation using perorganic acids was promoted by basic compounds, *e.g.*, Na_2CO_3 and $NaHCO_3$. (*53, 54*) Therefore, we think that basic character of hydrotalcites plays an important role in the above Baeyer-Villiger oxidation. Figure 1 shows a possible mechanism of the Baeyer-Villiger oxidation using hydrotalcites. Reaction of hydroxyl group on hydrotalcite surface with perbenzoic acid gives metal perbenzoate and H_2O. The perbenzoate attacks ketone to form a metal alkoxide intermediate which undergoes rearrangement to give lactone or ester, and benzoic acid. On the other hand, the reaction of perbenzoic acid with ketone gives a Criegee intermediate. The hydroxyl group promotes the rearrangement step of the Criegee intermediate. As mentioned in a previous section, $Mg_{10}Al_2(OH)_{24}CO_3$ showed the highest catalytic activity for the Baeyer-Villiger oxidation and had the largest quantity of the bases at the low strength range of 7.1-15.0 pK_{BH}. The above results indicate that this oxidation might favor the hydrotalcites having not strong basicity but weak one.

Introduction of transition metals of Fe, Ni, and Cu into the Brucite-like layer leads to increases in the catalytic activity for the Baeyer-Villiger oxidation. Clearly, in the oxidation using a combination system of molecular oxygen and benzaldehyde, the transition metals in the hydrotalcites promote the oxygen transfer step of the perbenzoic acid to the ketone because multi-metallic hydrotalcites catalyzed the Baeyer-Villiger oxidation using m-CPBA. At present, we have not mentioned clearly

Table VI. Baeyer-Villiger Oxidation Using Multi-Metallic Hydrotalcites [a)]

Substrate	R. Time(h)	Conv., Yield (%)			
		Mg-Al-Fe-CO_3 (Mg:Al:Fe=3:1:0.3)	Mg-Al-Ni-CO_3 (Mg:Al:Ni=3:1:0.3)	Mg-Al-Cu-CO_3 (Mg:Al:Cu=3:1:0.3)	Mg-Al-CO_3 (Mg:Al=3:1)
	5	92, 83 (17.8)	90, 78 (16.3)	60, 45 (9.4)	56, 46 (8.7)
	5	100, quantitative (21.5)	90, 90 (18.8)	73, 62 (13.0)	50, 49 (9.3)
	5	100, quantitative (21.5)	100, 85 (17.8)	94, 92 (19.3)	82, 70 (13.2)
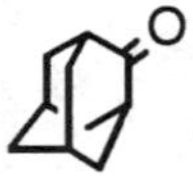	5	54, 54 (11.6)	100, quantitative (20.9)	100, quantitative (21.0)	41, 41 (7.8)
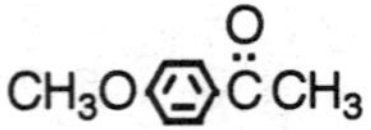 CH_3O-C₆H₄-C(O)CH_3	5	51, 51 (11.0)	68, 68 (14.2)	72, 72 (15.1)	51, 51 (9.6)
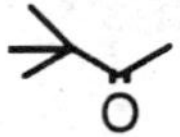	5	51, 45 (9.7)	75, 70 (14.6)	72, 66 (13.9)	25, 16 (3.0)

[a)] Reaction conditions: catalyst 0.025g, substrate 4 mmol, benzaldehyde 12 mmol, 1, 2-dichloroethane 20 ml, 40 °C. Turnover rate (moles of the product per molar unit cell of the hydrotalcite catalyst per hour) is in parentheses.

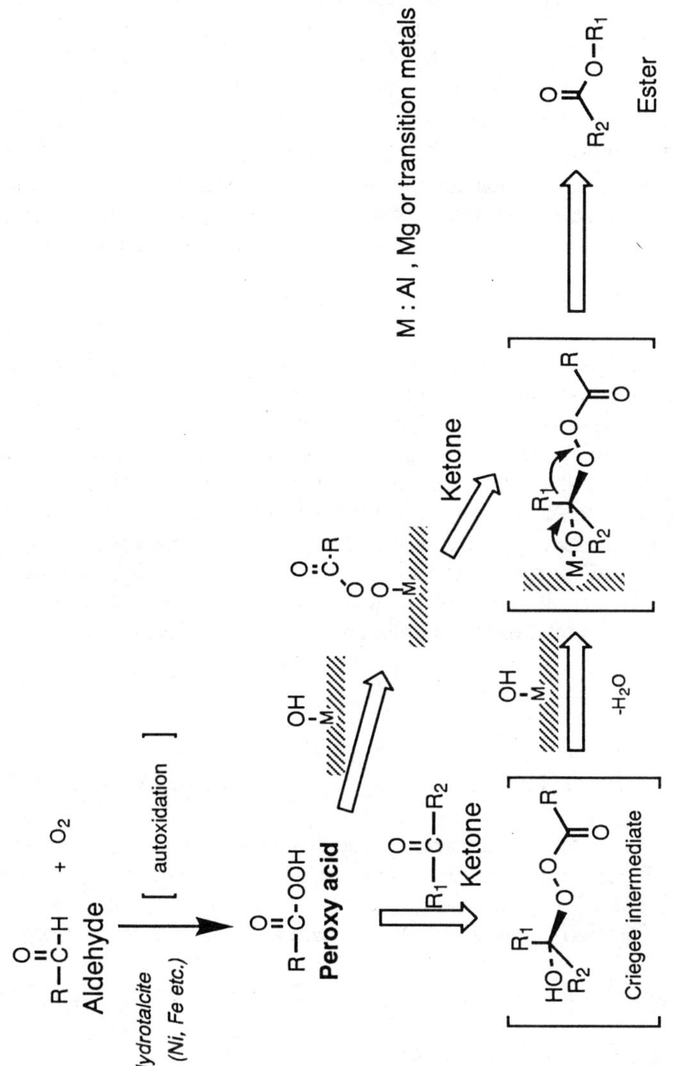

Figure 1. Possible Reaction Mechanism Using Hydrotalcite Catalyst

the role of transition metals in the Baeyer-Villiger oxidation. The transition metals might accelerate autoxidation of benzaldehyde to form perbenzoic acid. (*55*)

Since the hydrotalcites acted as heterogeneous catalysts, they could be expected to be reusable catalysts for the Baeyer-Villiger oxidation. In oxidation of cyclopentanone, the catalytic activities of the spent hydrotalcites were examined in two cases of $Mg_{10}Al_2(OH)_{24}CO_3$ and Mg-Al-Fe-CO_3 under the same conditions as fresh hydrotalcites. The spent $Mg_{10}Al_2(OH)_{24}CO_3$ hydrotalcite had an extremely low catalytic activity , but treatment of the hydrotalcite with aqueous Na_2CO_3 solution regenerated the activity to give a satisfactory yield of δ-valerolactone. On the other hand, oxidation with the spent Mg-Al-Fe-CO_3 hydrotalcite could give a high yield of δ-valerolactone even without the aqueous Na_2CO_3 treatment. This result shows that introduction of some transition metals into the Brucite-like layer not only increases the catalytic activity of hydrotalcites, but also prevents catalyst deactivation.

Conclusions

Hydrotalcites catalyze the Baeyer-Villiger oxidation of various ketones using a combination oxidant system of molecular oxygen and benzaldehyde to give high yields of lactones and esters. The catalytic reactions can be explained in terms of the basic character of the hydrotalcites. Use of multi-metallic hydrotalcites having Fe, Ni, or Cu in the Brucite-like layer can lead to higher yields of the oxidation products, which might be due to the cooperative action between the basic sites of the hydrotalcites and the transition metal sites. These results give insight for the preparation of highly functionalized hydrotalcite catalysts for organic syntheses. Since hydrotalcites act as heterogeneous catalysts, there are advantages of easy separation of the hydrotalcites from the reaction mixture after the oxidation and of reuse of the catalysts.

Literature Cited

1) Kaneda, K.; Haruna, S.; Imanaka, T.; Kawamoto, K. *J. Chem. Soc., Chem. Commun.* **1990**, 1467.
2) Mukaiyama, T.; Takai, T.; Yamada, T.; Rhode, O. *Chem. Lett.* **1990**, 1661.
3) Yamada, T.; Takai, T.; Rhode, O.; Mukaiyama, T. *Chem. Lett.* **1991**, 1.
4) Takai, T.; Yamada, T.; Rhode, O.; Mukaiyama, T. *Chem. Lett.* **1991**, 281.
5) Yamada, T.; Takai, T.; Rhode, O.; Mukaiyama, T. *Bull. Chem. Soc. Jpn.* **1991**, *64*, 2109.
6) Takai, T.; Hata, E.; Yamada, T.; Mukaiyama, T. *Bull. Chem. Soc. Jpn.* **1991**, *64*, 2513.
7) Irie, R.; Ito, Y.; Katsuki, T. *Tetrahedron Lett.* **1991**, *32*, 6891.
8) Yamada, T.; Imagawa, K.; Nagata, T.; Mukaiyama, T. *Chem. Lett.* **1992**, 2231.
9) Mukaiyama, T.; Yamada, T.; Nagata, T.; Imagawa, K. *Chem. Lett.* **1993**, 327.
10) Bouhlel, E.; Laszlo, P.; Levart, M.; Montaufier, M.-T.; Singh, G. P. *Tetrahedron Lett.* **1993**, *34*, 1123.
11) Hamamoto, M.; Nakayama, K.; Nishiyama, Y.; Ishii, Y. *J. Org. Chem.* **1993**, *58*, 6421.

12) Mizuno, N.; Tateishi, M.; Hirose, T.; Iwamoto, M. *Chem. Lett.* **1993**, 1839.
13) Mizuno, N.; Tateishi, M.; Hirose, T.; Iwamoto, M. *Chem. Lett.* **1993**, 1985.
14) Lopez, L.; Mastrorilli, P.; Mele, G.; Nobile, C. F. *Tetrahedron Lett.* **1994**, *35*, 3633.
15) Nagata, T.; Imagawa, K.; Yamada, T.; Mukaiyama, T. *Chem. Lett.* **1994**, 1259.
16) Mastrorilli, P.; Nobile, C. F. *Tetrahedron Lett.* **1994**, *35*, 4193.
17) Lassila, K.; Waller, F. J.; Werkheiser, S.; Wressell, A. L. *Tetrahedron Lett.* **1994**, *35*, 8077.
18) Reddy, M. M.; Punniyamurthy, T.; Iqbal, J. *Tetrahedron Lett.* **1995**, *36*, 159.
19) Yamada, T.; Takahashi, K.; Kato, K.; Takai, T.; Inoki, S.; Mukaiyama, T. *Chem. Lett.*, **1991**, 641.
20) Murahashi, S.; Oda, Y.; Naota, T. *Tetrahedron Lett.*, **1992**, *33*, 7557.
21) Bolm, C.; Schlinggloff, G.; Weickhardt, K. *Tetrahedron Lett.*, **1993**, *34*, 3405.
22) Bolm, C.; Schlinggloff, G. *J. Chem. Soc., Chem. Commun.* **1995**, 1247.
23) Yamada, T.; Rhode, O.; Takai, T.; Mukaiyama, T. *Chem. Lett.* **1991**, 5.
24) Bhatia, B.; Iqbal, J. *Tetrahedron Lett.* **1992**, *33*, 7961.
25) Murahashi, S.; Oda, Y.; Naota, T. *J. Am. Chem. Soc.* **1992**, *114*, 7913.
26) Murahashi, S.; Oda, Y.; Naota, T.; Komiya, N. *J. Chem. Soc., Chem. Commun.* **1993**, 139.
27) Mizuno, N.; Tateishi, M.; Hirose, T.; Iwamoto, M. *Chem. Lett.* **1993**, 2137.
28)Hata, E.; Takai, T.; Yamada, T.; Mukaiyama, T. *Chem. Lett.* **1994**, 1849.
29) Kalra, S. J. S.; Punniyamurthy, T.; Iqbal, J. *Tetrahedron Lett.* **1994**, *35*, 4847.
30) Anoune, N.; Lanteri, P.; Longeray, R.; Arnaud, C. *Tetrahedron Lett.* **1995**, *36*, 6679.
31) Imagawa, K.; Nagata, T.; Yamada, T.; Mukaiyama, T. *Chem. Lett.* **1995**, 335.
32) Kaneda, K.; Haruna, S.; Imanaka, T.; Hamamoto, M.; Nishiyama, Y.; Ishii, Y. *Tetrahedron Lett.* **1992**, 33, 6827.
33) Kaneda, K.; Ueno, S.; Imanaka, T.; Shimotsuma, E.; Nishiyama, Y.; Ishii, Y. *J. Org. Chem.* **1994**, 59, 2915.
34) Cavani,F.; Trifiro, F.; Voccari, A. *Catal. Today*, **1991**, *11*, 173.
35) Miyata, S.; Kumura, T.; Hattori, H.; Tanabe, K. *Nippon Kagaku Zasshi*, **1971**, *92*, 514.
36) Krow, G. R. *Organic Reaction*, John Wiley & Sons, New York, **1993**, *43*, 251.
37) Ogata, Y.; Sawaki, Y. *J. Am. Chem. Soc.* **1972**, *94*, 4189.
38) Kaneda, K.; Ueno, S.; Imanaka, T. *J. Chem.* Soc., *Chem. Commun.* **1994**, 797.
39) Kaneda, K.; Ueno, S.; Imanaka, T. *J. Mol. Catal. A: Chemical* **1995**, *102*, 135.
40) Perrin, D. D.; Amarego, W. L. F. *Purification of Laboratory Chemicals*, 3rd ed., Pergamon Press, Oxford, **1988**.
41) Reichle, W. T.; Kang, S. Y.; Everhordt, D. S. *J. Catal.*, **1986**, *101*, 352.
42) Drezdson, M. A. *Inorg. Chem.*, **1988**, *27*, 4628.
43) Miyata, S. *Clays Clay Miner.*, **1975**, *23*, 369.
44) Take, J.; Kikuchi, N.; Yoneda, Y. *J.Catal.*, **1971**, *21*, 164.
45) Luthy, C; Konstantin, P.; Untch, K. G. *J. Am. Chem. Soc.* **1978**, *100*, 6211.
46) Taylor, R. T.; Flood, L. A. *J. Org. Chem.* **1983**, *48*, 5160.

47) Matsubara, S.; Takai, K.; Nozaki, H. *Bull. Chem. Soc. Jpn.* **1983**, *56*, 2029.
48) Olah, G. A.; Yamato, T.; Iyer, P. S.; Trivedi, N. J.; Singh, B. P.; Pradesh, G. K. S. *Mater. Chem. Phys.* **1987**, *17*, 21.
49) Token, K.; Hirano, K.; Yokoyama, T.; Goto, K. *Bull. Chem. Soc. Jpn.* **1991**, *64*, 2766.
50) Jacobson, S. E.; Mares, F.; Zambri, P. M. *J. Am. Chem. Soc.* **1979**, *101*, 6938.
51) Brougham, P.; Cooper, M. S.; Cummerson, D. A.; Heaney, H.; Thompson, N. *Synthesis*, **1987**, 1015.
52) Hawthorne, M. F.; Emmons, W. D.; McCallum, K. S. *J. Am. Chem. Soc.* **1958**, *80*, 6393.
53) Krow, G. R.; Johnson, C. A.; Guare, J. P.; Kubrak, D.; Henz, K. J.; Shaw, D. A. Szczepanski, S. W.; Carey, J. T. *J. Org. Chem.*, **1982**, *47*, 5239.
54) Ehitesell, J. K.; Matthews, R. S.; Helbling, A. M. *J. Org. Chem.*, **1978**, *43*, 784.
55) Sheldon, R. A.; Kochi, J. K. *Metal-Catalyzed Oxidations of Organic Compounds*, Academic Press, New York, **1981**.

Chapter 23

The Oxidative Ammonolysis of Ethylene to Acetonitrile over γ-Al_2O_3-Supported Molybdenum Catalysts

I. Peeters, J. van Grondelle, and R. A. van Santen

Schuit Institute of Catalysis, Eindhoven University of Technology, P.O. Box 513, 5600 MB Eindhoven, Netherlands

The reaction of ethylene with ammonia, without gaseous oxygen, to acetonitrile over γ-Al_2O_3 supported molybdenum catalysts was studied. The effects of molybdenum loading and pretreatment on the catalytic activity were investigated. Experiments showed that the activity at the semi-steady-state is highly structure sensitive. Pretreated in oxygen, the catalyst is highly selective towards CH_3CN, with CO_x formed as side product. Pretreated in hydrogen, the catalyst is more active but less selective, with ethane formed as side product. Two mechanisms were deduced:

1) ammoxidation mechanism with consumption of lattice oxygen.

2) oxidative ammonolysis with coproduction of ethane, without lattice oxygen consumption.

The steady-state activity was independent of pretreatment and no oxygen containing products were observed, indicating that mechanism 1) can gradually change into 2) when removable lattice oxygen becomes depleted. The product distribution indicated that the mechanisms can be active simultaneously and separately. Mechanism 2 appeared to be operational on a MoO_2-like structure.

Acetonitrile is a well-known polar solvent. Although one of the more stable nitriles, it can be used as a reactant in a wide range of typical nitrile reactions, such as the synthesis of amides, amines and isocyanates.(*1* ,*2*) A multitude of different processes are known to produce acetonitrile, such as the decomposition of alkyl amines(*3* ,*4*) Also, reactions of ammonia with ethanol(*5*), acetic acid, acetic anhydride or thermal decomposition of several nitrogen containing compounds, and reactions of cyanogen, hydrocyanic acid or ammonia with hydrocarbons all yield acetonitrile(*6* ,*7*). The most extensively studied system is the

0097–6156/96/0638–0319$15.00/0

formation of acetonitrile from CO, H_2 and NH_3.(*6,8 ,9*) For this reason, catalysts are most frequently based on molybdenum(*10 ,11*) or iron oxides(*12 ,13*) on silica or alumina supports. The addition of silver, copper, rhodium(*10*), manganese and alkali earths(*8*) is mentioned to improve the selectivity to acetonitrile considerably. HCN has been suggested as the principal intermediate for this reaction.(*11*) So basically, acetonitrile is produced by catalytic decomposition of higher molecules, or by substitution of oxygen with nitrogen. CH_3CN and HCN are the main side products of the propylene ammoxidation over Bi/Mo catalysts (Sohio process) and in some plants acetonitrile is recovered and purified.(*1,2*)
Much research has been done on allylic (amm)oxidation of olefins over molybdenum containing catalysts.[14] However, little is known about vinylic (amm)oxidation.
Here we report the formation of CH_3CN from ammonia and ethylene as reactants over γ-Al_2O_3 supported molybdenum catalysts. To focus on the reactivity of lattice oxygen and to exclude surface reoxidation, oxidative ammonolysis experiments were done in the absence of gaseous oxygen. Two reaction mechanisms are proposed to explain the results. One is based on stoichiometric consumption of lattice oxygen and resembles the ammoxidation reaction mechanism, whereas the other is based on the hydrogenation of ethylene. Which mechanism is operative strongly depends on the catalyst pretreatment and the morphology created in time due to reaction.

Experimental

Catalysts were prepared by incipient wetness impregnation of a commercial γ-Al_2O_3 (Akzo, Ketjen CK-300, 200 m^2 g^{-1}, 0.6 ml g^{-1}, 250-500 μm) with aqueous solutions of ammonium heptamolybdate ($(NH_4)_6Mo_7O_{24}$), followed by drying overnight in air at 110 °C and heating in a flow of artificial air to 550 °C for 1h. Before cooling to reaction temperature, the catalyst was heated in a 5 vol% O_2 in helium flow to 550 °C for 24 h and, when reductive pretreatment was desired, subsequently heated in a 10 vol% H_2 in helium flow from room temperature to 650 °C for 24 h. The temperature ramp was set at 5 °C/min for all cases and the total flow was kept at 20 Nml/min. No indications were found of the formation of bronzes (H_xMoO_3) after a hydrogen treatment. Bronzes are known for their colour changes while forming, and the formation is reversible, so a heat treatment in an inert atmosphere will reform the corresponding oxide. After testing no bronzes were observed. Reactions were performed in a fixed bed plug flow reactor, containing 1.0 g of catalyst. The temperature window applied was 350 to 550 °C and incremental molybdenum loadings were tested (3, 5, 10 and 15 wt%, based on atomic Mo). To a helium flow of 20 Nml/min containing 0.75 vol% of ethylene, pulses of ammonia were added, yielding 1.1 vol%. In this way flow conditions were mimicked. The reactor effluent was monitored on-line by a quadrupole mass spectrometer (Balzers QMG-420) which operated at ionisation potential of 70 eV and inlet pressure of $5.0 \cdot 10^{-6}$ mBar. The Balzers software program "Quadstar 420 v3.05" was used in the multiple ion detection mode, where 16 different masses could be adjusted. The masses m/e= 2, 17, 18, 27, 30, 40 and 44 were used for identifying hydrogen, ammonia, water, ethylene, ethane, acetonitrile and carbon dioxide, respectively. Organic compounds

were quantified by introducing a sample into a Hewlett Packard 5890 A gas liquid chromatograph equipped with a Poraplot Q column, 10 m * 0.32 mm i.d., and flame ionisation detector (FID). Besides on-line sampling, off-line analysis could be performed by storing 15 samples in a 16 loop valve. Inorganic compounds, such as CO_x, were quantified by introducing a sample into a Porapak N column, 6 ft * 3x2mm (e.d. x i.d.) 80/100 mesh equipped with a thermal conductivity detector (TCD). The GLC was connected to a Nelson interface and a computer for spectra assimilation.
While collecting data, three important regions of the reaction could be distinguished, which are schematically shown in Figure 1:
1) the first 2 hours of a semi-steady-state, which was reached within 10 minutes after starting the reaction. Although the acetonitrile production was constant over this 2 hour period, the selectivity to other products was changing.
2) a transition period where strong changes in conversion and selectivity occurred, dependent on the catalyst.
3) after at least 20 hours of reaction when a *real* steady-state was reached.
A quasi turnover frequency (qTOF) was defined as moles acetonitrile produced per mole molybdenum per hour.
At various reaction times, samples for XPS analysis were taken. This was done by shutting the reactor after cooling the catalyst in helium. Subsequently the reactor was introduced into a nitrogen-filled glove box, where the XPS sample was prepared. The catalyst was crushed and mounted on an iron stub carrying an indium film. The sample was placed in a vessel in a nitrogen atmosphere for transport to the spectrometer. XPS spectra were measured on a VG-ESCALAB equipped with a standard dual X-ray source, of which the Al-Kα part was used. Spectra were fitted with the VG analysis software.[15] Charging was corrected for by using the C *1s* peak at 284.5 eV. The peak areas of the Mo $3d_{5/2}$ signals were used for calculating the percentage of Mo(IV) and Mo(VI). The ratio of Mo *3d* and Al *2p* areas was used as a tracer for changes in the dispersion of the molybdenum phase on the alumina surface.

Results and Discussion

Characteristic regions of the activity profile. Activity measurements were performed on catalysts with various molybdenum loading after an oxidative or a reductive pretreatment. Monitoring more than 20 hours time on stream, an unusual profile of acetonitrile formation was observed which is schematically shown in Figure 1. Three regions of interest are distinguished: a semi-steady-state, a transition and a steady-state period.

The Semi-Steady-State Period. The semi-steady-state is characterized by the constant formation of acetonitrile, while the selectivity to side products is changing. Considering the yield of acetonitrile versus reaction temperature a temperature (T_{opt}) was found where a maximum yield of acetonitrile was obtained. This T_{opt} shifted to a lower value at increasing Mo content for both pretreatments

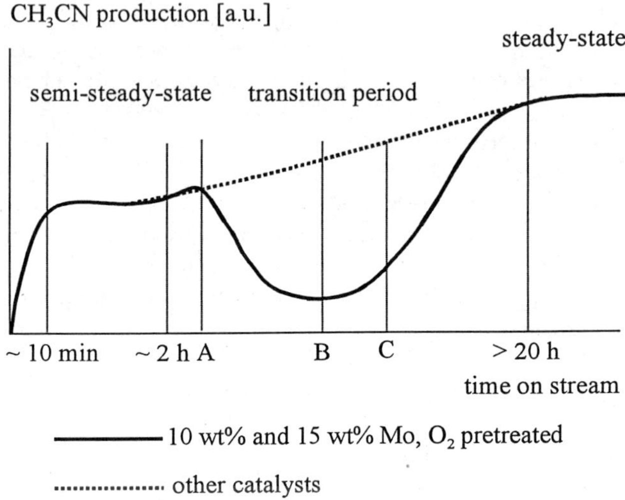

Figure 1: A schematic profile of the acetonitrile formation versus time on stream for various catalysts.

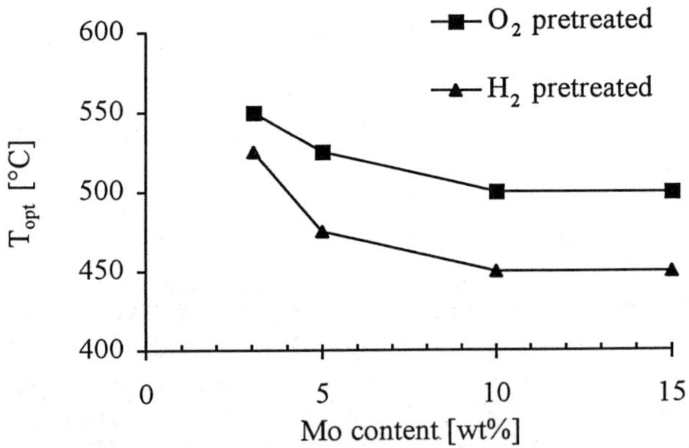

Figure 2: The temperature where maximum yield of acetonitrile is obtained versus molybdenum content for different pretreated catalysts at semi-steady-state condition (after 2 h time on stream).

(Figure 2). At temperatures exceeding T_{opt}, the complete dissociation of ammonia started to dominate, resulting in the desorption of nitrogen and thereby suppressing the desired reaction.

Figure 3 shows the ratio of the intensity of the XPS signals of Mo *3d* and Al *2p* for fresh catalysts and those which have been at various time on stream at 450 °C. Obviously, this ratio for the fresh catalysts increases with increasing molybdenum content (Figure 3; solid lines). Overall, this ratio of the oxygen pretreated catalysts is higher than the ones after a reductive treatment. Up to 5 wt% molybdenum, a similar increase is observed independent of the pretreatment applied. At higher molybdenum contents, however, the ratio for oxygen pretreated catalysts increased sharply compared to that of the hydrogen pretreated ones, which showed an almost linear behaviour. This observation is confirmed in the literature which states that molybdenum particles are better dispersed after calcination than after treating in hydrogen.[16]

So, with increasing molybdenum content the dispersion decreases, while hydrogen pretreatment induces an even stronger decrease. Considering that larger molybdenum particles are expected to be more active towards complete ammonia dissociation, it is explicable that this reaction starts to dominate at lower temperatures, shifting T_{opt} to a lower value with increasing molybdenum loading. It also follows that, for hydrogen pretreated catalysts T_{opt} is lower for the whole range considered (Figure 2; solid triangles).

Considering the optimal quasi turnover frequency ($qTOF_{opt}$) versus molybdenum content, the oxygen pretreated catalysts were less strongly affected by variation of the molybdenum content, and decreased slightly at molybdenum contents exceeding 10 wt% (Figure 4; solid squares). This indicates that the molybdenum added is well-spread on the alumina surface and accessible for reaction. Seemingly, the influence of dispersion effects starts to dominate at higher molybdenum content. This trend is also illustrated earlier in Figure 3.

The hydrogen pretreated catalysts with low molybdenum content initially had the best performance among all catalysts tested. At higher molybdenum content the $qTOF_{opt}$ decreased sharply (Figure 4; solid triangles), due to the lower dispersion of the hydrogen pretreated catalysts.

In summary, the semi-steady-state activity is highly structure sensitive. When oxygen pretreated, a higher yield of acetonitrile is obtained at higher molybdenum content (lower dispersion), whereas for hydrogen pretreated catalysts this concerns a low molybdenum content (high dispersion). This causes in Figure 4 the point of intersection at approximately 7 wt% molybdenum loading. Apparently the complete dissociation of ammonia, resulting in the desorption of nitrogen, is favoured on reduced larger molybdenum particles.

Transition Period. During the transition period strong fluctuations were observed in both activity and selectivity. All hydrogen pretreated catalysts and the oxygen pretreated catalysts with 5 wt% Mo or less showed a gradual increase in the formation of acetonitrile, until an upper limit was reached (Figure 1; dotted line). During this increase, the formation of ethane and hydrogen increased, while the

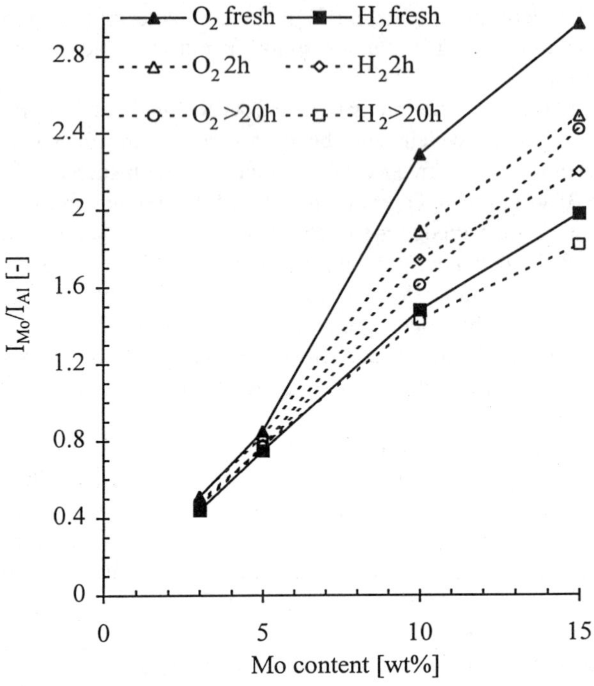

Figure 3: Ratios of the intensity of the XPS signals of Mo *3d* and Al *2p* of different catalysts at various time on stream.

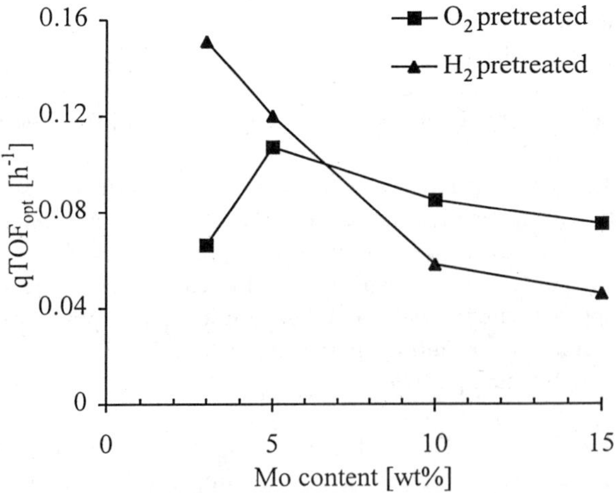

Figure 4: The maximum quasi turnover frequency versus molybdenum content for different pretreated catalysts in the semi-steady-state (after 2 h time on stream).

formation of CO_x and water decreased, resulting in an overall increase of the activity.

Only with the 10 wt% and 15 wt% Mo catalysts pretreated in oxygen, the yield of acetonitrile passed through a minimum (Figure 1; solid line). Although a strong increase in activity is observed, the selectivity to acetonitrile drops sharply, resulting initially in an overall slight increase of the yield (Figure 1; A). During the period of low acetonitrile production the conversion of ethylene and ammonia increased to almost 100%, while products were methane, ethane, nitrogen, hydrogen and carbon monoxide (Figure 1; B). Despite the fact that the conversion slowly decreased, the formation of acetonitrile slowly increased. This was due to the reduction of methane, nitrogen and carbon monoxide formation (Figure 1; C).

XPS analysis of different stages of reaction of the 10 wt% Mo catalyst showed clearly an increase of the Mo(IV) fraction (Figure 5; black bars). A fresh catalyst, containing 100% Mo(VI), was not very active, but highly selective to acetonitrile (Figure 5; after ~ 1.3 h). Before the yield started to decrease, little reduction of molybdenum was observed with XPS (after ~ 2.5 h). Evidently, due to the higher molybdenum content, the presence of a large amount of still removable surface oxygen in a partly reduced catalyst (due to reaction) leads to a very high activity, but only towards the side reactions. These side reactions are terminated at depletion of removable lattice oxygen. Apart from the fact that the start of the decrease of the yield of acetonitrile occurred several hours later, the oxygen pretreated 15 wt% Mo catalyst reacted similar to its 10 wt% analogue: A, B and C in Figure 5 represent 6, 8.5 and 11.5 hours of reaction respectively.

Due to the lower Mo content and to the poor reducibility of smaller particles, these catalysts contain less removable lattice oxygen. Apparently, the reductive treatment of the 10 wt% and 15 wt% Mo catalysts has removed enough lattice oxygen to suppress a larger part of the side reactions, resulting in also a gradual increase of the formation of acetonitrile, similar to the catalysts with lower molybdenum content.

Steady-State Period. Compared to the semi-steady-state, significant increases in catalytic activity were observed over at least 20 hours time on stream at 450 °C. No oxygen containing products were detected in the reaction stream and the product distribution was constant, so a *real* steady-state was reached. This is different from the ammoxidation of propylene. Aykan [17] observed a fast decline in catalytic activity with increasing reduction of the catalyst in the absence of gaseous oxygen. When using different ratios of bismuth and molybdenum, the 30% MoO_3 on silica appeared to be a very poor catalyst. Experiments where acrylonitrile was introduced to the catalyst showed that a low activity to decomposition of this desired product gave a high selectivity to acrylonitrile. Amorphous SiO_2 and MoO_3 were found to be inert with respect to acrylonitrile. Only after complete reduction to MoO_2 the catalyst showed initially a high activity towards conversion of acrylonitrile to CO_2, acetonitrile and propionitrile which decreased very rapidly due to large amounts of carbon deposition.

The point where steady-state was reached coincided with the absence of oxygen containing components in the product stream. Comparing the qTOF of the semi-

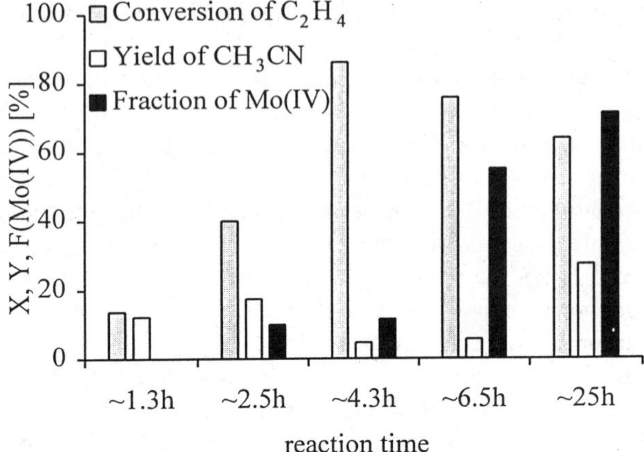

Figure 5: Reactivity profile of the oxygen pretreated 10 wt% Mo catalyst with corresponding percentage Mo(IV) obtained by XPS.

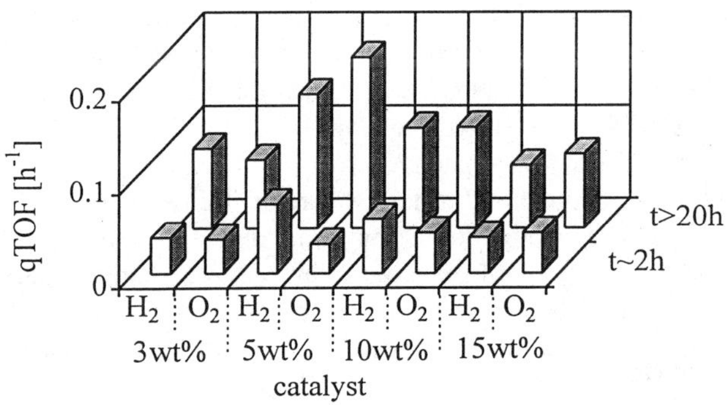

Figure 6a: The quasi turnover frequency at semi-steady-state and steady-state conditions at 450 °C of various catalysts.

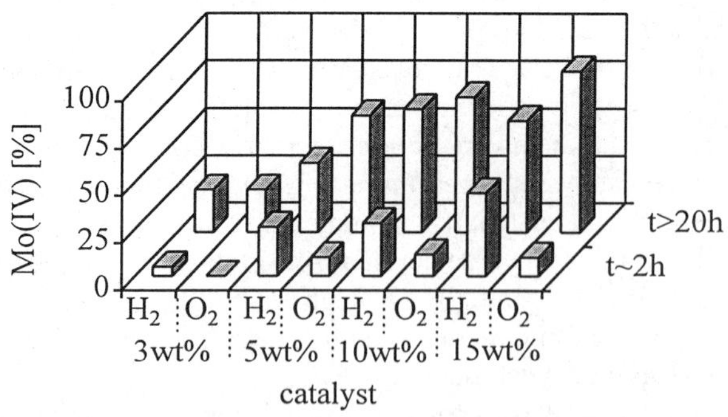

Figure 6b: The area percentage of the $3d_{5/2}$ XPS signal of Mo(IV) at ~228.5-229.5 eV at semi-steady-state and steady-state conditions at 450 °C of various catalysts.

steady-state and the steady-state, the largest increase is observed when the catalysts were pretreated with oxygen (Figure 6a). Overall, when reaching the steady-state, the oxygen pretreated catalysts reacted similarly to the hydrogen pretreated ones, considering the qTOF and product distribution.

Samples taken at various time on stream showed different ratios of Mo(IV)/Mo(VI) in XPS. Figure 6b depicts the percentage of Mo(IV) in the period of semi-steady-state and in the steady-state of the reaction for the different pretreated catalysts with various molybdenum content. Again, it can be seen that large molybdenum particles are easier to reduce, because the amount of Mo(IV) increased with increasing loading. But it appeared that the hydrogen pretreatment did not remove all reactive oxygen, since in all cases the amount of Mo(IV) had increased after steady-state was reached. The increase of the yield of acetonitrile can not be attributed only to the increase of the amount of Mo(IV). Although the oxygen pretreated 5 wt% Mo catalyst yielded the highest qTOF, the largest amount of Mo(IV) was observed on the oxygen pretreated 15 wt% Mo catalyst. Evidently, a well-dispersed calcined precursor is needed to obtain the best performance in the steady-state, where both Mo(IV) and Mo(VI) species are present. At higher molybdenum loading the activity in the steady-state becomes independent of the pretreatment. This suggests the formation of a stable molybdenum structure during reaction. Fresh, hydrogen pretreated, catalysts with higher loading showed the Mo $3d_{5/2}$ peak at 228.5 eV. After a reaction period until steady-state was reached this peak shifted to 229.5 eV. Catalysts with lower loading showed already this Mo $3d_{5/2}$ peak at 229.5 eV after hydrogen pretreatment, suggesting a MoO_2-like structure.(*15*) An XRD spectrum of the hydrogen pretreated 15 wt% Mo catalyst which had reached the steady-state showed clearly the presence of crystalline MoO_2. This in contrast with the fresh hydrogen pretreated catalyst, which showed no crystalline phase.

When submitted to reaction, an increase of the C *1s* peak was observed suggesting that coke formation occurred. Since the catalytic activity showed no indications of deactivation, this carbon could be removed by the reaction stream, or even act as a reaction intermediate.

Proposed reaction mechanisms. It is expected that the formation of the main side-products is a strong function of the pretreatment. Water and CO_x were dominant on oxygen pretreated catalysts. Because of the formation of water, the formation of acetonitrile on an oxidic catalyst surface is favoured thermodynamically. Hydrogen pretreated catalysts formed mainly ethane and hydrogen. It should be remembered that the formation of acetonitrile and hydrogen directly from ethylene and ammonia is thermodynamically unfavourable (ΔG^0(427 °C) ~ 10 kJ/mol). This in contrast with the hydrogenation of ethylene: ΔG^0(427 °C) ~ -50 kJ/mol. Since ethane is always the main side product for hydrogen pretreated catalysts and catalysts in the steady-state period (~50% selectivity), this exothermic reaction probably delivers the energy needed to drive the reaction. The following two mechanisms are proposed (Scheme 1):

It is obvious that deactivation occurs in mechanism (1), because the catalyst acts as a reactant. Apart from the fact that lattice oxygen is not replenished, this mechanism is

O_2 pretreated (1)

$$C_2H_4 + NH_3 + MoO \longrightarrow CH_3CN + H_2O + Mo\square$$

H_2 pretreated (2)

$$2C_2H_4 + NH_3 \xrightarrow{\text{Mo catalyst}} CH_3CN + H_2 + C_2H_6$$

Scheme 1

similar to the ammoxidation mechanism proposed by Grasselli *et al.*(*14*). In mechanism (2) no deactivation is expected due to the fact that the catalyst is not altered by reaction. It is therefore suggested that the two mechanisms occur on different sites formed by different molybdenum structures.

Since conditions were found where besides acetonitrile, both oxygen containing products (water, CO_x) and non-oxygen containing products (ethane, hydrogen) were formed, it is concluded that the two mechanisms can occur simultaneously. The semi-steady-state condition where this happened showed a strong structure sensitivity.

In summary, it appears that during reaction a calcined catalyst, where mechanism (1) is dominant, changes into a catalyst where mechanism (2) dominates after removal of lattice oxygen. This transition proceeds gradually for molybdenum contents of 5 wt% or less. At higher molybdenum contents only after reductive treatment a gradual transition is observed. When calcined, the transition proceeds via a minimum in the formation of acetonitrile, due to a highly active period where side reactions dominate. Surface oxygen is removed by reaction until depletion, reaching a steady-state. This way the molybdenum phase is converted into a particular stable structure, acting as a real catalyst. The exact nature of this structure is suggested to be MoO_2-like, and is independent of the pretreatment applied at higher molybdenum content.

Conclusions

The oxygen pretreated catalysts have initially a high selectivity to acetonitrile and water with total combustion as side reaction. At higher molybdenum content, the yield of acetonitrile passes through a minimum, while the activity goes through a maximum. A hydrogen pretreated catalyst exhibits a gradual increase in activity and is less selective. The major products, besides acetonitrile, are hydrogen and ethane.

Two mechanisms are suggested:

1) one based on the ammoxidation mechanism with the consumption of lattice oxygen.

2) the other is based on oxidative ammonolysis, i.e. without consumption of lattice oxygen. For thermodynamic considerations a synergetic formation of ethane and acetonitrile is suggested.

The mechanisms can occur separately or simultaneously, depending on the amount of removable lattice oxygen. XPS showed only Mo(IV) and Mo(VI) species. The fraction of Mo(IV) increases with increasing Mo content for both pretreatments, due to the reducibility of larger Mo particles. Freshly calcined catalysts are highly selective, attributed to Mo(VI) sites present on a MoO_3-like structure. Catalysts in the steady-state, containing reduced molybdenum, are highly active, but less selective. This is due to Mo(IV) sites, which enhance more effective N-H and C-H activation. For this reason the selectivity is also a strong function of the dispersion. Large, reduced molybdenum particles are too active towards complete ammonia dissociation, resulting in termination of the selective reaction. Although, due to the reaction the catalyst becomes more reduced, a large amount of removable lattice oxygen still present at higher molybdenum loading is very active, resulting in formation of methane, ethane, nitrogen, hydrogen and CO_x.

Finally, in the steady-state a reduced molybdenum structure is obtained which is not altered by reaction anymore, and acts as a real catalyst. This MoO_2-like structure can only be formed by the removal of lattice oxygen under reaction conditions.

Literature Cited

(1) *Encyclopaedia of Chemical Technology*, Nitriles, 3rd Ed, **1978**, Vol. 15, pp 888-909.

(2) *Ullmann's Encyclopaedia of Industrial Chemistry*, Nitriles, **1991**, 5th Ed, Vol. A17, pp 363-376.

(3) Xu, B.-Q., Yamagushi, T., Tanabe, K. *Chem. Lett.* **1987**, pp 1053.

(4) Xu, B.-Q., Yamagushi, T., Tanabe, K. *Chem. Lett.* **1988**, pp 281.

(5) Reddy, B.M., Manohar, B. *J. Chem. Soc.-Chem. Comm.* **1993**, pp 234.

(6) Olivé, G., Olivé, S. *US Patent 4 179 462* **1979**.

(7) Catani, R., Centi, G. *J. Chem. Soc.-Chem. Comm.* **1991**, pp 1081.

(8) Gambell, J.W., Auvil, S.R. *US Patent 4 272 451* **1981**.

(9) Auvil, S.R., Penquite, C.R. *US Patent 4 727 452* **1981**.

(10) Hamada, H., Kawuhara, Y., Matsuno, Y., Wakabayashi, K. *Sekiyu Gakkaishi* **1987**, *30 (3)*, pp 188.

(11) Tatsumi, T., Kunitomi, S., Yoshiwara, J., Maramutsu, A., Tominaga, H. *Cat. Lett.* **1989**, pp 223.

(12) Hummel, A.A., Bandani, M.V., Hummel, K.E., Del gass, W.N. *J. Catal.* **1993**, *139*, pp 392.

(13) Eshelman, L.M., Delgass, W.N. *Catal. Today* **1994**, *21*, pp 229.

(14) Grassellli, R.K., Burrington, J.D. *Adv. Catal.* **1987**, *87*, pp 363.

(15) De Jong, A.M., Borg, H.J.,Van IJzendoorn, L.J., Soudant, V.G.F.M., De Beer, V.H.J., Van Veen, J.A.R., Niemantsverdriet, J.W. *J. Phys. Chem.* **1993**, *97*, pp 6477.

(16) Grünert, W., Stakheev, A.Y., Mörke, W., Feldkraus, R., Anders, K., Shpiro, E.S., Minachev, K.M. *J. Catal.* **1992**, *135*, pp 269.

(17) Aykan, K. *J. Catal.* **1968**, *12*, pp 281.

Chapter 24

Catalytic Cooperation via Spillover of Oxygen: Dehydration–Dehydrogenation of 2-Butanol over SnO_2–MoO_3 Catalysts

E. M. Gaigneaux[1], D. Herla, P. Tsiakaras[2], U. Roland[3], P. Ruiz, and B. Delmon

Unité de Catalyse et Chimie des Matériaux Divisés, Université Catholique de Louvain, Place Croix du Sud 2/17, B–1348 Louvain-la-Neuve, Belgium

Strong synergetic phenomena in the dehydration-dehydrogenation of sec-butyl alcohol were observed at very low temperature using mechanical mixtures of separately prepared SnO_2 and MoO_3. The principal effects were an increase in conversion as well as increases in yields and selectivities for butene and methyl-ethyl ketone. Characterization of the solids before and after catalytic tests by X-ray Photoelectron Spectroscopy coupled with Transmission Electron Microscopy, Raman Spectroscopy and X-ray Diffraction analysis excluded the possibility of an *in situ* formation of a mixed oxide or mutual contamination. Attempts for preparing true mixed Sn/Mo oxides have been unsuccessful as predicted from the literature, and catalysts prepared by incipient decomposition of homogeneous precursor, which should exhibit maximum contamination, have exhibited much lower activities than the corresponding mechanical mixtures. Taking into account these and previous results with SnO_2 + Sb_2O_4 and MoO_3 + Sb_2O_4 mixtures, the origin of the observed synergy can be explained by a "Remote Control Mechanism" : SnO_2 acts as a donor of spillover oxygen species (Oso) and MoO_3 as an acceptor of Oso. The role of Oso is to create selective sites on MoO_3 and to protect them from deactivation.

When synergetic effects are observed between two oxide phases in a selective oxidation reaction, the existence of a mixed phase or its *in situ* formation during the catalytic test is a common way to explain the observed improvement of

[1]Please see Acknowledgments, page 345.
[2]On leave from Department of Chemical Engineering, University of Thessaloniki, Greece
[3]On leave from University of Dresden, Germany

0097–6156/96/0638–0330$15.00/0

the catalytic performances compared to those observed with the pure oxides. It has been argued that, even if present in very small amounts in the outer surface of the catalyst, the supposed mixed phase could present a higher activity than the pure oxides *(1)*.

Nevertheless, several other hypotheses have been proposed in order to explain these synergetic effects. Among those, a promising one is the "Remote Control Mechanism" (RCM) concept. Remote Control involves the activation of molecular oxygen on one of the two oxides ("donor"), to form a mobile oxygen species, namely "spillover oxygen" (O_{SO}). This migrates onto the surface of the other phase ("acceptor") where it creates and/or regenerates selective active sites *(2-4)*. As a result, the performances of the biphasic catalyst are improved, presenting higher yields and selectivities for partial oxidation products, extended lifetime and enhanced resistance to deactivation by coke deposition *(5)*. All those changes are the consequence of the improvement of the surface properties of the acceptor due to the action of the donor through the spillover species.

The present work corresponds to studies of synergetic effects observed at *low temperatures*. In these conditions, the formation of new phases and mutual contamination between the oxides are minimized and the synergetic effects are consequently more easily explained. We have already presented the results of some studies dealing with the dehydration-dehydrogenation of 2-butanol to produce methyl-ethyl-ketone and butene, over Sb_2O_4-MoO_3 and Sb_2O_4-SnO_2 biphasic catalysts. The catalysts were prepared by mechanical mixtures of the oxides prepared separately. The reaction temperature was 190°C and 240°C. Strong synergetic effects were observed. The absence of any indication of the existence of a new contaminated phase led to the conclusion that the RCM was playing an important role in the reaction : Sb_2O_4 acts as "O_{SO} donor", thus improving the activities of the other oxides, which were "O_{SO} acceptors" *(6)*.

Okamoto et al obtained strange results in the dehydration-dehydrogenation of 2-butanol on SnO_2 - MoO_3 catalysts. It was observed a change in the selectivities for butene and methyl-ethyl-ketone with the composition (Mo/Mo+Sn) in the surface of the bimetallic catalysts *(7)*. These catalysts were co-precipitated, but no oxide containing Mo and Sn in the same structure was mentioned. The origin of the selectivity change was thus difficult to explain.

In the present work, we investigate the same reaction, but the catalytic system chosen consisted of mechanical mixtures of SnO_2 and MoO_3.

As both oxides have been previously reported to exhibit an “acceptor” behaviour *(2-6)*, it was interesting to see if, even in this case, the RCM could operate and, if so, how it would operate. On the other hand, SnO_2 has already been reported to be able to act as “O_{SO} donor”, nonetheless weaker than Sb_2O_4 *(3-4)*.

Much attention has been paid in this work to the detection of mixed phase, supposing that this could be formed during the preparation of the catalysts or during the catalytic reaction. For this, we used several characterization techniques, either bulk or surface sensitive.

Very few mentions are made in the literature concerning the existence of mixed oxide phases containing Sn and Mo*(8,9)*. Among these, only the compound SnO_2.2 MoO_3 hse been the object of a deep crystallographic study. This compound is reported to be unstable below 500 °C, decomposing into the pure Mo and Sn oxides *(9)*. This is in complete agreement with several papers dealing with the SnO_2-MoO_3 system. Solid solutions of Mo in SnO_2 can be obtained using different preparation procedures, but no new bimetallic oxide phases were mentioned *(10-12)*. In the present investigation, we attempted to prepare some tin-molybdenum mixed phases corresponding to the Sn:Mo = 1:2 composition using the citrate method *(13)*. This method has been shown to be particularly efficient homogeneously interdispersing several metals in the same phase. The preparation starts from a homogeneous bimetallic precursor obtained by simultaneous gelification. The objective of this part of the work was to evaluate the possible role of the thus prepared phases, supposed to be as much "contaminated" by the other elements as possible in the activities observed for the corresponding mechanical mixtures.

In the same context some catalytic tests have also been performed during long times of reaction (10 times longer than usual) in order to check the stability of the catalysts. Much attention was especially placed on the possibilities of interdiffusion of one metal into the phases containing the other metal (formation of solid solution).

Experimental

Catalyst Preparation

Pure oxides

Both pure oxide phases were prepared separately. SnO_2 was obtained by precipitation of an acidic $SnCl_2$ aqueous solution with ammonia. The chloride anions were washed out of the resulting tin hydroxide using distilled water. The obtained solid was dried at 110 °C for 20 hours and calcined at 600 °C for 20 hours and at 900 °C for 16 hours.

MoO_3 was synthesized starting from an aqueous solution of ammonium heptamolybdate complexed with an adequate quantity of oxalic acid. The mixture was stirred at 40 °C until a homogeneous solution was obtained. After having removed the solvent under vacuum, the obtained solid was dried overnight at 80 °C, decomposed at 300 °C for 20 hours and calcined at 400 °C for 20 hours.

Both phases were shown by X-ray Diffraction (XRD) to be respectively cassiterite and molybdite *(14)*. Specific surface area measurements (SBET) gave values of 3.7 m^2/g and 7.4 m^2/g, for SnO_2 and MoO_3 respectively.

Mechanical mixtures

Mechanical mixtures (MM) of different compositions (R_m = 0.05, 0.25, 0.5, 0.75 - see Equation [1]) were prepared by vigourously mixing a suspension of adequate amounts of both finely ground pure oxides in n-pentane. After removing the solvent under vacuum at room temperature, the catalyst was gently dried overnight at 80 °C.

$$R_m = \frac{\text{mass } MoO_3}{\text{mass catalyst}} \quad [1]$$

Before being tested, the pure oxide phases, SnO_2 (R_m=0) and MoO_3 (R_m=1), were treated according to exactly the same procedure as the mechanical mixtures. All the catalysts were prepared starting from the same batches of pure oxides.

Attempt to prepare Sn/Mo mixed oxide phase, and artificially "possibly contaminated catalysts" - Citrate method

The attempt to prepare a Sn/Mo mixed oxide phase with a composition Sn:Mo = 1:2 was performed using a variant of the citrate method*(13)*. A quantity of ammonium heptamolybdate tetrahydrate was complexed with an equivalent amount of citric acid monohydrate in the smallest possible volume of water. The same procedure was achieved with $SnCl_2$. $2H_2O$. Adequate volumes of these solutions were mixed together in order to reach the desired Sn:Mo ratio. The resulting dark blue solution was thereafter stirred overnight at room temperature, and then concentrated under vacuum at 30 °C. The obtained viscous residue was maintained at 110 °C during 20 hours at 50 mbar, so forming a solid cake, hereafter called "precursor". This was manually finely ground and separated into three parts. Each part was respectively calcined during 8 hours at 450 °C (Sample I), 550 °C (Sample II) and 700 °C (Sample III) and washed with distilled water in order to remove chloride anions.

Figure 1 shows the XRD patterns obtained for the 3 samples. None of the samples exhibited any of the peaks distinguishing the SnO_2.2 MoO_3 phase from the pure oxides *(14)*. On the other hand, the sample III exhibited all the peaks corresponding to a mechanical mixture of SnO_2 + MoO_3. The intensities of the peaks were those of the mechanical mixture with a composition Sn:Mo = 1:2 (corresponding to R_m=0.66). The same peaks were present on the pattern of sample II but with lower intensities. Sample I only exhibited the features of SnO_2 with low intensities.

Table I presents the specific areas measured for samples I to III.

Table I. Specific surface areas values measured for the mixed Sn/Mo precursors calcined at 450 °C, 550 °C and 700 °C

Sample	Specific area
Sample I (450 °C)	3.3 m^2/g
Sample II (550 °C)	24.2 m^2/g
Sample III (700 °C)	5.0 m^2/g

Considering these results, it is concluded that, as predicted from the literature, the true mixed oxide phase cannot be formed at low temperature *(9-12)*. The calcination of the mixed Sn/Mo precursor triggers the independant crystallisation of pure SnO_2 and MoO_3. The SBET of sample III has a value similar to that of the corresponding mechanical mixture (R_m=0.66, SBET = 6.0 m^2/g), meaning that the sizes of the crystallites are similar in the two samples. In sample I, SnO_2 has just started to form small crystallites, while sample II is mainly constituted of MoO_3 crystallites and SnO_2 bigger crystals. The fact that the MoO_3 crystallites are small explains the high SBET values measured for these samples as well as the low XRD intensities.

Starting from the precursor, which associates homogeneously Sn and Mo, and considering the series of samples calcined at increasingly higher temperatures, this undergoes a progressive decontamination of each oxide (SnO_2 and MoO_3) from the foreign cation (Mo and Sn, respectively). The decontamination of the system SnO_2-$MoO_3$was already reported in papers dealing with "impregnated catalysts" *(15)*. However, the possibility that the samples I to III could be, at least partially, constituted of a solid solution of Mo in the SnO_2, as suggested in several papers *(10-12)*, could not be excluded. In this case, it is admitted that the amount of solid solution in the samples would decrease with the temperature of calcination.

Samples I to III must thus be considered provisionally as constituted of SnO_2 and MoO_3 oxides containing as much contamination as possible.

Catalytic Activity Measurement.

Catalytic tests were performed in a fixed bed reactor. The partial pressure of 2-butanol was 176 mmHg in a 90 ml/min flow of air. We used 500 mg of catalyst for each test, with a granulometry between 500 and 800 μm. Preliminary experiments in the test conditions showed that no diffusion limitation occurred. Standard tests were run at 190 °C during 3 hours of steady state. Some additional reactions were carried out at 150 °C with different batches of catalysts. The amount

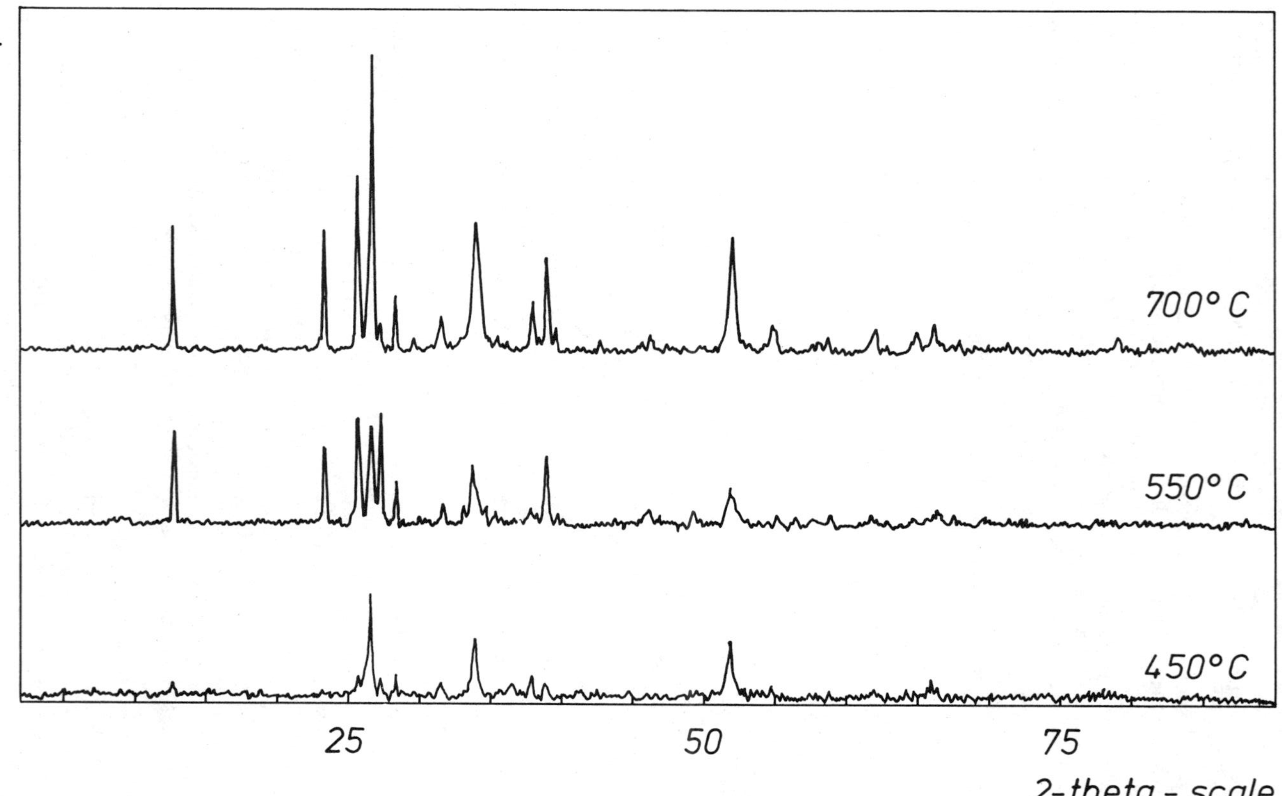

Figure 1. XRD patterns of mixed Sn/Mo phase precusor calcined at 450 °C, 550 °C and 700 °C.

of unreacted 2-butanol and the products of the reaction, methyl-ethyl-ketone (MEK) and butene (But) were measured at the reactor outlet by on-line gas chromatography.

Catalytic activity was expressed in terms of % conversion of 2-butanol (%C = number of moles of 2-butanol converted per 100 moles of 2-butanol introduced), % yields ($\%Y_{MEK}$ and $\% Y_{But}$ = number of moles of MEK or But produced per 100 moles of 2-butanol introduced) and % selectivities ($\%S_{MEK}$ and $\%S_{But}$ = number of moles of MEK or But produced per 100 moles of 2-butanol converted).

For each mechanical mixture, theoretical values for these expressions have been calculated on the basis of the properly averaged sum of the activities measured for the pure oxides, as described in equations 2. These theoretical values are representative of the activity of a biphasic catalysts where the constituting phases perform the catalytic reaction without mutual interaction (namely, as if they were alone in the reactor) assuming, as a first approximation, zero-order reactions.

Theoretical Conversion for the MM (composition = R_m) = $\%C_{th}^{R_m}$

$$R_m * \text{Conversion of pure } MoO_3 + (1 - R_m) * \text{Conversion of the pure } SnO_2 \quad [2a]$$

Theoretical Yield (But or MEK) for the MM (composition = R_m) = $\%Y_{th}^{R_m}$

$$R_m * \text{Yield of pure } MoO_3 + (1 - R_m) * \text{Yield of the pure } SnO_2 \quad [2b]$$

Theoretical Selectivity (But or MEK) for the MM (composition = R_m) = $\%S_{th}^{R_m}$

$$\%Y_{th}^{R_m} / \%C_{th}^{R_m} \quad [2c]$$

No further normalisation of the observed activity with the specific areas, or with the number of surface exposed atoms of the active phases on the particles of the catalysts was made. As all the mechanical mixtures were prepared from the same batches of pure oxides, these values are directly correlated with the mass of each phases in the catalysts.

Two additional tests have been performed with pure MoO_3 and the mechanical mixture with R_m=0.25 during 30 hours in the same conditions of reaction as described before.

Characterization.

Before and after the tests at 190 °C, the mechanical mixtures were characterized by Transmission Electron Microscopy (TEM), X-ray Diffraction (XRD), BET specific surface area measurement (SBET), Photoelectron Spectroscopy (XPS) and Raman Spectroscopy.

TEM investigations were performed on a JEOL microscope - Model TEMSCAN 100C. The accelerating voltage was fixed at 100kV. The crystallites of each oxide present in the mixtures were identified using Energy Dispersive X-ray Spectroscopy (EDS) with a Kevex X-ray Energy Spectrometer 5100 interface.

XRD was performed on a Kristalloflex Siemens D5000 diffractometer using the $K_{\alpha 1,2}$ radiation of Cu (λ=1.5418 Å) between 2θ angles from 2° to 90°.

SBET was measured on a Micromeritics Flowsorb II taking the single point approximation of the BET equation for the adsorption of N_2 at 77 K. A theoretical value of the SBET is calculated for the mechanical mixtures on the basis of a linear relation with the masses of each pure oxides in the mixture.

XPS analysis was carried out on a VG MKII spectrometer with a Mg anode (Mg K_{α}=1253.6 eV). Contamination carbon C_{1s} peak, taken as a reference for the measurement of kinetic energies, was set at 284.8 eV. Particular attention was placed on the C_{1s}, $Mo_{3d5/2}$, $Sn_{3d5/2}$ and O_{1s}. The different elements were quantified using the Wagner sensitivity factors.

A Bruker RFS100 spectrometer was used for Raman analysis. The power of the Nd-YAG excitation laser was 15 mW in order to avoid thermal radiation, fluorescence, and thermal degradation of the samples. Under these conditions, previous analysis of impregnated catalysts (0.5 to 2 monolayers of SnO_2 on MoO_3 supports) had shown Raman spectroscopy to be a very sensitive technique *(16)*.

Results

Catalytic Results.

Mechanical mixtures

Table II summarizes the results of the catalytic tests performed at 190 °C. MoO_3 is highly active, producing principally butene, although it also produces some MEK. SnO_2 is nearly inactive. It produces only small amounts of MEK, probably with a high selectivity (mentioned as nearly 100 % in the Table). Strong synergetic effects were observed for the different mechanical mixtures, especially when considering the % conversion of 2-butanol and the % yields for MEK and butene. For high content of MoO_3, an increase in the selectivity of MEK and butene was also observed.

Table II. Observed and theoretical values (the bold figures in parentheses were calculated assuming no synergy) of the % conversion of 2-butanol (%C), % yields (%Y_{MEK}, %Y_{But}) and % selectivites (%S_{MEK}, %S_{But}) for MEK and butene, for the different mechanical mixtures at 190 °C. Due to the low activity, selectivity of pure SnO_2 is an approximation (~ 100%).

Rm	%C	%Y_{MEK}	%Y_{But}	%S_{MEK}	%S_{But}
0 (SnO_2)	~1.0	~1.0	0	~100	-
0.05	24.3	5.7	5.8	25	23.9
	(3.3)	**(1.0)**	**(0.8)**	**(30.3)**	**(24.2)**
0.25	32.1	3.9	10.5	12.6	32.6
	(12.7)	**(1.1)**	**(4.2)**	**(8.7)**	**(33.1)**
0.5	40.9	3.4	15.1	8.7	37.5
	(24.45)	**(1.25)**	**(8.4)**	**(5.1)**	**(34.4)**
0.75	44.4	2.4	15.9	5.5	36.2
	(36.2)	**(1.4)**	**(12.6)**	**(3.9)**	**(34.8)**
1 (MoO_3)	47.9	1.5	16.8	3.1	35

Although presenting a lower activity (conversion of about 10%), the catalysts tested at 150 °C exhibited even stronger cooperative effects.

For all the catalysts (excepted pure SnO_2), an induction time of approximately 1 hour was observed. For the pure MoO_3, decreases of the conversion (about 10%) and yield in butene (about 5%) in time were observed. The yield in MEK remained constant. For the mechanical mixtures, increases of the conversion (about 15%) and of the yield in butene (about 7%) were noted. After the induction period, the activities remained constant even after 30 hours of reaction.

Samples prepared by citrate method

Table III presents the catalytic results obtained with the samples prepared by the citrate method (Samples I to III).

Table III. Observed values of the % conversion of 2-butanol (%C), % yields (%Y_{MEK}, %Y_{But}) and % selectivites (%S_{MEK}, %S_{But}) for MEK and butene, for the samples prepared y the citrate method at 190 °C. The activities measured for the mechanical mixtures with similar Sn:Mo ratios are also presented.

Sample	%C	%Y_{MEK}	%Y_{But}	%S_{MEK}	%S_{But}
Sample I (450 °C)	15.2	3.7	3.6	24.7	23.8
Sample II (550 °C)	73.8	6.7	24.5	9.1	33.3
Sample III (700 °C)	43.9	4.4	16	10.2	36.4
MM R_m=0.5	40.9	3.4	15.1	8.7	37.5
MM R_m=0.75	44.4	2.4	15.9	5.5	36.2

Sample I exhibited low conversion and yields both for MEK and butene. Sample II presented the highest conversion and yields, while sample III showed intermediates values. Conversely the selectivities for both butene and MEK were the highest of the series. Samples II and III exhibited very similar selectivities.

In comparison with the mechanical mixtures with Sn:Mo ratios similar to those of the samples prepared by the citrate method (Sn:Mo = 1:2, i.e. R_m=0.66), sample I exhibited much lower conversion of 2-butanol, and lower yield and selectivity in butene. On the other hand, the selectivity in MEK was much higher. Conversion and yields of sample II were higher than the ones of the corresponding MM. The selectivity in MEK was also a slightly higher, but the selectivity in butene remained in the same range. The activity of sample III was similar to the ones of the MM.

Characterization Results

Transmission Electron Microscopy

Figures 2 a, b, c and d shows micrographs of, respectively, pure SnO_2 and pure MoO_3 (both having been submitted to the same treatment as those used for preparing mechanical mixtures), the mechanical mixture with R_m=0.05 and the mechanical mixture with R_m= 0.25. Pure tin oxide particles have a spherical shape with an average diameter of 50 nm. Pure molybdenum oxide particles exhibited the typical hexagonal prismatic morphology. The size of the crystallites were in the range 200 nm and 500 nm. In the mechanical mixtures, the shapes and sizes of the crystallites of both phases remained unchanged, showing that no degrading abrasive effect of one oxide on the other occured during the mixture. No autoaggregation of the cristallites was observed in any sample. Conversely, no deflocculation phenomena were observed in the mechanical mixtures when compared with the pure phases.

The characterization of the catalysts after the tests did not reveal any changes when comparing with the corresponding samples before the reaction.

X-ray diffraction

X-ray diffraction patterns of the pure oxide phases submitted to the same treatment as that used for preparing the mechanical mixtures exhibited exactly the same features as the non treated oxides. Perfect fits were obtained with the ASTM patterns of cassiterite for SnO_2, and molybdite for MoO_3. The patterns of the mechanical mixtures presented, in proportions corresponding to the composition, all

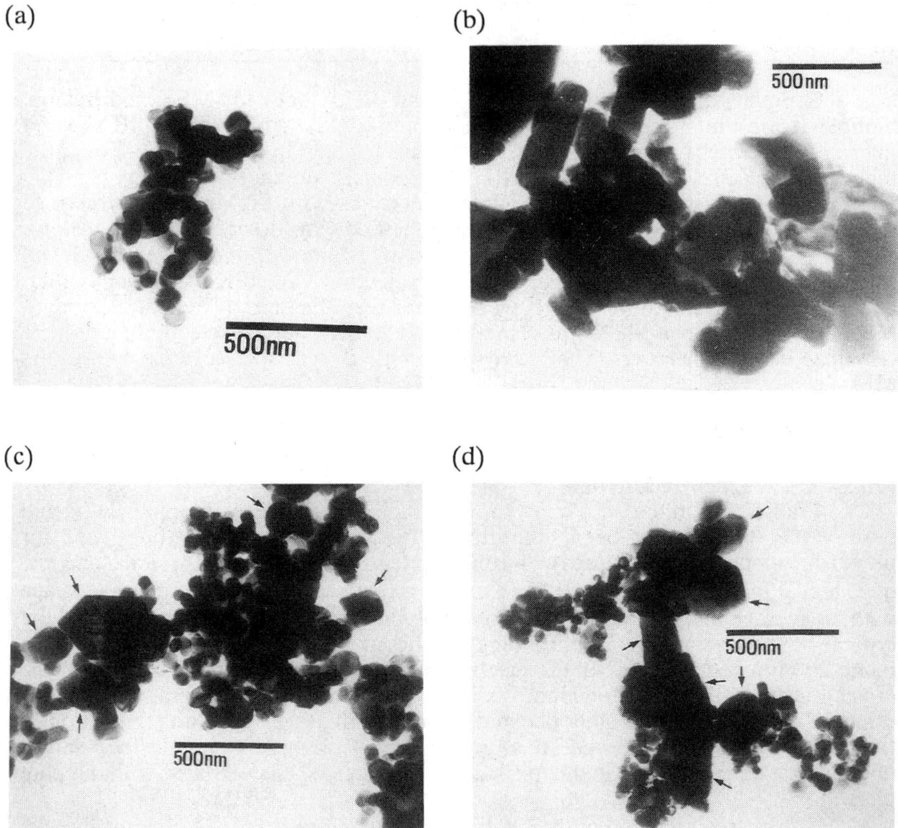

Figure 2. TEM micrographs of pure tin oxide (a), pure molybdenum oxide (b) after mechanical mixture procedure, mechanical mixtures with R_m=0.05 (c) and R_m=0.25 (d) before catalytic tests. (Arrows indicates some MoO_3 crystallites)

the features of both oxides without any shift or disappearance of peaks, nor appearance of new peaks. After the tests (3 hours and 30 hours), th e patterns of the used catalysts were absolutely identical to the ones before the tests. Figure 3 shows XRD patterns of the MM with R_m=0.75 before and after the catalytic test.

Specific surface area

Table IV gives a comparison of the SBET for the catalysts before, theoretically calculated and after the tests.

Table IV. Specific surface area values of the catalysts (i) before the tests, (ii) theoretically calculated and (iii) after the catalytic tests

R_m	Before test	Theoretical	After test
0 (SnO_2)	3.7 m^2/g	-	3.8 m^2/g
0.05	3.9 m^2/g	3.9 m^2/g	8.2 m^2/g
0.25	4.4 m^2/g	4.6 m^2/g	8.3 m^2/g
0.5	5.9 m^2/g	5.6 m^2/g	8.8 m^2/g
0.75	6.9 m^2/g	6.5 m^2/g	8.0 m^2/g
1 (MoO_3)	7.4 m^2/g	-	9.0 m^2/g

The SBET measured for the mixtures before the tests are very close from the calculated theoretical values.

When tested alone, the SBET of MoO_3 increases. On the contrary, when tested alone, the SBET of SnO_2 remains unchanged.

The general trend for the mixtures is an important increase of the SBET value after the test. This phenomenon, already reported in previous publications, is due to coke formation on the surface of the catalyst.

Photoelectron spectroscopy

For both mechanical mixtures and pure oxides, the binding energies of the $Mo_{3d5/2}$ and $Sn_{3d5/2}$ bands appeared to be characteristic of Mo^{6+} in MoO_3 and Sn^{4+} species in SnO_2, respectively. No significant differences of binding energies were observed when comparing the catalysts before and after the tests.

Table V presents the main atomic ratios calculated from the XPS data.

Table V. Main atomic ratios calculated from the XPS data. Values in ( ) were obtained after the catalytic tests (3hours). Values in [ ] were obtained after 30 hours of catalytic tests.

R_m	C/Mo	C/Sn	Mo/Sn
0(SnO2)	-	1.2 **(0.7)**	0 **(0)**
0.05	21.8 **(2.8)**	1.4 **(2.0)**	0.1 **(0.7)**
0.25	5.6 **(2.6) [3.5]**	1.9 **(4.6) [8.2]**	0.3 **(2.1) [2.3]**
0.5	3.1 **(1.5)**	3.2 **(6.6)**	1.0 **(4.3)**
0.75	2.0 **(1.7)**	5.8 **(21.8)**	2.9 **(13.2)**
1(MoO3)	0.3 **(1.6) [3.0]**	-	-

The coke content on pure SnO_2 was lower after the test, while it increased on pure MoO_3. When considering the mechanical mixtures, the general trend was an increase of the C/Sn ratio coupled with a decrease of the C/Mo ratio after the catalytic test, meaning that the coke deposition in the mixtures took place

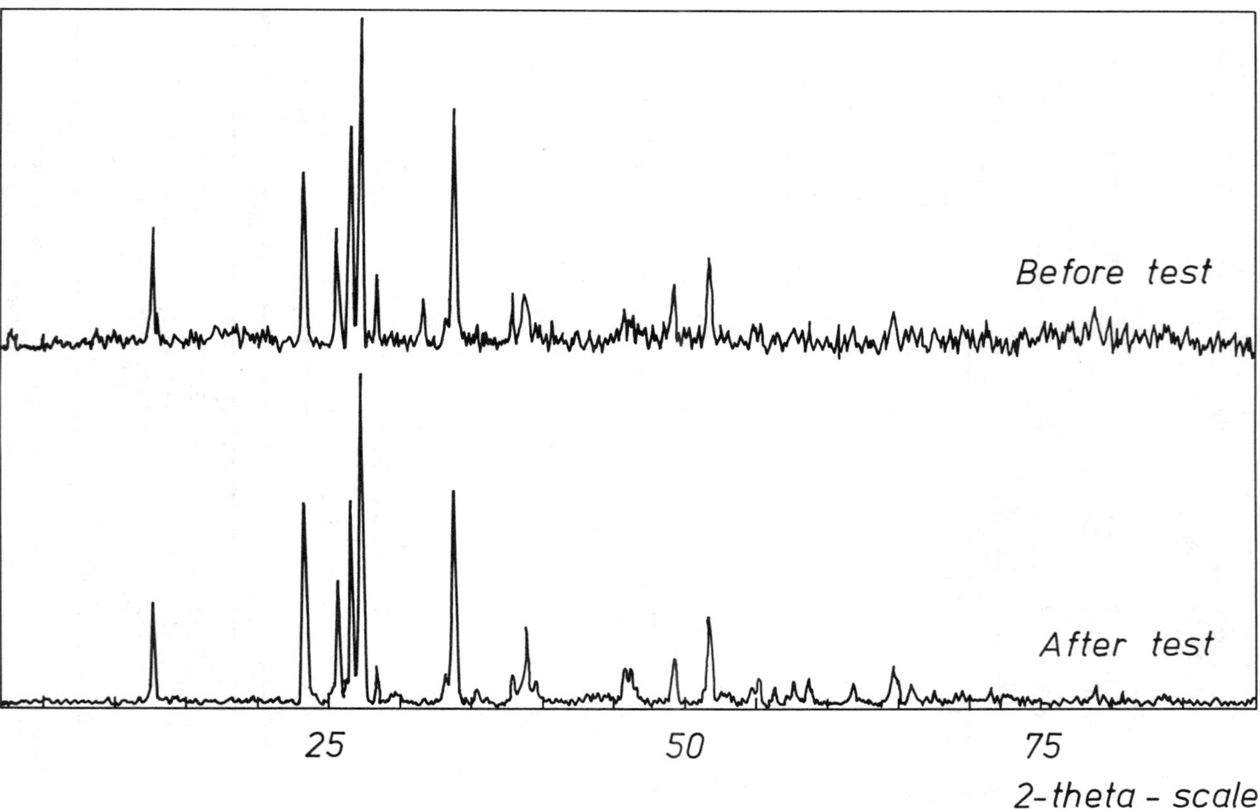

Figure 3. XRD patterns of MM with R_m=0.75 before and after the catalytic test (3 hours) at 190 °C.

preferentially or exclusively on the SnO_2 rather than on the MoO_3 surface. This is consistent with the apparent Mo/Sn ratio which was observed to be higher after the tests than before, because less photoelectrons of Mo are stopped by the coke overlayer while more of those originating from Sn are.
After 30 hours of catalytic tests, in comparison with 3 hours of reaction, the coke level had increased both on the Mo and on the Sn in the MM, and on the Mo in pure MoO_3. Nevertheless, the Mo/Sn ratio had remained constant. This evidenced that the decrease of the Mo/Sn during the test is an effect of a coke deposition occuring with different rates on each oxides rather than a migration of one metal into the oxide phase of the other.

Raman spectroscopy

Pure MoO_3 exhibited a spectrum very similar to that of a reference commercial molybdenum oxide, presenting intense bands at 157, 284, 666, 818 and 995 cm^{-1}.(Fig. 4) Pure SnO_2 exhibited a typical tin oxide spectrum with main bands at 85, 475 and 633 cm^{-1}. Nevertheless, the intensities of the bands were very weak. Consequently, in the mechanical mixtures containing more than 5% wt of MoO_3, only the features typical of this last oxide were observed.

After the tests, the catalysts presented all the bands exhibited before. Nevertheless, a slight broadening and a decrease in the intensity of the bands was noted. This could be attributed to the coke deposition on the surface of the catalysts (demonstrated by results of other characterization techniques), the increase of the coke level corresponding to an increase of the disruption of the signal due to the "black body" radiation. No additional bands or shifts were detected. Figure 4 shows Raman spectra for the pure MoO_3, and for mechanical mixture (R_m=0.5) before and after the test.

Discussion

In situ formation of a contaminated phase

The formation of a contamination phase between the pure oxides during the reaction would have led to modifications of the characteristics of the catalysts when comparing spectra before and after the tests. No such modifications were observed, even after a long time of reaction, either by "bulk sensitive" (XRD) or "surface sensitive" techniques (XPS and Raman). It must be concluded that no mutual contamination takes place between the pure oxides during the reaction. The only restriction is that the results are limited by the sensitivity of the techniques used.

Owing to this restriction, it could however be argued that some contamination could take place between the two oxides, but in so small amounts that they would remain undetected by the characterization methods used. Some of the results obtained in the present investigation make this hypothesis very unlikely.

Role of a true bimetallic oxide and of solid solution in the catalytic activity

First, the attempt to prepare a true bimetallic oxide phase has been unsuccessful. This, and several other tentatives related in the literature, confirm that the mixed phase is not stable at low temperature, and in particular when in the presence of oxygen, which is the case in the conditions of reaction. It is therefore highly improbable that such a contamination phase be formed, even in small quantities, during the preparation of the mechanical mixtures (no thermal treatment at temperatures higher than 80 °C) or during the catalytic test.

Second, the tests performed with the samples which should suffer maximum contamination (or even contain solid solutions of Mo in the SnO_2) show that this

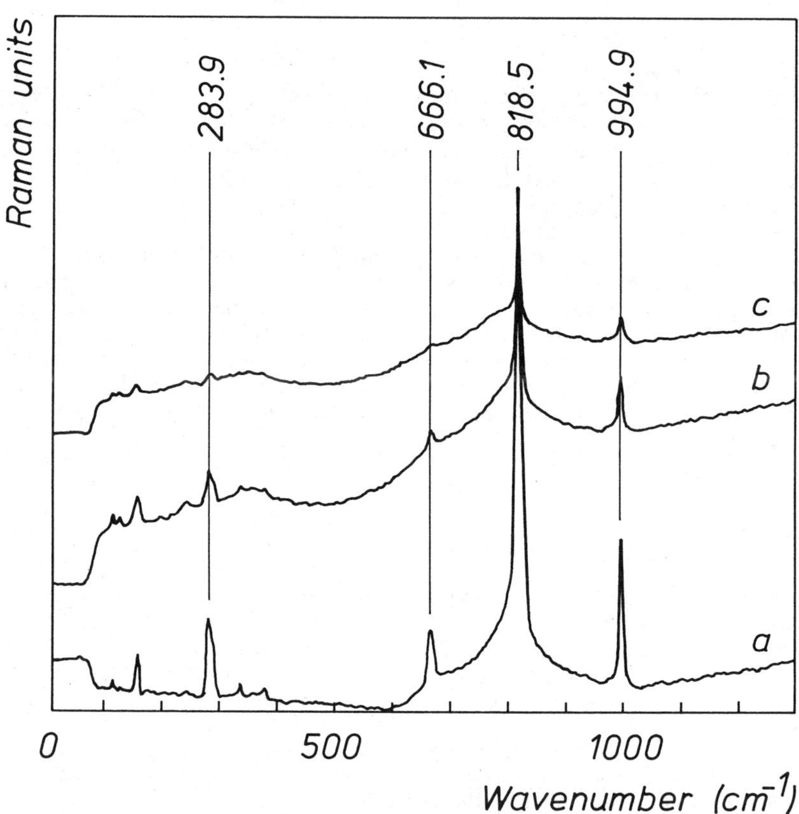

Figure 4. Raman spectra for the pure MoO_3 (a), and for mechanical mixture (R_m=0.5) before (b) and after (c) the test.

type of contamination, if it existed, could not explain the performances of the mechanical mixtures, and in particular the synergies. Indeed, sample I (the one whose preparation made the most difficult the segregation of the elements contained in the precursor) exhibits conversions and yields, in particular in butene, which are very modest compared to those of mechanical mixtures which have similar Sn:Mo compositions and specific areas. The higher selectivity in butene is probably explained by the fact that the conversion is much lower than for sample II and III, and the higher selectivity in MEK is very likely explained by the fact that SnO_2 (which produces only MEK) is the only crystalline phase in the sample. Considering sample II, the very high activity exhibited comes from its high specific area, as MoO_3 (which is the major compound of the catalyst) has just started to form small crystallites. This also explains the increase of the selectivity in butene (MoO_3 produces mainly butene) and the decrease of the selectivity in MEK (SnO_2 is no more the only crystalline phase and the size of the SnO_2 crystallites has increased), in comparison with sample I. In sample III, crystallites of SnO_2 and MoO_3 are more segregated. The fact that sample III presents performances very similar to those of the MM seems totally consistent with the above described data, because the two metals have completely segregated to form a mixture of pure simple oxides, as shown by the XRD.

According to the hypothesis that the solid solution would be selective, and that sample II should contain more solid solution than sample III, sample II should thus be more selective than sample III. This is not observed. This thus shows undoubdtedly that the solid solution is not the selective element of the catalysts.

Synergy between two separate phases

The synergetic effects between SnO_2 and MoO_3 in the dehydration-dehydrogenation of 2-butanol at low temperature cannot be explained by mutual contamination. Another explanation is necessary. A logical explanation of the cooperative effects has to be based on the existence of *two separate phases in intimate contact.*

Possible physical changes of the catalyst particles

It could be argued that the synergetic effects could be a consequence of the mechanical actions to which the oxide particles were subjected during the mixing procedure. This effect would mainly concern MoO_3, which is the most active phase, under the influence of SnO_2. SnO_2 would trigger a decrease of the size of the MoO_3 crystallites and/or some deflocculation of the MoO_3 aggregates, with, as a consequence, an increase of the number of active atoms of Mo exposed on the surface of the catalysts. This is very unlikely because the reference samples were subjected to the same mechanical treatment as the mixtures. None of the techniques used detects any effect. TEM does not reveal such changes in the morphology of the particles, either during the preparation of the mixtures, or during the catalytic tests. Measured SBET values for the MM are the same as the theoretically calculted ones. The aforementioned hypothesis must thus be discarded.

Existence of a Remote Control Mechanism

On the other hand, the observed results can be easily explained by the existence of a "Remote Control Mechanism (RMC)", due to the migration of "spillover oxygen" species.

The present results were unexpected. As mentioned in the introduction, we recently carried out the same reaction, under similar experimental conditions, over mechanical mixtures of α-Sb_2O_4 with MoO_3 and α-Sb_2O_4 with SnO_2 biphasic catalysts, and found similar synergetic effects. We were led to the conclusion that both oxides, in the presence of α-Sb_2O_4, behave as acceptors. However, former studies dealing with other oxidations (or reactions in the presence of oxygen as oxygen aided dehydrations) had shown that MoO_3 was not an "O_{so} donor", and that SnO_2, although not as effective as Sb_2O_4, exhibited donor properties*(3-4)*.

In the present case, it is suggested that SnO_2 very likely plays the role of the "donor", while MoO_3 would act as the "O_{so} acceptor". This is supported by several experimental observations. The principal effect was an increase in the yields and selectivities for butene. As i) butene is not formed on pure SnO_2 and ii) no butene formation was reported in our previous study over Sb_2O_4-SnO_2 catalysts, we must conclude that, even in the presence of O_{so}, SnO_2 does not present active sites selective for butene formation. Consequently, the effect of the O_{so} present in the SnO_2-MoO_3 system is probably to create active sites selective for butene on MoO_3. Another argument for that is that α-Sb_2O_4, which is a typical donor, produces similar effects when mixed with MoO_3.

These results can recieve easily an explanation. Dehydration of 2-butanol to butene needs acids sites. We have shown previously that in mechanical mixtures of MoO_3 and α-Sb_2O_4, O_{so} produced by this last oxide reacts with the surface of MoO_3 to produce, in a way which we have not yet fully clarified, new acid sites (Brönsted sites) *(17)*.

These observations strongly support the conclusion that SnO_2 acts as the O_{so} donor while MoO_3 plays as the acceptor. The result is an increase in the acidic sites necessary for dehydration.

The results concerning the second reaction, namely the oxidative dehydrogenation of sec-butanol to MEK, are also consistent with an effect of Oso. Only a small amount of MoO_3 in the mixture increases significantly the amount of MEK. This shows that the increase of MEK is due principally to the presence of MoO_3. The results do not demonstrate unequivocally that new sites for the formation of MEK are also created on MoO_3 by O_{so}. But the formation of MEK by oxidative dehydrogenation is an oxido-reduction process, and is similar to other similar reactions of alcohols we have investigated. In all these cases, we found that O_{so} increased the selective oxidative dehydrogenation of alcohols *(18)*. It is therefore logical to suppose that the process needs superficial oxygen species to regenerate adequately the active sites which accidentally become deactivated during the reaction : O_{so} coming from SnO_2 help to maintain those sites in an active and selective state.

Both effects, that concerning acid sites and that involved in the oxidative dehydrogenation, correspond to a remote control mechanism (RCM).

The RCM is also consistent as an explanation for the very high activity observed with sample II (the one constitued of small SnO_2 crystals with crystallites of MoO_3). As mentioned, part of the phenomenum comes very likely from the high specific surface of the sample. But in addition the crystallites of both phases are in extremely good contact in sample II, because they have just seggregated from the same precursor. In this case, the Oso formed on SnO_2 can easily migrate onto the surface of the close MoO_3, so triggering more efficiently the creation and the regeneration of the selective sites.

On the other hand, the induction periods observed for mechanical mixtures also find a logical explanation in the RCM. In the first moments of the reaction, the main action of Oso will be to generate the selective sites on MoO_3. This is probably made by triggering a reconstruction of the surface to structure permitting the selective coordinations of Mo atoms *(19)*. As the amount of Oso produced by SnO_2 is relatively small (SnO_2 is not a very efficient donor), the time needed to reach the equilibrium would be relatively long.

The coke content in pure oxides and their mixtures gives a new indication concerning the intervention of O_{SO}.

We have shown previously that coke formation on MoO_3 decreases significantly in the presence of a donor of O_{SO}. From Tables IV and V, it is observed that coke is formed on MoO_3 when MoO_3 is alone. Conversely, in the presence of SnO_2, the coke is not formed (or formed in smaller amounts) on MoO_3 but preferentially formed on SnO_2. However, SnO_2 alone does not form coke.

These observations set new questions. It is well established that coke is burned off from the surface of the catalysts by the action of O_{SO}. It turns out that SnO_2 accumulates coke, which should be removed by O_{SO} from SnO_2 itself. We have no strong argument for clarifying this problem. There are strong indications with MoO_3 that coke elimination and formation of acid Brönsted sites are parallel phenomena *(17)*. A possible explanation of the different behaviour of MoO_3 and SnO_2 with respect to the ability of O_{SO} to remove coke from their surface would be the absence of adequate sites on SnO_2 (sites permitting the same parallel process as with MoO_3). On the other hand, we have also observed that, according to experimental conditions, superficial oxygen can flow from an oxide to another. Another possible explanation would be that superficial oxygen (in particular O_{SO}) could flow too easily from SnO_2 to MoO_3 because of a too strong free energy gradient, thus letting the surface of SnO_2 in a too reduced state and promoting its ability to accumulate coke. But these are mere speculations which should be verified experimentally.

Conclusion

Strong synergetic effects have been observed at low temperatures in the dehydration-dehydrogenation of 2-butanol over mechanical mixtures of SnO_2 and MoO_3: important enhancement of the conversion, yields and selectivities for both butene and methyl-ethyl-ketone have been observed.

No phase different from the pure oxides are formed during the reaction. Solid solutions of Mo in SnO_2 have been shown to be inactive. No physical changes of the catalysts particles have been observed. Consequently, the observed cooperative effects seem to be explained by the Remote Control Mechanism. SnO_2 would play the role of a very weak O_{SO} donor, MoO_3, acting as acceptor of O_{SO}; O_{SO} on MoO_3 creates and/or regenerates selective sites active for the production of butene and MEK.

This is one of the first instances where a Remote Control process is shown to occur at temperatures below 200°C.

Acknowledgements

The support of the Région Wallonne (Belgium) in the frame of an "Action Concertée" is gratefully acknowledged. Authors also thank the Fonds National de la Recherche Scientifique (Belgium) for the fellowship awarded to Eric Gaigneaux.

Literature cited

1) J. Nilsson, A. Landa-Canovas, S. Hansen and A. Andersson, American Chemical Society Annual Meeting. Symposium on "Catalysis and Photocatalysis on Metal Oxides".Chicago, Illinois,USA, August 21-25, **1995**.

2) L.T. Weng, Y.L. Xiong, P. Ruiz and B. Delmon, Tokyo Conference Catal. Sci. Technol., Tokyo, **1990**.
Catalytic Science and Technology, vol. 1, 207, 1991, Kodansha Ltd., Japan

3) L.T. Weng, P. Ruiz and B. Delmon, Studies in Surface Science and Catalysis, "New Developments in Selective Oxidation by Heterogeneous Catalysis", P. Ruiz and B. Delmon (eds), vol 72, **1991**, 399
4) L.T. Weng and B. Delmon, *Appl. Catal. A: General*, 81, **1992**, 141
5) L.T. Weng, L. Cadus, P. Ruiz and B. Delmon, *Catal. Today.* 11, N° 4, **1992** (J.J. Spivey, ed.) p.455.
6) E.M. Gaigneaux, P. Tsiakaras, D. Herla, L. Ghenne, P. Ruiz and B. Delmon, American Chemical Society Annual Meeting, Symposium on "Catalysis and Photocatalysis on Metal Oxides", Chicago, Illinois, USA, August 21-25, **1995**.
7) Y. Okamoto, K. Oh-Hiraki, T. Imanaka and S. Teranishi, *J. Catal.*, 71, **1981**, 99.
8) S.V. Gerei, K.M. Kholyavenko, I.M. Baryshevskaya, N.A. Chervukhina and V.I. Lazukin. in "Metody Issled. Katalizatorov i Katalitich. Reaktsii, Sb" (Novosibirsk : Sibirsk. Otd. Akad. Nauk SSSR) 2, **1965**, 334.
9) V.V. Safonov, N.V. Porotnikov and N.G. Chaban. *Russian Journal of Inorganic Chemistry*, 28 (3), **1983**, 462.
10) F.J. Berry and C. Hallett, *Inorganica Chimica Acta*, 98, **1985**, 135.
11) F.J. Berry and C. Hallett, *Inorganica Chimica Acta*, 98, **1985**, L69.
12) F.J. Berry and C. Hallett, *J. Chem. Soc. Dalton Trans.*, **1985**, 451.
13) P. Courty, A. Ajot, C. Marcilly and B. Delmon, *Powder Technology*, 7, **1973**, 21.
14) ASTM - XRD standard patterns n° 36-1240, 35-0609, 41-1445.
15) E.M. Gaigneaux, D. Herla, L. Ghenne, U. Roland, P. Tsiakaras, P. Ruiz and B. Delmon. to be submitted
16) E.M. Gaigneaux, U. Roland, P. Ruiz and B. Delmon. to be submitted.
17) Z. Bing, T. Machej, P. Ruiz and B. Delmon, *J. Catal.*, 132, **1991**, 183.
18) R. Castillo, P.A. Awasarkar, Ch. Papadopoulo, D. Acosta and P. Ruiz. in New Developments in Selective Oxidation II (V. Cortes Corberan and S. Vic Bellon, eds), Elsevier Science B.V., Amsterdam, **1994**. p 795-802.
19) E.M. Gaigneaux, P. Ruiz and B. Delmon. 5th European Workshop Meeting on Selective Oxidation by Heterogeneous Catalysis, 6-7 Nov, 1995, Berlin (Germany). submitted in Catal. Today.

Chapter 25

Interaction of CO_2 with ZnO Powders of Different Microcrystalline Surfaces

A. Guerrero-Ruiz[1] and I. Rodriguez-Ramos[2,3]

[1]Instituto de Catálisis y Petroleoquimica, Consejo Superior de Investigaciones Cientificas, Campus Universitario de Cantoblanco, 28049 Madrid, Spain
[2]Departamento de Quimica Inorgánica, Universidad Nacional de Educación a Distancia, Senda del Rey, s/n, 28040 Madrid, Spain

The surface structure of several ZnO powder samples has been studied by temperature programmed desorption (TPD) of adsorbed CO_2. The isotopic distribution of evolved CO_2 species was followed when ^{18}O labelled carbon dioxide was adsorbed. The CO_2 species formed on the ZnO surface were studied by infrared spectroscopy on the samples with higher surface area. It is found that TPD profiles of CO_2 adsorbed on ZnO are affected by the chemical nature of the surface. On the other hand, the interaction of $C^{18}O_2$ with polar and nonpolar planes of ZnO is very different as showed by TPD. After $C^{18}O_2$ adsorption an abundant $C^{16}O_2$ desorption takes place. The extension of the oxygen exchange between CO_2 and the lattice oxygen of ZnO is related to the surface structure of the oxide and also to the CO_2 adsorbed species.

Zinc oxide is an industrially important oxide that is widely employed as a catalyst. It is used for hydrogenation, dehydrogenation, dehydration and oxidation, and also is an essential component of methanol synthesis catalysts. Therefore, considerable effort has been devoted to the study of its surface properties. Zinc oxide crystallizes in a wurtzite-type structure. Microcrystalline powder of ZnO is usually made up of hexagonal prisms, where the (0001) and $(000\bar{1})$ polar faces are located perpendicular to the c-axis while the apolar ones $(10\bar{1}0)$ or $(11\bar{2}0)$ are parallel to it. Studies on zinc oxide have demonstrate that its surface chemistry is dependent on the presence of point defects such as anion vacancies, step or line defects, and the atomic arrangement of the surface planes (*1-4*). It is well known that following different preparation routes the morphology of ZnO samples can be modified. Also catalytic properties (i.e. selectivity) change depending on the exposed faces. The catalytic performance of zinc oxide for the selective oxidation of light alkanes (C_1-C_3) is greatly affected by the catalyst preparation method (*5*). The same dependence has

[3]Corresponding author

0097–6156/96/0638–0347$15.00/0

been found for the oxidative dehydrogenation of propylene on zinc oxide (*6*). The results are explained in terms of the oxidation-reduction state of the catalyst surface. On the other hand, studies on CO oxidation using single crystals of ZnO have shown that the catalytic activity depends on the crystalline plane (*7*). So it is interesting to have experimental methods to detect the atomic arrangement of catalyst surface planes. The main difficulty in these studies lies in the limited surface area of metal oxide catalysts and techniques of high sensitivity are needed.

Carbon dioxide is frequently used as a probe molecule for investigating the basic properties of metal oxide surfaces (*8-10*) using classical methods such as temperature programmed desorption (TPD) or infrared spectroscopy (IR). The TPD experiment with detection of desorbed gases by mass spectrometry is a very sensitive method to obtain information about the chemical nature of the surface sites exposed on a catalyst. Additional insight can be obtained by the study of the temperature programmed desorption of adsorbed $C^{18}O_2$ (*11, 12*). Here the isotopic exchange between $C^{18}O_2$ and $Zn^{16}O$ can provide information about the structure of adsorbed CO_2 and about the surface acid-base properties. Moreover, the desorption temperatures combined with the isotope distribution reflect the variety of adsorption sites and their oxygen exchange reactivity.

In this paper, we report a detailed TPD study of carbon dioxide chemisorption on several ZnO powder samples differing in origin and surface area. The effect of CO_2 surface coverage on the TPD profiles and the isotope distribution of evolved C^xO_2 species when $C^{18}O_2$ was adsorbed has been studied. Also, an infrared study of adsorbed CO_2 on higher surface area samples has been performed. Direct observation of the morphology of the different samples was made by scanning electron microscopy.

Experimental

Four ZnO samples were studied. One of them is commercially available: ZnO-A (Analar, from B.D.H. Chemicals Ltd.), obtained by combustion of Zn. The specific surface area determined by the BET method, using nitrogen adsorption at 77 K, was 3.6 m^2g^{-1} (*13*). The three others were prepared by the thermal decomposition of zinc oxalate (ZnO-ox), zinc hydroxide (ZnO-h) and zinc hydroxycarbonate (ZnO-hc). Surfaces areas of these samples were 18.6 m^2g^{-1} for ZnO-ox, 0.2 m^2g^{-1} for ZnO-h and 4.2 m^2g^{-1} for ZnO-hc.

For the temperature programmed desorption (TPD) experiments, the ZnO samples (0.1-1.0 g) were placed in an adsorption vessel and pretreated under vacuum at 773 K for 1 h. Then the sample was heated in oxygen at 873 K to remove any residual carbonate from the surface and again treated at 773 K for 1 h under vacuum. After cooling to room temperature, a known amount of CO_2 or $C^{18}O_2$ was introduced into the vessel and contacted with ZnO for 30 min. Once the gas phase was evacuated at room temperature, the TPD was run at a heating rate of 10 $K\cdot min^{-1}$. The desorbed gases were analysed by a quadrupole mass spectrometer (Balzers QMG 421C). The ion current of the various products and the temperature of the sample were simultaneously collected in a personal computer. The ^{18}O–labelled carbon dioxide was supplied by Isotec Inc., and its isotopic purity was 95%.

For the IR experiments, a ZnO disk was placed in an "in situ" IR cell, and pretreated as described above for the TPD experiments. An amount of CO_2 (100 Torr) was introduced into the cell. The gas phase was removed at room temperature and the sample with the adsorbed CO_2 was heated in vacuum, increasing the temperature by 100 K increments up to 573 K. After evacuating the sample at each step for 15 min, the sample was cooled to room temperature and an IR spectrum was recorded on a Nicolet 5ZDX FTIR spectrometer with a resolution of 4 cm^{-1}. The IR spectra of the adsorbed species were obtained by subtracting the spectrum of the clean ZnO sample (background) from the spectrum obtained after adsorption.

The morphological study of the samples was carried out by scanning electron microscopy (SEM), using a Jeol JSM-35C electron microscope. The samples were pressed into wafers and then coated with a carbon film.

Results and Discusion

Scanning electron micrographs of typical crystallites of the different ZnO samples are shown in Figure 1. The ZnO-hc sample presents hexagonal crystals which average size is 0.6 μm (Figure 1.a). The ZnO-ox shows an amorphous morphology (Figure 1.b). The ZnO-h sample is made of rectangular crystallites which sizes are between 0.3 and 1.8 μm (Figure 1.c). Finally, the ZnO-A sample has hexagonal crystals which size is 200 Å (Figure 1.d). Thus, the selected ZnO powders have different surface areas and morphologies. This indicates that the proportion of crystal planes and surface defects are different for each sample. So ZnO-A sample exhibits a higher proportion of lateral faces in comparison with the ZnO-hc sample, and in agreement with other authors (*10*). Consequently, the various chemical properties of the different surface sites can be studied by examining the adsorption and desorption of CO_2 and $C^{18}O_2$.

The adsorption of CO_2 on the ZnO samples with higher surface areas was monitored by IR spectroscopy. Figure 2 shows the spectra for CO_2 adsorption on ZnO-hc and ZnO-A samples. It is seen that CO_2 adsorption on ZnO samples leads to very similar spectra with a multiplicity of species. The bands at 1575-95 and 1340-45 cm^{-1} can be assigned to the antisymmetric and symmetric vibrations of bidentate carbonate species. The broad peaks at 1630, 1419 and 1227 cm^{-1} can be ascribed to hydrocarbonate species formed through the adsorption of CO_2 on hydroxyl groups (*8*). These are reversibly adsorbed at room temperature and tend to disappear with time. The bands at 1515 and 1390 cm^{-1} are assigned to unidentate carbonates and the 1540 cm^{-1} peak corresponds to carboxylates (*8, 9*).

The multiplicity of species formed is due to the existence of different adsorption sites. The ZnO presents three natural surfaces: the Zn-polar (0001) plane, the O-polar $(000\bar{1})$ plane, and the non-polar prismatic $(10\bar{1}0)$ plane where both Zn and O ions are on the same plane. On the other hand, in powders, the number of surface step defects, $(n0\bar{n}1)$ and $(n0\bar{n}\bar{1})$ surfaces can be high. In addition to surface defects of steps and kinks, oxide surfaces often possess anion vacancies because of preferential removal of oxygen during vacuum activation processes. The IR results can be explained as the formation of: unidentate carbonates on O^{2-} ions from the $(000\bar{1})$ plane (*8*), bidentate carbonates on $Zn^{2+}O^{2-}$ ion-pair sites from either the non-polar $(10\bar{1}0)$ surface or the stepped surfaces (*8*), and carboxylates on Zn^{2+} ions either

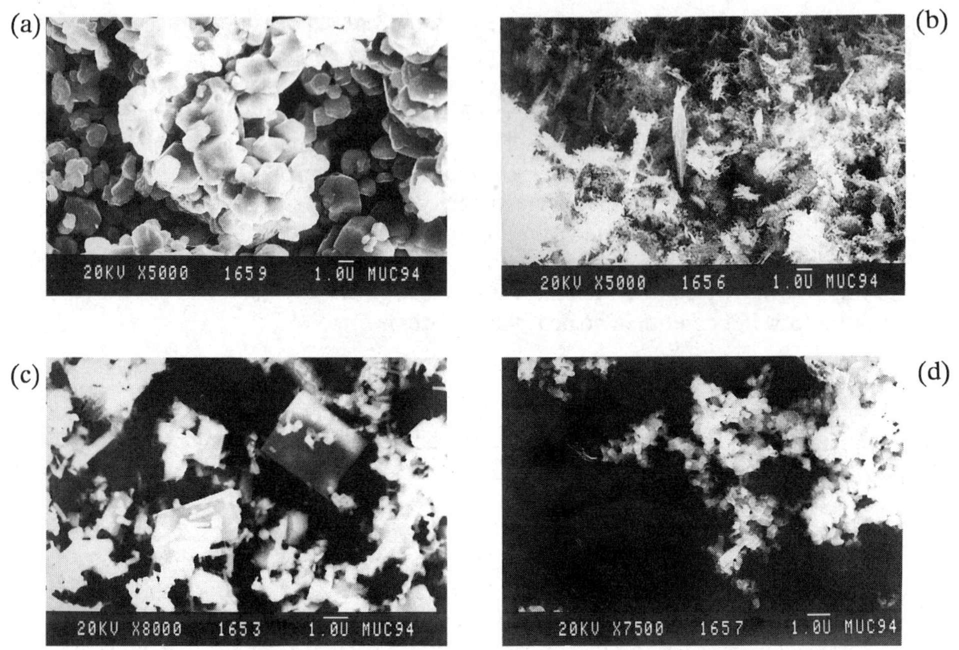

Figure 1.- Scanning electron micrograph of: a) ZnO-hc, b) ZnO-ox, c) ZnO-h and d) ZnO-A samples.

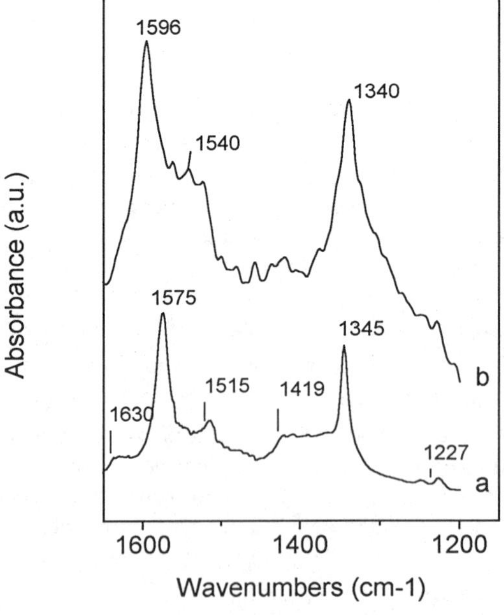

Figure 2.- Infrared spectra of CO_2 adsorbed on: a) ZnO-A and b) ZnO-hc samples.

from the (0001) plane (*8*) or from the (10$\bar{1}$0) plane with adjacent anion vacancies (V_{Os})(*2*).

Although the IR spectra showed in Figure 2 are essentially the same, CO_2 adsorption on the ZnO-hc sample seems to give rise to bands with a different relative intensity. So, bands at 1515 and at 1540 cm^{-1} are relatively more important in the case of the ZnO-hc sample. Considering that the former characterizes unidentate carbonate species and the latter carboxylate species, it appears that the polar faces, respectively (000$\bar{1}$) and (0001), and/or (10$\bar{1}$0) with oxygen vacancies (V_{Os}) in the latter, are more important on the ZnO-hc than on the ZnO-A sample.

Figure 3 shows the IR spectra of adsorbed CO_2 on ZnO-A, which were obtained after heating the sample with CO_2 adsorbed at different temperatures for 15 min in vacuum. At room temperature the above described peaks were observed. These peaks did not change by evacuation at room temperature, with the exception of those corresponding to hydrocarbonate species, which tend to disappear with time. By elevating the temperature to 373 K and 473 K, the spectra changed. The peaks at 1575 and 1345 cm^{-1}, wich were assigned to a bidentate carbonate, disappeared and those at 1470 and 1341 cm^{-1} increased. These latter bands, associated with species formed at high temperature, may be assigned to polydentate carbonates (*8*). Also, Göpel and coworkers (*2*) suggest the formation of carbonate-like complexes on the (10$\bar{1}$0) surface, which are stable up to 500 K. The peaks at 1515 and 1390 cm^{-1} stemming from unidentate carbonates remained unchanged. As the temperature was raised to 573 K and above no peaks were observed.

To have a deeper insight about the nature and chemical properties on the surface sites of the different ZnO samples, the adsorption of CO_2 was studied by TPD. TPD profiles for CO_2 desorption from ZnO samples are remarkably affected by the surface coverage. For lower surface coverage ($\sim$0.25 μmol CO_2/m^2 ZnO), as shown in Figure 4.a for the ZnO-A sample, desorption peaks at temperatures as high as 550 K can be identified. For higher surface coverage ($\sim$2.5 μmol CO_2/m^2 ZnO), the main observed peaks correspond to those of lower desorption temperatures (Figure 4.b). Finally, when the ZnO surface is saturated with CO_2, for instance by contacting the ZnO sample with a CO_2 pressure of 30 Torr (Figure 4.c), an increase of the peak intensity at lower desorption temperature ($\sim$410 K) was observed.

From Figure 4 the desorbed CO_2 can be divided into three regions in terms of desorption temperature. The first region ranges from room temperature to 430 K, the second one from 430 to 520 K, and the third above 520 K. To compare the different ZnO sample their CO_2-TPD profiles at a surface coverage of 0.25 μmol/m^2 and saturation are given in Figures 5 and 6, respectively. In general the desorption peaks despicted agree well with those observed by other authors (*3,10*). Nevertheless the peak at 540 K was not detected for zinc oxide single crystal samples (*10*), very probably because in powders the number of edge positions could be much higher.

The appearance of three regions in the TPD profiles suggests the existence of three kinds of adsorption sites, differing in adsorption strength for CO_2. On the basis of the IR studies of the CO_2 chemisorbed over the ZnO samples, the species associated with region I can be unidentate carbonate and/or bidentate carbonate, situated on the (000$\bar{1}$) plane and/or on the (10$\bar{1}$0) plane, respectively (*8, 14*). Moreover, oxygen vacancies act as specific sites for strong CO_2 chemisorption (*2*).

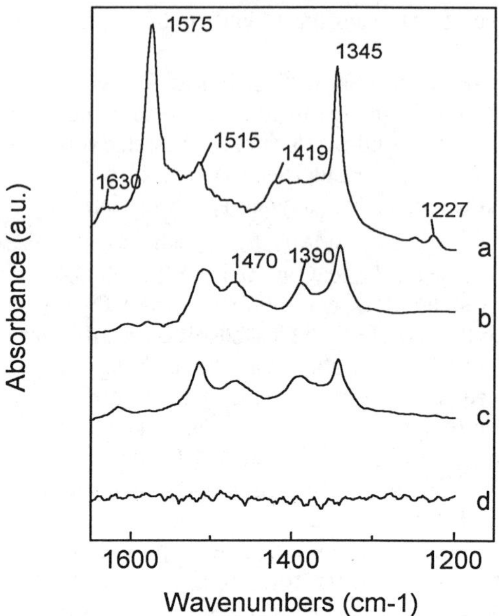

Figure 3.- Infrared spectra of CO_2 adsorbed on ZnO-A. The CO_2 was adsorbed at room temperature and desorbed by heating at different temperatures under vacuum: a) room temperature, b) 373 K, c) 473 K and d) 573 K.

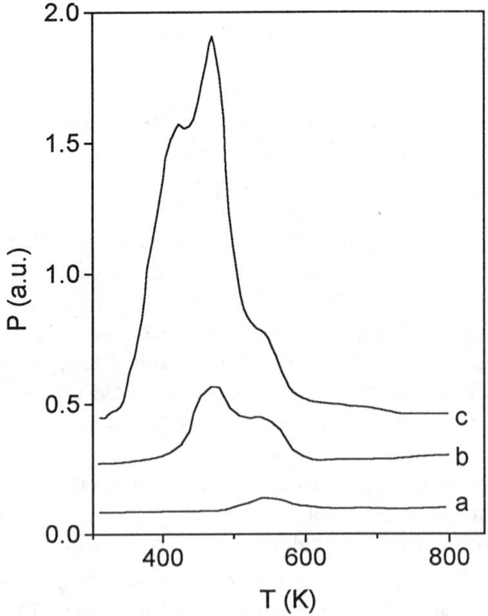

Figure 4.- TPD profiles for CO_2 adsorbed on ZnO-4. Amount of CO_2 introduced: a) 0.25 μmol CO_2/m^2, b) 2.5 μmol CO_2/m^2, and c) saturation.

So region II can be assigned to carbonate species associates to oxygen vacancies located on lateral faces(step or edges) of ZnO. Finally region III can be assigned to carboxylates, which are found on the $(10\bar{1}0)$ with adjacent anion vacancies (*2*) and/or on the (0001) plane (*3*). The former, which desorb at 540 K, are formed on ZnO samples with higher surface area in which the number of surface defects is larger. The latter, which appear at 670-700 K, are the strongest adsorbed CO_2 species on the ZnO samples.

It is interesting to note that different CO_2-TPD profiles are obtained depending on the ZnO origin. It is ease to deduce from Figure 5 that the ZnO-hc sample exposes polar (0001) faces in higher quantities than the other ZnO samples. This finding agrees with the IR results (Figure 2) and SEM observations of these samples. On the other hand, the density of defects and therefore the reducibility of the lateral faces seems to be higher in ZnO-ox and ZnO-A than in other samples (regions II and III), since they form species associated with surface oxygen vacancies to a greater extent. To obtain this information the ZnO surface was not saturated with CO_2, because under this latter conditions region I becomes the principal one (Figure 6).

To investigate the CO_2 desorption in regions II and III in more detail, small amounts of $C^{18}O_2$ were adsorbed on the ZnO samples and subjected to TPD measurements. Figures 7 and 8 show the TPD profiles for each isotopically labelled CO_2, at surface coverages of 2.5 and 0.25 $\mu mol/m^2$, respectively. It can be observed that for all desorption regions the oxygen-exchange reaction between $C^{18}O_2$ and the oxide surfaces is vesy extensive. At desorption temperatures above 500 K (region III), a nearly total oxygen exchange seems to take place (Figure 8). So, the desorbed CO_2 consisted of $\sim 90\%$ $C^{16}O_2$ and $\sim 10\%$ $C^{16}O^{18}O$, with no apreciable amount of $C^{18}O_2$. As discussed above region III can be assigned to carboxylates on $(10\bar{1}0)$ with adjacent anion vacancies or on polar (0001) faces, both species with the two oxygen atoms of the CO_2 molecule bound to the ZnO surface.

A slightly different picture is found in region II. As in region III the desorption of $C^{18}O_2$ is absent, but a considerable fraction of $C^{16}O^{18}O$ was observed. This species, $C^{16}O^{18}O$, being dominant for the ZnO-ox sample. The adsorbed species was a carbonate, which is bound to the surface with oxygen vacancies through two bonds. If these two bonds are equivalent the desorbed CO_2 would be a 1:1 mixture of $C^{16}O^{18}O$ and $C^{18}O_2$ upon TPD and $C^{16}O_2$ should not have been included. Therefore, for the description of CO_2 in region II and III, processes other than simple adsorption-desorption of CO_2 on two ZnO sites are involved.

To explain the isotopic distribution of region II and III, a multiple oxygen exchange between CO_2 and the lattice oxygen of the oxide has to occur. Different mechanisms have been proposed. First, repetitive adsorption-desorption of CO_2 molecules has been suggested (*15*). This latter was observed in the presence of gas-phase CO_2 ($>$1KPa). However, in our experiments after the dosage of 0.25 μmol CO_2/m^2 the partial pressure of CO_2 was negligible and its occurrence is excluded. Other mechanisms proposed (*16,17*) contemplate the formation of a rolling three-fold coordinated carbonate, but this species would result in the desorbed CO_2 containing at least 17% $C^{18}O_2$. Finally, it has been proposed (*12*) that extensive oxygen exchange results from adsorbed CO_2 migrating over the surface without leaving the surface. During the migration over the surface, the adsorbed CO_2 species undergoes O exchange with the surface O atoms.

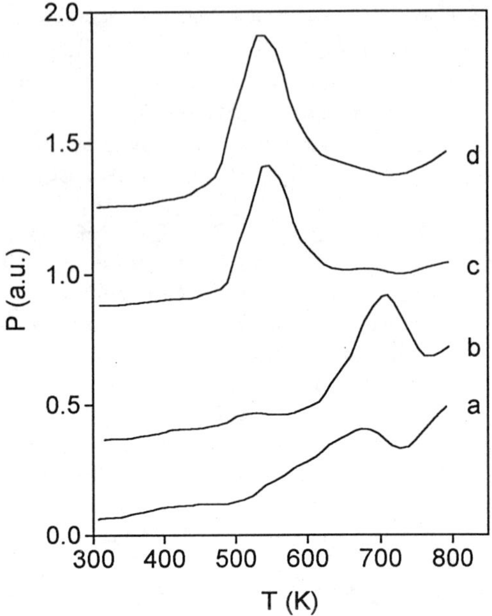

Figure 5.- TPD profiles for CO_2 adsorbed on: a) ZnO-h, b) ZnO-hc, c) ZnO-A and d) ZnO-ox. CO_2 concentration: 0.25 μmol CO_2/m^2.

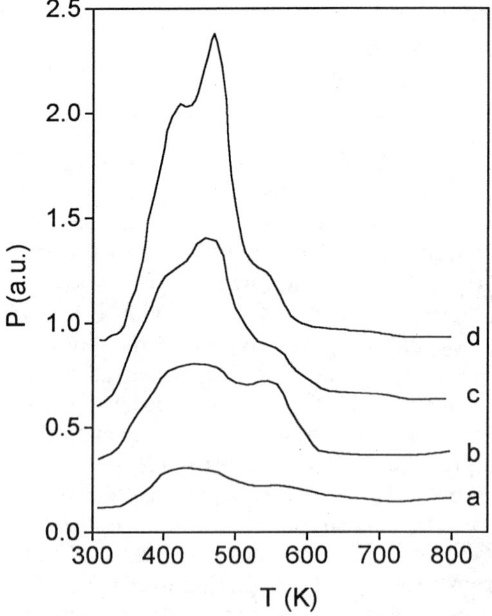

Figure 6.- TPD profiles for CO_2 adsorbed on: a) ZnO-h, b) ZnO-ox, c) ZnO-hc and d) ZnO-A. CO_2 concentration: saturation.

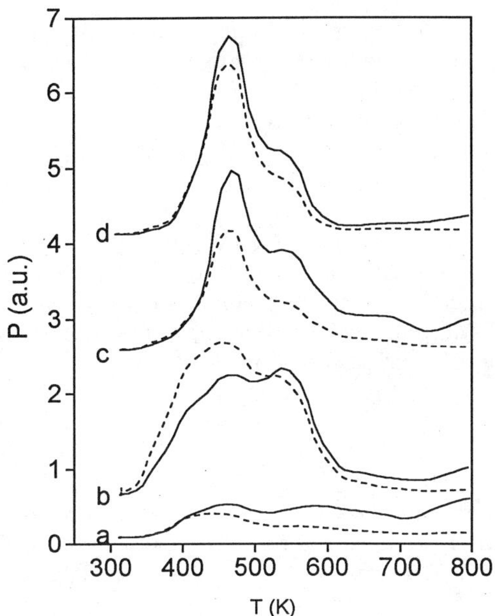

Figure 7.- TPD profiles for $C^{18}O_2$ adsorbed on: a) ZnO-h, b) ZnO-ox, c) ZnO-hc and d) ZnO-A. — $C^{16}O_2$, ---$C^{18}O^{16}O$. CO_2 concentration: 2.5 μmol/m^2.

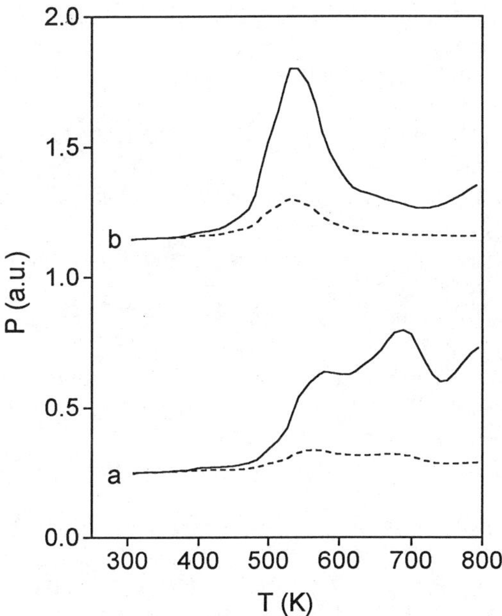

Figure 8.- TPD profiles for $C^{18}O_2$ adsorbed on: a) ZnO-hc and b) ZnO-ox. — $C^{16}O_2$, ---$C^{18}O^{16}O$. CO_2 concentration: 0.25 μmol/m^2.

Comparison of CO_2 desorption in region II for the different ZnO samples (Figure 7) shows that the migration process depends on the surface structure of the ZnO. So, differences in the probability of migration results in differences in the isotopic distribution in the desorbed CO_2.

Conclusions

The results obtained allow us to say that the temperature programmed desorption of the adsorbed probe molecule CO_2, with and without ^{18}O-labelled, provide information about the surface structure of microcrystalline ZnO samples with very different surface area. The TPD profiles reflect the variety of adsorption sites of the sample. So, the proportion of different adsorbing faces, the presence of surface defects such as steps, kinks and anion vacancies can be determined.

Acknowledgments

This work was supported by the CICYT (Spain) under Project MAT93-0477. We wish to thank to Prof. J. Rouquerol for kindly supplying the ZnO-A sample. Also, thanks are due to Dr. F.J. Perez-Trujillo of Univ. Complutense (Madrid) for SEM measurements.

Literature Cited

(1) Göpel, W., Brillson, L.J., Brucker, C.F. *J. Vac. Sci. Technol.* **1980,** *17*, 894.
(2) Göpel, W., Bauer, R.S. and Hansson, G. *Surf. Sci.* **1980,** *99*, 138.
(3) Cheng, W.H. and Kung, H.H. *Surf. Sci.* **1982,** *122*, 21.
(4) Akhter, S., Lui, K. and Kung, H.H. *J. Phys. Chem.* **1985,** *89*, 1958.
(5) Wada, K., Yoshida, K., Takatani, T. and Watanabe, Y. *Appl. Catal. A: General* **1993,** *99*, 21.
(6) Spinicci, R. and Tofanari, A. *Appl. Catal.* **1981,** *1*, 387.
(7) Amigues, P. and Teichner, S. J. *Discusion Faraday Soc.* **1966,** *41*, 362.
(8) Saussey, J., Lavalley, J.C. and Bovet, C. *J.Chem. Soc. Faraday Trans. 1* **1982,** *78*, 1457.
(9) Chauvin, C., Saussey, J., Lavalley, J.C. and Djega-Mariadassou, G. *Appl. Catal.* **1986,** *25*, 59.
(10) Bowker, M., Houghton, H., Waugh, K.C., Giddings, T. and Green, M. *J. Catal.* **1983,** *84*, 252.
(11) Mckenzie, A.L., Fishel, C.T. and Davis, R.J. *J. Catal.* **1992,** *138*, 547.
(12) Tsuji, H., Shishido, T., Okamura, A., Gao, Y., Hattori, H. and Kita, H. *J. Chem. Soc. Faraday Trans.* **1994,** *90*, 803.
(13) Grillet, Y., Rouquerol, F. and Rouquerol, J. *Thermochimica Acta* **1989,** *148*, 191.
(14) Little, L.H. *Infrared Spectra of Adsorbed Species;* Academic Press: New York, NY, 1966.
(15) Ito, T. *J. Chem. Soc., Faraday Trans. 1* **1982,** *78*, 1603.
(16) Yamagisawa, Y., Shimodama, H. and Ito, A. *J. Chem. Soc., Chem. Commun.* **1992,** 610.
(17) Takaishi, T. and Endoh, A. *J. Chem. Soc., Faraday Trans. 1* **1987,** *83*, 411.

Chapter 26

Alkene Oxidation on Pd (100): Total, Not Partial

Xing-Cai Guo and Robert J. Madix[1]

Department of Chemical Engineering, Stanford University, Stanford, CA 94305

Partial alkene oxidation has previously been observed on Ag(110) and Rh(111) under ultrahigh vacuum conditions. Atomic oxygen on Ag(110) reacts with alkenes through acid-base reactions or epoxidations, whereas on Rh(111) alkenes can be oxidized to ketones. On Pd(100), a neighbor to both Ag and Rh in the periodic table, we detected no partial oxidation products for O(a) coverages up to 1 O/Pd for any of the alkenes studied including ethylene, propene, 1-butene, and 1,3-butadiene; only total oxidation (combustion) products were observed. We found that strong bonding of alkenes to Pd(100) invalidates the application of gas phase acidity concepts as on Ag(110) and that efficient O-H bond formation on Pd(100) prevents ketone formation as on Rh(111). Two different combustion pathways were identified, namely, direct (oxygen-activated) combustion and indirect (decomposition and oxidation) combustion.

Alkene oxidation has been investigated and is relatively well understood on Ag(110) (*1*) and Rh(111) (*2*) under ultrahigh vacuum (UHV) conditions. Atomic oxygen on Ag(110) reacts facilely with alkenes through an acid-base reaction (proton abstraction from the acidic C-H bonds) or epoxidation (oxygen cyclic addition to the C=C double bonds)(*1*). Ethylene does not react with O(a) on Ag(110) under UHV because it desorbs before it can react (*3*). Propene is totally oxidized (or combusted) because of its most acidic allylic H (*4,5*). Partial oxidation occurs for 1-butene to form 1,3-butadiene via acid-base reaction; oxygen addition to butadiene yields 2,5-dihydrofuran which can be further converted to furan and maleic anhydride (*5*).

In contrast, no acid-base reaction occurs on Rh(111)-p(2×1)-O (0.50 O/Rh). On this surface alkenes with a vinylic C-H bond can be oxidized to ketones, e.g. propene to acetone, butene and butadiene to the corresponding methyl ketones (*2*). It has been proposed that direct oxygen addition to 2-C occurs to form an oxametallacycle, and subsequent β-H elimination and 1-C hydrogenation form ketone. During these reactions

[1]Corresponding author

0097–6156/96/0638–0357$15.00/0

allylic C-H bonds are not broken. Inhibition of dehydrogenation by O(a) is thought to be the key for partial oxidation of alkenes on Rh(111)-p(2×1)-O, because at lower O(a) coverages only combustion takes place.

In the periodic table palladium is next to both Ag (on its right) and Rh (on its left), which makes one wonder what mechanism will be followed by alkene oxidation on Pd. Does acid-base reaction occur as on Ag(110)? Does oxygen atom attack the C atoms of alkene molecule to form a C-O bond as on Rh(111)? Are there any partial oxidation products from alkene oxidation on Pd? With these questions in mind we have carried out alkene oxidation experiments on Pd(100). We have studied a wide range of O(a) coverages up to 1 monolayer (1 ML = 1 O/Pd) using temperature-programmed reaction spectroscopy (TPRS) under UHV conditions. The alkenes studied include ethylene, propene, 1-butene, and 1,3-butadiene. Reactions on both clean and O(a)-covered Pd(100) surfaces were investigated. Briefly, no partial oxidation products were detected for O(a) coverages up to 1 O/Pd for any of the alkenes studied; only combustion products were observed. However, two different combustion pathways have been identified, namely, *direct (oxygen-activated) combustion* and *indirect (decomposition and oxidation) combustion*. Generally, combustion occurs at temperatures below which dehydrogenation would be complete on the clean surface.

Results

Evolution of Product Molecules H_2, H_2O, CO, and CO_2. For the purpose of calibration we first determined the temperature programmed desorption (TPD) or TPR spectra of the product molecules H_2, H_2O, CO, and CO_2. Desorption of H_2 starts at 300 K and reaches maximum at 355 K. This experiment sets a reference temperature for desorption-limited H_2 evolution. H_2O evolution from H(a) adsorbed on Pd(100)-p(2×2)-O starts at 195 K, showing a desorption maximum at 285 K. For lower O(a) coverages H_2O evolves in the same temperature range. H_2O evolution in this temperature range is rate-limited by OH disproportionation, since desorption-limited H_2O evolves below 200 K from Pd(100) (*6*). Desorption-limited CO evolution from CO(a) on clean Pd(100) is around 500 K. CO adsorption on Pd(100)-p(2×2)-O leads to CO_2 evolution at 335 K. CO_2 evolution is rate-limited by CO(a)+O(a) reaction, because CO_2 rapidly desorbs at this surface temperature. Reaction-limited CO and CO_2 evolution from reactions of O(a) and C(a) on Pd(100) occurs at 625 and 500 K, respectively.

Dehydrogenation of Ethylene, Propene, 1-Butene, and 1,3-Butadiene . Alkenes were adsorbed on clean Pd(100) at a surface temperature of 120 K with variable exposures. Dehydrogenation of alkenes yields adsorbed hydrogen H(a) on the surface, which desorbs associatively as H_2. At low exposures no molecular alkene desorbs. All alkenes on the surface dehydrogenate, yielding H(a) and C(a). Even at very low coverages all four alkenes exhibit two H_2 evolution peaks: one at 350 K, and the other above 400 K. These results indicate that each alkene dehydrogenates in two major steps, one occurring around 300 K and the other above 400 K. As coverage increases further, a fraction of the adsorbed alkene molecules start to desorb. For butadiene at saturation coverage, self-hydrogenation forms 0.02 butene/Pd. No hydrogenation was detected for ethylene, propene, and butene.

Table I. Total Adsorption Coverage, Molecular Desorption Yield, and Dehydrogenation Yield of Alkenes on the Clean Pd(100) Surface for Saturation Exposures at 120 K.

Alkene Molecule	Desorption Yield (alkene/Pd)	Dehydrogenation Yield		Adsorption Coverage (alkene/Pd)
		(H_2/Pd)	(alkene/Pd)	
ethylene	0.15	0.19	0.10	0.25
propene	0.13	0.32	0.11	0.24
butene	0.12	0.36	0.09	0.21
butadiene	0.12	0.38	0.13	0.27

Table I lists the total adsorption coverages and the dehydrogenation yields for saturation alkene exposure at 120 K. The saturation coverage for simple alkenes is in the range from 0.25 to 0.21 alkene/Pd, decreasing as the molecular chain increases from ethylene to propene to butene. From butene to butadiene there is an increase of 8 kcal/mol in adsorption energy as estimated from the high temperature desorption peak (225 K for butene and 350 for butadiene). The dehydrogenation yields are similar for simple alkenes (0.10 alkene/Pd), but slightly more dehydrogenation occurs for butadiene (0.13 butadiene/Pd).

Oxidation of Ethylene, Propene, 1-Butene, and 1,3-Butadiene. Adsorption of ethylene, propene, 1-butene, and 1,3-butadiene was studied on the atomic oxygen-covered Pd(100)-p(2×2)-O surface at 120 K with a variable alkene exposure. Under these conditions TPR spectra heating up to 1070 K yielded no partial oxidation products; only total oxidation (combustion) products H_2O and CO_2 (and CO when alkene is in excess) were found. Even at higher O(a) coverages (up to 1 O/Pd) combustion was the only reaction detected. The product yields for the saturation alkene exposure are listed in Table II.

Table II. Yield (molecules/Pd or atoms/Pd) of Alkene Combustion on the Pd(100)-p(2×2)-O Surface for Saturation Exposures at 120 K.

Alkene Molecule	Reaction Products						Re-acted Alkene	De-sorbed Alkene	Ad-sorbed Alkene
	H_2O	H_2	CO_2	CO	O(a)	C(a)			
ethylene	0.02	0.02	0.04	0.00	0.13	0.00	0.02	0.14	0.16
propene	0.20	0.07	0.01	0.03	0.00	0.23	0.09	0.14	0.23
butene	0.19	0.20	0.01	0.04	0.00	0.36	0.10	0.14	0.24
butadiene	0.12	0.19	0.03	0.07	0.00	0.30	0.10	0.13	0.23

Very little ethylene reacts with O(a) on Pd(100)-p(2×2)-O, and there is no indication of direct reactions. H_2O evolution near 350 K is rate-limited by ethylene dehydrogenation on Pd sites unaffected by adsorbed O and subsequent reactions between H(a) and O(a), as the peak temperature for water evolution is well above that from H(a)+O(a) reaction on the clean surface. The CO_2 evolution at 530 K is apparently due to O(a) reacting with totally dehydrogenated C(a) atoms, as it closely resembles that from O(a)+C(a) reaction. No CO is formed from ethylene on Pd(100)-p(2×2)-O, since O(a) is in excess by 0.13 O/Pd. Of 0.16 ethylene/Pd adsorbed at saturation, which is considerably smaller than on clean Pd(100), 0.14 ethylene/Pd desorbs molecularly in a broad peak below 370 K. O(a) on Pd(100) seems to block ethylene adsorption *and* inhibit ethylene dehydrogenation.

Propene, butene, and butadiene have very similar reacted coverages (0.10 molecule/Pd) and adsorbed coverages (0.23 molecule/Pd) for saturation exposures at 120 K (Table II). In fact, these quantities on Pd(100)-p(2×2)-O are comparable to those on clean Pd(100) (cf. Table I). Clearly O(a) does not block adsorption or inhibit reactions of propene, butene, and butadiene – unlike the case of ethylene. The TPRS spectra on Pd(100)-p(2×2)-O are also similar for propene, butene, and butadiene. At low coverages *all* H atoms of the alkene molecule react with O(a) to form H_2O below 400 K; no H_2 evolves from the surface. This result is in sharp contrast to the behavior on clean Pd(100), where reaction-limited H_2 evolves above 400 K in the second stage of dehydrogenation. Apparently, O(a) facilitates the C-H bond scission of higher alkenes on Pd(100), at least subsequent to the first dehydrogenation step.

For propene and butene there are two H_2O peaks which develop simultaneously as coverage increases, indicating that there are two stages in O(a)+alkene reactions. The low temperature H_2O peak at 280 K coincides with that from H(a)+O(a) reaction on Pd(100), suggesting that the H_2O evolution is not rate-limited by C-H bond scission, but by recombination. However, the high temperature H_2O peak above 300 K (340 K for propene and 350 K for butene) is clearly rate-limited by C-H bond scission. As coverage increases, both H_2O peaks shift down in temperature and a shoulder appears above 400 K. To further distinguish the two water evolution processes within the alkene, experiments were performed using deuterated alkenes. In these experiments alkene exposures were low enough so that there was no evolution of hydrogen or its isotopes. All alkene molecules reacted with O(a) to form water. Experiment with $D_2C{=}CD{-}CD_3$ shows that Deuterated propene reacts similarly as non-deuterated propene. Unlike the results for Ag(110)(*3*), there is no significant deuterium kinetic isotope effect. Qualitatively, in the low temperature peak there is little C-D bond breaking for $H_2C{=}CH{-}CD_3$ and $D_2C{=}CH{-}CH_2{-}CH_3$. This result implies that the low temperature H_2O originates mainly from the vinylic 2-H atoms of the 1-alkene. Quantitative distributions of H (or D) in the two pathways to water are listed in Table III.

The distribution remains approximately constant as the coverage varies. As Table III shows, of six H atoms in $H_2C{=}CH{-}CH_3$ one H goes to low temperature H_2O, suggesting cleavage of one C-H bond, while the remaining five appear in the high temperature H_2O. This ratio also holds for the corresponding routes of $D_2C{=}CD{-}CD_3$. For $H_2C{=}CH{-}CD_3$ one C-H bond reacts at low temperature, while the two C-H bonds and three C-D bonds react at high temperature. For $H_2C{=}CH{-}CH_2{-}CH_3$ and $D_2C{=}CH{-}CH_2{-}CH_3$ the low temperature reaction occurs with two C-H bonds. The two C-D bonds in D_2C=CH-CH_2-CH_3 all react at high temperature. Taken these together, it is clear that the vinylic C-H bond in propylene is predominantly cleaved at low temperature while the allylic or methylenic C-H bonds react at high temperature. The exact origin of the two hydrogens in 1-butene cannot be specified, however.

Table III. Distributions of H or D in Water Evolution in the Low Temperature Range and in the High Temperature Range for Propene and 1-Butene on Pd(100)-p(2x2)-O. The Numbers Shown Are Rounded from Those in the Parentheses.

	Hydrogen (H)			Deuterium (D)		
	Low	High	Total	Low	High	Total
H_2CCHCH_3	1(1.2)	5(4.8)	6	0(0.0)	0(0.0)	0
D_2CCDCD_3	0(0.1)	0(0.0)	0	1(1.2)	5(4.8)	6
H_2CCHCD_3	1(0.8)	2(2.2)	3	0(0.2)	3(2.8)	3
$H_2CCHCH_2CH_3$	2(2.1)	6(5.9)	8	0(0.0)	0(0.0)	0
$D_2CCHCH_2CH_3$	2(2.2)	4(3.8)	6	0(0.1)	2(1.9)	2

Formation of CO and CO_2 from propene, butene, and butadiene on Pd(100)-p(2×2)-O represents a minor reaction channel compared to H_2O formation (Table II). Consequently, a large amount of C(a) is left on the surface. Two CO_2 peaks emerge as coverage increases, which appear to coincide with the high temperature H_2O peaks. As proven by isotope coadsorption experiments, these CO_2 evolution peaks are from neither CO(a)+O(a) reaction nor C(a)+O(a) reaction. They must be due to direct O(a) reactions with alkene intermediates following C-H bond scission, forming C-O bonds with an activation energy lower than that of C(a)+O(a) reaction but higher than that of CO(a)+O(a) reaction. This suggestion is further supported by the appearance of desorption-limited CO at 450 K. Although it occurs at 500 K on clean Pd(100), desorption-limited CO evolution is shifted down by 50 K with coadsorbed alkenes as shown by CO^{18} coadsorption experiments. The desorption-limited CO evolution at 450 K also indicates that the C-O bond forms below 400 K. The high temperature CO peak around 600 K is from O(a) reaction with C(a) on the surface.

Discussion

Reactivity of the vinylic C-H bond on Pd(100)-p(2×2)-O seems not to be governed by the gas phase acidity, which applies well on Ag(110). The allylic C-H bond of propene is the most acidic (ΔH_{acid} = 391 kcal/mol), but it is less reactive than the vinylic C-H bond. We also found that norbornene (bicyclo[2.2.1]-2-heptene) combusts easily, despite the fact that its gas phase acidity ΔH_{acid} = 402 kcal/mol) is similar to that of ethylene (ΔH_{acid} = 406 kcal/mol). On the other hand, initial C-H bond scission is not induced by O addition to the C atom of alkene as it is proposed to occur on Rh(100)-p(2×1)-O.

Oxygen addition to C atoms does occur on Pd(100)-p(2×2)-O, but apparently only to the C atoms of alkene intermediates after the C-H bond reaction with O(a). In addition, efficient H_2O formation and evolution on Pd(100) make hydrogenation impossible, as no self-hydrogenation of butadiene occurs on Pd(100)-p(2×2)-O, although it occurs on clean Pd(100). Hydrogenation is an essential step in the proposed mechanism of ketone formation from alkene on Rh(100)-p(2×1)-O (2), where H_2O evolves at a higher temperature than ketone evolution. In conclusion, efficient reactions between O(a) and the C-H bonds may prevent ketone formation on Pd, while strong bonding of alkene to the surface prevents the application of gas phase acidity concepts to alkene oxidation on Pd. This is why Pd is a good catalyst for alkene total oxidation rather than partial oxidation.

Summary

The oxidation reactions of ethylene, propene, 1-butene, and 1,3-butadiene have been studied on atomic oxygen-covered Pd(100)-p(2x2)-O surface. It was found that O(a) inhibits adsorption and reaction of ethylene, whereas O(a) does not inhibit the adsorption of higher alkenes and also activates their reactions. At low alkene coverages, all C-H bonds react facilely with O(a), forming water below the temperature of complete dehydrogenation on the clean surface. Isotope experiments show that the low-temperature water evolution at 280 K results predominately from O(a) reactions with the vinylic C-H bonds of 1-alkenes. Low-temperature CO_2 evolution and desorption-limited CO evolution indicate that O(a) directly attacks the C atoms of dehydrogenated alkene intermediates. However, efficient water formation prevents O-containing intermediates from hydrogenating and producing partial oxidation products.

Acknowledgment

The authors gratefully acknowledge the support of the National Science Foundation (NSF CTS-9300188-001).

Literature Cited

(*1*) Madix, R. J.; Roberts, J. T. in *Surface Reactions*, Madix, R. J., Ed.; Springer: Berlin, 1994; pp 5-53.
(2) Xu, X.; Friend, C. M. *J. Am. Chem. Soc.* **1991**, *113*, 6779.
(*3*) Barteau, M. A.; Madix, R. J. *J. Am. Chem. Soc.* **1983**, *105*, 344.
(*4*) Roberts, J. T.; Madix, R. J.; Crew, W. W. *J. Catal.* **1993**, *141*, 300.
(*5*) Roberts, J. T.; Capote, A. J.; Madix, R. J. *Surf. Sci.* **1991**, *253*, 13.
(*6*) Stuve, E. M.; Jorgensen, S. W.; Madix, R. J. *Surf. Sci.* **1984**, *146*, 179.

NOVEL OXIDANTS

Chapter 27

Catalytic Methane Oxidation at Low Temperatures Using Ozone

W. Li and S. Ted Oyama[1]

Department of Chemical Engineering, Virginia Polytechnic Institute and State University, 133 Randolph Hall, Blacksburg, VA 24061

Methane oxidation using ozone at low temperatures (< 673 K) was studied both in the gas phase and with catalysts. Gas phase reactions were significant even at low temperatures, however, catalysts could improve the reactivity both under oxygen-rich and oxygen-deficient conditions. MgO was found to be active for the conversion of methane to CO and CO_2 at oxygen-rich conditions, while Li/MgO catalyst promoted the production of formaldehyde at oxygen-deficient conditions. A selectivity to HCHO over 90% was achieved with a 4% methane conversion at 650 K on a Li/MgO catalyst, while no detectable conversion of methane was observed using oxygen as the oxidant. Methane oxidation to form formaldehyde was found to occur at the same temperature range when ozone decomposition happened, which suggested that the active oxygen species for methane oxidation were formed by ozone decomposition. Values of the ratio of converted CH_4/converted O_3 above unity clearly indicated the involvement of a chain reaction mechanism.

The direct oxidation of methane to methanol or formaldehyde has been receiving considerable attention due to its fundamental as well as practical significance. Oxygen is the most commonly used oxidant. However, severe conditions (>400 °C) are usually necessary to achieve detectable conversion of methane, and high selectivities to the desired oxygenates occur only at small (<2%) methane conversion levels (*1-7*). Consequently, alternative oxidants, i.e. O_3, N_2O etc. (*8-15*) have also been studied. As methane is the most stable hydrocarbon, its interaction with O_3 is particularly interesting because ozone is the second strongest oxidizing agent, only next to fluorine *(16)*. Hutchings and co-workers (*11,12*) studied the effect of ozone versus oxygen on the methane oxidation reaction. Ozone was found to be more active and selective than oxygen at temperatures ≤ 400 °C. Hydrogen, carbon monoxide, and carbon dioxide were the only products observed. Two crucial questions were unanswered in that

[1]Corresponding author

0097–6156/96/0638–0364$15.00/0

study. First, ozone is thermodynamically unstable, and it decomposes to oxygen both thermally and catalytically. Hence ozone can be consumed by reaction with methane as well as by decomposition to oxygen. The role of ozone decomposition in relation to its reaction with methane needs to be clarified. Second, because of the strong oxidizing ability of ozone, contributions from gas phase reactions cannot be neglected even at low temperatures. The importance of these homogeneous reactions needs to be quantitatively resolved.

In this paper, we report a detailed study of methane oxidation at relatively low temperatures (≤ 400 °C) using ozone or oxygen as the oxidizing agents. Due attention has been given to establish the role of gas phase reactions. The ozone and methane concentrations before and after the reaction were monitored to determine the stoichiometry of the reaction and the possible involvement of chain reactions. MgO and Li/MgO were used as catalysts for this reaction.

Experimental

The catalysts used in this study were MgO and 5 wt% Li/MgO. MgO (Alfa, 99.85%) was treated with distilled water to form a paste, which was dried at 120°C and then calcined at 500°C for 6 hours. The Li/MgO catalyst was prepared by a similar procedure described in ref. 17. An aqueous solution of LiOH (Aldrich, 99.9+%) was added to MgO to form a paste. Then the catalyst was dried and calcined as the MgO catalyst. The surface area of the catalysts was measured by nitrogen adsorption using the BET method. Methane oxidation was carried out in a fixed-bed flow reactor system using 100 mg of catalysts. Ozone was produced by passing oxygen (Air Products, 99.6%), dried with a gas purifier (Alltech, Drierite and Molecular Sieve 5A), through an ozone generator (OREC, Model V5-0), and its concentrations before and after the reaction were monitored by an ozone analyzer (H-1, IN-USA). In this study, the initial ozone concentration was always lower than 2 mol%. Methane (Air Products, 99.99%) and helium (Air Products, 99.999%) were also dried using the gas purifiers, and their flow rates were controlled by mass flow controllers (Brooks 5850E). The flow rate of the ozone and oxygen mixture was set by a needle valve (Nupro), and monitored by a mass flow meter (Aalborg, GFM17). The product compositions were determined by an on-line gas chromatograph (SRI, 8610) with a TCD and an FID detector. The gas phase reaction was evaluated with a reactor loaded with quartz chips (18-20 mesh), using oxygen-rich and oxygen-deficient conditions. The flow rates for these conditions are listed in Table I.

Table I. Flow Rates for the Oxygen-Rich and Oxygen-Deficient Conditions

Gas	Oxygen-Rich		Oxygen-Deficient	
He	0	0	510 cm^3min^{-1}	379 $\mu mol\ s^{-1}$
CH_4	12 cm^3min^{-1}	8.9 $\mu mol\ s^{-1}$	60 cm^3min^{-1}	45 $\mu mol\ s^{-1}$
$O_2(O_3)$	800 cm^3min^{-1} (1.5 mol%O_3)	595 $\mu mol\ s^{-1}$ (8.9 $\mu mol\ s^{-1}$)	30 cm^3min^{-1} (2 mol% O_3)	22 $\mu mol\ s^{-1}$ (0.4 $\mu mol\ s^{-1}$)

Under all the reaction conditions studied, no methane conversion was observed when oxygen alone was used as the oxidizing agent. Methane conversion was calculated by summing all of the products, and the converted CH_4/converted O_3 ratio was calculated as (CH_4 conversion x CH_4 flow rate)/(O_3 Conversion x O_3 flow rate). During the experiments, special caution was taken due to the high toxicity of ozone. The reactor system was carefully leak-tested, and the effluent gas was properly vented.

Results and Discussions

Gas Phase Reaction. The role of the gas phase reaction in methane oxidation by ozone was studied using a reactor filled with quartz chips (18-20 mesh). The quartz chips used had a very low specific surface area (< 0.2 m^2/g). The gas phase reaction was carried out under both oxygen-rich and oxygen-deficient conditions. As shown in Figure 1, significant gas phase reactions between ozone and methane were observed even at low temperatures, especially in an oxygen-rich atmosphere (Figure 1a). Under these conditions, the main product was CO_2 (Figure 1b). A substantial amount of CO was produced only at temperatures higher than 550 K. Interestingly, the reaction rate was observed to decrease with increasing temperature. This can be understood as arising from the decreasing availability of ozone in the gas phase. With increasing temperature, ozone decomposes faster and is less available to react with methane. Thus, the observed kinetic behavior is a combination of the kinetics of ozone reaction with methane and of ozone decomposition to oxygen.

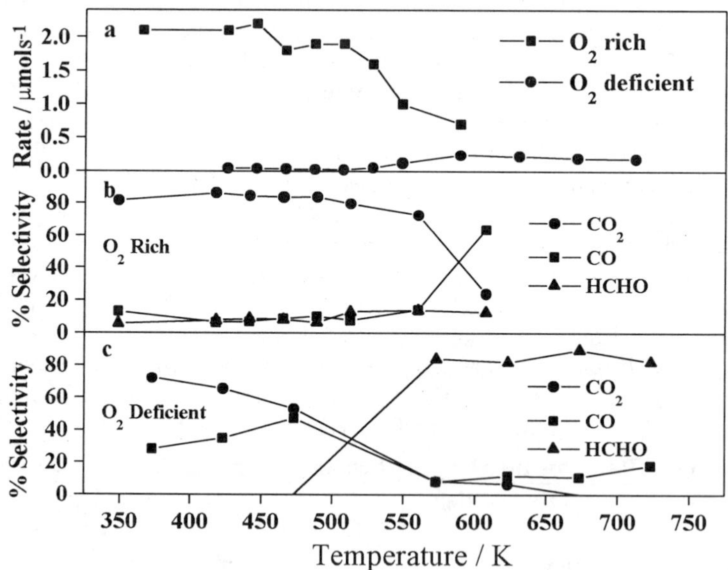

Figure 1. Gas phase reactivity under both O_2 rich and O_2 deficient conditions

At the oxygen-deficient conditions, CO and CO_2 were the only products at low temperatures (Figure 1c). However, the amount of formaldehyde produced increased sharply at about 523 K, and a very high selectivity to formaldehyde was observed at higher temperatures. At even higher temperatures (>750 K), the further oxidation of formaldehyde became important, and CO formation increased.

Notice that the selectivities are different from those reported by Hutchings and coworkers (*11,12*). The reason is probably that the contact time used in this study (< 0.05 s) is much shorter than that used in their study (> 5 s). It is well accepted that selectivity depends strongly on the contact time in partial oxidation reactions.

To better understand the relationship between ozone decomposition and its reaction with methane, we also carried out ozone decomposition with no methane under oxygen-deficient condition. Interestingly, the ozone decomposition was found to follow the same temperature profile as the reaction between ozone and methane (Figure 2). The reaction rates of both reactions were found to follow Arrenhius equation (Figure 3) within the temperature range of 473 to 523 K. Their activation energies were determined to be 74 $kJmol^{-1}$ for methane oxidation, and 80 $kJmol^{-1}$ for ozone decomposition. This suggests that the active oxygen species for methane conversion are probably formed by ozone decomposition.

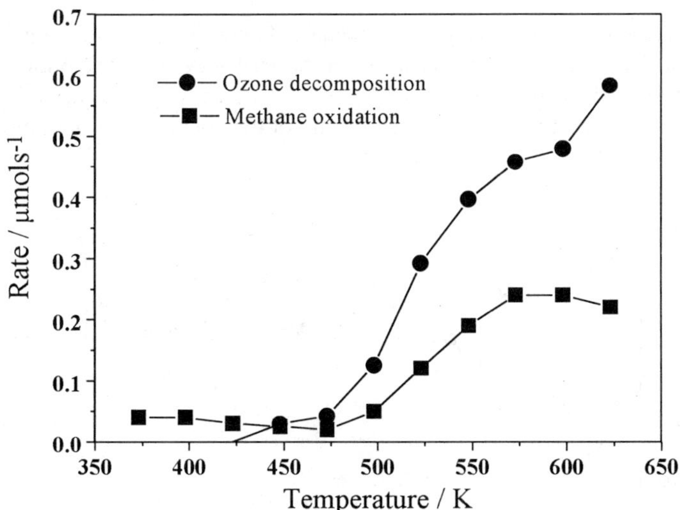

Figure 2. Comparison of methane oxidation rate and ozone decomposition rate in the gas phase under O_2 deficient condition

The mechanism of methane reaction with ozone in the gas phase was postulated to go through the following steps (*18*):

(1) the decomposition of ozone to form an atomic oxygen species and an oxygen molecule,
(2) the reaction between methane and the atomic species to form methyl radicals,
(3) the further oxidation of methyl radicals to methanol and/or formaldehyde.

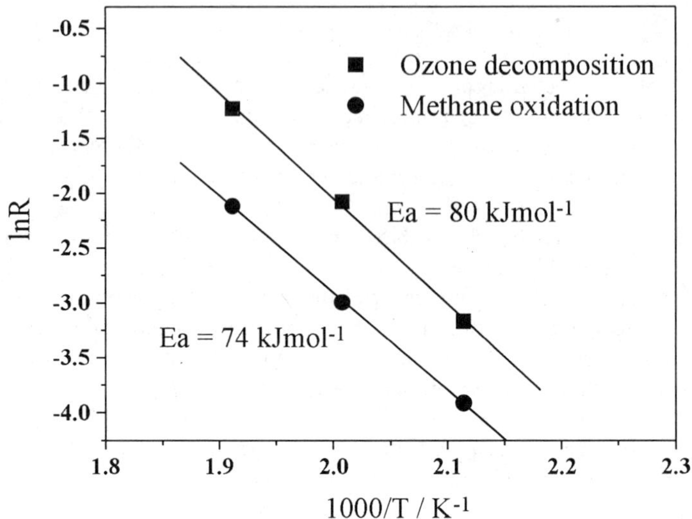

Figure 3. Arrenhius plots for ozone decomposition and methane oxidation rates

Our results are consistent with this mechanism, however, there exists a possible involvement of a chain reaction mechanism. The measurement of inlet and outlet ozone concentrations allowed calculation of converted CH_4/converted O_3 ratio (Figure 4). This ratio gives information on the stoichiometry of the reaction as well as insights

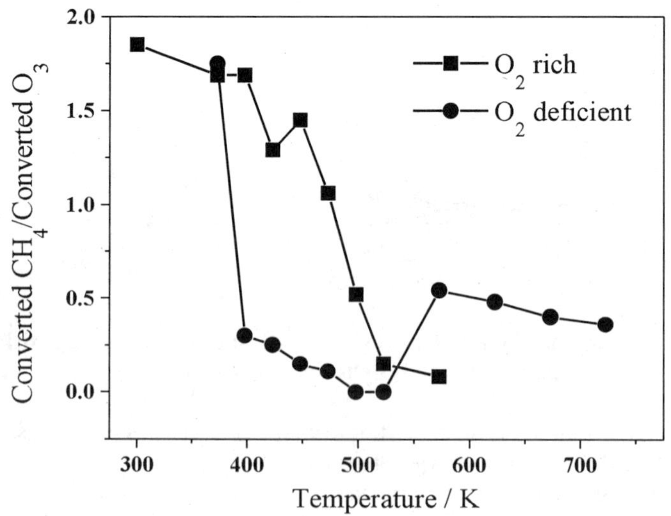

Figure 4. Converted CH_4 / Converted O_3 ratio in the gas phase reaction

for the reaction mechanism. The ratio was found to be larger than 1 under certain conditions, which indicates that a chain reaction mechanism may be involved. The ratio for oxygen-rich conditions was found to be considerably higher than that for oxygen-deficient conditions, which suggests that molecular oxygen is involved in the chain propagation reaction.

Reaction over Catalysts. MgO and 5 wt% Li/MgO were used to study the effect of catalysts on the reaction between ozone and methane.

MgO. Under oxygen-rich conditions, MgO showed considerably higher reactivity than quartz for methane conversion, especially at lower temperatures (Figure 5a). However, the selectivities were comparable to those over quartz (Figure 5b). On the other hand, the gas phase reaction dominated under oxygen-deficient conditions (Figure 6a, b). Both the activity and the selectivities over the catalyst showed no difference from those over quartz. Therefore, under the latter conditions the reaction occurred mainly in the gas phase instead of on the MgO surface.

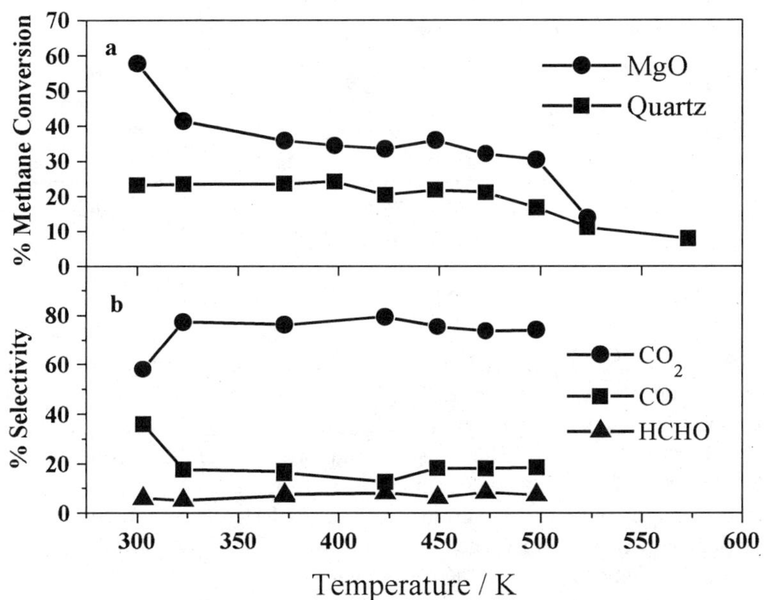

Figure 5. Reactivity of MgO under O_2 rich condition

Notice that the selectivities under oxygen-deficient conditions were remarkably different with increasing temperature. At lower temperatures, the main products were CO and CO_2, while formaldehyde became the predominant product at higher temperatures. The production of CO and CO_2 at lower temperatures suggests that they are not produced by the further oxidation of formaldehyde, even though the oxidation of formaldehyde to CO is observed at higher temperature (> 650 K). There exist two reaction pathways: one to formaldehyde, which can be further oxidized to CO, and a separate one to CO and CO_2. The production of formaldehyde increased sharply around 523 K. This large temperature dependence indicated a high activation energy for the oxidation of methane to produce formaldehyde.

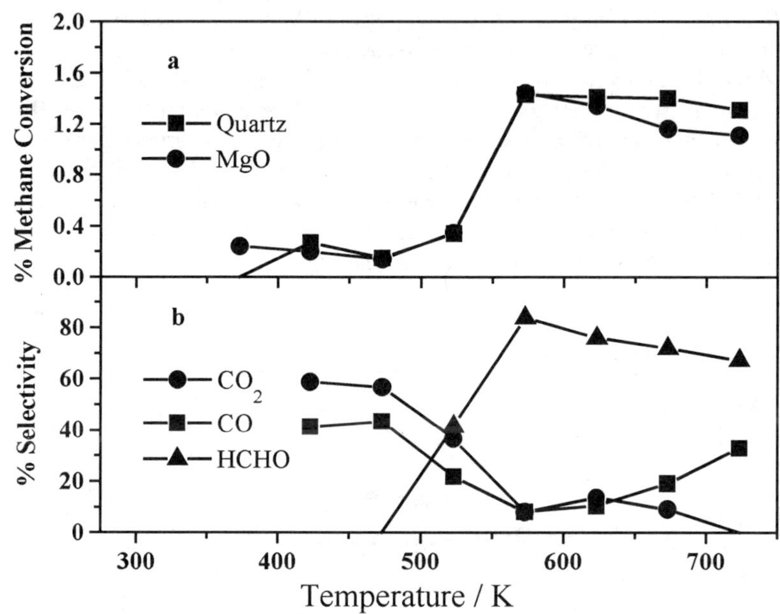

Figure 6. Reactivity of MgO under O_2 deficient condition

Li/MgO. Under oxygen-rich conditions, Li/MgO showed similar reactivity to MgO. The methane conversion over Li/MgO was much lower than that over MgO, however, their areal rates were comparable because of the lower surface area of the Li/MgO catalyst (Figure 7a, b). CO_2 was the main product. Only at temperatures higher than 550 K, was a substantial amount of CO formed.

Under oxygen-deficient conditions Li/MgO showed higher activity and selectivity than MgO. A 4% methane conversion was achieved with the selectivity to formaldehyde exceeding 90% at 650 K (Figure 8a, b).

Li/MgO is known as an effective catalyst for the oxidative coupling of methane, which involves the generation of methyl radicals and its coupling to C_2H_6 (*19*). Surface O^- species (Li^+O^-) are proposed as the active centers. When ozone is used as the oxidizing agent, O^- species and methyl radicals could be generated at much lower temperatures on the Li/MgO catalyst surface as well as in the gas phase.

The reaction of methyl radicals with metal oxides has been studied by Lunsford and coworkers (*20*). CH_3OH and HCHO are formed by the reaction of methyl radicals with metal oxides at temperatures lower than 773 K. It was concluded that high selectivity and yield to oxygenates are possible to achieve if CH_4 could be

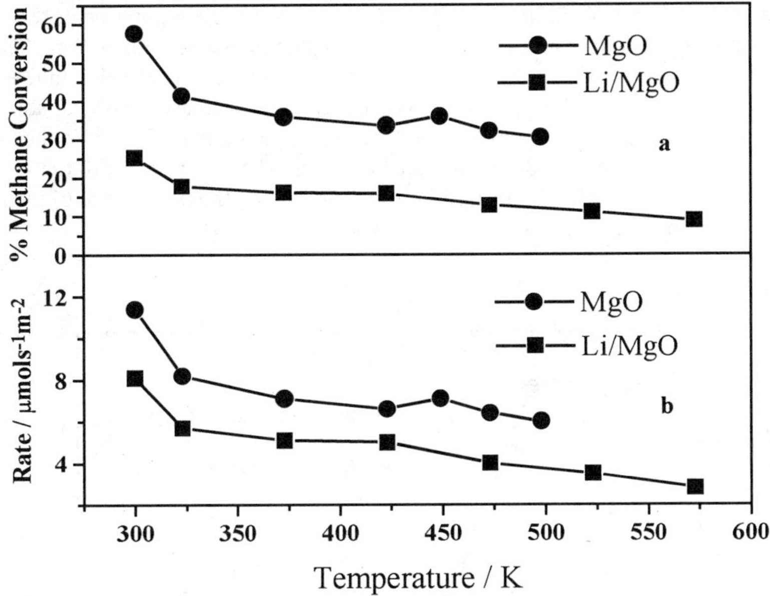

Figure 7. Comparison of activities of MgO and Li/MgO under O_2 rich condition

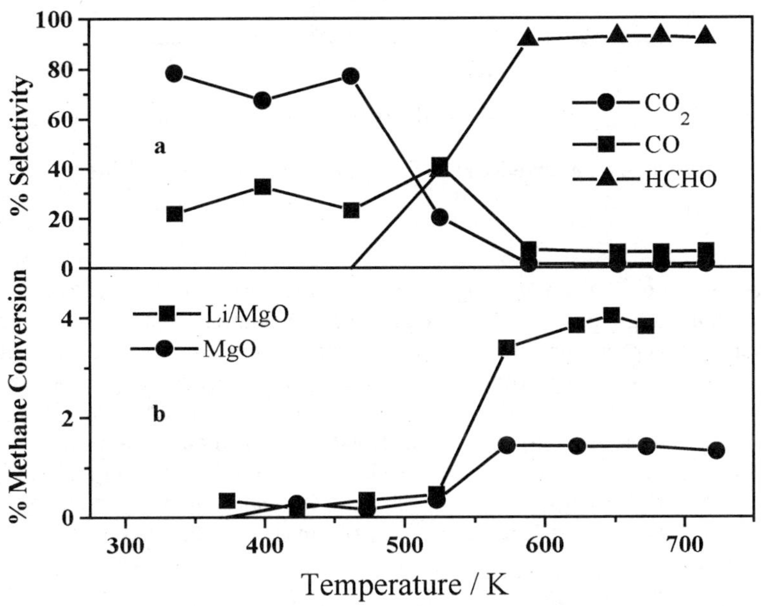

Figure 8. Reactivity of Li/MgO under O_2 deficient condition

converted to $CH_3\bullet$ radicals at moderate temperatures. When ozone is used as the oxidant, active oxygen species can be produced through ozone decomposition. The formed oxygen species can react with methane to generate methyl radicals. In the experiments here the formation of HCHO was observed to increase sharply at about 523 K, which is the thermal decomposition temperature of ozone. Therefore the results are consistent with a radical reaction mechanism shown earlier, with the calculation of converted CH_4/converted O_3 ratio confirming the occurrence of the chain reaction mechanism (Figure 9).

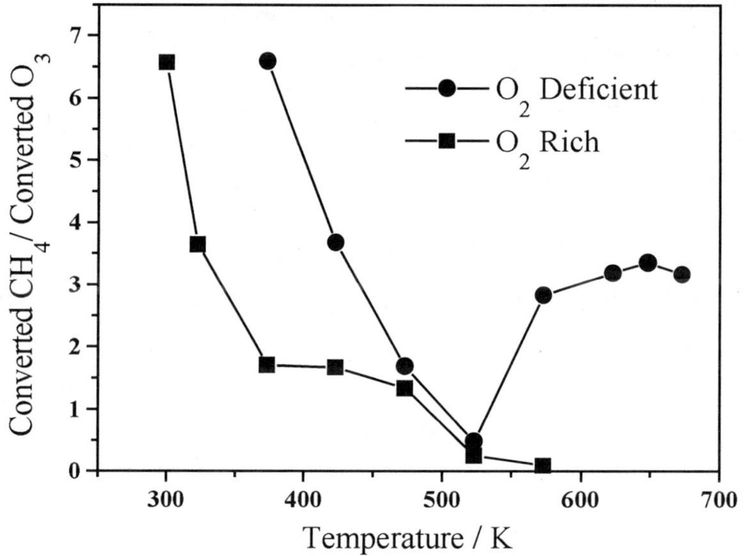

Figure 9. Converted CH_4 / Converted O_3 ratio over Li/MgO

Conclusions

The direct oxidation of methane using ozone at low temperatures (< 673 K) was studied both in the gas phase and over MgO, Li/MgO catalysts. Although significant reaction in the gas phase was observed, catalysts could improve the reactivity both under oxygen-rich and oxygen-deficient conditions. MgO was found to be active for the conversion of methane to CO and CO_2 at oxygen-rich conditions, while Li/MgO catalyst promoted the production of formaldehyde at oxygen-deficient conditions. A 4% methane conversion with a selectivity to HCHO over 90% was achieved at 650 K on a Li/MgO catalyst using ozone as the oxidant. Our calculations of converted CH_4/converted O_3 ratio clearly indicate the involvement of a chain reaction mechanism.

Acknowledgments

We gratefully acknowledge the financial support for this work by the Director, Division of Chemical and Thermal System of the National Science Foundation, under Grant CTS-9311876.

Literature Cited

(1) Gesser, H. D.; Hunter, N. R. In *Methane Conversion by Oxidative Processes*; Wolf, E.E., Ed.; Van Nostrand Reinhold: New York, 1992, 403.
(2) Spencer, N. D. *J. Catal.* **1988**, 109, 187.
(3) Barbaux, Y.; Elamrani, A. R.; Payne, E.; Gengembre, L.; Bonnelle, J. P.; Grzybowska, B. *Appl. Catal.* **1988**, 44, 17.
(4) Spencer, N. D.; Pereira, C. J. *J. Catal.* **1989**, 116, 399.
(5) Parmaliana, A.; Frusteri, F.; Mezzapica, A.; Miceli, D.; Scurrell, M. S.; Giordano, N. *J. Catal.* **1993**, 143, 262.
(6) Smith, M. R.; Ozkan, U. S. *J. Catal.* **1993**, 141, 124.
(7) Smith, M. R.; Zhang, L.; Driscoll, S. A.; Ozkan, U. S. *Catal. Lett.* **1993**, 19, 1.
(8) Liu, H.-F.; Liu, R.-S.; Liew, K. Y.; Johnson, R. E.; Lunsford, J. H. *J. Am. Chem. Soc.* **1984**, 106, 4117.
(9) Khan, M. M.; Somorjai, G. A. *J. Catal.*, **1985**, 91, 263.
(10) Zhan, K. J.; Khan, M. M.; Mak, C. H.; Lewis, K. B.; Somorjai, G. A. *J. Catal.* **1985**, 94, 501.
(11) Hutchings, G. J.; Scurrell, M. S.; Woodhouse, J. R. *Appl. Catal.* **1988**, 38, 157.
(12) Hutchings, G. J.; Scurrell, M. S.; Woodhouse, J. R. *Methane Conversion*, Bibby, D. M.; Chang, C. D.; Howe, R. F.; Yurchak, S., Eds.; Elsevier: Amsterdam, 1988.
(13) Gesser, H. D.; Hunter, N. R.; Das, P. A. *Catal. Lett.* **1992**, 16, 217.
(14) Otsuka, K.; Wang, Y. *Chem. Lett.* **1994**, 1893.
(15) Otsuka, K.; Wang, Y. *Catal. Lett.* **1994,** 24, 85.
(16) *Ullmann's Encyclopedia of Industrial Chemistry*, Vol. A18, 350, VCH, 1991
(17) Ito, T.; Wang, J.-X.; Lin, C.-H.; Lunsford, J. H. *J. Am. Chem. Soc.*, **1985**, 107, 5062.
(18) Toby, S.; Toby, F. S. *J. Phys. Chem.*, **1989**, 93, 2453.
(19) Lunsford, J. H. *Angew. Chem. Int. Ed. Engl.*, **1995**, 34, 970.
(20) Pak, S.; Rosynek, M. P.; Lunsford, J. H. *J. Phys. Chem.* **1994**, 98, 11786.

Chapter 28

Alkane Oxidation on Vanadium Silicalite Compared to Titanium Silicalite

T. Tatsumi, Y. Hirasawa, and J. Tsuchiya

Engineering Research Institute, Faculty of Engineering, University of Tokyo, 2–11–16 Yayoi, Bunkyo-ku, Tokyo 113, Japan

Silicates containing Ti and V, TS-2 an VS-2, have been synthesized using tetrabutylammonium cation as the template. Contrasted with TS-2, VS-2 gives appreciable terminal oxidation products in the oxidation of hexane and allylic oxidation products in the oxidation of hexenes. The relative rate of 2-hexene/1-hexene epoxidation is much lower on VS-2 than on TS-2. These findings indicate the radical character of the active oxygen species on VS-2. Spin trapping experiments have revealed that the primary hexyl radical is generated in the VS-2-H_2O_2-hexane system and that no alkyl radical is observed with the TS-2 system. For both TS-2 and VS-2, hydroperoxy radicals are trapped by nitrones. It is proposed that in the oxidation of alkanes the oxidation of internal carbons and that of terminal carbons proceed by different mechanisms.

The Ti analogs of ZSM-5 (TS-1) (*1*) and ZSM-11 (TS-2) (*2*) catalyze the oxidation of simple alkanes at mild temperatures (*3-5*). Similar materials containing V have also been prepared (*6-8*) and exhibit interesting oxidation properties: the oxyfunctionalization with hydrogen peroxide of the primary carbon atom leading to the formation of primary alcohols and aldehydes was observed only with the vanadium silicalite VS-2 (*8*) and not with TS-1 or TS-2 (*3-5*). A comparison has been made of the oxidation efficiency of Ti and V silicalites and it has been postulated that a predominantly heterolytic pathway is operative in the case of TS-2 and that the principal pathway is homolytic in the case of VS-2 (*9*).

The purpose of this study is to synthesize and characterize silicalites containing Ti and V and to investigate the similarities and differences in the metal environments and reaction pathway in the oxidation of hydrocarbons. Attention has been also focused on identifying the reactive radical intermediates. Although direct observation should generally provide the most reliable information of the radical, many radicals cannot be observed directly by ESR. The technique of spin trapping has been developed to detect and identify radicals too short lived or too scarce for direct ESR observation. In this approach, a transient radical reacts with a spin trap to produce a persistent radical, a spin adduct. The ESR spectrum of the adduct is characteristic of the trapped radical and can help in identifying the transient species.

0097–6156/96/0638–0374$15.00/0

Experimental

Catalysts. Titanium silicalite (TS-2) was prepared by the method reported by Reddy and Kumar (*2*). Vanadium silicalite (VS-2) was prepared as follows: 54 g of tetrabutylammonium hydroxide (TBAOH) (40% aqueous solution, TCI) was added slowly under nitrogen atmosphere at 273 K to 136.5 g of tetraethyl orthosilicate (TEOS) (TCI) dissolved in 2-propanol (45 g). To the resultant mixture, a solution of 2.1 g of vanadium trichloride in 30 g of deionized water was added slowly together with 177 g of TBAOH (6.1 wt%) under vigorous stirring. Then 60 g of TBAOH (10 %) was added and the mixture was stirred for 90 min. Then the temperature was raised to 323-333 K and the solution was stirred at that temperature for 2 h. The clear green liquid was then transferred into a teflon flask, placed into an autoclave and heated at 433 K under static conditions for 4 days. After crystallization, the white solid materials were recovered by filtration, washed with water and dried overnight at 383 K. The template was removed by calcining the solid at 763 K in air for 6 h.

Characterization. X-ray diffraction (XRD) powder patterns were collected on a Rigaku RINT 2400 X-ray diffractometer. Fourier-transform infrared spectra were recorded on a Perkin Elmer 1600 spectrometer. Raman spectra were obtained on a Nihon Bunko NR-1800 spectrometer. Electron spin resonance spectra were recorded at room temperature on a JEOL JM-EFIX spectrometer at 9.3 GHz (X-band). UV-visible diffuse reflectance spectra were recorded on a Hitachi 340 spectrometer.

Reactions. The oxidation of hexane and the epoxidation of 1-hexene and 2-hexene were performed in a flask with 30% aqueous H_2O_2. The products were analyzed on a Shimadzu GC 14 gas chromatograph equipped with a Nukol capillary column for alkane oxidation and an OV-17 capillary column for alkene epoxidation.

Spin trapping. The spin traps 2-methyl-2-nitrosopropane (MNP), phenyl-*N-tert*-butyl nitrone (PBN) and α-4-pyridyl 1-oxide *N-tert*-butyl nitrone (POBN) were used as purchased. To detect radicals in the presence of hexane, 0.3 ml of 30% aqueous H_2O_2 was added to 1.25 ml of hexane in 4.0 ml of acetonitrile. To the resultant mixture, 4 mg of MNP and 16 mg of catalyst was added and the mixture was shaked for 30 seconds. To detect radicals in the absence of hexane, 20 mg of catalyst was added to the mixture of 1.0 ml of 30% aqueous H_2O_2 and 10 mg of PBN or POBN was added and the resultant mixture was shaked for 30 seconds. An aliquot of the supernant was transferred to an ESR cell and the ESR spectrum recorded. Highly dispersed Mn^{2+} on MgO was used as a marker.

Results and discussion

Structure of TS-2 and VS-2. XRD data show that TS-2 and VS-2 samples are highly crystalline and their patterns closely matched with those reported for MEL structures (*8*). However, TEM observation suggests that these materials are agglomerates of very small crystallites of MFI structure, not the MEL structure expected from the template (*10*). The absence of (110) reflection in their XRD pattern is also indicative of the MFI structure. The apparently singlet peak at $2\theta = 45°$ would be due to the coalescence resulting from line broadening of the (0 10 0), (0 8 4), (10 0 0) and (8 0 4) reflections.

In the ESR spectrum of the as-synthesized VS-2, the anisotropic hyperfine splitting (8-fold) caused by ^{51}V nucleus is very well resolved without the presence of appreciable superimposed broad singlet, indicating the atomic dispersion and immobility of the V^{4+} species. The *g*-values ($g_{\parallel} = 1.935$, $g_{\perp} = 1.994$) and hyperfine coupling constants ($A_{\parallel}$ = 188 G, $A_{\perp}$ = 72 G) are typical of V^{4+}=O complexes with square pyramidal coordination (*11*). The high dispersion is preserved after calcination at 763 K and subsequent photoreduction with H_2 at 77 K (*12*).

The IR spectrum of calcined VS-2 exhibits a medium intensity band at 965 cm^{-1}, although it shows no band around this wavenumber as synthesized. The Raman spectrum of VS-2 also shows an intense peak at 965 cm^{-1}. Photoluminescene studies indicates the presence of tetrahedrally coordinated V-oxide species; a vibrational fine structure owing to the vanadyl band was clearly observed (*12*). The energy gap between the (0 → 0) and (0 → 1) transition bands of the V=O vibration was found to be *ca.* 965 cm^{-1} in good agreement with the IR and Raman V=O stretching bands. This energy gap is slightly different from that of the vanadium oxide species highly dispersed on Vycor glass or silica (1035 cm^{-1}), suggesting the presence of some electronic perturbation owing to the neighboring OH group in the zeolite lattice. Although there is an IR band at 960-970 cm^{-1} in the TS-1 and TS-2 samples, no evidence for the Ti=O structure is obtained from their phosphorescence spectra. This IR band is assigned to Si-O^- defect or Si-OH associated with the incorporation of Ti into the zeolite framework (*13-15*).

Oxidation Reaction. The catalytic activities of these materials are shown in Table I. In the oxidation of hexane, VS-2 gives appreciable terminal oxidation products as well as products formed by the oxidation of internal secondary carbons, in contrast with TS-2, which gives only internal oxidation products. These results are in agreement with the previously reported ones (*8,9*). Since the relative rate of 2-hexene/1-hexene epoxidation is much lower on VS-2 than on TS-2, it is shown that VS-2 generates active oxygen species of less electrophilic character. This finding and formation of allylic oxidation products on VS-2 suggest radical character of the active oxygen species.

Table I. Oxidation of Hexane and Hexenes with Aqueous H_2O_2

Catalyst	Reactant	Turnovers (mol/mol-Ti or V)	Terminal Oxidation (%)	Allylic Oxidation (%)	Turnover Rate (x 10^{-3} s^{-1})
TS-2	Hexane	20	0	-	2.8
TS-2	1-Hexene	14	-	0	1.9
TS-2	2-Hexene	74	-	0	10.3
VS-2	Hexane	4.4	32	-	0.61
VS-2	1-Hexene	15	-	51	2.1
VS-2	2-Hexene	20	-	31	2.8

TS-2 (Si/Ti = 85) 50 mg, H_2O_2 (30% aq.) 2.5 ml, substrate 2.5 ml, 333 K, 2 h.
VS-2 (Si/V = 58) 50 mg, H_2O_2 (30% aq.) 1.0 ml, substrate 3.9 ml, acetonitrile 12.5 ml, 333 K, 2 h.

Spin trapping is a valuable tool for the study of free radical processes. Two kind of spin traps have been developed, nitrone and nitroso compounds. Nitroso compounds, such as MNP, can provide considerably more information than nitrones as the radical to be trapped adds directly to the nitroso nitrogen.

When MNP is added to the hexane oxidation system catalyzed by VS-2, we have obtained an ESR spectrum consisting of a triplet (1:2:1) of triplets (Figure 1). This signal is identified as originating from the MNP-$CH_2(CH_2)_4CH_3$. This clearly shows that the $CH_3(CH_2)_4CH_2\bullet$ radical is generated in the VS-2-H_2O_2-hexane system. If secondary hexyl radical such as $CH_3(CH_2)_3CH(CH_3)\bullet$ had been trapped instead of $CH_3(CH_2)_4CH_2\bullet$, the ESR signal would have been doublet of triplets.

In spite of the predominant oxidation of secondary carbon atoms of hexane in the case of VS-2, no MNP-*sec*-hexyl radical adduct is observed. There are two explanations for this finding. One explanation is that, owing to the transition state shape selectivity in the restricted space in the zeolite channel, the interaction of *sec*-hexyl radical with MNP may be hindered. The other explanation is that only the terminal oxidation proceeds by a homolytic mechanism; a distinct homolytic mechanism involving free radicals operates for this terminal oxidation whereas a kind of heterolytic mechanism seems to operate in the oxidation of secondary carbon atoms. In line with this hypothesis, as shown in Figure 2, the ESR signal assignable to MNP-radical adduct was hardly observed with the TS-2-H_2O_2-hexane system which gives only the products resulting from the oxidation of secondary carbons.

Postulated Reaction Mechanism. While oxygen-centered radical adducts of MNP are quite unstable, nitrones can be used for the study of oxygen-centered radicals. To detect the reactive species generated directly from H_2O_2 in the presence of VS-2, PBN and POBN were employed. The ESR spectrum of the PBN adduct is shown in Figure 3. Its coupling constants, $A_N = 14.4$ G and $A_H = 2.63$ G, suggest that the trapped radical is $\bullet$OOH (*16*). For the PBN-OH radical adduct in aqueous solution, A_N larger than 15.2 G is expected (*16*). This means that the VS-2-H_2O_2 system is clearly different from the Fenton system, in which the reactive species is revealed to be the hydroxy radical by using PBN (*17*) and POBN (*18*). The coupling constants of the POBN adduct, $A_N = 13.8$ G and $A_H = 1.66$ G, also support this conclusion. The POBN- OH radical adduct should have $A_N = ca.$ 15 G (*16*). Similar spectra are obtained with the reaction of H_2O_2 and TS-2 in the presence of the spin trapping agents (Figure 4).

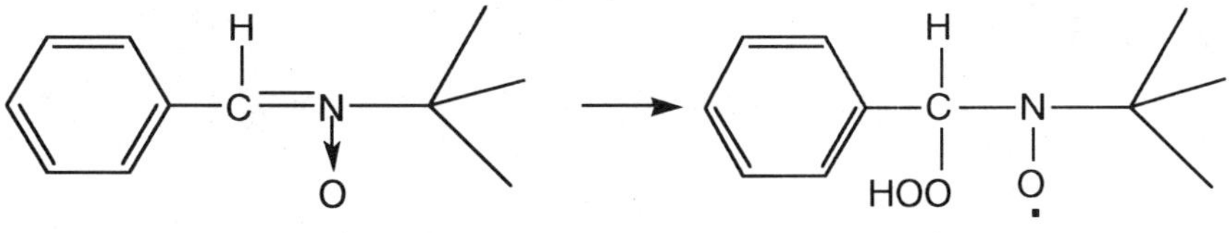

Thus, it has been revealed that the $\bullet$OOH radical is generated and is the major product both in the VS-2-H_2O_2system and in the TS-2-H_2O_2 system. Since $\bullet$OOH radical is relatively stable, it is unlikely that the $\bullet$OOH radical abstracts the H$\bullet$ radical from an alkane. Therefore it is not unreasonable that no alkyl radical is detected in the TS-2-H_2O_2-hexane system. Considering that VS-2 has vanadyl groups, the following reaction scheme is proposed. On addition of aqueous H_2O_2,

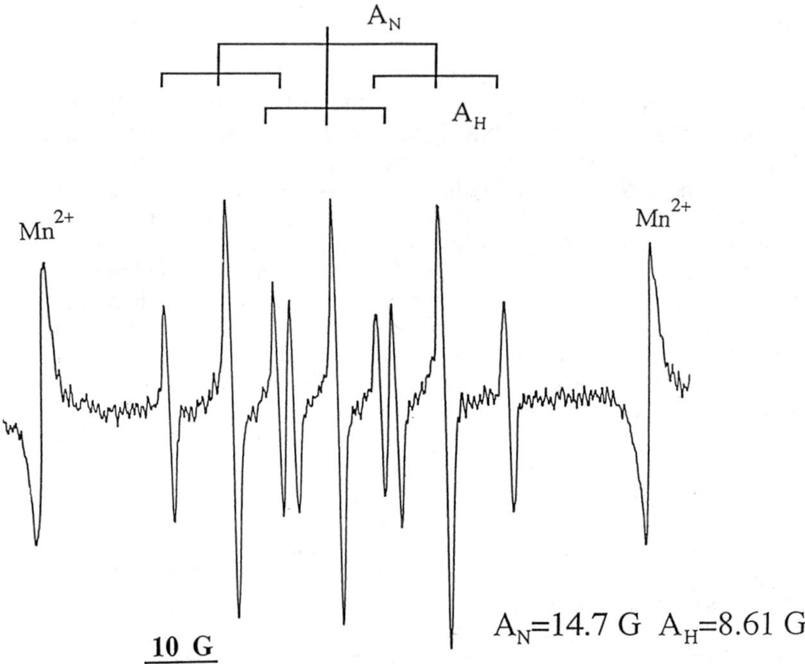

Figure 1. ESR spectrum of spin adduct of MNP in the hexane oxidation by aqueous H_2O_2 catalyzed by VS-2.

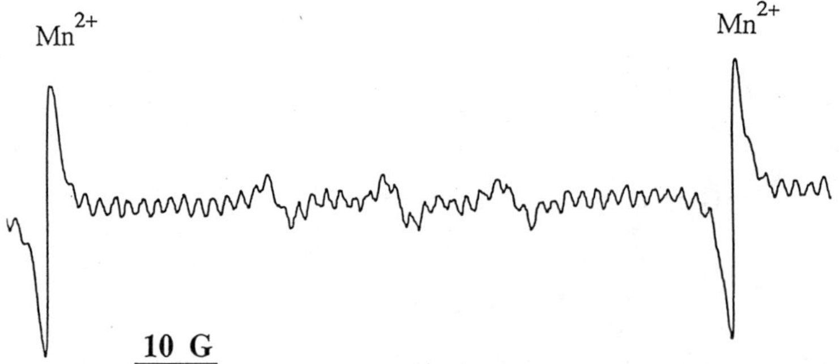

Figure 2. ESR spectrum of spin adduct of MNP in the hexane oxidation by aqueous H_2O_2 catalyzed by TS-2.

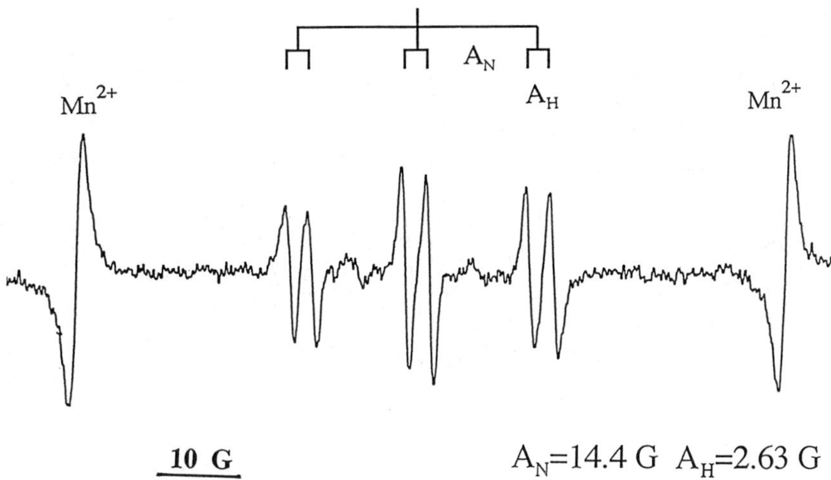

Figure 3. ESR spectrum of spin adduct of PBN in the reaction of VS-2 with aqueous H_2O_2 .

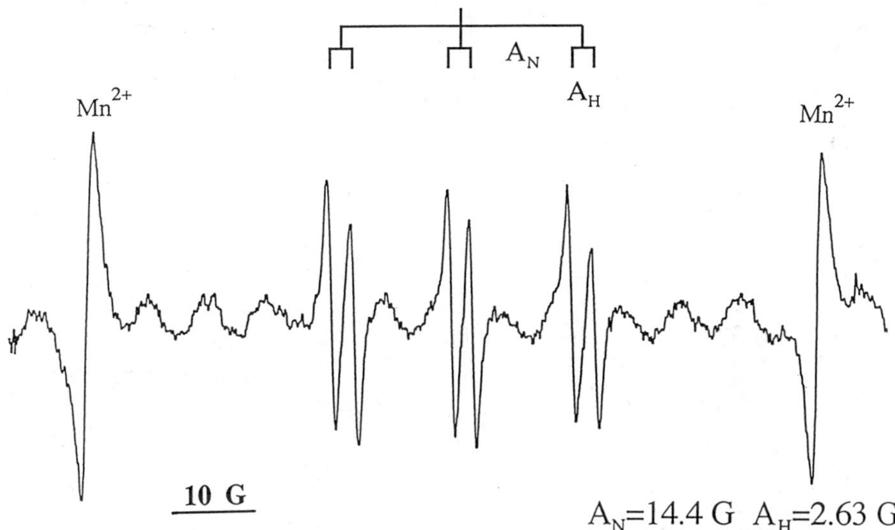

Figure 4. ESR spectrum of spin adduct of PBN in the reaction of TS-2 with aqueous H_2O_2 .

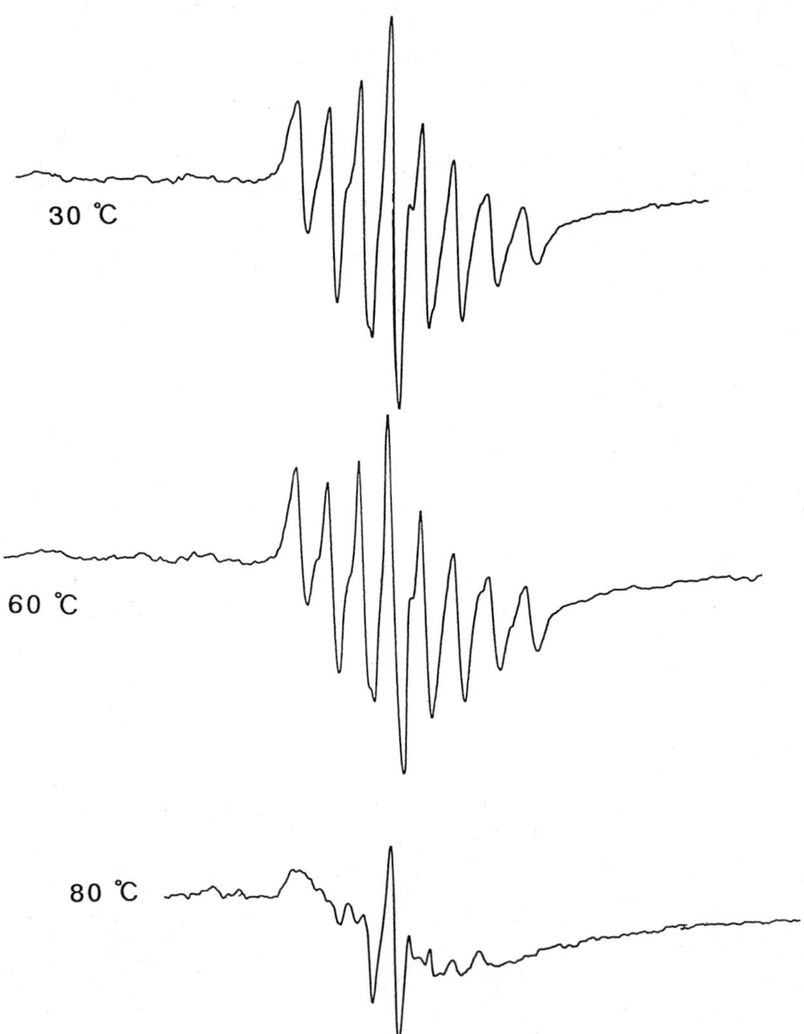

Figure 5. ESR spectrum of VS-2 contacted with aqueous H_2O_2 .

the IR band at 965 cm^{-1} assignable to the vanadyl group disappears, while it is virtually unchanged on addition of water. Although the calcined VS-2 is ESR-silent, the typical spectrum of V^{4+} reappears when VS-2 is contacted with H_2O_2. The ESR spectra of the VS-2 which has been calcined, treated with H_2O_2 and evacuated at increasing temperatures are shown in Figure 5. The spectra are only partly anisotropic at room temperature. Evacuation of the H_2O_2-adsorbing sample at temperatures higher than 343 K reduces the ESR signal intensity, indicating reoxidation of V^{4+} to V^{5+}.

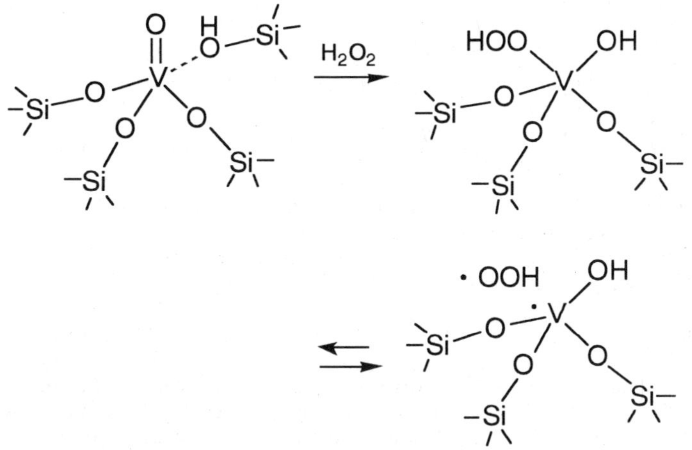

As shown below, we propose that the formation of V^{4+}-OO • concomitant with the release of water might be responsible for the oxidation of primary carbons of alkanes. However, we could not detect such species, probably because they are too short-lived.

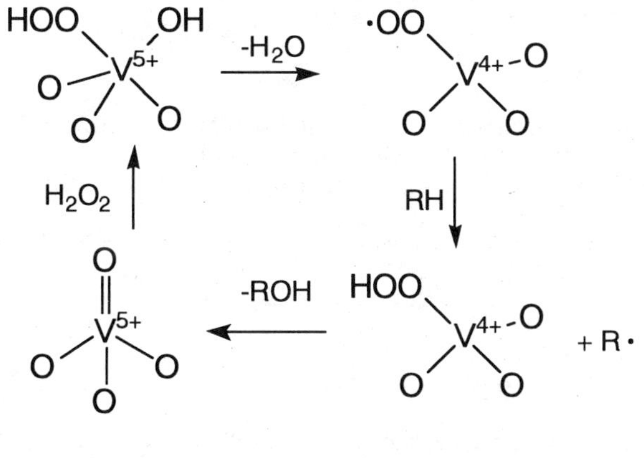

This scheme is similar to the one proposed by Hari Prasada Rao et al. (*19*). Since only the primary hexyl radical is observed, only the oxidation of terminal carbons may be envisaged to occur via this pathway; the oxidation of internal carbons might proceed by a different mechanism, which may be common to both VS-2 and TS-2 systems.

Since titanyl groups are absent on TS-2 and most Ti is considered to be surrounded by three SiO and one OH groups (*14,15*) we can depict the following scheme. Although the redox potential of the Ti^{4+}/Ti^{3+} couple suggests that Ti^{4+} is hardly reducible, it can be expected that the highly sensitive ESR measurements enabled us to detect the $\cdot$ OOH radical trapped by the nitrones.

$$\text{HO}(\text{O})_3\text{Ti}^{4+} \xrightarrow[-\,H_2O]{H_2O_2} \text{HOO}(\text{O})_3\text{Ti}^{4+} \rightleftharpoons \cdot\text{OOH}\ (\text{O})_3 \cdot\text{Ti}^{3+}$$

It is conceivable that oxidation of internal carbons of alkanes on both TS-2 and VS-2 proceeds via a mechanism involving a metal coordinated OOH group (*5, 20*) while the electrophilicity of this group should be influenced by the difference in the electron-withdrawing nature between Ti^{4+} and V^{4+} .

Conclusions

A comparison of oxidation reactions on VS-2 and TS-2 shows that the active oxygen species on VS-2 have more radical character than those on TS-2. Spin trapping experiments have revealed that the primary hexyl radical is generated in the VS-2-H_2O_2-hexane system, in contrast with the TS-2 system where no alkyl radical is observed. The formation of hydroperoxy radicals is observed both for TS-2 and VS-2. It is proposed that the oxidation of internal carbons of alkanes on VS-2 and TS-2 proceeds by a mechanism involving a metal coordinated OOH group and that V^{4+}-OO $\cdot$ is the active species for oxidation of terminal carbons of alkanes on VS-2.

Acknowledgment

We thank Prof. M. Anpo for measuring photoluminescence spectra of TS-2 and VS-2.

Literature Cited

1. Taramasso, M.; Perego, G.; Notari, B., U.S. Patent 4,410,501 (1983).
2. Reddy, J.S.; Kumar, R.; Ratnasamy, P., *Appl. Catal.*, **1990**, *58*, L1.
3. Tatsumi, T.; Nakamura, M.; Negishi, S.; Tominaga, H., *J. Chem. Soc. Chem. Commun.*, **1990**, 476.
4. Huybrechts, D.R.C.; DeBryuker, L.; Jacobs, P.A., *Nature*, **1990**, *345*, 240.
5. Clerici, M.G., *Appl. Catal.*, **1991**, *68*, 249.
6. Rigutto, M.S.; Van Bekkum, H., *Appl. Catal.*, **1991**, *58*, L1.
7. Centi, G.; Perathoner, S., Trifilo, F., Aboukais, A.; Aissi, C.F.; Guelton, M., *J. Phys. Chem.*, **1993**, *96*, 123.
8. Hari Prasada Rao, P.R.; Ramaswamy, A.V.; Ratnasamy, P., *J. Catal.*, **1992**, *137*, 225.
9. Ramaswamy, A.V.; Sivasanker, S.; Ratnasamy, P., *Microporous Mater.*, **1994**, *2*, 451.

10. Tatsumi, T., Koyano, K.A.,Terasaki, O., unpublished data.
11. Kucherov, A.B.; Slinkin, A.A. ; *Zeolites,* **1987**, *7*, 583.
12. Anpo, M.; Zang, S.; Yamashita, H.; Hirasawa, Y.; Tatsumi, T., Preprints of the 7th Japan-China-USA Symposium on Catalysis, 1992, p.55.
13. Camblor, M.A.; Corma, A.; Perez-Pariente, J., *J. Chem. Soc. Chem. Commun.*, **1993**, 557.
14. Deo, G.; Turek, A.M.; Wachs, I.E.; Huybrechts, D.R.C.; Jacobs, P.A., *Zeolites*, **1993**, *13*, 365.
15. Khouw, C.B.; Davis, M.E., *J. Catal.*, **1995**, *151*, 77.
16. Buettner, G.R., *Free Radical Biology & Medicine*, **1987**, *3*, 259.
17. Schaich, K.M.; Borg, D.C., In *Autoxidation in Food and Biological Systems*, M.G. Simic, M. Karel, Ed., Plenum Press, New York, NY, 1980, pp 45-70.
18. Lain, C.S.; Piette, L.H., *Biochem. Biophys. Res. Commun.*, **1977**, *78*, 51.
19. Hari Prasada Rao, P.R.; Ramaswamy, A.V.; Ratnasamy, P., *J. Catal.*, **1993**, *141*, 604.
20. Khouw, C.B.; Dartt, C.B.; Labinger, J.A.; Davis, M.E., *J. Catal.*, **1994**, *149*, 195.

Chapter 29

Retardation of Carbon Deposition in CO_2–CH_4 Reaction on Metal Sulfide Catalysts

Toshihiko Osaki

National Industrial Research Institute of Nagoya, Hirate-cho, Kita-ku, Nagoya 462, Japan

The CO_2-oxidation (reforming) of methane on MoS_2 and WS_2 catalysts was compared to that on a Ni/SiO_2 catalyst using a continuous flow technique. The sulfide catalysts had lower activity than the Ni/SiO_2 catalyst, however in contrast to Ni catalyst, deactivation due to carbon deposition was not observed on the sulfide catalysts. The direct decomposition of CH_4[g] to C[s] and $2H_2$[g] was much less significant on the sulfides than on Ni, however, the dissociation of CO_2[g] to CO[g] and O[ads] occurred to a greater extent on the sulfide surface. The rate was expressed by the following equations: $r = kP_{CH4}^{0.54\sim0.76}P_{CO2}^{-0.09\sim-0.10}$ on MoS_2 and WS_2 and $r = kP_{CH4}^{-0.30}P_{CO2}^{0.16}$ on Ni/SiO_2, respectively. The rate equations suggest that during steady-state reaction, CO_2 species are abundant on the surface of the sulfide catalysts, while CH_4 species are abundant on Ni. This is probably the cause for the retardation of carbon deposition on the sulfide catalysts and for their low activity for this reaction.

Much attention has been paid to the CO_2-oxidation (reforming) of methane from the view point of producing valuable synthesis gas (CO and H_2, syngas) (*1-2*). The conversion of methane into syngas has been usually carried out by the H_2O-reforming reaction (*3*), however, there has been a need to operate at lower H/C ratio to obtain the optimum H_2/CO ratio for the synthesis. This can be achieved by using CO_2 alone or by combining CO_2 with H_2O as oxidant for the reforming reaction.

Attention has been given to the CO_2-CH_4 reaction from another point of view. Because both CO_2 and CH_4 are greenhouse effect gases, their emission in the atmosphere causes global warming. Among the gases, released the quantity of CO_2 is enormously large. In order to decrease the emission of CO_2, great efforts have been paid to the separation of CO_2 from exhaust gas and its utilization (*4*). Therefore, the CO_2-oxidation of methane can be considered as an appropriate reaction because both harmful gases can be converted to useful gases.

Most of the previous work on the CO_2-oxidation of methane has concerned with supported Ni catalysts because the reaction can be regarded as related to the steam-oxidation of methane (*5*). One of the major challenges in the commercialization of the

0097–6156/96/0638–0384$15.00/0

CH_4-CO_2 reaction is the development of a high-performance catalyst which can efficiently produce syngas without the formation of carbon. Carbon deposition commonly leads to catalyst deactivation and reactor blocking. The deposition has been controlled by the use of excess carbon dioxide and by the addition of alkaline metals to the supported Ni catalysts to accelerate the oxidation of carbon by carbon dioxide and also to retard the coking (*6-9*). A serious carbon build-up commonly reported with acidic-supported catalysts tends to disappear as basic supports are employed (*10-11*). Another way for large-scale applications is the use of a partly sulfur poisoned Ni catalyst as in the *SPARG process* in which sulfur blocks active sites for the nucleation of carbon (*12*).

Although the catalyst performance of Ni has been extensively studied, little work has been reported on metal sulfides of Mo and W. It was reported that on Ni, the activity for the methanation of CO significantly decreased with time, however, not so significantly on MoS_2 (*13*). In the present paper, the catalyst performance for CH_4-CO_2 reaction and the carbon deposition due to the decomposition of methane were investigated on MoS_2 and WS_2 catalysts and compared to that on Ni/SiO_2. It was hoped that the sulfides of Mo and W would exhibit high performance for the CO_2-oxidation of CH_4 without deactivation due to carbon deposition. The results are discussed in terms of the abundance of adsorbed species on the catalyst surface.

Experimental

Catalyst and Reagents

MoS_2 and WS_2 catalysts were prepared by decomposing $(NH_4)_2MoS_4$ and $(NH_4)_2WS_4$ (both from Aldrich, purity; both more than 99.9 %), respectively, in a stream of H_2. For comparison, a 20 wt % Ni/SiO_2 catalyst was prepared by impregnating a SiO_2 (GL Science, surface area; 401.9 m^2/g) support with an aqueous solution of $Ni(NO_3)_2$, followed by drying and subsequent calcination at 773 K for 3 h.

Hydrogen was purified by passage through a silica gel column where the small amount of oxygen present were converted to water. Helium was purified by passage through a titanium metal sponge heated at 1073 K, and then through a molecular sieve trap. Both CO_2 (ca. 99.9%) and CH_4 (> 99.9 %) were used without further purification.

Apparatus and Procedure. The CH_4-CO_2 reaction was carried out using a conventional flow type microreactor. After reducing the catalysts in flowing hydrogen at 723 K for 3 h, helium (40 ml/min) was passed over the catalysts at 723 K for 0.5 h. Then an equimolar mixture of CH_4 and CO_2 (28 ml/min) was introduced onto the catalyst. The catalyst weight was 5.1 mg, 305.1 mg, and 359.0 mg for Ni/SiO_2, MoS_2, and WS_2, respectively. The reaction was carried out under the conditions of a differential reactor. Effluent gases were analyzed by gas chromatography using separation columns (Shimadzu GC-14A).

The dependence of methane conversion on reaction time was investigated by reducing the catalyst (catalyst weight, 93.5 mg for Ni/SiO_2 and 500 mg for MoS_2) in flowing hydrogen at 873 K for 3 h, passing helium over the catalyst at 873 K for 0.5 h, followed by an equimolar mixture of CH_4 and CO_2 (total flow rate, 28 ml/min and 60 ml/min for MoS_2 and Ni/SiO_2, respectively) at 873 K for 6 h

The rate of carbon deposition was investigated by thermal gravimetric analysis (TGA, Shimadzu DT-20B). After the reduction of a catalyst with hydrogen followed by purging with helium at 773 K, CH_4 (40 ml/min) was introduced onto the catalyst

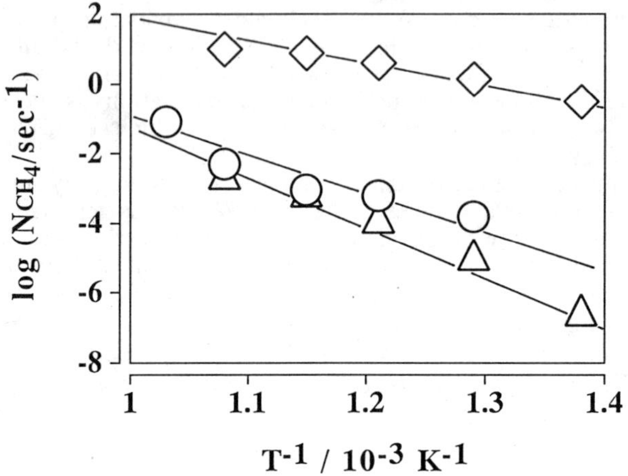

Fig. 1 Arrhenius plots for the rate of methane conversion in steady-state CH_4-CO_2 reaction on Ni/SiO_2 (◇), on MoS_2 (○), and on WS_2 (△) by use of a continuous flow technique at atmospheric pressure.

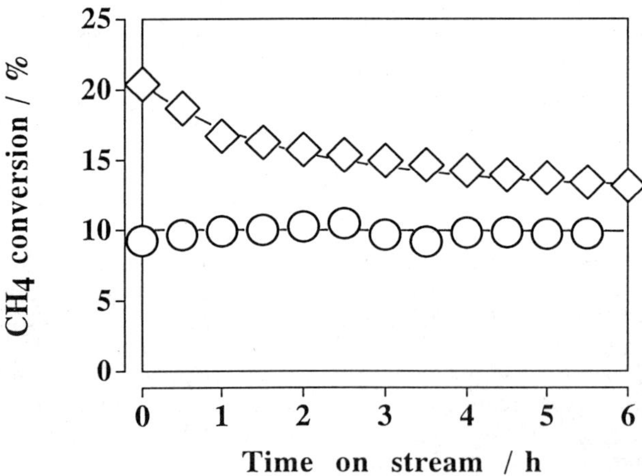

Fig. 2 Deactivation of a catalyst in the CO_2-reforming of CH_4 on Ni/SiO_2 (◇) and on MoS_2 (○). Reaction temperature = 873 K. W/F = 0.582 g h mol^{-1} for Ni/SiO_2 and 6.67 g h mol^{-1} for MoS_2.

and the catalyst weight was continuously measured. The weight of the catalyst used was 25.5 mg for Ni/SiO_2, 16.5 mg for MoS_2, and 50.7 for WS_2.

The dissociation of carbon dioxide on a catalyst was investigated by a conventional pulse technique with a thermal conductivity detector (Shimadzu GC-3BT). After reducing a catalyst with hydrogen (catalyst weight was 100.5 mg and 224.6 mg for Ni/SiO_2 and WS_2, respectively) at 973 K for 3 h, CO_2 (0.54 ml) was pulsed onto the catalyst for 15-16 times at 973 K in a helium carrier gas. The amount of CO_2 dissociatively adsorbed on the catalyst was determined by the amount of CO generated.

The dependence of reaction rate on partial pressures was investigated under differential reactor conditions. After reducing a catalyst followed by purging with helium at 973 K, a mixture of CH_4, CO_2, and helium (total flow rate, 54 ml/min) was introduced onto the catalyst. The catalyst weights were 5.1 mg, 305.1 mg, and 280.3 mg for Ni/SiO_2, MoS_2, and WS_2, respectively.

Characterization of catalysts. The BET surface area of the prepared sulfide catalysts was measured with an N_2 adsorption apparatus (Carlo Erba Sorptomatic 1800). X-ray diffraction analysis of the prepared sulfide catalysts was carried out using a Rigaku RAD-1VC (CuKα, 30 kV, 30 mA). Chemical analysis of the prepared sulfide catalysts was carried out using a sulfur-carbon analyzer and an oxygen-nitrogen analyzer (LECO). The amount of CO adsorbed on the catalysts was measured using a conventional pulse microreactor at 298 K in a flow of helium carrier gas.

Results

Characterization of catalysts and amount of CO adsorbed on catalysts

X-ray diffraction patterns of the sulfide catalysts prepared from $(NH_4)_2MoS_4$ and $(NH_4)_2WS_4$ showed broad diffraction peaks due to molybdenum disulfide and tungsten disulfide, respectively. It was also found that the ratio: S/Mo (W) of the prepared sulfide was close to the stoichiometric value of ca. 2 by chemical analysis. The BET surface area of the sulfide catalysts were 34.2 m^2/g and 63.8 m^2/g for MoS_2 and WS_2, respectively, which were much larger than those of commercially obtained MoS_2 (0.90 m^2/g) and WS_2 (1.34 m^2/g).

The amount of CO adsorbed on the catalysts was; 41.0, 15.0, and 38.6 μmol/g for Ni/SiO_2, MoS_2, and WS_2 catalyst, respectively. Even on the sulfide catalysts, a considerable amount of CO was adsorbed.

CH_4-CO_2 reaction under steady-state reaction conditions. Fig. 1 shows Arrhenius plots for the rate of methane conversion on sulfide catalysts and on Ni/SiO_2 catalyst in the initial period of time. The rates are displayed on the basis of turnover frequency (N_{CH4} / s^{-1}), where methane conversion rate is divided by the number of adsorbed CO on the catalyst. On the sulfide catalysts, selective oxidation of methane to CO was observed, while no ethane nor higher hydrocarbon was obtained. As shown in Fig. 1, the sulfide catalysts showed lower activity than the Ni/SiO_2. Between the sulfide catalysts, MoS_2 exhibited a little higher activity than WS_2, while the apparent activation energy seemed a little larger on WS_2 than on MoS_2.

Results of continuous CH_4-CO_2 reaction for 6 h are shown in Fig. 2. As shown, Ni/SiO_2 gradually lost its activity with time; the methane conversion decreased from ca. 20 % to ca. 13 % over a period of 6 h at the described conditions. This can be

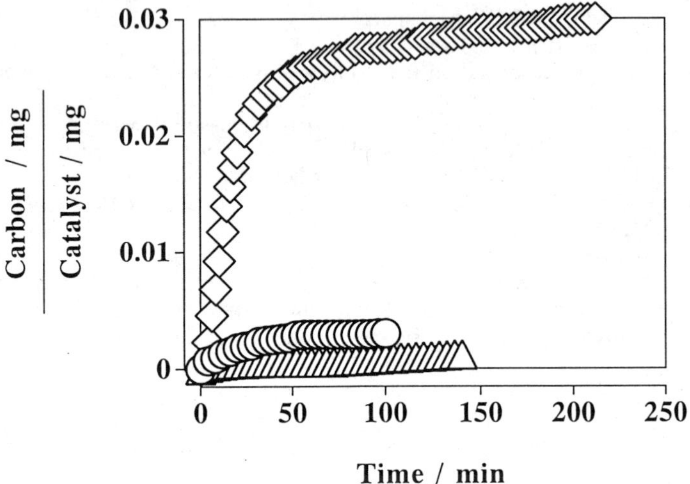

Fig. 3 Formation of carbon due to the decomposition of methane in the absence of CO_2 on Ni/SiO_2 (◇), on MoS_2 (○), and on WS_2 (△) at 773 K.

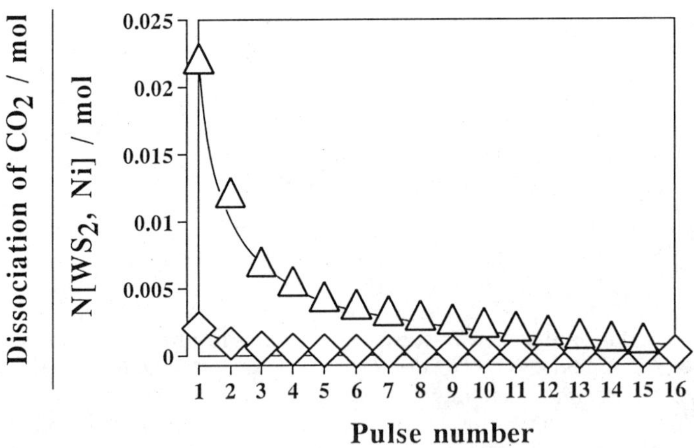

Fig. 4 Dissociation of CO_2 on Ni/SiO_2 (◇) and WS_2 (△) at 973 K.

ascribed to the deposition of carbon on the Ni surface due to the decomposition of methane, viz.

$$CH_4[g] \text{ -----> } C[s] + 2H_2[g] \quad (1)$$

On the other hand, the deactivation of the catalyst was not observed on the sulfide catalyst; the methane conversion was kept about 10 % for 6 h under the conditions, although the activity was lower than on the supported Ni catalyst.

Carbon deposition and dissociation of CO_2 on catalyst surface. The activity for methane decomposition in the absence of CO_2 was examined by TGA. Fig. 3 shows the results on the sulfides and on the supported Ni. The amount of carbon deposited on the catalyst is shown on the basis of deposited carbon weight (mg) per catalyst weight (mg) used. When methane was continuously flowed over the catalyst, the catalyst weight increased with time, indicating the decomposition of methane to produce carbon and hydrogen. As shown, supported Ni exhibited significantly high activity for the decomposition of methane, while the sulfide catalysts showed much lower activity.

The activity for dissociation of CO_2 in the absence of methane was investigated by a conventional pulse technique. When CO_2 was pulsed onto the catalyst in a flowing carrier gas of helium, the production of CO was observed, indicating that CO_2 was dissociatively adsorbed to give CO[g] and O[ads] on the catalyst surface, viz.

$$CO_2[g] \text{ -----> } CO[g] + O[ads] \quad (2)$$

Fig. 4 shows the results on the sulfide and the supported Ni catalyst. The amount of CO_2 dissociated was compared on the basis of mole of dissociated CO_2 (mol) per mole of catalyst (mol) used. As shown, the amount of dissociated CO_2 decreased with pulse number, indicating that the active sites on the catalyst were gradually filled with the adsorbed O[ads]. The amount of CO_2 dissociated was much larger on the sulfide catalysts than on the Ni catalyst.

Dependence of reaction rate on partial pressure of CH_4 and CO_2. The dependence of the reaction rate on the partial pressure of CH_4 and CO_2 was investigated to determine which adsorbed species were more adsorbed on the catalyst surface. The rate for the CH_4-CO_2 flow reaction is represented by the following equation:

$$r_i = k_i \, P_{CH_4}{}^x \, P_{CO_2}{}^y \quad (3)$$

where r_i and k_i are the reaction rate and the rate constant, respectively, and p_j is the partial pressure of CH_4 and CO_2. Fig. 5 shows the dependence of the reaction rate on the partial pressure. From the slope of the straight line obtained, the kinetic orders, x and y, can be calculated, and are summarized in Table 1. The Ni/SiO_2 catalyst used in our experiment gave almost similar results to those already reported on Ni/SiO_2 (*14*) and preliminarily on Ni/Al_2O_3 (*15*). The reaction order on Ni is zero or negative with respect to the partial pressure of methane, while positive with respect to that of CO_2. On sulfide catalysts, on the other hand, the orders are positive with respect to the partial pressure of methane, while negative with respect to that of CO_2.

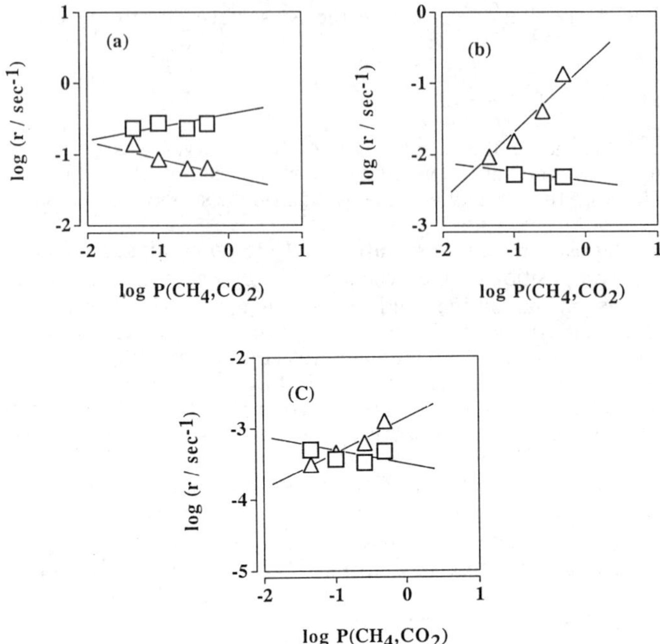

Fig. 5 The dependence of the reaction rate on the partial pressure of CH_4 and CO_2 for the CO_2-reforming of CH_4 on Ni/SiO_2 (a), on MoS_2 (b), and on WS_2 (c) at 973 K. △: log r vs. log P(CH_4), □: log r vs. log P(CO_2).

Table 1. Kinetic orders for the rate equation: $r = k\, P_{CH4}{}^{x}\, P_{CO2}{}^{y}$ in CH_4-CO_2 reaction

Catalyst	x	y	Temperature K	Partial pressure[a] atm
MoS_2	0.76	-0.10	973	0.26 - 0.74
WS_2	0.54	-0.09	973	0.26 - 0.74
Ni/SiO_2	-0.30	0.16	973	0.26 - 0.74
Ni/SiO_2[b]	0.02 - 0.05	0.5 - 0.6	823-973	ca. 0.15 - 0.60
Ni/Al_2O_3[c]	-0.27	0.21	1073	0.11 - 0.56

[a]Examined Range of partial pressure for CH_4 and CO_2.
[b]Y. Sakai et al., see reference (*14*).
[c]Our preliminary data, see reference (*15*).

Discussion

Selectivity for methane oxidation reaction by carbon dioxide

It has been proposed for the CO_2-reforming of methane that the following steps are involved, viz.

$$CO_2[g] \Leftrightarrow CO[ads] + O[ads] \quad (4)$$
$$CH_4[g] \Leftrightarrow CH_x[ads] + (4-x)/2\ H_2[g] \quad (5)$$
$$CH_x[ads] + O[ads] \Leftrightarrow CO[ads] + x\ H[ads] \quad (6)$$
$$CO[ads] \Leftrightarrow CO[g] \quad (7)$$
$$x\ H[ads] \Leftrightarrow x/2\ H_2[g] \quad (8)$$

The surface reaction between CH_x[ads] and O[ads] gives CO[ads] and H[ads] as shown in equation 6, from which both CO and H_2 are produced. It is reported that the formation of C_2H_6 was observed in the initial stages of reaction on a Rh catalyst, indicating the coupling of CH_x[ads] on the Rh surface (*16*). On sulfide catalysts, on the other hand, no ethane, ethylene, acetylene, nor other higher hydrocarbon were produced during the steady-state reaction, suggesting that the sulfide catalysts are selective for the conversion of methane to CO by carbon dioxide.

In the continuous CH_4-CO_2 reaction, carbon deposition due to the decomposition of methane was significant on supported Ni catalysts. The deposited carbon accumulated on the Ni surface, and resulted in a gradual deactivation of the catalyst as shown in Fig. 2. On the sulfide catalysts, on the other hand, the contribution of equation 1 seems small since little coking was observed. Therefore, it can be concluded that sulfide catalysts are more selective because of lack of coking from CH_4, although the activity was lower than that of supported Ni catalysts.

Strongly adsorbed species on catalyst surface. The formation of carbon from methane could involve intermediate hydrocarbon species, CH_x[ads], produced through the consecutive elimination of hydrogen atoms:

$$CH_4[g] \longrightarrow CH_4[ads] \longrightarrow CH_3[ads] \longrightarrow CH_2[ads] \longrightarrow CH[ads] \longrightarrow C[s] \quad (9)$$

Each adsorbed species including surface carbon, C[s], react with CO_2 (or O[ads]) to give CO (and H_2). From the results in Fig. 3, supported Ni seems to have a strong affinity for CH_4, causing the dehydrogenation of methane step by step to deposit carbon. On the sulfide catalysts, on the other hand, there seems to be not so strong on affinity for methane. This must be one of the causes for the suppression of carbon deposition on the sulfide catalysts in addition to the their low activity for this reaction.

It is reported that the dehydrogenation of 2-propanol proceeds preferentially on reduced MoS_2, while dehydration proceeds on partially oxidized MoS_2. The active sites of reduced and partially oxidized MoS_2 catalysts are assumed to be coordinatively unsaturated sites and acidic sites formed by reduction and partial oxidation of MoS_2, respectively (*17*). During the CH_4-CO_2 reaction, a part of the active sites, the coordinatively unsaturated sites on the sulfide catalysts might be oxidized by CO_2 to give acidic sites. This is evidenced by the finding that CO_2 pulse onto sulfide catalysts produced significantly large amounts of CO as was shown in Fig. 4.

The most abundantly adsorbed species on the catalysts is indicated by the dependence of the reaction rate on the partial pressures of CH_4 and CO_2. As shown in Table 1, the reaction order on the Ni/SiO_2 catalyst is negative with respect to CH_4

partial pressure while positive with respect to CO_2 partial pressure. The result indicates that CH_4 is strongly adsorbed species on the Ni surface. On the sulfide catalysts, on the other hand, the reaction order is positive with respect to CH_4 partial pressure while negative with respect to CO_2 partial pressure, suggesting that CO_2 species are strongly adsorbed on the sulfide surface.

The kinetic orders on other supported Ni catalysts are also shown in Table 1 for comparison. Although the reaction conditions are significantly different for each set of experiments, the reaction order with respect to CH_4 partial pressure is zero or negative, while positive with respect to CO_2 partial pressure. From the data on the supported Ni catalysts, it is found that CH_4 rather than CO_2 is more strongly adsorbed on Ni and that the results obtained in our experimental conditions are in qualitative agreement with those reported earlier, despite the large differences in conditions.

These findings indicate that during the steady-state CH_4-CO_2 reaction, CH_4 species are abundant on the Ni/SiO_2 surface, while CO_2 species are abundant on the sulfide surface. The speculation is substantiated by the results shown in Figs. 3 and 4. The retardation of deactivation due to carbon deposition on sulfide catalysts can be ascribed to the abundantly adsorbed CO_2 species.

From the results shown in Figs. 3, 4, and 5, it seems for the CH_4-CO_2 reaction that the first step on the sulfide catalysts is the dissociative adsorption of CO_2[g] on the active site, the coordinatively unsaturated site, to give CO[ads] and O[ads]. Whereas, the coordinatively unsaturated site accepts O[ads] from CO_2[ads], being changed to an acidic site, viz.

$$\frac{CO_2[\text{ads}]}{\text{coordinatively unsaturated site}} \Longleftrightarrow \frac{CO[\text{ads}]}{\text{acidic site (O[ads])}}$$

On Ni catalysts, on the other hand, the first step must be the decomposition of methane to give CH_x[ads] and (4-x)/2 H_2. Since carbon deposition is caused by the decomposition of methane, covering the catalyst surface with adsorbed CO_2 (or O[ads]) should give favorable conditions for the suppression of coking, which results in no deactivation as shown in Fig. 2. Our preliminary investigation also showed that the promotion of Ni/Al_2O_3 by alkaline metal salts changed the reaction order from negative to positive with respect to CH_4 partial pressure, while positive to negative with respect to CO_2 pressure (*15*). The addition of alkaline metal salts also decreased the activity of supported Ni for the direct decomposition of methane (*15*). The results on alkali-promoted Ni catalyst are very similar to the present findings on sulfide catalysts, which lead us to speculate that the sulfide catalysts play the same role as alkali-promoted Ni catalysts.

It is found that the sulfide catalysts retard the coking reaction, while still maintaining an adequate, although reduced, activity for the CO2-reforming of methane. A similar effect was observed by adding H_2S to the reactant gases of CH_4 and CO_2 on a nickel catalyst in a process, called the *SPARG process* (*12*). In this process, the sulfur inhibits the rate of carbon formation more than that of the reforming reaction on the nickel. The effects are explained by assuming that a larger ensemble is involved in the formation of carbon than in the reforming reaction. In our present study, although the active sites on MoS_2 and/or WS_2 is still not identified, the sulfide catalysts used seem to exhibit a similar tendency to the partly sulfur poisoned nickel catalyst. Further study is necessary for clarification.

The carbon deposition deactivates a catalyst, however, the ease of CH_4 decomposition on a catalyst appears to be important to obtain high activity for methane conversion reactions. In fact, the activity difference between the sulfide catalysts and the supported Ni catalyst in the CH_4-CO_2 reaction must be ascribed to the relative ease of adsorption and decomposition of CH_4 on the catalyst surfaces. Therefore, the promotion of CH_4 adsorption, which may lead to not only high activity but also significant coking, must be reconciled with the retardation of carbon formation for developing a practical catalyst for industry.

Conclusions

The CO_2-reforming of methane was investigated on sulfide catalysts of Mo and W by comparison to a supported Ni catalyst. Little carbon formation on the sulfide catalysts occurred in contrast to the supported Ni catalyst. Rate equations for both catalysts were determined. From the dependence of reaction rate on the partial pressures of CH_4 and CO_2, it was deduced that the sulfide surface was covered with adsorbed CO_2[ads] (or O[ads]) during the steady-state CH_4-CO_2 reaction. This must be one of the causes for the retardation of carbon deposition in addition to their low activity for this reaction.

Acknowledgment

I am grateful to Dr. Toshiaki Mori of National Industrial Research Institute of Nagoya for the helpful discussions.

Literature Cited

(1) Richardson, J. T.; Paripatyadar, S. A. *Appl. Catal.* **1990**, 61, 293.
(2) Perera, J. S. H. Q.; Couves, J. W.; Sankar, G.; Thomas, J. M. *Catal. Lett.* **1991**, 11, 219.
(3) Rostrup-Nielsen, J. R. In *Catalysis, Science and Technology*; Anderson, J. R.; Boudart, M., Ed.; Springer, Berlin, **1984**, Vol. 5; 1.
(4) *Proc. Symp. on Chemical Fixation of Carbon Dioxide*, Nagoya **1991**.
Proc. Int. Conf. on Carbon Dioxide Removal, Amsterdam **1992**.
Proc. Int. Conf. on Carbon Dioxide Utilization, Bari **1993**.
Proc. Int. Conf. on Carbon Dioxide Removal, Kyoto **1994**.
Proc. Int. Conf. on Carbon Dioxide Utilization, Oklahoma **1995**.
(5) Gadalla, A. M.; Sommer, M. E. *Chem. Eng. Sci.* **1989**, 44, 2825.
(6) Sacco, A.; Geurts, Jr., F. W. A. H.; Jablonski, G. A.; Lee, S.; Gately, R. A. *J. Catal.* **1989**, 119, 322.
(7) Gadalla, A. M.; Bower, B. *Chem. Eng. Sci.* **1988**, 43, 3049.
(8) Gadalla, A. M.; Sommer, M. E. *J. Am. Ceram. Soc.* **1989**, 72, 683.
(9) Yamazaki, O.; Nozaki, T.; Omata, K.; Fujimoto, K. *Chem. Lett.* **1992**, 1953.
(10) Mizuhara, Y.; Miyashita, Y.; Fujita, T.; Ishihara, T.; Takita, Y. *Preprint of Spring Annual Meeting of Chem. Soc. Jpn.*, **1992**, 3C441.
(11) Ashcroft, A. T.; Cheetham, A. K.; Green, M. L. H.; Vernon, P. D. F. *Nature*, **1991**, 352, 225.
(12) Dibbern, H. C.; Olesen, P.; Rostrup-Nielsen, J. R.; Tottrup, P.B.; Udengaard, N. R. *Hydrocarbon Processing*, **1986**, 65, 71.
(13) Saito, M.; Anderson, R. B. *J. Catal.*, **1980**, 63, 438.
(14) Sakai, Y.; Saito, H.; Sodesawa, T.; Nozaki, F. *React. Kinet. Catal. Lett.*, **1984**, 24, 253.
(15) Sakuma, K.; Horiuchi, T.; Fukui, T.; Osaki, T.; Mori, T. *Preprint of Spring Annual Meeting of Chem. Soc. Jpn.*, **1995**, 2H628.
(16) Solymosi, F.; Kutsan, Gy.; Erdohelyi, A. *Catal. Lett.* **1991**, 11, 149.
(17) Sugioka, M.; Kimura, F. *J. Japan Petrol. Inst.* **1985**, 28, 306.

Chapter 30

Short-Chain Alkane Activation

An Investigation of SO_2-Promoted Propane Oxidation over Pt (111) and Pt–AlO_x Model Systems

Karen Wilson[1], Christopher Hardacre[2], and Richard M. Lambert[1]

[1]Department of Chemistry, University of Cambridge, Lensfield Road, Cambridge CB2 1EW, England
[2]School of Chemistry, Queen's University of Belfast, Belfast BT9 5AG, Northern Ireland

SO_2 chemisorption on oxygenated Pt(111) enormously enhances the dissociative chemisorption and subsequent combustion of propane. This activation of the metal surface is induced by an adsorbed sulfoxy species which is formed > 220 K. In the absence of adsorbed SO_2 the sticking probability of propane is immeasurably small. However in the presence of SO_2, the precursor-mediated initial sticking probability rises from ~0.02 at 300 K to ~ 0.15 at 160 K. XPS and HREELS measurements identify the active species as SO_4^{2-}, and also demonstrate the consumption of SO_4 during the oxidation reaction. Coincident CO_2 and SO_2 formation suggest decomposition of a complex reaction intermediate: this is supported by isotope experiments involving CO adsorption onto $SO_2/^{18}O_2$ precovered Pt(111). Propane oxidation over sulphated AlO_x films on Pt(111) is also reported, with increased activity for submonolayer oxidised Al films being observed compared to clean Pt(111).

The activation of short chain alkanes is often the rate limiting step in catalytic oxidation of hydrocarbons [1]. In particular, the slow oxidation of alkanes, namely propane and methane, is a technological problem faced by the automotive exhaust catalyst industry [2]. Reactor studies of Pt/Al_2O_3 and $Pt\text{-}Rh/Al_2O_3/CeO_2$ catalysts have shown that the inclusion of SO_2 in the gas feed significantly promotes the oxidation of propane [3,4], whilst inhibiting reactions of CO, NO and propene. It has also been observed that in the presence of SO_2, the activity of Pt/Al_2O_3 catalysts for propane oxidation becomes independent of Pt loading. In the absence of SO_2, the activity usually increases with Pt loading, a result which suggests the formation of different active sites on inclusion of SO_2 in the gas feed. Infra-red analysis of these catalysts following exposure to O_2 and SO_2, indicates the presence of SO_4^{2-} on the support, which is thought to be involved in the activation process. Enhanced C-H bond dissociation is often attributed to catalyst acidity [5]; for example homogeneous Pt(III) solutions display increased oxidative capacity in the presence of sulphuric acid.

0097–6156/96/0638–0394$15.00/0

In the case of heterogeneous catalysis, the effects of acid strength have been investigated for propane oxidation over Pt/ZrO_2, Pt/SiO_2 and Pt/Al_2O_3 [6]. Following sulphonation, the increase in acid strength of Pt/ZrO_2 is more than that observed for Pt/Al_2O_3, however only Al_2O_3 supported catalyts display the SO_2 promotional effect. Effects due to support acidity do not therefore explain the activation of alkane oxidation by SO_2, so there appears to be little fundamental understanding of the mechanism involved.

Recently we have shown that under UHV conditions SO_2 can promote propane dissociation over oxygenated Pt(111) at 300 K [7]. This dissociative chemisorption occurs only in the presence of coadsorbed O_2 and SO_2 , which is striking in view of the fact that clean Pt(111) is inactive towards the dissociative chemisorption of propane [8]: only molecular physisorption occurs at 100 K.

We report here the results of a study of propane adsorption/oxidation in the presence of SO_2 over Pt(111). Particular attention is given to the identity of the active sulphur species to understand the likely mechanism of SO_2 induced propane activation.

Experimental

Experiments were performed using two different UHV chambers. Chamber 1 [9] was used for all TPRS and Al deposition experiments, and is equipped with a RFA for LEED/Auger measurements and a multiplexed mass spectrometer. XPS and HREELS experiments were performed on Chamber 2 [10], which was a VSW 12" ARIES system equipped with a HA-100 hemispherical analyser, Mg K_α X-ray source, and a HA-300 HREELS analyser. In both chambers the sample could cooled to 160K. HREELS measurements were taken in the specular direction (45°) using a primary beam energy of 9.6eV.

Crystal cleaning was achieved by Ar^+ bombardment and 800K oxygen treatment. Gases used in these experiments were CO (99.97%); O_2 (99.995%); C_3H_8 (99.995%) and SO_2 (99.98%) which were supplied by M.G.Distillers. Al deposition was achieved by means of a heated ceramic crucible filled with an Al wire melt (99.99% Goodfellow Metals), the design of which has been described in detail elsewhere [11]. Al uptakes were performed by following the attenuation of the 44 eV Pt Auger transition inorder to ascertain the monolayer point.

No S (152eV) Auger data are reported, due to the susceptibility of SO_2 and SO_x species to electron stimulated desorption/dissociation [12]. Post-reaction Auger spectroscopy did not show any detectable amounts of residual S or C remaining after the propane oxidation experiments.

Results

The interaction of propane over clean and oxygen pre-covered Pt(111) was initially studied at 300 K. As may be predicted from earlier studies of hydrocarbon adsorption [8], no adsorption occurred over the clean or oxygen pre-covered surface at either

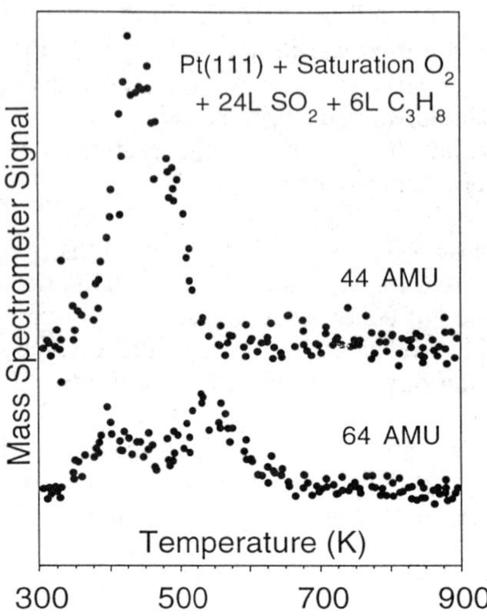

Figure 1: TPR following 300K adsorption of 6L C_3H_8 on Pt(111) precovered by saturation O_2 and 24L SO_2. (Reproduced with permission from ref. 7. Copyright 1995 American Chemical Society)

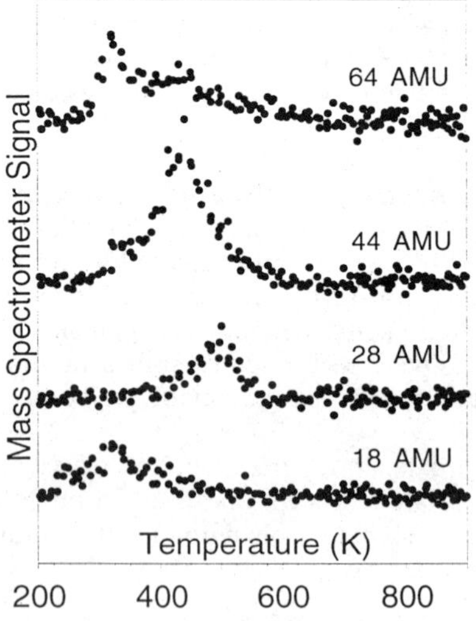

Figure 2: TPR following 160K adsorption of 6L C_3H_8 on Pt(111) precovered by saturation O_2 and 24L SO_2 (dosed at 300K). (Reproduced with permission from ref. 7. Copyright 1995 American Chemical Society)

temperature. Likewise, over SO_2 pre-covered Pt(111), no adsorption of C_3H_8 was observed at any temperature. However, stepwise exposure of Pt(111) at 300K to (i) O_2 then (ii) SO_2 produced a surface active for dissociative adsorption of C_3H_8, with CO_2 being observed at 420K in the subsequent TPR spectrum (Figure 1).

We examined other saturated hydrocarbons ranging from C_1 to C_7 for their oxidation activity over Pt(111), *n* - butane exhibited similar promotional behaviour to propane in the presence of both chemisorbed oxygen and SO_2. However with *n*-heptane, oxidation was possible in the presence of chemisorbed oxygen alone. In this instance SO_2 appears to behave as an extra source of $O_{(a)}$, with an increased amount of CO being formed during the reaction. For CH_4 and C_2H_6, no promotional effect of SO_2 was observed, in good agreement with conventional catalytic studies; no beneficial effect of SO_2 was observed for methane oxidation over Pt-Rh/CeO_2/Al_2O_3 [4].

Reactively-formed H_2O desorbs from Pt(111) $\leq$ 300 K, hence in order to study the oxidation reaction in its entirety the 300 K O_2/SO_2 pre-treated crystal was cooled to 160 K prior to C_3H_8 adsorption. The resulting TPR spectrum is shown in Figure 2, with H_2O, CO_2 and CO desorbing at 320 K, 430 K & 500 K respectively. A low temperature shoulder in the 44 amu desorption is also visible at 320K, the temperature at which CO oxidation occurs over Pt(111) [13]. However when the Pt(111)/$O_{(a)}$ system was exposed to SO_2 at 160 K, *no subsequent adsorption of C_3H_8 occurred* (Figure 3), suggesting that there is a threshold temperature for formation of the surface species responsible for activation of Pt with respect to C_3H_8 adsorption. In order to determine this threshold temperature, SO_2 was adsorbed on the Pt(111)/$O_{(a)}$ surface at 160 K, then annealed to 220 K, 250 K, 300 K, 400 K and 600 K prior to C_3H_8 exposure at 160 K. The degree of C_3H_8 oxidation was found to increase sharply (Figure 4) following annealing the surface to 250 K, and continued to increase with annealing temperature, before passing through a maximum at 300 K. No C_3H_8 adsorption was observed following 600 K pre-treatment. The trend in CO and CO_2 production with annealing temperature indicates that the active SO_x species is stable up to 400K; however by 600 K loss of SO_x by desorption has occurred, and C_3H_8 adsorption is no longer possible. In all instances, no residual carbon was detectable by AES or XPS, indicating complete combustion of the hydrocarbon.

Additional information about reaction intermediates and mechanism was acquired by XPS and HREELS. The Pt sample was first exposed to O_2 then SO_2 at 160 K; XP and HREEL spectra of the surface species were then recorded as a function of annealing temperature. Figure 5 shows the resulting XPS, in which the S(2p) state is observed at 170.7 eV, approximately the same energy as that for SO_2 on clean Pt [14]. However annealing to 250 K and 300 K, resulted in the binding energy increasing to 171.8 and 172.3 eV respectively. This is consistent with oxidation of SO_2 to an SO_x species (x>2). On Ag(110) [15], the S(2p) binding energy of SO_3 and SO_4 were reported to be 0.7 eV and 2.5 eV higher than SO_2 respectively. We observe a final shift of 1.6 eV, which is therefore attributed to SO_4 formation.

For SO_2 on Pt(111) three vibrational states are observed at 524, 910 and 1240 cm^{-1} which correspond to the deformation, symmetric and asymmetric modes of SO_2 [16]. On the O_2 saturated surface (Figure 6) the deformation mode is very weak, and the symmetric and asymmetric modes are observed at 855 and 1250 cm^{-1}

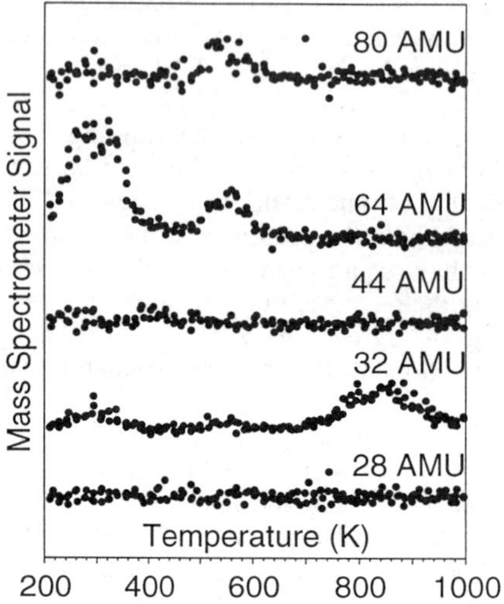

Figure 3: TPR following 160K adsorption of 6L C_3H_8 on Pt(111) pre-covered by saturation O_2 and 24L SO_2 (dosed at 160K).

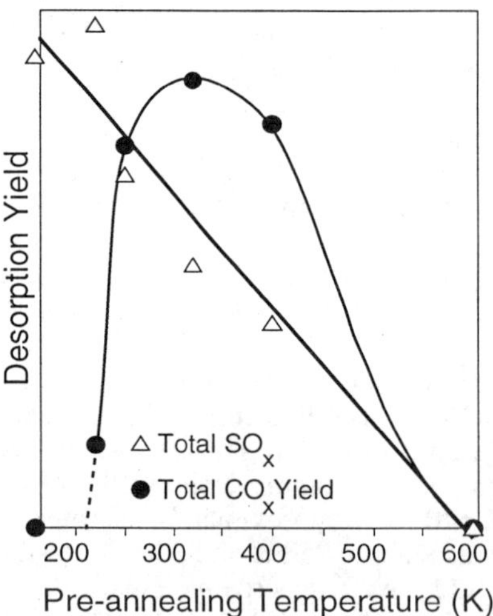

Figure 4: Oxidation product yield following 160K C_3H_8 adsorption over Pt(111) pre-covered by saturation O_2 (at 300K) and 24L SO_2 (at 160K) as a function of annealing temperature of the O_2/SO_2 adsorbate over layer. (Reproduced with permission from ref. 7. Copyright 1995 American Chemical Society)

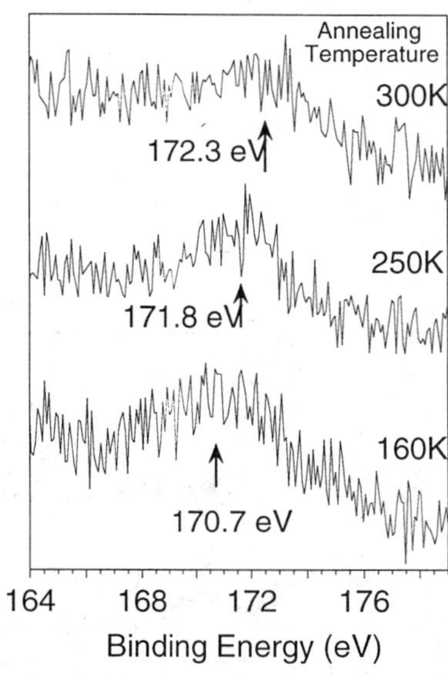

Figure 5: S(2p) XPS 160K O_2+SO_2 annealed to 250K and 300K

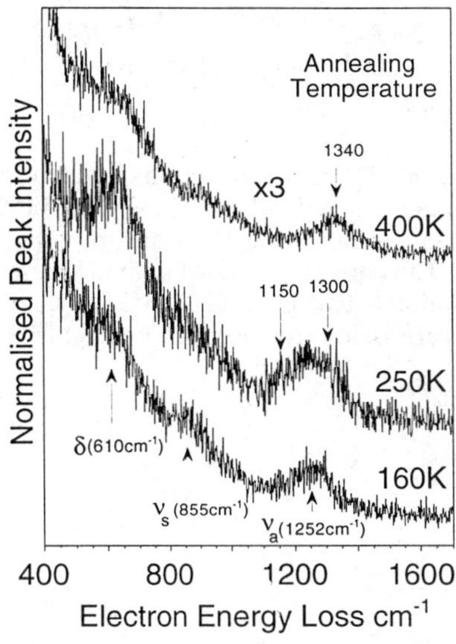

Figure 6: HREELS of 160K O_2 + SO_2 annealed to 250K and 400K

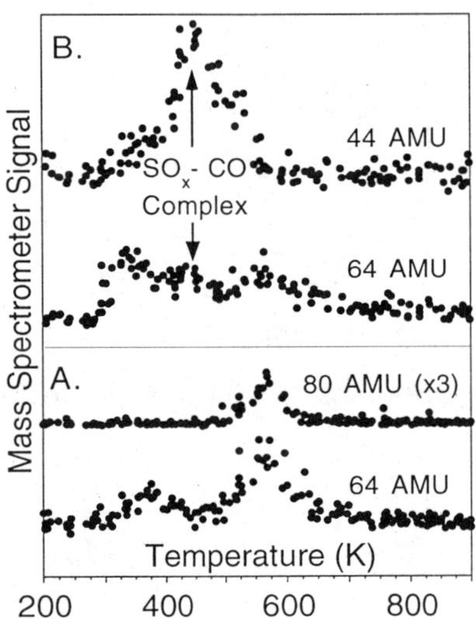

Figure 7: Comparison of 64 amu desorption obtained following co-adsorption of saturation O_2 and 24LSO_2 (dosed at 300K then cooled to 160K), with that obtained following 160K adsorption of 3L C_3H_8 on the O_2/SO_2 overlayer. The corresponding 44 amu desorption is also shown to illustrate the correlation between the 420K SO_x peak and CO_2. (Reproduced with permission from ref. 7. Copyright 1995 American Chemical Society)

respectively. Following annealing to 250K, the intensity of the 610 cm^{-1} state increases, and there is broadening of the asymmetric loss which can be attributed to two new states around 1150 and 1300 cm^{-1}. The stretching and deformation modes of SO_4 range from 1290-856 cm^{-1}, and 660-493 cm^{-1} respectively [17]. The corresponding modes for SO_3 are observed in the range 1110-940 cm^{-1} and 670-461 cm^{-1}: this implies that the observed changes in HREELS can only be accounted for by formation of adsorbed SO_4. When the sample is annealed to 450K, the precursor to SO_3 desorption is isolated on the surface (see Figure 3). Thus at this temperature, the surface species that desorbs as SO_3 can be isolated, and we observed an associated asymmetric stretch at 1340 cm^{-1} . From the frequency data quoted, this value is obviously too high for SO_3, confirming that the species must be SO_4. However 1340 cm^{-1} is high even for SO_4, although frequency shifts of this magnitude may be produced if there is a change in coordination to the surface, e.g. from mono-dentate to bi-dentate, and indeed such transitions have been observed for SO_3 on Ag(110) [18]. That is, as free surface sites are produced, the interaction of another S-O bond with the surface should be facilitated, producing a bi-dentate chemisorbed SO_4. Desorption of SO_3 would result from cleavage of one of the surface-coordinated S-O bonds.

Significant modification of the 64 amu desorption occurs following adsorption of C_3H_8 on the O_2/SO_2 precovered surface. Figure 7 shows a comparison of the 64 amu desorption following adsorption at 160 K of (i) $O_{(a)}$ and SO_2 alone and (ii) in the presence of C_3H_8. In the absence of C_3H_8 two 64 amu desorption states are observed at 370 K and 550 K. This high temperature state is coincident with a peak in the 80 amu spectrum (Figure 7A) and is ascribed to SO_3, in agreement with previous workers [19]. Following reaction with C_3H_8, (Figure 7B) substantial changes in the 64 amu TPR occur: the 550 K state is attenuated and a new state at 420 K appears. This 550 K 64 amu state continued to decrease with increasing C_3H_8 exposure, and was no longer observed under partial oxidation conditions (i.e. when CO desorption is observed e.g. Figure 2). There is a strong correlation between the intensities of the coincident 420 K 64 amu and 44 amu features, Figure 8. These observations suggest complex formation between the promoting SO_4 species and the hydrocarbon either as it adsorbs, or decomposes. Evidence for the formation of such a CO-SO_4 complex decomposing to form CO_2 is found in CO oxidation over Pt(111) precovered by $^{18}O_2$ and $S^{16}O_2$. Figure 9 shows the resulting TPR spectrum from which $C^{18}O^{16}O$ (46 amu) is seen to desorb at 310 K - the expected desorption temperature for CO_2 by the Pt/O system. Two 44 amu states are also observed at 370 K and 500 K which correspond to oxidation using ^{16}O from SO_2. The 370 K state is coincident with the 64 amu desorption, suggesting decomposition of a common intermediate (CO-SO_4). It should be noted that desorption of CO_2 at 500 K would result from CO oxidation by residual O(a) remaining on the surface after the CO-SO_4 complex has decomposed to CO_2 and SO_2 + O(a).

The consumption of SO_4 by C_3H_8 can be followed using HREELS as shown in Figure 10. Pt(111) was exposed to O_2 and SO_2 at 300 K, cooled to 160 K prior to C_3H_8 exposure, then annealed to 400 K. In the control spectrum (with no C_3H_8 exposure), the losses at 1325 and 925 cm^{-1} are assigned to the asymmetric and symmetric stretching modes of SO_4. If the O(a)/SO_2 covered surface is exposed at 160 K to C_3H_8 prior to annealing to 400 K, these SO_4 loss features are significantly attenuated. For low C_3H_8 exposures (<1L), the SO_4 modes are still visible, which is

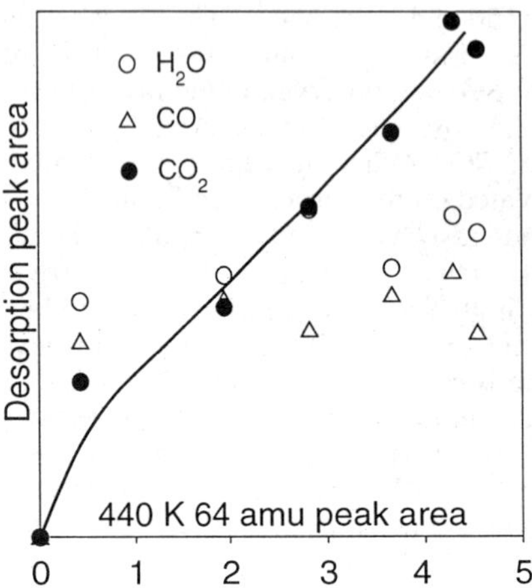

Figure 8: Total CO_x yield as a function of the 420K 64amu desorption yield. Increase in this 64amu state intensity was obtained by increasing the O_2 exposure (at 300K) followed by 24L SO_2 (at 300K) and 6L C_3H_8. Reproduced with permission from ref. 7. Copyright 1995 American Chemical Society)

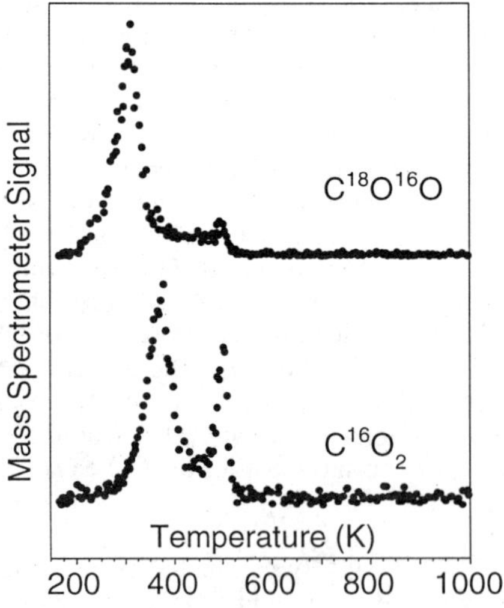

Figure 9: TPR following 160K CO adsorption over Pt(111) pre-covered by $^{18}O_2$ and SO_2 at 300K.

consistent with the TPD observation that SO_3 still desorbs under these conditions. However for larger exposures, complete extinction of the SO_4 modes occurs, again in accord with the TPD results which show no SO_3 desorption. We have thus demonstrated that adsorbed SO_4 plays a direct role in triggering the dissociative chemisorption of C_3H_8 on Pt.

Having shown that under UHV conditions SO_2 promotes propane chemisorption and oxidation over unsupported Pt, we now address the question "what is the role of alumina in the conventional catalyst during SO_2 promoted alkane oxidation?". To investigate this, Al films of varying thicknesses were deposited on the Pt(111) single crystal and oxidised (100 L O_2, 300 K), prior to performing the TPR experiment with O(a), SO_2 and C_3H_8 chemisorbed at 300 K. Oxidation by O_2 at 300 K yields AlO_x films characterised by an Al(KLL) Auger transition at 58 eV [20]; exposure to SO_2 shifts the Auger emission to 53 eV indicating complete oxidation and the formation of Al^{3+}. The TPR yields of CO_2 and CO are shown as a function of as-deposited Al film thickness in Figure 11. In the presence of submonolayer oxide films the yield of CO is increased compared to the clean metal surface, whilst the CO_2 yield decreases. The enhanced oxidative capacity of the submonolayer films is clearly demonstrated by the total CO_x yield which passes through a maximum at ~1ML Al, decreasing for higher initial Al loadings. This enhancement in oxidation capacity induced by the presence of AlO_x on Pt suggests that the former acts to enhance the population of SO_4 on the surface of the model catalyst.

Discussion

Promotion of propane oxidation has been studied only over supported Pt catalysts [3,4,21], the results being interpreted in terms of support-mediated effects due to the presence of sulphate on the oxide phase; this view is at least consistent with infra-red observations which show that sulphated alumina adsorbs propane [3]. However, the present results clearly demonstrate that in the presence of chemisorbed oxygen, SO_2 dramatically enhances the chemisorption and oxidation of propane on platinum *in the absence of any effects due to a support phase.*

There is some uncertainty in the literature regarding the nature of SO_2 adsorption on clean Pt(111) and no detailed information is available about the co-adsorption of O_2 and SO_2. Wassmuth *et al* concluded that SO_2 adsorbs dissociatively at 160 K forming SO_a and O_a, which undergo recombinative desorption at 300 K [22,23,24]. In the light of their HREELS and XPS data, White *et al* [16] argue in favour of molecular adsorption at 130 K, followed at 300 K either by desorption as SO_2, or dissociation to form adsorbed S, SO and SO_4. We have studied SO_2 [14] adsorption over both clean and O_2 exposed Pt(111) and conclude that for low coverages SO_2 dissociates to SO + O, with non-dissociative adsorption at higher coverages. In contrast to the work of White *et al*, we find that formation of SO_4 requires prior exposure of the surface to O_2. In the presence of oxygen a threshold temperature of >220 K is observed for the formation of the hydrocarbon activating species: formation of this sulphate species at 250 K gives rise to new vibrational modes at 1150 and 1300 cm^{-1}. (For SO_2 bonded through oxygen as in $SbF_5.SO_2$ [25], ν_s and ν_a are observed at 1100 and 1323cm^{-1} respectively. This supports our assignment of the

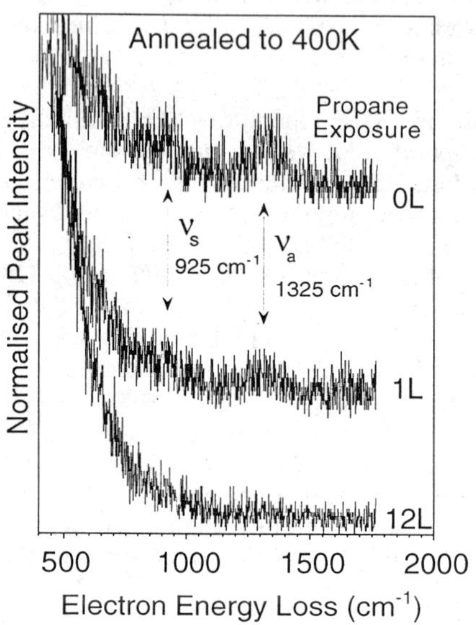

Figure 10: HREELS of O_2+SO_2 dosed at 300K and cooled to 160K prior to C_3H_8 exposure then annealing to 400K

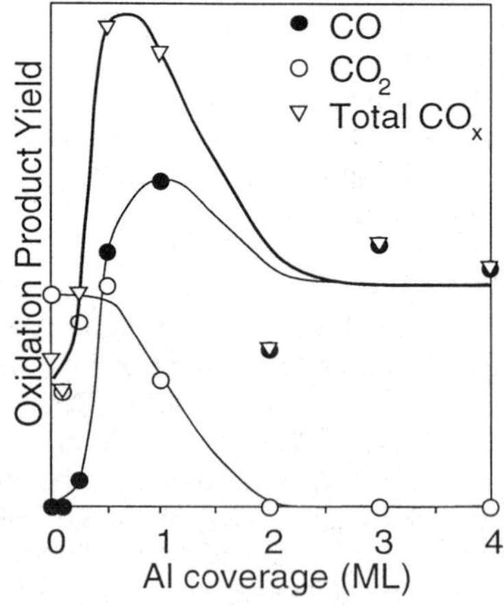

Figure 11: Reactivity of AlO_x films on Pt(111) towards propane oxidation in the presence of SO_2

1150 cm^{-1} and 1300 cm^{-1} losses to SO_4 bonded to the surface through oxygen). The observed changes in the HREELS suggest that upon annealing the SO_4 species is also bound through oxygen to Pt. The increase in frequency of the SO_4^{2-}(a) asymmetric stretch which occurs upon annealing to 450K indicates an increase in the S-O bond order. This is consistent with a change in bonding geometry from monodentate to bidentate. For SO_4^{2-} the net negative charge would be localised near the surface S-O bonds [18], thus in the bidentate geometry the terminal S-O bond order would be ~2. The transformation is shown in scheme 1.

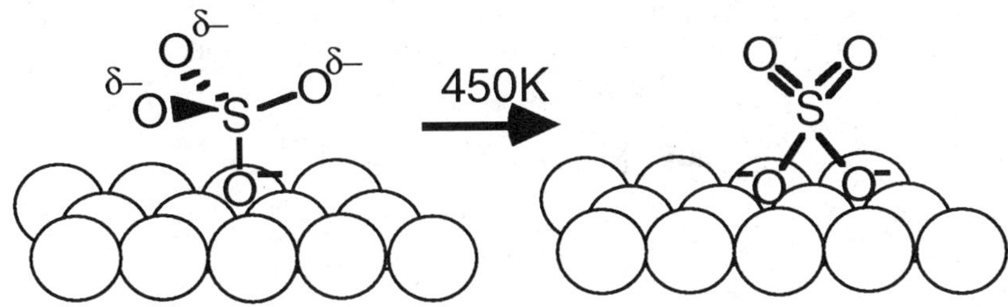

Scheme 1

The 560 K SO_2^+ peak (Figure 7B) and the associated coincident 80 amu signal (SO_3^+); (Figure 7A) are ascribed to SO_3 desorption. In these experiments, SO_2 was dosed at 300 K, which is above the dissociation temperature for molecular SO_2 [16]. Therefore as the 370 K 64 amu feature is *not* associated with a coincident SO_3^+ peak, we assign it to SO_2 desorption resulting from O_a + SO_a recombination. In the presence of adsorbed propane (Figure 7B) this feature is masked by a 330 K 64 amu peak: since C_3H_8 chemisorption almost certainly involves H-abstraction, we tentatively suggest that the 330 K SO_2^+ peak is due to adsorbed HSO_x which undergoes decomposition to SO_2. (In this connection it should be noted that Leung *et al* have suggested that SO_2 and H_2S can react to form H_2SO_3 over Cu(100) [26].) Furthermore, this is consistent with the pronounced decrease in SO_3 desorption induced by the presence of co-adsorbed propane which strongly suggests the occurrence of a reaction between the hydrocarbon and the precursor to SO_3 desorption (Figure 7). Our HREELS data (Figure 6) identify this precursor as SO_4, and in Figure 10 it can clearly be seen that following coadsorption of C_3H_8 the sulphate stretches are significantly diminished. This result therefore confirms that sulphate species are consumed in the oxidation of propane and accounts for the reduction in SO_3 desorption. The correlation between the integrated intensity of the 440 K SO_2^+ feature and the CO_2 desorption yield (Figure 8) points to the formation of a complex between the SO_4 species and either the hydrocarbon fragments or reactively formed CO. Note that the CO_2 peak exhibits a shoulder at ~320 K which corresponds to oxidation of CO_a by O_a [13] indicating that some of the reactively-formed CO is adsorbed directly onto the metal surface and not incorporated into a SO_4-containing complex. Strong evidence for the formation of a CO-SO_4 complex is shown in Figure 9, which demonstrates that there are two distinct routes to CO oxidation by ^{18}O(a) and $S^{16}O_2$.

CO desorption (Figure 2) is observed only under oxygen-lean conditions (i.e. in the absence of the 560 K SO_2^+ peak due to SO_3 desorption). This CO peak occurs at the same temperature (500 K) as that observed during TPR oxidation of ethene and propene on Pt: therefore it probably results from the oxidation of adsorbed carbon atoms[27,28,29].

Our observations may be rationalised in terms of the following scheme.

1. Initial dissociative adsorption of propane involving H abstraction by SO_4. The surface intermediate resulting from this process has not yet been identified, however there are two plausible possibilities:

$$SO_4 + C_3H_8 \rightarrow C_3H_y + HSO_4$$

Scheme 2

In scheme 2, H abstraction yields HSO_4 and a hydrocarbon fragment in a single step, both of which chemisorb on the Pt surface. An alternative explanation involves the formation of an alkyl sulphate (scheme 3), which undergoes pyrolysis on heating [30].

$$C_3H_7\text{-O-}\underset{\substack{|\\O}}{\overset{\substack{O\\|}}{S}}\text{-O-H} \rightarrow H_2C{=}CH.CH_3 + SO_2 + HOH + O(a)$$

Scheme 3

The olefin formed during the decomposition step would be rapidly dehydrogenated [31,28] and oxidised on Pt(111) to yield CO and H_2O. The CO formed may then be further oxidised by surface oxygen to form CO_2 at 320K, or incorporated in the CO-SO_4 complex as already discussed.

2. Coincident desorption of SO_2 and CO_2 at 440 K indicates decomposition of a surface complex. No coincident H_2O desorption occurs, suggesting that the complex does not contain any H, i.e. is of the form CO-SO_4. This could decompose according to:

$$\mathbf{CO\text{-}SO_4 \rightarrow CO_2 + SO_2 + O_a \quad \sim 440\ K}$$

At high propane coverages (oxygen-lean system) incomplete oxidation occurs. No SO_3 formation is observed as all the SO_4 has been consumed, hence the associated 560 K amu desorption feature is no longer observed. CO is then produced by the oxidation of surface carbon by residual O(a) at the characteristic temperature associated with this process, as observed during catalytic hydrocarbon oxidation [27,28,29].

When the reaction was performed over the Pt(111)/AlO_x model catalyst, an enhanced yield of oxidation products was observed for submonolayer AlO_x films. This can be rationalised in the following terms. Pt(111)/Submonolayer AlO_x behaves as an efficient bifunctional catalyst. Oxygen, SO_2 and propane chemisorption occur on the metal, with propane also adsorbing on the sulphated AlO_x [3]. Enhanced oxidation activity then results from spillover of adsorbed propane on the support to Pt sites, where adsorbed O(a) facilitates catalytic oxidation. The reduced activity with increased AlO_x loading can simply be understood in terms of there being fewer bare Pt sites, which is indeed confirmed by CO titration of these surfaces [32]. The reduced surface concentration of O(a) is also demonstrated by the transition from total oxidation (CO_2 formation) to partial oxidation (CO only) as the Al thickness increases.

In this study we have shown that SO_2 is capable of promoting propane oxidation over Pt alone, and submonolayer AlO_x films further enhance this activity. We are now able to explain some of the observations in the original reactor study [3,4]. Sulphate species were only identified on the support. However, this is hardly surprising, because with the real catalyst, the infra red spectrum would be dominated by sulphate species on the support, obscuring any contribution from the metal component. Recall that for Pt/Al_2O_3 [3] the presence of SO_2 makes the catalytic activity invariant with Pt loading. This is understandable in the light of our UHV data and the bifunctional behaviour proposed above.

Conclusions

1. At 300K, adsorption of propane occurs over Pt(111) pre-covered by O(a) and SO_2. The promotional effect of SO_2 on hydrocarbon oxidation can thus be observed over Pt alone and does not require Al_2O_3 to stabilise the adsorbed SO_x species.

2. There is a threshold temperature of ~220K for the formation of the active SO_x species.

3. HREELS and XPS measurements enable us to identify the SO_x species as SO_4. The direct involvement of SO_4 in propane oxidation has also been demonstrated.

4. SO_2 promotion of oxidation also occurs for C_3-C_7 hydrocarbons, but not for methane and ethane.

5. 64 and 44 amu TPR spectra suggest formation of a CO-SO_x complex during the reaction by SO_4 scavenging reactively formed CO. CO oxidation over $^{18}O_2/S^{16}O_2$ precovered Pt(111) show that there are two distinct routes to CO_2 formation, supporting the proposal of a SO_4-CO complex.

6. Submonolayer films of AlO_x further enhance the promotional effect of SO_2 compared to clean Pt(111). This may be rationalised in terms of bifunctional catalytic behaviour.

References

1. Kemball.C.; *Advances in Catalysis*, **1959**,*11*, 223
2. Otto,K.; Andino,J.M.; Parks,C.L.; *J.Catal.* ,**1991**, *131* , 243
3. Yao,H.C.; Gandhi,H.S.; Stephien,H.K.; *J.Catal.*, **1981**, *67* , 231
4. Ansell,G.P.; Golunski,S.E.; Hatcher, H.A.; Rajaram,R.R.; *Catal.Lett.*, **1991**, *11*, 183
5. Shilov, A.E.; "Activation of Saturated Hydrocarbons by Transition Metal Catalysts" - Reidel Publishing Co.
6. Hubbard,C.P.; Otto,K.; Gandhi,H.S.; Ng,K.Y.S.; *J.Catal.*, **1993**, *131*, 484
7. Wilson,K.; Hardacre,C.; Lambert, R.M.; *J.Phys.Chem.*, **1995**, *99*, 13755
8. McMaster, M.C.; Arumainayagam, C.R.; Madix, R.J.; *Chemical Physics*, **1993**, *177*, 461
9. Hardacre,C.; Roe,G.M.;Lambert,R.M.; *Surf.Sci.* , **1995**, *326* , 1
10. Horton,J.H.; Moggridge,G.D.; Ormerod,R.M.; Kolobov,A.V.; Lambert,R.M.; *Thin Solid Films*, **1994**, *237* , 134
11. Wytenburg,W.J.; Lambert,R.M.; *J.Vac.Sci.&Tech.*, **1992**, *10*, No.(6), 3598
12. Wassmuth,H.-W.; Ahner,J.; Hofer,M.; Stolz,H.; *Prog.Surf.Sci.*, **1993**, *42*, 257
13. Gland,J.L.; Kollin,E.B.; *J.Chem.Phys.*, **1983**, *78(2)*, 963
14. Wilson,K.;Hardacre,C.; Baddeley,C.J.; Leudeke,J.;Woodruff,D.P.; Lambert,R.M.; *manuscript in preparation*
15. Outaka,D.A; Madix,R.J.; *Surf.Sci.*, **1984**,*137*, 242
16. Sun,Y.M.; Sloan,D.; Alberas,D.J; Kovar,M.; Sun,Z.J.; White,J.M.; *Surf.Sci.*, **1994**, *319*,
17. Nakamoto,K., *Infrared and Raman Spectra of inorganic and Coordination Compounds* , Wiley, New York
18. Outaka, D.A.; Madix, R.J.; Fisher,C.B.; Maggio,C.D.; *J.Phys.Chem* , **1986**, *90* , 4051
19. Astegger, St.; Bechtold,E.; *Surf,Sci.*, **1982**, *122*, 491
20. Ko,C.S.; Gorte,R.J.; *Surf.Sci.*, **1992**, *155*, 296
21. Gandhi,H.S.; Shelef,M.; *Applied Catalysis*, **1991**,*77*,175
22. Hofer,M.; Hillig,S.; Wassmuth,H.W.; *Vacuum.*, **1990**,*41* , 102
23. Kohler,U.; Wassmuth,H.W.;*Surf.Sci.*, **1982**, *117*, 668
24. Kohler,U.; Wassmuth,H.W.;*Surf.Sci.*, **1983**, *126*, 448
25. Moore,J.W.; Baird,H.W.; Millar,H.B.; *J.Am.Chem.Soc*, **1968**, *90*, 1358
26. Leung, K.T.; Zhang, X.S.; Shirley, D.A.; *J.Phys. Chem* , **1989**, *93*, 6164
27. Steiniger,H.; Ibach,H.; Lehwald,S.; *Surf.Sci.*, **1982**, *117*, 685
28. Wilson,K.; Hardacre,C.; Lambert,R.M.;*manuscript in preparation*
29. Schafer,L.; Wassmuth,H.W.; *Surf.Sci.*, **1989**, *208*, 55
30. Kice,J.L; *in"The Chemistry of Organic Sulphur Compounds"(Ed.Kharasch,N.; Meyers,C.Y)* **1965**, *2*, 128
31. Avery,N.R.; Sheppard, N.S.; *Proc.Roy.Soc.Lond.*, **1986**, *A405*, 1
32. Wilson,K.; Hardacre,C.; Lambert,R.M. *manuscript in preparation*

Chapter 31

Selective Photooxidation of Small Hydrocarbons by O_2 with Visible Light in Zeolites

Hai Sun, Fritz Blatter, and Heinz Frei[1]

Structural Biology Division, Calvin Laboratory, Lawrence Berkeley National Laboratory, 1 Cyclotron Road, Berkeley, CA 94720

Small alkenes, alkanes, or alkyl substituted benzenes loaded with O_2 gas into cation-exchanged zeolite Y react upon irradiation with visible light to yield corresponding carbonyl products at very high selectivity. Alkyl (alkenyl) hydroperoxides are formed as intermediates, in the case of isobutane as the final product. This was observed when monitoring the reactions in situ by Fourier-transform infrared spectroscopy. Experiments were typically run at room temperature, in some cases at zeolite temperatures as low as -100°C to elucidate mechanisms. Chemistry was induced by light from a tungsten lamp. Frequently, studies were also conducted with the emission of an Ar ion or cw dye laser in order to determine the visible wavelengths responsible for the reaction. Diffuse reflectance spectra revealed a visible absorption tail which originates from a hydrocarbon•O_2 collision complex. It is attributed to the hydrocarbon•O_2 charge-transfer absorption whose onset is shifted from the UV into the visible region by the very high electrostatic field of the zeolite (shifts of the order of 1.5 to 3 eV). Quantum efficiencies are in the region 10-30%, and selectivities remain high even upon conversion of more than 50% of the hydrocarbon loaded into the zeolite matrix. Many of the reactions studied are of commercial importance: toluene to benzaldehyde, propylene to acrolein or propylene oxide, isobutane to *t*-butyl hydroperoxide, cyclohexane to cyclohexanone, ethane to acetaldehyde.

Oxidation by O_2 is the single most important process for the conversion of abundant hydrocarbons to oxygenated derivatives such as organic building blocks for the manufacture of plastics and synthetic fibers, and industrial intermediates for the synthesis of fine chemicals (*1-6*). In large-scale synthesis, the use of molecular oxygen as oxidant is dictated primarily by economic factors. Yet, autoxidation of small hydrocarbons is inherently unselective, whether conducted in the gas or liquid phase, or whether catalyzed

[1]Corresponding author

0097–6156/96/0638–0409$15.00/0

by transition metals or not (*1-4*). One reason is that the desired products such as alcohols or carbonyls are more easily oxidized by O_2 than the parent hydrocarbon. Overoxidation can only be minimized by keeping conversions low (at a few percent). Another factor is diversion of the radical chain reaction leading to the primary product (alkyl or alkenyl hydroperoxide) by termination steps which result in the formation of oxy radicals. This is especially a problem in the case of olefin oxidations. The highly reactive oxy radicals can undergo several competing reactions that lead to a multitude of products. Hence, oxidations by O_2 exhibit most often little chemo- or regioselectivity. A major challenge in the field of hydrocarbon + O_2 chemistry is, therefore, to find reaction paths that afford the primary product with high selectivity at high conversion.

We have developed a method that affords partial oxidation of small alkenes, alkanes, and substituted aromatics by O_2 at very high selectivity. The approach is based on photoexcitation of hydrocarbon•O_2 pairs in a large-pore zeolite (faujasite). The 3-dimensional network of molecular-size cages of the latter offer a natural environment for the formation of hydrocarbon•O_2 collisional pairs at high concentration (Fig. 1). The key to selectivity is a low-energy reaction path that is opened up by a very strong stabilization of the hydrocarbon•O_2 excited charge-transfer state by the high electrostatic field of the zeolite cage. The stabilization causes a red-shift of the charge-transfer absorption from the UV into the visible region. Access to this low-energy excited state, coupled with the positional constraint imposed by the zeolite nanocage furnishes a new, tightly controlled reaction path for small hydrocarbon + O_2 systems.

Experimental

Self-supporting zeolite wafers of 5-10 mg (1.2 cm diameter) were placed in a miniature infrared or UV-Vis vacuum cell (*7-14*). For infrared measurements, the cell was mounted inside a variable temperature vacuum system (Oxford Model DN1714 or DN1724). The zeolite was dehydrated by heating the cell to 200°C for 12-15 h while evacuating with a turbomolecular pump. The reactants were subsequently loaded from the gas phase into the zeolite. The loading level was adjusted by the gas pressure and the zeolite temperature. Both laboratory-synthesized and commercial zeolite NaY (Aldrich) were used. Alkaline-earth exchanged zeolite Y (BaY, CaY) was prepared by repeated ion-exchange of NaY at 90°C in 0.5 M solution of the corresponding chloride salt (*15*). The degree of exchange (ICP) was typically 95% or better.

Photochemistry was monitored in situ by Fourier-transform infrared spectroscopy using a Bruker Model IFS 113 or IR 44 instrument. Zeolite Y is transparent in the infrared except for the region 1200-920 cm^{-1} and below 800 cm^{-1}. For photolysis, a prism-tuned cw Ar ion laser (Coherent Model Innova 90) or the emission of a tungsten lamp was used (equipped with a UV cut-off filter). The light beam was expanded to cover the entire zeolite pellet. Experiments were conducted at temperatures between 100 K and room temperature. UV-Vis spectra were recorded by the diffuse reflectance method using an integrating sphere (Shimadzu Model 2100 equipped with a Model ISR-260 sphere).

Results

Our work has focused on the partial oxidation of small olefins, aromatics, and alkanes. A partial overview of the reactions studied thus far is given in Scheme 1. We will present, in turn, one or two examples for each class of hydrocarbons.

Toluene to Benzaldehyde Conversion. Upon loading of toluene (5 Torr) and O_2 (760 Torr) into a BaY or CaY matrix, a continuous absorption with a tail extending into the visible region is observed (Fig. 2) (*9*). Alkali or alkaline earth exchanged zeolite Y has no optical absorption in this spectral range (*16*). The band appears only when the hydrocarbon and O_2 are simultaneously present in the zeolite. The absorption can be reversibly removed by pumping off the oxygen gas, which constitutes direct evidence that it originates from the hydrocarbon•O_2 complex. Use of the diffuse reflectance method is required because of the strong scattering behavior of the zeolite pellet (type Y crystallites have a size of about one micron).

The onset of the lowest energy absorption of toluene•O_2 contact complexes in the oxygen-saturated liquid lies in the UV region around 370 nm (*17*). It originates from excitation of the hydrocarbon•O_2 charge-transfer state. By contrast, the diffuse reflectance spectra indicate onset of the toluene•O_2 absorption in BaY and CaY at 500 and 600 nm, respectively.

Speculating that the very large red shift of the hydrocarbon•O_2 absorption is caused by high electrostatic fields inside the cage (*18*), we have determined experimentally the magnitude of the field in the exchanged zeolite Y samples used in our work (*19*). The method consists of measuring the induced infrared fundamental absorption of O_2 or N_2, or the symmetric stretch vibration of CH_4 loaded into the matrix. These infrared-forbidden transitions become active in the presence of an electrostatic field, and the band intensity is proportional to the square of the field (*20*). The method has previously been used by Cohen de Lara in the study of zeolite A (*21*). Fig. 3 shows the induced infrared absorption of O_2 at 1550 cm^{-1} upon loading of the gas into a NaY or BaY pellet. Pressures were adjusted so as to obtain 1.5 molecules O_2 per supercage on average. These intensity measurements revealed very high electrostatic fields. For example, in NaY at -50°C the molecules experience a field of 0.3 $VÅ^{-1}$, or 0.9 $VÅ^{-1}$ in BaY (*19*).

When shining green or blue light on zeolite BaY loaded with toluene and O_2, benzaldehyde and H_2O grew in under concurrent depletion of toluene. Fig. 4 shows an infrared difference spectrum after photolysis with 488 nm light from an Ar ion laser at room temperature (400 mW cm^{-2}, 3 h). The same result was obtained when using the visible emission of a tungsten lamp. Yields were independent of the light source and simply reflect the number of light quanta absorbed by the reactants. The use of monochromatic laser light had the advantage of furnishing the wavelength-dependence of product yields. The positive bands agree completely with the spectrum of an authentic sample of benzaldehyde in BaY. Aside from a shoulder at 1640 cm^{-1} which is due to H_2O coproduct, no other infrared absorption grew in even upon prolonged photolysis. This signals completely selective oxidation of toluene to benzaldehyde by O_2. Observation of some thermal growth of benzaldehyde *after* toluene + O_2

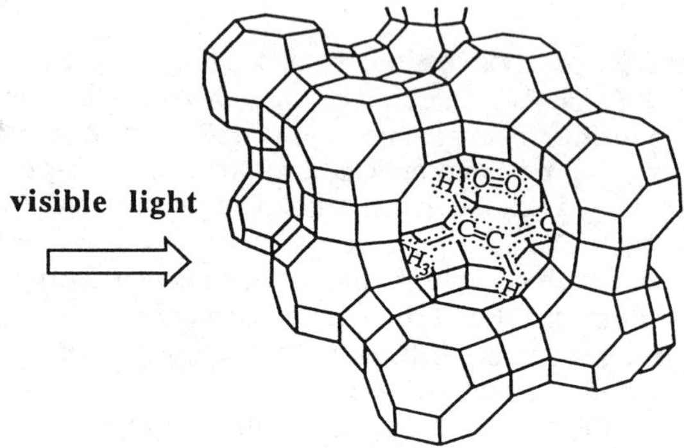

Figure 1. Photo-induced reaction of hydrocarbon•O_2 collisional pairs inside a zeolite Y supercage.

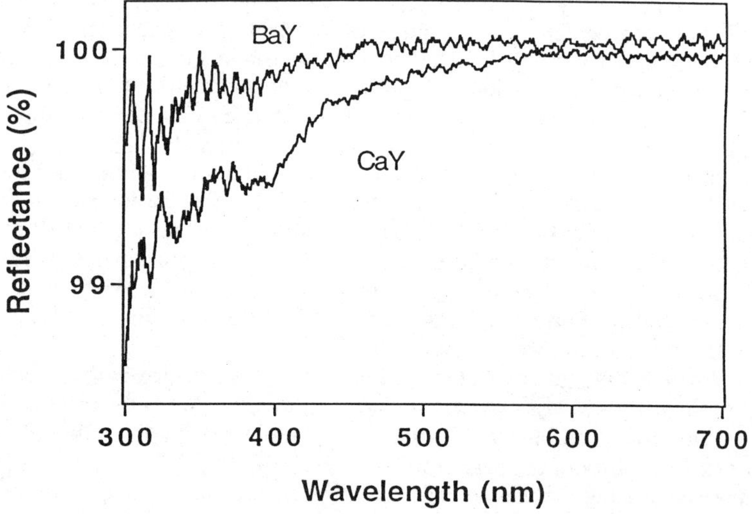

Figure 2. Toluene•O_2 diffuse reflectance spectra in zeolite BaY and CaY.

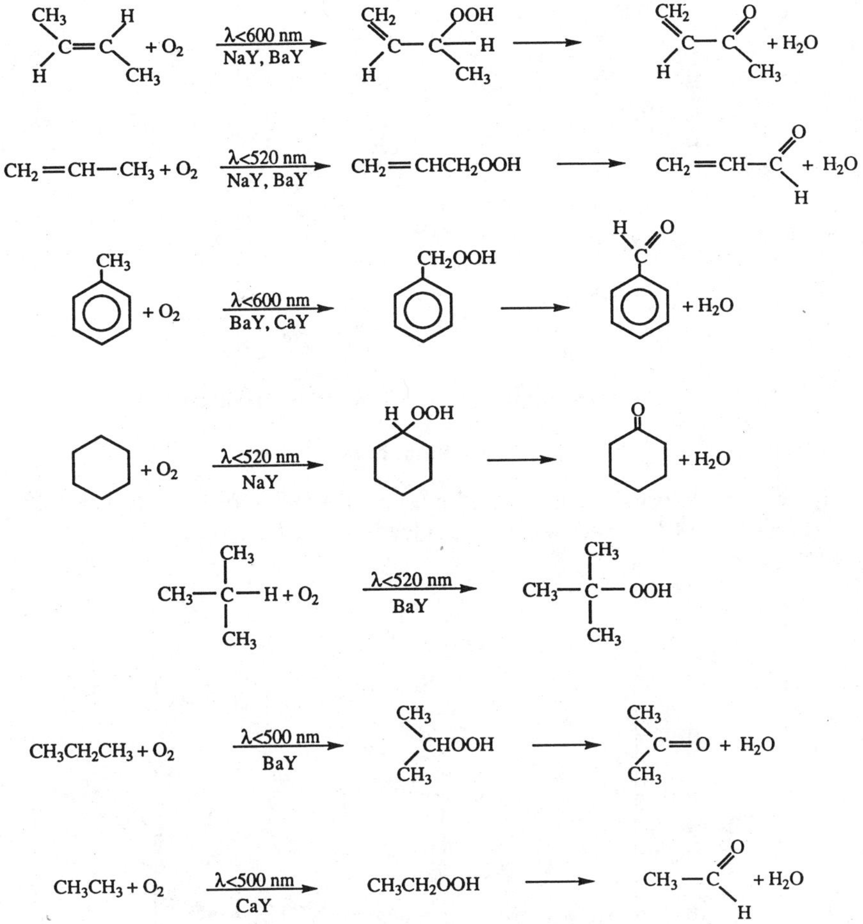

$$CH_3CH{=}CHCH_3 + O_2 \xrightarrow[\text{NaY, BaY}]{\lambda<600\ \text{nm}} CH_2{=}CH{-}CH(OOH){-}CH_3 \longrightarrow CH_2{=}CH{-}C(=O){-}CH_3 + H_2O$$

$$CH_2{=}CH{-}CH_3 + O_2 \xrightarrow[\text{NaY, BaY}]{\lambda<520\ \text{nm}} CH_2{=}CHCH_2OOH \longrightarrow CH_2{=}CH{-}C(=O){-}H + H_2O$$

$$C_6H_5CH_3 + O_2 \xrightarrow[\text{BaY, CaY}]{\lambda<600\ \text{nm}} C_6H_5CH_2OOH \longrightarrow C_6H_5CHO + H_2O$$

$$C_6H_{12} + O_2 \xrightarrow[\text{NaY}]{\lambda<520\ \text{nm}} C_6H_{11}OOH \longrightarrow C_6H_{10}O + H_2O$$

$$CH_3{-}C(CH_3)_2{-}H + O_2 \xrightarrow[\text{BaY}]{\lambda<520\ \text{nm}} CH_3{-}C(CH_3)_2{-}OOH$$

$$CH_3CH_2CH_3 + O_2 \xrightarrow[\text{BaY}]{\lambda<500\ \text{nm}} (CH_3)_2CHOOH \longrightarrow (CH_3)_2C{=}O + H_2O$$

$$CH_3CH_3 + O_2 \xrightarrow[\text{CaY}]{\lambda<500\ \text{nm}} CH_3CH_2OOH \longrightarrow CH_3{-}C(=O){-}H + H_2O$$

Scheme 1. Examples of reactions studied.

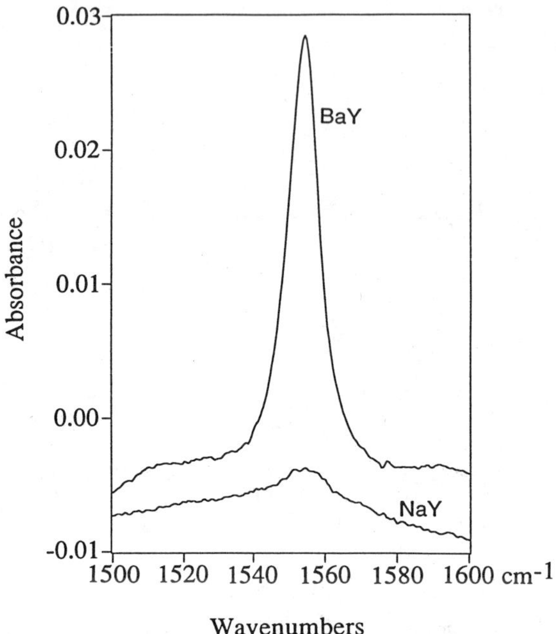

Figure 3. Electrostatic field-induced infrared absorption of O_2 in zeolite NaY and BaY at -100°C. Each supercage is loaded with 1.5 molecules on average.

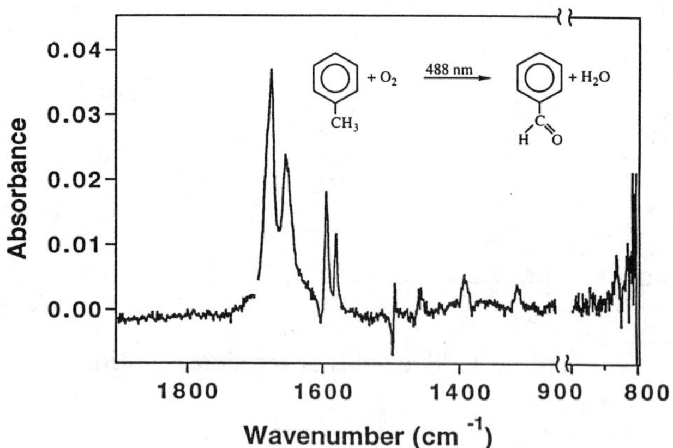

Figure 4. Infrared difference spectrum of zeolite BaY loaded with toluene and O_2 taken upon photolysis with Ar ion laser irradiation at 488 nm (400 mW cm^{-2}, 3 h).

photolysis is consistent with the formation of benzyl hydroperoxide as a reaction intermediate (*9*). Photooxidation of toluene to benzaldehyde in zeolite CaY proceeded without side reaction as well. However, the rate of reaction was faster, presumably reflecting the higher extinction of the visible toluene•O_2 absorption band (Fig. 2).

Propylene to Acrolein and Propylene Oxide. Selectivity is a particular challenge in the case of autoxidation of propylene and other small olefins, as mentioned in the introduction (*1,4,6,22*). We have studied visible light induced reactions of O_2 with all methyl substituted ethylenes.

Irradiation of propylene and O_2-loaded zeolite BaY at room temperature with green or blue light of an Ar ion laser, or a tungsten lamp, induced partial oxidation of the olefin (*12*). Readily identified products in the room temperature zeolite are acrolein and allyl hydroperoxide. Since the hydroperoxide is stable at -100°C, experiments at this temperature allowed us to find out about the details of the reaction path. Fig. 5 shows spectra of such a photochemical run at low temperature. The top trace presents the infrared spectrum recorded after loading of 3 Torr propylene and 700 Torr O_2 gas into BaY (corresponding to 4 olefin and 4 oxygen molecules per supercage) (*18*). The infrared difference spectrum upon 488 nm photolysis (Fig. 5b) shows allyl hydroperoxide as the main product (characteristic modes: C=C stretch at 1640 cm^{-1}, COH bending mode at 1343 cm^{-1}). Identification of this product is based on ^{18}O and D isotope frequency shift, and a good agreement with literature infrared data (*12*). The growth of some propylene oxide is signaled by absorptions at 1490 cm^{-1} (CH_2 bending mode), 1268 cm^{-1} (CC stretching absorption), and 820 cm^{-1} (CO stretching mode). This assignment was confirmed by recording spectra of an authentic propylene oxide sample in BaY. Aside from allyl hydroperoxide (87%) and propylene oxide (13%) only a small trace of acrolein was observed at -100°C (C=O stretch at 1670 cm^{-1}).

Warm-up experiments following photo-accumulation of allyl hydroperoxide at -100°C allowed us to elucidate the path leading to acrolein and propylene oxide, and establish ways to control the formation of the two final oxidation products. In one experiment, the remaining reactants were removed from the zeolite by evacuation. The infrared spectrum of Fig. 5c shows that allyl hydroperoxide rearranges quantitatively to acrolein upon warm-up to room temperature. Hence, the acrolein produced upon photooxidation of propylene by O_2 at ambient temperature stems from dehydration of allyl hydroperoxide intermediate (Scheme 2). On the other hand, when conducting the warm-up of allyl hydroperoxide in the presence of excess propylene, growth of propylene oxide is observed. This signals formation of the epoxide by a secondary thermal reaction in the zeolite in which allyl hydroperoxide transfers an O atom to propylene, presumably under concurrent formation of allyl alcohol (*12*). Our observation that the allyl hydroperoxide rearrangement to acrolein shows a steep temperature dependence while epoxidation of propylene does not, opens up a means to manipulate the aldehyde to epoxide branching ratio by the zeolite temperature. Variation of the propylene loading level gives an additional handle on the acrolein/propylene oxide branching. The most important result, however, is the unprecedented selectivity in terms of the allyl hydroperoxide intermediate (>98% at ambient

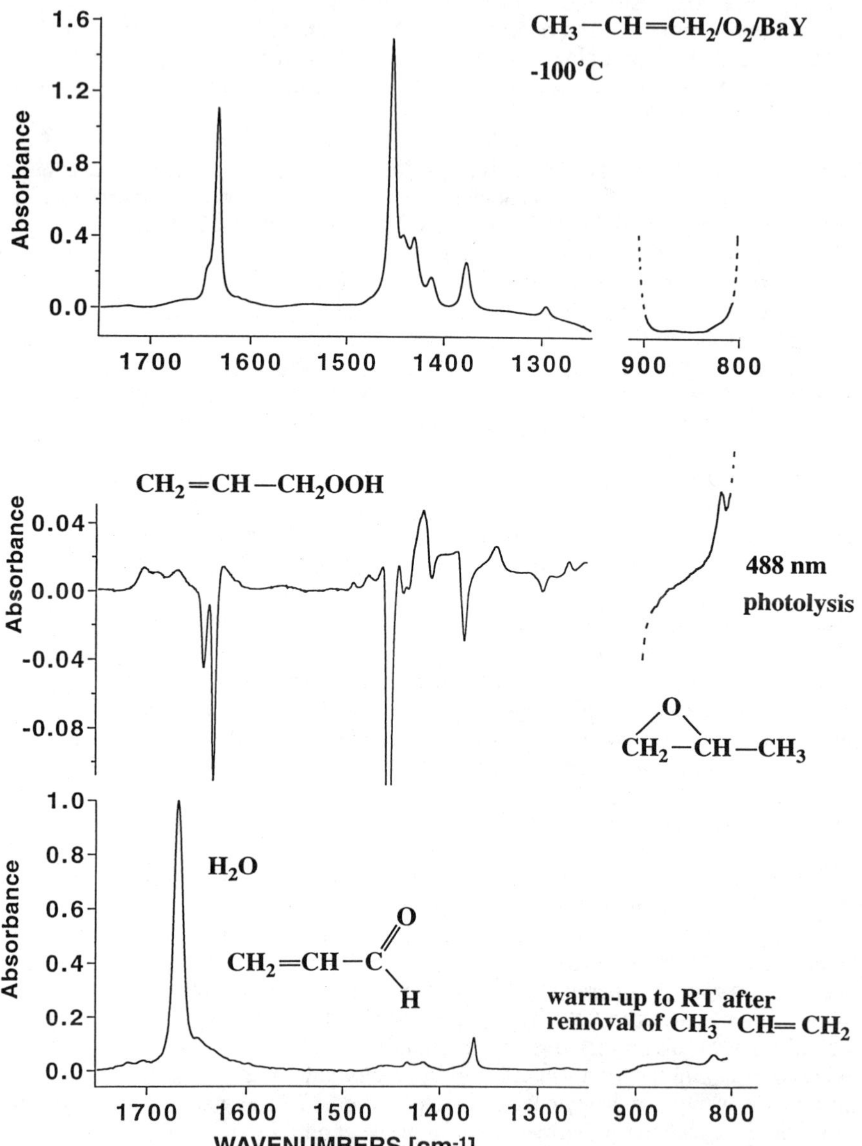

Figure 5. Visible light-induced reaction of propylene with O_2 in zeolite BaY at -100°C monitored by FT-IR spectroscopy. Top: Difference spectrum before and after loading of propylene and O_2. Middle: Difference spectrum following irradiation at 488 nm (400 mW cm^{-2}) for 300 minutes. Bottom: Difference spectrum following warm-up of the photolysis product allyl hydroperoxide to room temperature.

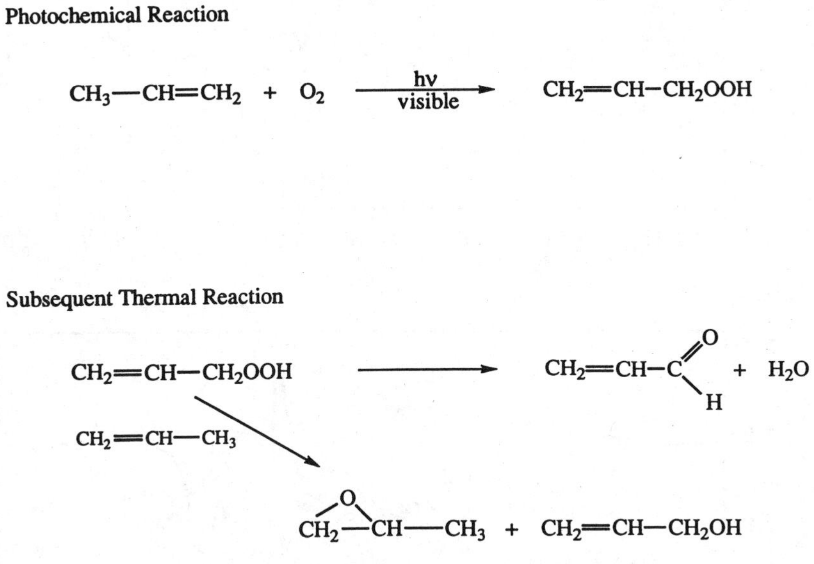

Photochemical Reaction

$$CH_3{-}CH{=}CH_2 + O_2 \xrightarrow[\text{visible}]{h\nu} CH_2{=}CH{-}CH_2OOH$$

Subsequent Thermal Reaction

$$CH_2{=}CH{-}CH_2OOH \longrightarrow CH_2{=}CH{-}CHO + H_2O$$

$$CH_2{=}CH{-}CH_2OOH \xrightarrow{CH_2{=}CH{-}CH_3} \underset{\text{(epoxide, O bridging } CH_2 \text{ and } CH)}{CH_2{-}CH{-}CH_3} + CH_2{=}CH{-}CH_2OH$$

Scheme 2. Reaction scheme for propylene oxidation.

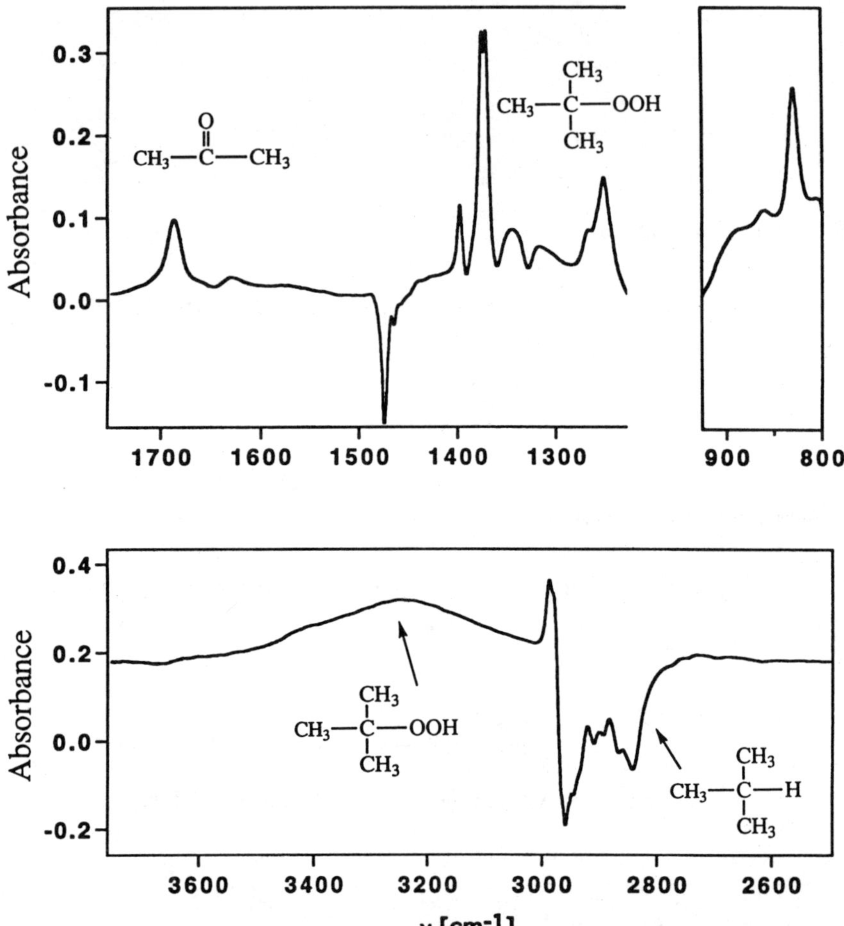

Figure 6. Infrared difference spectrum upon prolonged irradiation at 488 nm of BaY loaded with isobutane and O_2. Conversion of the initially loaded alkane is 57%.

temperature, >99.8% at -100°C) (*12*). The selectivity is undiminished even upon consumption of 20% of the propylene loaded into the zeolite. Note that since the strong scattering of the visible photolysis light restricts penetration to the front section of the zeolite, the conversion of propylene in the irradiated section of the pellet is substantially larger. On the basis of diffuse reflectance spectroscopy of the visible propylene•O_2 charge-transfer absorption and the measured infrared product growth, a rather high reaction quantum yield of 20% was estimated (*12*).

Selective Oxidation of Isobutane and Ethane. Particularly interesting is the finding that even small alkanes can be partially oxidized by oxygen in cation-exchanged zeolite Y under visible light. The corresponding alkyl hydroperoxides and carbonyl compounds are produced with very high selectivity at high conversion of the hydrocarbon.

When loading isobutane (1.6 Torr) and O_2 (900 Torr) into zeolite BaY and irradiating with visible light, *t*-butyl hydroperoxide grew in at 98% selectivity (*13*). As in all hydrocarbon oxidations studied thus far, the linear dependence of the yield with photolysis light intensity indicated a single photon process. Fig. 6 shows the infrared difference spectrum upon photochemical conversion of more than half (57%) of the isobutane loaded into the matrix. All positive bands originate from *t*-butyl hydroperoxide growth except for the small absorption at 1686 cm^{-1}, which indicates the formation of 2% acetone. $(CH_3)_2C{=}O$ and CH_3OH (which was observed when accelerating the *t*-butyl hydroperoxide rearrangement at 50°C (*13*)) are established thermal products of $(CH_3)_3COOH$ (*23*). Hence, the trace amount of acetone and methanol in the zeolite is secondary thermal products of the hydroperoxide. The absorption of the isobutane•O_2 complex in BaY was observed by diffuse reflectance spectroscopy (*13*). While the band was weak because only a fraction of the reactant pairs are probed by visible light in the highly scattering pellet, it was sufficiently strong to allow a quantum efficiency estimate of 15% (blue and green photons).

An interesting aspect of the synthesis of *t*-butyl hydroperoxide in the zeolite is the opportunity of in situ use for olefin epoxidation. This allows us to avoid accumulation of the important but hazardous oxidizer in bulk quantities. After photochemical synthesis of the hydroperoxide in BaY, the remaining oxygen and isobutane were pumped off and *trans*-2-butene was loaded into the zeolite. Growth of *trans*-2,3-epoxybutane and *t*-butanol was observed in the dark at room temperature only a few minutes after adsorption of the olefin. Infrared spectroscopic analysis showed that no *cis*-epoxide was formed upon conversion of as much as one third of the butene. Similarly, complete stereospecificity was obtained when conducting the epoxidation reaction with *cis*-2-butene (Scheme 3) (*13*).

Ethane was found to react with O_2 in zeolite CaY under irradiation with blue light when several hundred Torr C_2H_6 and one atm of O_2 were loaded into the dehydrated pellet (*24*). Fig. 7 shows the infrared product growth after 3 hours of photolysis at room temperature with the 488 nm line of an Ar ion laser (500 mW cm^{-2}). Acetaldehyde (1707, 1419, 1357 cm^{-1}) and H_2O (3400, 1650 cm^{-1}) are the only observed products. Not even a trace of carbon dioxide was produced. The latter could have easily been detected by its very strong infrared

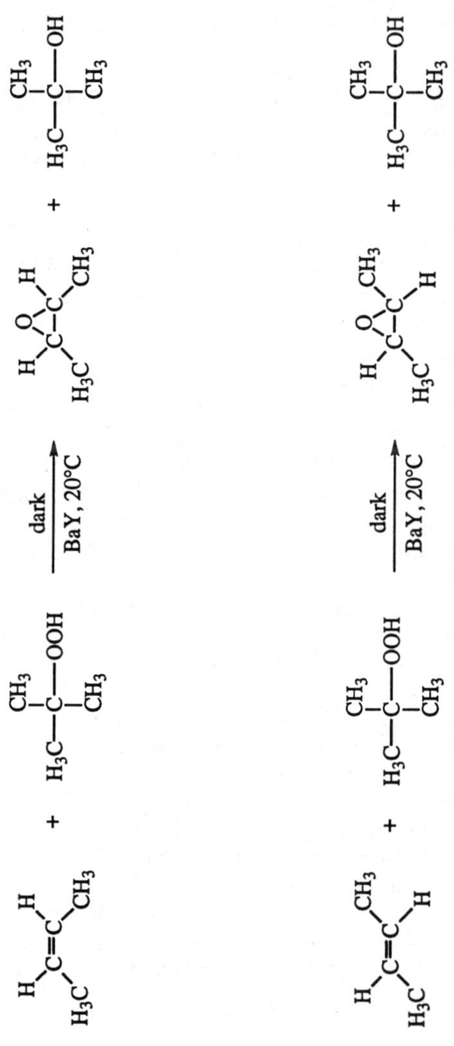

$$cis\text{-}CH_3CH{=}CHCH_3 + (CH_3)_3C{-}OOH \xrightarrow[\text{BaY, 20°C}]{\text{dark}} cis\text{-}2,3\text{-epoxybutane} + (CH_3)_3C{-}OH$$

$$trans\text{-}CH_3CH{=}CHCH_3 + (CH_3)_3C{-}OOH \xrightarrow[\text{BaY, 20°C}]{\text{dark}} trans\text{-}2,3\text{-epoxybutane} + (CH_3)_3C{-}OH$$

Scheme 3. Stereospecific thermal epoxidation of 2-butene at room temperature in zeolite BaY.

absorption at 2340 cm^{-1}. The same selective oxidation of ethane was observed when using a tungsten lamp instead of the laser source. We conclude that photooxidation of ethane by O_2 in zeolite CaY to acetaldehyde under blue light is completely selective. In contrast to the isobutane and cyclohexane•O_2 systems, the diffuse reflectance method was not sensitive enough to reveal the visible tail of the ethane•O_2 absorption. This presumably reflects the 0.9 eV higher ionization potential of ethane compared to isobutane.

Discussion

Electronic Absorption of Hydrocarbon•O_2 in Zeolite Y. We can conceive of two possibilities for the appearance of an electronic transition in the visible spectral region when coadsorbing a hydrocarbon and O_2 on the zeolite. One is an oxygen-enhanced triplet absorption of the hydrocarbon, the other a hydrocarbon•O_2 charge-transfer absorption. The assignment to a triplet state absorption can be ruled out because the lowest triplet states of small alkanes, alkenes, and alkyl-substituted benzenes in conventional phase lie in the UV region (*25*). According to numerous studies reported in the literature (e.g., phosphorescence work by Ramamurthy (*26*)), spectral shifts of triplet states are negligible compared to the red shifts of hydrocarbon•O_2 absorptions by the zeolite that we encountered here. By contrast, assignment to the well-known hydrocarbon•O_2 charge-transfer transition (*27*) is strongly supported by the observed dependence of the photochemical reaction threshold on the ionization potential of the hydrocarbon. The onset of photolysis shifts monotonically to higher energies (shorter wavelengths) with increasing ionization potential of the hydrocarbon. For example, 2,3-dimethyl-2-butene, *trans* (or *cis*)-2-butene, and propylene•O_2 complexes have photolysis thresholds in NaY at 760, 600, and 460 nm, reflecting ionization potentials of 8.30 eV (2,3-dimethyl-2-butene), 9.13 eV (*trans, cis*-2-butene), and 9.73 eV (propylene), respectively (*8,12*). In a plot of the reaction threshold energy versus hydrocarbon ionization potential, not all data points lie on a straight line, however. This is because the reaction onset depends not only on the absorption cross section, but also on the quantum efficiency. The latter is influenced by electronic and steric factors that vary among olefins, alkanes, and aromatics. Note that no maximum of the absorption band is discernible, very similar to most hydrocarbon•O_2 charge-transfer absorptions in the gas phase and solution (*17,28*).

The large cages of zeolite Y (so-called supercages shown in Fig. 1) are approximately spherical and have a diameter of 13 Å. Window openings between these cages are 8 Å across. The walls of the supercage carry a formal negative charge of 7 due to the presence of Al in the zeolite lattice (Si/Al ≈ 2.5). This charge resides on the framework oxygen atoms and is counterbalanced by 7 alkali, or 3-4 alkaline earth cations per cage (*18,29*). In the case of Na^+ exchanged zeolite Y approximately half of the cations are located in the supercage while the rest resides in the smaller sodalite and hexagonal cages (*30*). The much larger Ba^{2+} ions have difficulty entering the smaller cavities and hence occupy mostly sites in the supercages (*31*). The electric shielding of these cations in the supercages, where the hydrocarbon and O_2 molecules reside, is poor. It is this lack of shielding by the framework oxygens that gives rise to the high electrostatic fields in the supercage of cation

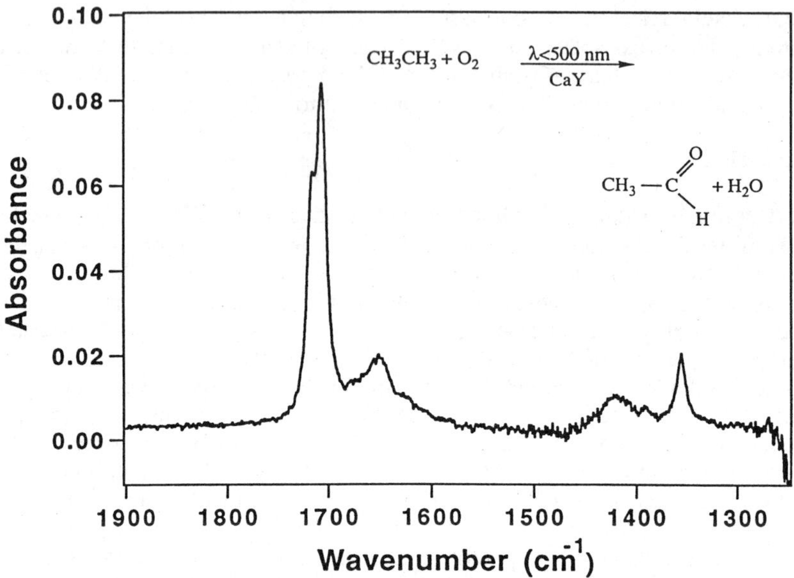

Figure 7. Infrared difference spectrum showing growth of acetaldehyde and H_2O upon irradiation of ethane and O_2 loaded zeolite CaY at 488 nm.

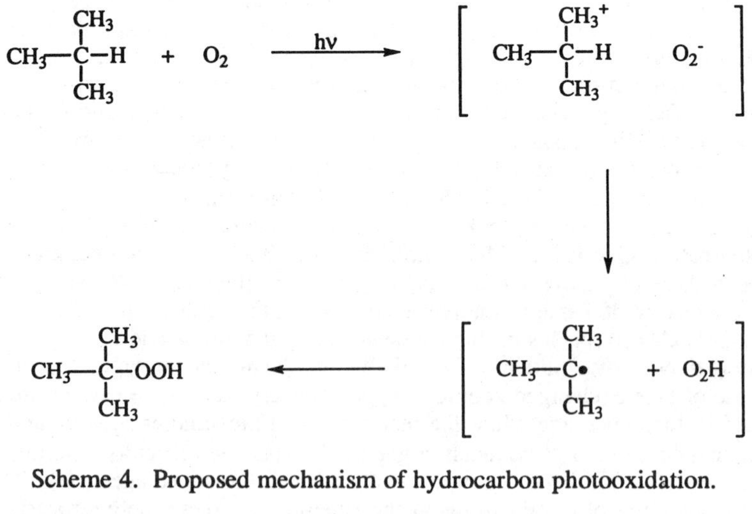

$$(CH_3)_3C{-}H + O_2 \xrightarrow{h\nu} [(CH_3)_3C{-}H^{+} \; O_2^{-}]$$

$$\downarrow$$

$$(CH_3)_3C{-}OOH \longleftarrow [(CH_3)_3C\bullet + O_2H]$$

Scheme 4. Proposed mechanism of hydrocarbon photooxidation.

exchanged zeolite Y that we observed with the induced infrared measurements. Such fields have also been estimated by other experimental methods (e.g. effect on hyperfine coupling constants of guest radicals in zeolites by ESR spectroscopy) (*32,33*), and were predicted by model calculations (*18*).

Among the potential energy terms describing the interaction of the hydrocarbon•O_2 charge-transfer complex with the zeolite cage (electrostatic, induction, dispersion, repulsion), the electrostatic field-dipole interaction ($-\mu \bullet E$) is expected to be most strongly affected by excitation of the charge-transfer state. Photoexcitation is accompanied by the development of a large dipole across the reactant complex because of the charge separation. Assuming a distance of 4 Å between the alkane$^+$•O_2^- (alkene$^+$•O_2^-, toluene$^+$•O_2^-) charge centers and an electrostatic field of 0.3 to 0.9 VÅ^{-1}, we calculate a dipole stabilization between 1.2 and over 3 eV. We attribute the large red shifts of the hydrocarbon•O_2 charge-transfer bands to this electrostatic field effect. While a quantitative prediction of the charge-transfer state energy and the shape of the absorption band will require a quantum chemical treatment of the hydrocarbon•O_2-zeolite interaction, this estimate nonetheless shows that the expected dipole stabilization is in the range of the observed red shift of the absorption onset.

The assignment of the visible absorption of hydrocarbon•O_2 complexes in zeolites in terms of a charge-transfer transition is strongly supported by our observation that the threshold photon energy to chemical reaction (and the onset of the optical absorption) are sensitive to the magnitude of the electrostatic field. For example, the photolysis onset of the 2,3-dimethyl-2-butene + O_2 reaction shifts from 760 nm to below 600 nm when replacing zeolite NaY by high-silica faujasite (*7*). The latter has the same framework structure as zeolite Y, but most supercages contain no free charge (Si/Al > 100). Hence, electrostatic fields are much lower than in NaY, although the zeolite environment is still very polar. Going in the opposite direction, substitution of Na^+ by Ba^{2+} results in an increase of the electric field in the supercage from 0.3 to 0.9 VÅ^{-1} (Results Sect.). Accordingly, the onset of the hydrocarbon•O_2 absorption shows a substantial shift to the red. For example, the threshold of the diffuse reflectance spectrum of *trans*-2-butene•O_2 shifts by 100 nm (4000 cm^{-1}) from NaY to BaY (*10*).

Mechanism and Selectivity. The main steps of the proposed mechanism are common to all visible light-induced alkane, alkene, and arene + O_2 reactions encountered thus far. Scheme 4 shows the path proposed for the isobutane oxidation. The initial step following excitation of the charge-transfer state is most likely a proton transfer from the isobutane radical cation to O_2^-. Alkane (alkene, toluene) radical cations involved in the reactions presented in this paper are spectroscopically established transients (*34*). Lifetimes of some of these species with respect to deprotonation have been reported for the room temperature liquid and are of the order of a few nanoseconds or less (*35*). We expect the proton transfer to be even more efficient in the presence of a base like O_2^-. Hydrocarbon radical cations are known to be highly acidic in general (*36*). We consider the efficient proton transfer of the charge-transfer pair as the main reason for the rather high quantum yields to reaction because it furnishes a path that is competitive with back electron transfer. The latter process suppresses useful chemistry of charge-transfer states in many cases. The alkyl and hydroperoxy radical so produced are expected to undergo cage recombination to yield the observed alkyl hydroperoxide. In the case of hydroperoxides with an α C-H group, heterolytic thermal rearrangement results in the formation of the corresponding carbonyl compound. The latter is an established mechanism in acidic solution (*22*).

Several factors contribute to the tight control of the photochemical hydrocarbon oxidations observed in this work. The principal factor is the very strong stabilization of the excited hydrocarbon•O_2 charge-transfer state by the electrostatic field of the zeolite cage. The charge-transfer state can be accessed by low-energy visible instead of the more energetic UV photons, with the result that primary products emerge with minimal excess energy. This, coupled with the confinement imposed by the zeolite cage, prevents random radical coupling reactions and homolytic fragmentation of the primary products. Specifically, no homolytic OO bond rupture of the hydroperoxide intermediates occurs which would lead to the formation of OH and alkoxy radicals. These are the two species that destroy the selectivity upon thermal autoxidation of small hydrocarbons (*1*). Moreover, the use of visible light insures that no photodissociation or further oxidation of hydroperoxide or carbonyl products by O_2 takes place. The main reason in the case of alkene and toluene systems is that the ionization potential (IP) increases upon partial oxidation of the hydrocarbon. For example, IP of propylene is 9.73 eV, compared to 10.10 eV for acrolein or 10.22 eV for propylene oxide; IP of toluene is 8.82 eV, compared to 9.52 eV for benzaldehyde (*37*). The result is that the onset of the charge-transfer absorption of the product•O_2 complexes lies at shorter wavelengths than for the alkene•O_2 or toluene•O_2 systems, which renders secondary photooxidation of the primary products unlikely. Hence, the method of hydrocarbon oxidation by O_2 with visible light is inherently stable against further reaction with oxygen. Overoxidation is a major obstacle to selectivity in the case of thermal catalytic autoxidation (*1-6*). Only for light alkanes like propane and ethane are the ionization potentials of the corresponding carbonyl products (acetone, acetaldehyde) lower than that of the parent hydrocarbon (*37*). However, we have not observed overoxidation under visible light irradiation in these cases thus far (*24*).

Conclusions

Very high selectivities have been achieved upon oxidation of olefins, alkanes, and alkyl substituted benzenes by O_2 in zeolites under visible light. For example, toluene to benzaldehyde oxidation by O_2 without overoxidation or side reactions is unprecedented. Cobalt(III) catalyzed autoxidation of toluene in solution used currently in an industrial process for benzaldehyde synthesis lacks this selectivity, mainly because of continued oxidation of the aldehyde to benzoic acid (*1,2*). Yet, toluene to benzaldehyde conversion by oxygen is an important process. The aldehyde serves as an intermediate in the manufacture of agrochemicals, fragrances, and other specialty chemicals (*38*). Among the olefin photooxidations demonstrated thus far, the propylene to acrolein and propylene oxide conversion is perhaps the most interesting one from an applications standpoint. While there is a commercially important method for propylene to acrolein oxidation by O_2 using a Bi molybdate catalyst (*39*), no selective transformation of propylene to propylene oxide by oxygen through thermal catalysis has been reported thus far. Selective propylene epoxidation schemes still use H_2O_2 or organic hydroperoxides as O donors (*1*). A recent breakthrough in the area of epoxidation with aqueous H_2O_2 are Ti silicalite catalysts (*40*).

The products of all alkane photooxidations studied thus far are of commercial importance. For example, *t*-butyl hydroperoxide is a widely used oxidizing agent, even in large-scale processes like propylene epoxidation (*1,22,41*). Cyclohexanone is an intermediate in the manufacture of nylons (*2*) as well as a variety of fine chemicals (*38*). Acetaldehyde and acetone are both bulk chemicals. Yet, existing thermal autoxidation processes are very unselective. Practical selectivities ($\approx$ 70%) of desired hydroperoxide (*t*-butyl hydroperoxide) or carbonyl compounds (cyclohexanone, acetone, acetaldehyde) can only be obtained at low conversion (below 10%) (*1,2*). Even then, carbonyls are often co-produced along with the corresponding alcohols. By contrast, the new method presented here demonstrates that these alkane oxidations to the corresponding hydroperoxides or carbonyls can be accomplished at very high selectivity even at high conversion of the reactants. In the case of the isobutane oxidation an additional attractive feature is that the hydroperoxide product can be used in situ for subsequent epoxidation reactions. Especially intriguing is the completely selective ethane to acetaldehyde conversion. Thermal catalytic methods produce substantial amounts of CO_2 without exception. By contrast, none was observed in our experiments. It may open up ethane, a constituent of natural gas, as a new feedstock for the aldehyde (instead of the currently used ethylene). More generally, the photochemical oxidations by O_2 in zeolites constitute a new and highly selective method for activation of alkane C-H bonds (*3*).

Successful upscaling of our experiments with micromolar quantities reported here requires a reduction of the scattering of photolysis light, and choice of operating conditions that allow continuous desorption of the products from the zeolite host. The preferred solution for the light scattering problem would be translucent zeolite membranes. Such membranes have been reported very recently for pentasil-type zeolites (ZSM-5) (*42*). Although release of small oxygenated hydrocarbons from zeolite Y by polar organic solvents is routine (*43*), a solvent-free method would be preferable. Use of a carrier gas and modestly elevated temperatures may be sufficient to effect desorption of the polar products at acceptable rates.

Acknowledgment

This work was supported by the Director, Office of Energy Research, Office of Basic Energy Sciences, Chemical Sciences Division, of the U. S. Department of Energy under Contract No. DE-AC03-76SF00098.

Literature Cited

[1] Sheldon, R. A.; Kochi, J. K. *Metal-Catalyzed Oxidation of Organic Compounds*; Academic Press: New York, 1981.

[2] Parshall, G. W.; Ittel, S. *Homogeneous Catalysis*; 2nd ed.; Wiley: New York, 1992.
[3] *Activation and Functionalization of Alkanes*; Hill, C. L., Ed.; Wiley: New York, 1989.
[4] Sheldon, R. A.; Dakka, J. *Catal. Today* **1994**, *19*, 215.
[5] Lyons, J. E.; Parshall, G. W. *Catal. Today* **1994**, *22*, 313.
[6] Dartt, C. B.; Davis, M. E. *Ind. Eng. Chem. Res.* **1994**, *33*, 2887.
[7] Blatter, F.; Frei, H. *J. Am. Chem. Soc.* **1993**, *115*, 7501.
[8] Blatter, F.; Frei, H. *J. Am. Chem. Soc.* **1994**, *116*, 1812.
[9] Sun, H.; Blatter, F.; Frei, H. *J. Am. Chem. Soc.* **1994**, *116*, 7951.
[10] Blatter, F.; Moreau, F.; Frei, H. *J. Phys. Chem.* **1994**, *98*, 13403.
[11] Blatter, F.; Sun, H.; Frei, H. U. S. Patent filed, 1995.
[12] Blatter, F., Sun, H., and Frei, H. *Catal. Lett.* **1995**, *35*, 1.
[13] Blatter, F.; Sun, H.; Frei, H. *Angew. Chem. Int. Ed. Engl. (Chemistry)*, in press.
[14] Sun, H.; Blatter, F.; Frei, H. *J. Am. Chem. Soc.*, submitted.
[15] Bellat, J. P.; Simonot-Grange, M. H.; Jullian, S. *Zeolites* **1995**, *15*, 124.
[16] Engel, S.; Kynast, U.; Unger, K. K.; Schüth, F. In *Zeolites and Related Microporous Materials, Studies in Surface Science and Catalysis*; Weitkamp, J.; Karge, H. G.; Pfeifer, H.; Hölderich, W., Eds.; Elsevier: Amsterdam, 1994, Vol. 84; p. 477.
[17] Chien, J. C. W. *J. Phys. Chem.* **1965**, *69*, 4317.
[18] Breck, D. W. *Zeolite Molecular Sieves: Structure, Chemistry, and Use*; Wiley: New York, 1974.
[19] Blatter, F.; Frei, H. *J. Phys. Chem.*, to be submitted.
[20] Condon, E. U. *Phys. Rev.* **1932**, *41*, 759.
[21] Barrachin, B.; Cohen de Lara, E. *J. Chem. Soc., Farad. Trans. 2* **1986**, *82*, 1953.
[22] Sheldon, R. A. In *The Chemistry of Functional Groups - Peroxides*; Patai, S., Ed.; Wiley: New York, 1983; Ch. 6.
[23] Hiatt, R. In: *Organic Peroxides*; Swern, D., Ed.; Wiley: New York, 1971, Vol. 2; p. 81.
[24] Sun, H.; Blatter, F.; Frei, H., to be submitted.
[25] Murov, S.L.; Carmichael, I.; Hug, G. L. *Handbook of Photochemistry*; 2nd ed.; Marcel Dekker: New York, 1993.
[26] Ramamurthy, V. In *Photochemistry in Organized and Constrained Media*; Ramamurthy, V., Ed.; VCH Publishers: New York, 1991; Ch. 10.
[27] Mulliken, R. S.; Pearson, W. B. *Molecular Complexes*; Wiley: New York, 1969.
[28] Davis, K. M. C. In *Molecular Association*; Foster, R., Ed.; Academic Press: New York, 1975, Vol. 1; p. 151.
[29] Introduction to Zeolite Science and Practice, Studies in Surface Science and Catalysis; Van Bekkum, H.; Flanigen, E. M.; Jansen, J. C.; Eds.; Elsevier: Amsterdam ,1991, Vol. 58.
[30] Spackman, M. A.; Weber, H. P. *J. Phys. Chem.* **1988**, *92*, 794.
[31] Liu, S. B.; Fung, B. M.; Yang, T. C.; Hong, E. C.; Chang, C. T.; Shih, P. C.; Tong, F. H.; Chen, T. L. *J. Phys. Chem.* **1994**, *98*, 4393.
[32] Coope, J. A. R.; Gardner, C. L.; McDowell, C. A.; Pelman, A. I. *Mol. Phys.* **1971**, *21*, 1043.
[33] Sugihara, A.; Shimokoshi, K.; Yasimori, I. *J. Phys. Chem.* **1977**, *81*, 669.

[34] Iwasaki, M.; Toriyama, K.; Nunome, K. *J. Am. Chem. Soc.* **1981**, *103*, 3591.
[35] Sauer, M. C.; Werst, D. W.; Jonah, C. D.; Trifunac, A. D. *Radiat. Phys. Chem.* **1991**, *37*, 461.
[36] Hammerich, O.; Parker, V. D. *Adv. Phys. Org. Chem.* **1984**, *20*, 55.
[37] *CRC Handbook of Chemistry and Physics*; 53rd ed.; Weast, R. C., Ed.; The Chemical Rubber Co.: Cleveland, 1972; p. E-62.
[38] Szmant, H. H. *Organic Building Blocks of the Chemical Industry*; Wiley: New York, 1989.
[39] Grasselli, R. K.; Centi, G.; Trifiro, F. *Appl. Cat.* **1990**, *57*, 149.
[40] Notari, B. *Stud. Surf. Sci. Cat.* **1987**, *37*, 413.
[41] Indictor, N.; Brill, W. F. *J. Org. Chem.* **1965**, *30*, 2074.
[42] Kiyozumi, Y.; Maeda, K.; Mizukami, F. *Workshop on Cluster Science*; National Institute for Advanced Interdisciplinary Research: Tsukuba, Japan, 1995.
[43] Thibault-Starzyk, F.; Parton, R. F.; Jacobs, P. A. In *Zeolites and Related Microporous Materials, Studies in Surface Science and Catalysis*; Weitkamp, J.; Karge, H. G.; Pfeifer, H.; Hölderich, W., Eds.; Elsevier: Amsterdam, 1994, Vol. 84; p. 1419.

Chapter 32

Photocatalytic Destruction of Automobile Exhaust Emissions

P. D. Kaviratna and C. H. F. Peden[1]

Pacific Northwest National Laboratory, P.O. Box 999, Richland, WA 99352

Hydrocarbons, carbon monoxide, and nitrogen oxides contained in automobile exhaust emissions are among the major atmospheric air pollutants. During the first few minutes of a cold start of the engine, the emission levels of unburned hydrocarbon and CO pollutants are very high due to the inefficiency of the cold engine and the poor activity of the catalyst at lower temperatures. Therefore, it is necessary to provide an alternative approach to deal with this specific problem in order to meet near-term regulatory requirements. Our approach has been to use known photocatalytic reactions obtainable on semiconducting powders such as titanium dioxide. In this paper, we describe our recent studies aimed at the photocatalytic oxidation of unburned hydrocarbons in automobile exhaust emissions. Our results demonstrate the effective destruction of propylene into water and carbon dioxide. The conversion was found to be a strong function of the propylene flow rate. The reaction rate was studied as a function of time, humidity and temperature. The effect of the power of the UV source on conversion is also discussed.

A number of technologies exist for the removal or destruction of dilute levels of organic contaminants in air in enclosed living or working environments, off-gas treating of water treatment facilities, and emission control in process and manufacturing plants. Since most of these common contaminants are oxidizable, a chemical oxidation process is in principle a viable alternative. However, because of the diluted levels of organics, combined with large volumes of air to be treated, it is necessary that the process be extremely fast and energy efficient.

[1]Corresponding author

0097–6156/96/0638–0428$15.00/0

Moreover, it would be desirable if the oxygen in the air could be used as the oxidant, so that would not be necessary to employ additional chemicals. Novel processes based on photocatalyzed reactions with semiconducting powders such as titanium dioxide show great promise for rapid efficient destruction of environmental pollutants that have proven difficult or expensive to treat using established remediation methods (*1-3*). For example, the ability of ultraviolet illuminated titanium dioxide to affect the partial or complete oxidation of gaseous paraffins, olefins and alcohols is well documented (*4-6*). Most efforts to apply photoassisted reactions to the destruction of environmental contaminants have focused on the purification of aqueous solutions in slurry reactors using relatively intense levels of ultraviolet light (*7-15*). However, the photodegradation of gaseous organic pollutants has gained attention in recent years (*15-22*).

Ground level ozone, formed by photochemical reactions of nitrogen oxides (NO_X) and hydrocarbons in the air, is one of the major air pollutants of concern. Automobiles contribute nearly one-third of the total hydrocarbon emissions. Therefore, reducing hydrocarbon emissions from automobiles has become a center of attention by the government regulators. The problem is particularly severe in heavily populated urban areas. Modern automobiles sold in the US are equipped with computer-controlled three-way catalytic converters to reduce the emissions of NO_X as well as CO and hydrocarbons. However, both hydrocarbon and carbon monoxide emission during the first few minutes of a cold start of the engine is still much higher than the permitted emission levels of these pollutants. This is a direct result of the inefficiency of the cold engine as well as the poor activity of the catalyst at lower temperatures. Therefore, it is necessary to either improve the activity of the catalyst in automobile catalytic converters or develop new catalysts superior to those currently being used. This work is aimed at assessing the potential of a photocatalytic process for the reduction of unburned hydrocarbons, carbon monoxide and oxides of nitrogen in automobile exhaust emissions (*23*). At this stage, the emphasis is given to the particular problem of unburned or partially burned hydrocarbon emissions in automobile exhaust during cold-start. An attempt is made to identify some of the factors that control the photocatalytic destruction of these pollutants. In particular, we report here a determination of the reaction rate as a function of gas flow (space velocity), the stability of the rate over time, the effect of impurity water-vapor concentration, and the dependence of the reaction rate on UV flux.

Experimental

As shown in Figure 1 the photocatalytic reactor system used for these studies is composed of a mass flow controller (Tylan General), a UV lamp (Aquafine SP-2), a catalytic reactor, and a sampling valve. The catalytic reactor is made out of titania coated quartz capillary tubes assembled in a 3/8" quartz tube and connected to a gas flow system made out of 1/8" stainless steel tubing. The capillaries are coated both inside and outside with a titania film by a sol-gel process to ensure the maximum contact area of the catalyst. After calcination in air, Raman spectroscopy confirmed that the films were anatase TiO_2 (*24*). The test gas contains 300 ppm propylene (C_3H_6) in dry, carbon dioxide free air. This particular hydrocarbon and its concentration is considered to be prototypical for an automobile exhaust composition. Gas samples were analyzed by using a Hewlett Packard 5890 Series II gas chromatograph equipped with a TCD detector, and Porapak Q and molecular sieve packed columns.

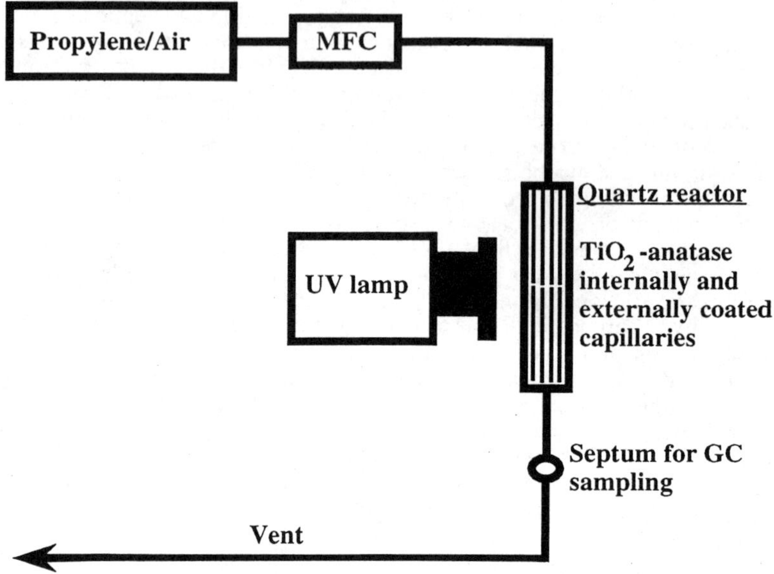

Figure 1. A schematic illustration of the photocatalytic reactor system.

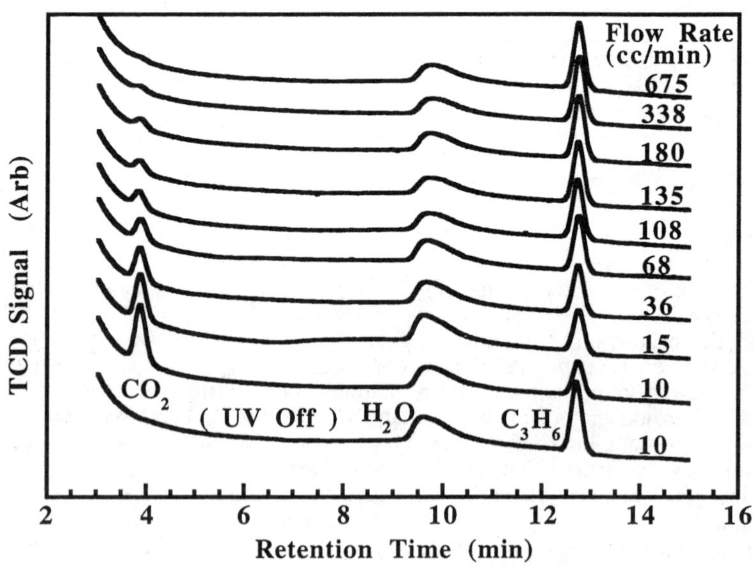

Figure 2. Gas chromatographic traces of the products generated in the photochemical destruction of propylene in the presence of a UV light source.

Results and Discussion

Traces from the gas chromatograph (GC) at different propylene/air flow rates are shown in Figure 2. The GC peak around 4 minutes retention time is due to the presence of carbon dioxide, the peak around 10 minutes indicates the presence of water vapor, and the peak around 13 minutes is due to the reactant, propylene. Each GC trace shown in Figure 2 is obtained after an equilibration time of one hour at each flow rate. A blank experiment is shown in the bottom GC trace, and was obtained with a flow rate of 10 cc/min with the UV source turned off.

In the absence of UV illumination, no reaction is observed. However, a GC peak corresponding to the product of reaction, carbon dioxide, is present upon irradiation. Concurrently a decrease in the GC peak corresponding to propylene is observed. Importantly, note the absence of any other reaction product from incomplete combustion, such as carbon monoxide or partially oxidized hydrocarbon species. Thus, the overall reaction proceeds according to equation 1:

$$C_3H_6 + 9/2\, O_2 \dashrightarrow 3\, CO_2 + 3\, H_2O \quad (1)$$

The appearance of carbon dioxide and the decrease in propylene clearly depend on the flow rate of the propylene/air mixture as illustrated in Figure 3. At the lowest flow rates studied here, greater than 50% of propylene is converted to carbon dioxide and water. However, at higher flow rates, the propylene conversion rate drops dramatically due to the decrease in residence time. Based on the volume of the reactor containing catalyst, we estimate that the highest flow rates studied here (≈900 cc/min) correspond to a space velocity of approximately 6800 hr^{-1}.

We found the photocatalytic propylene oxidation activity to be remarkably stable over time as shown in Figure 4. The data plotted in this figure were obtained with a flow rate of 36 cc/min. During the entire reaction time the propylene conversion rate was steady demonstrating the photostability of the catalyst. Experiments carried out for still longer times (> 24 hours, not shown in Figure 4) showed no change in the reaction rate. In fact, the same set of titania-coated capillary tubes have been used for over a year under a variety of reaction conditions without any loss in activity.

Another important consideration is the effect of water vapor on the photocatalytic reactivity of titania because of the potential for relatively high concentrations of water in automobile exhaust, particularly in high humidity climates. It has also been reported in the literature (*19*) that while the presence of water in the reactant gas mixture is critical for the gas-phase photocatalytic destruction of chlorohydrocarbons using titania, high water-vapor concentrations lead to catalyst deactivation. Figure 5 depicts the effect of water vapor concentration on photocatalytic propylene conversion rates. In the figure, the diamond symbols represent the conversion of propylene when the reaction is carried out with the as-received propylene gas mixture. The open squares indicate the activity of titania if this gas mixture is first passed through an isopropanol/dry ice trap designed to remove impurity water. The open circles are for reactions carried out after saturating the propylene gas mixture with water vapor by bubbling the gas through a room temperature water reservoir. It is clear from these data that the presence of water vapor is not critical for the photocatalytic destruction of hydrocarbons, nor does water poison the catalyst.

Finally, the effect of the UV light source power on the photocatalytic activity was studied. There are two competing issues of concern here. First, sufficient source power is needed to deliver UV photons to all of the catalyst in the reactor. However, a practical design will ultimately be constrained by the power

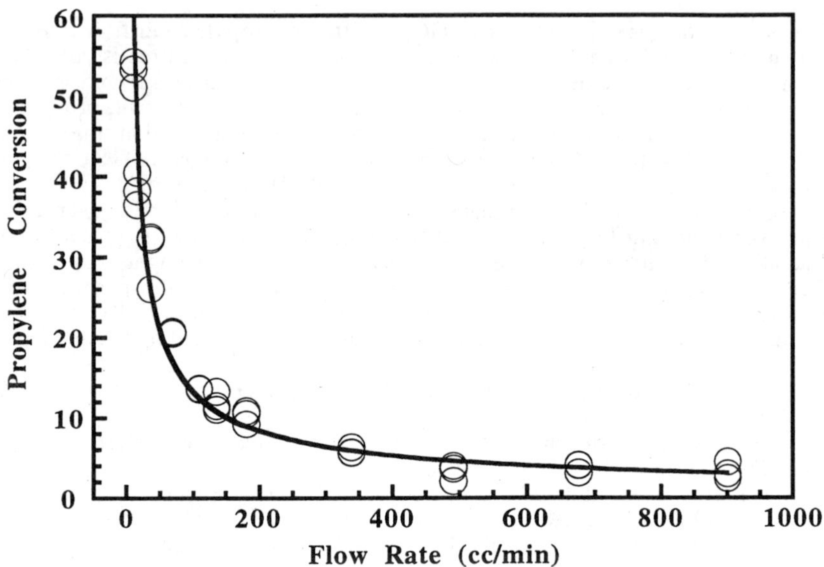

Figure 3. Flow rate dependence of propylene conversion.

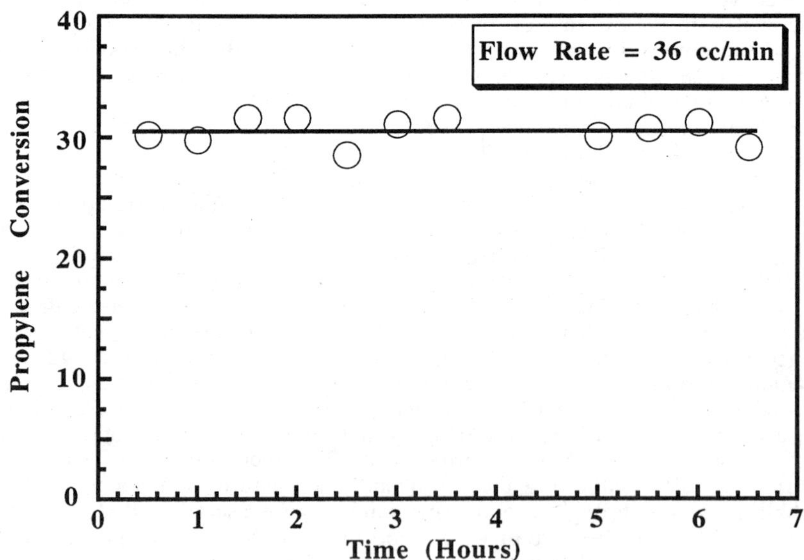

Figure 4. Time dependence of propylene conversion at a flow rate of 36 cc/min.

deliverable by the battery. Our current reactor design uses very low (< 100 watts) power and can be scaled up well within the current available power. Of more immediate concern is an accurate assessment of reaction rates (space velocities) that will not be obtained if some of the catalyst is not irradiated. The data plotted in Figure 6 shows a strong dependence of propylene conversion rates on the UV flux at least at low power. However, at higher power levels, the conversion seems to level off with further increases in power. This result may indicate that the reaction rates (space velocities) we measured at the highest power levels used here represent the intrinsic activity of the present catalyst in our reactor.

Conclusions

In the present studies, we have demonstrated that a prototypical automobile exhaust hydrocarbon, propylene, can be completely and efficiently combusted to carbon dioxide and water by a photochemical process over a titania catalyst. Importantly, we have also shown that the catalyst does not deactivate over time, and that the propylene conversion rates are insensitive to atmospheric water vapor even at relatively high concentrations.

We compared our measured reaction rates (space velocities) to those typically desired in present 3-way catalytic converters. The current catalyst system needs to handle a total gas flow rate of at least 4.4×10^5 cc/min. Because of the need to minimize both the amount of precious metal used in the three-way catalyst as well as the weight of the converter, space velocities on the order of 20,000-50,000 hr^{-1} with destruction efficiencies of 80-90% or better are desirable for the current systems. To date, we have demonstrated reaction rates that are considerably below these levels. For example, we find that for space velocities of about 1000 hr^{-1}, we obtain propylene conversion rates of 10-20%. However, there are a number of (thus far) unexplored areas of improvement that could dramatically improve these numbers. For example, neither the catalyst geometry (*e.g.*, surface area) or the catalyst composition have been optimized. Recent reports have demonstrated that small additions of a second oxide material to titania leads to greatly enhanced photocatalytic oxidation rates (*12-14*). Additionally, small amounts of precious metal on titania can lead to significant increases in electron transfer rates and, thus, photoreactivity (*14*). We are currently exploring these and other potential improvements to the photocatalytic activity of our system.

It is important to keep in mind, however, that the "targets" described above for the 3-way catalysts (> 80% destruction at space velocities > 20,000 hr^{-1}) are likely to be much higher than necessary for the particular system described here. First, as discussed above, these targets are dictated to some extent by the need to minimize the precious metal content in the catalyst while the titania photocatalyst may not require any precious metal. Thus, a larger reactor may be possible to overcome lower activities. Perhaps even more important to consider is the fact that this approach is designed to deal with the specific problem of cold-start hydrocarbon emissions. The current EPA Federal Test Procedure (FTP) makes an integrated measure of emissions over a lengthy (several minutes) engine cycle from cold-start through several accelerations, decelerations and cruising modes. A large fraction of the hydrocarbon emissions occur during the cold-start period and any reduction during this period can result in significantly more flexibility in the operating envelope (*e.g.*, air/fuel ratio) for the remainder of the FTP test.

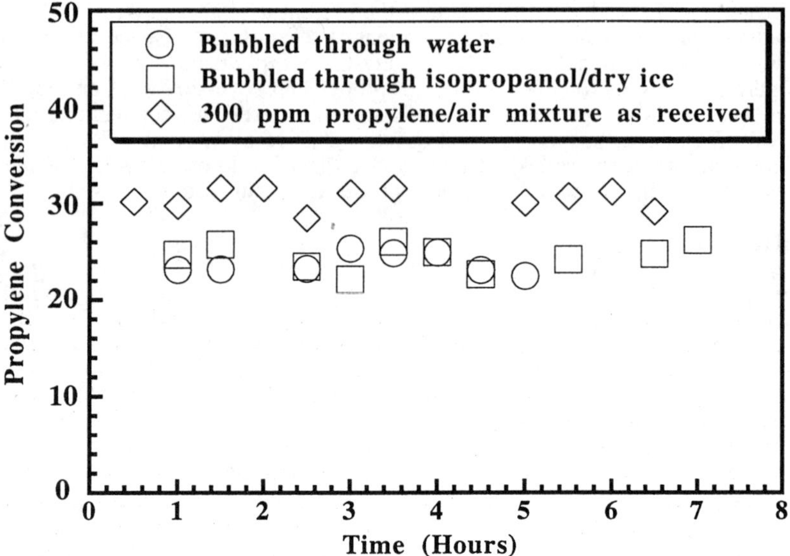

Figure 5. Effect of humidity on propylene conversion in the photocatalytic destruction process.

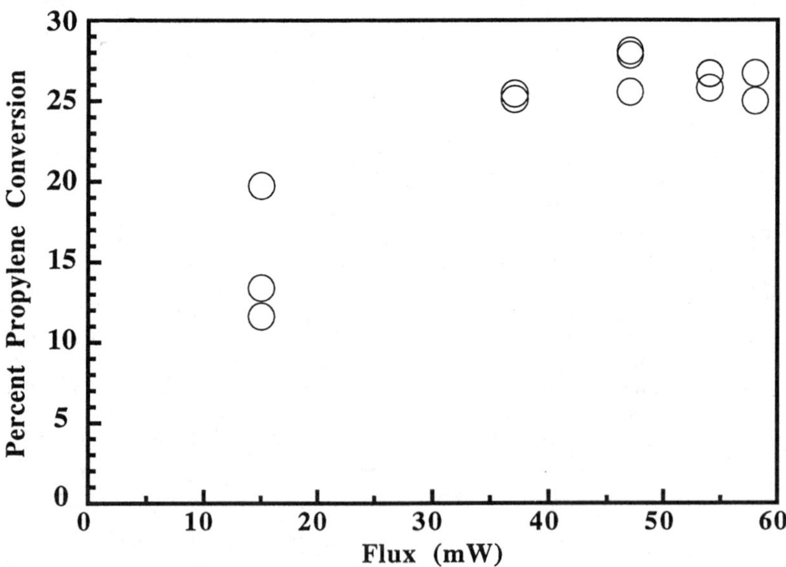

Figure 6. Propylene conversion as a function of the UV lamp source power.

Pacific Northwest National Laboratory is a multiprogram national laboratory operated for the U.S. Department of Energy by Battelle Memorial Institute under contract DE-AC06-76RLO 1830.

Acknowledgments

This work was supported by Laboratory Technology Transfer Funds provided by the U. S. Department of Energy's (DOE) Office of Energy Research. One of us (PDK) wishes to thank the Associated Western Universities, Inc., Northwest Division (AWU-NW) for a post-doctoral fellowship. AWU-NW is supported by the grant DE-FG06-89ER-75522 with the U. S. DOE.

Literature Cited

(1) Frank, S.; Bard, A. *J. Phys. Chem.* **1977,** *81*, 1484.
(2) Ollis, D. *Environ. Sci. Tech.* **1983,** *17*, 628.
(3) Gratzel, C.; Jirousek, M.; Gratzel, M. *J. Mol. Catal.* **1987,** *39*, 347.
(4) McLintock, I.; Ritchie, M. *Trans. Faraday Soc.* **1965,** *61*, 1007.
(5) Pichat, P.; Hermann, J.; Courbon, H.; Disdier, J.; Mozzanega, M. *Can. J. Chem. Eng.* **1982,** *60,* 27.
(6) Blake, N.; Griffin, G. *J. Phys. Chem.* **1988,** *92*, 5697.
(7) Pruden, A.; Ollis, D. *J. Catal.* **1983,** *60*, 404.
(8) Ollis, D.; Hsiao, C.; Budiman, L.; Lee, C. *J. Catal.* **1984,** *88*, 89.
(9) Matthews, R. *J. Catal.* **1988,** *113*, 549.
(10) Sabate, J.; Anderson, M. A.; Kikkawa, H.; Edwards, M.; Hill, C. G., Jr. *J. Catal.* **1991,** *127*, 167.
(11) Brezova, V.; Stasko, A. *J. Catal.* **1994,** *147*, 156.
(12) Papp, J.; Soled, S.; Dwight, K.; Wold, A. *Chem. Mater.* **1994,** *6*, 496.
(13) Cui, H.; Dwight, K.; Soled, S.; Wold, A. *J. Solid State Chem* **1995,** *115*, 187.
(14) Hoffmann, M. R.; Martin, S. T.; Choi, W.; Bahnemann, D. W. *Chem. Rev.* **1995,** *95*, 69.
(15) Augugliaro, V.; Loddo, V.; Palmisano, L.; Schiavello, M. *J. Catal.* **1995,** *153*, 32.
(16) Formenti, M.; Juillet, F.; Meriaudeau, P.; Teichner, S. J. *Chem. Tech.* **1971,** *1*, 680.
(17) Dibble, L. A.; Raupp, G. B. *Catal. Lett.* **1990,** *4*, 345.
(18) Dibble, L. A.; Raupp, G. B. *Environ. Sci. Technol.* **1992,** *26*, 492.
(19) Raupp, G. B.; Junio, C. T. *Appl. Surf. Sci.* **1993,** *72*, 321.
(20) Peral, J.; Ollis, D. F. *J. Catal.* **1992,** *136*, 554.
(21) Sampath, S.; Uchida, H.; Yoneyama, H. *J. Catal.* **1994,** *149*, 189.
(22) Tada, H.; Honda, H. *J. Electrochem. Soc.* **1995,** *142*, 3438.
(23) Janata, J.; McVay, G.L., Peden, C. H. F.; Exarhos, G. J. patent pending.
(24) Exarhos, G. J. unpublished results.

Indexes

Author Index

Affiliation Index

Subject Index

N

R